THE BOOK OF ORCHIDS

THE BOOK OF ORCHIDS

A LIFE-SIZE GUIDE TO SIX HUNDRED SPECIES
FROM AROUND THE WORLD

MARK CHASE
MAARTEN CHRISTENHUSZ
TOM MIRENDA

THE UNIVERSITY OF CHICAGO PRESS

Chicago

PROFESSOR MARK CHASE is a senior research scientist at the Royal Botanic Gardens, Kew. He is also an adjunct professor at the School of Biological Sciences at the University of London and at the School of Plant Biology at the University of Western Australia, and a fellow of the Linnean Society and the Royal Society. He co-edited *Genera Orchidacearum* and has contributed to more than 500 publications on plant science.

DR. MAARTEN CHRISTENHUSZ is a botanist-consultant who has worked for the Finnish Museum of Natural History in Helsinki, the Natural History Museum (London) and the Royal Botanic Gardens, Kew. Maarten was initiator of the journal *Phytotaxa* and is deputy editor of the *Botanical Journal of the Linnean Society*. He has written about a hundred scientific and popular publications.

TOM MIRENDA is the Orchid Collection Specialist at the Smithsonian Institution in Washington, D.C. He frequently lectures on orchid ecology and conservation in the US and abroad, and is a columnist for *Orchids*, the magazine of the American Orchid Society.

The University of Chicago Press, Chicago 60637
© The Ivy Press Limited 2017
All rights reserved. No part of this book may be used or reproduced in any manner whatsoever without written permission, except in the case of brief quotations in critical articles and reviews. For more information, contact the University of Chicago Press, 1427 E. 60th St., Chicago, IL 60637.
Published 2017.
Printed in China

26 25 24 23 22 21 20 19 18 17   1 2 3 4 5

ISBN-13: 978-0-226-22452-7 (cloth)
ISBN-13: 978-0-226-22466-4 (e-book)
DOI: 10.7280/chicago/9780226224664

Library of Congress Cataloging-in-Publication Data

Names: Chase, Mark W., 1951– author. | Christenhusz, Maarten J. M., 1976– author. | Mirenda, Tom, 1957– author. | Ivy Press.
Title: The book of orchids : a life-size guide to six hundred species from around the world / Mark Chase, Maarten Christenhusz and Tom Mirenda.
Description: Chicago : The University of Chicago Press, 2017. | "This book was conceived, designed, and produced by Ivy Press." Identifiers: LCCN 2016039433 | ISBN 9780226224527 (cloth : alk. paper) | ISBN 9780226224664 (e-book)
Subjects: LCSH: Orchids.
Classification: LCC QK495.O64 C359 2017 | DDC 584/.4— dc23 LC record available at https://lccn.loc.gov/2016039433

The views expressed in this work are those of the authors and do not necessarily reflect those of the publisher or the Royal Botanic Gardens, Kew.

This book was conceived, designed, and produced by
Ivy Press
Ovest House, 58 West Street
Brighton, BN1 2RA
United Kingdom
www.quartoknows.com

Publisher SUSAN KELLY
Creative Director MICHAEL WHITEHEAD
Editorial Director TOM KITCH
Art Director WAYNE BLADES
Commissioning Editor STEPHANIE EVANS
Senior Project Editor JOANNA BENTLEY
Copy-editor JOHN ANDREWS
Designers JANE MCKENNA & GINNY ZEAL
Illustrator DAVID ANSTEY
Picture Researcher ALISON STEVENS
Assistant Editor JENNY CAMPBELL

JACKET IMAGES
Maarten Christenhusz: *Orchis militaris*
Norbert Dank: *Gomesa forbesii*
Lourens Grobler: *Acineta superba*
Eric Hunt: *Anguloa virginalis, Oncidium nobile, Prosthechea mariae*
Malcolm M Manners: *Sacoila lanceolata*
Herbert Stärker: *Brownleea coerulea*
Swiss Orchid Foundation / G Meyer: *Aganisia cyanea*; / R Parsons: *Calanthe sieboldii*; / Rogier Van Vugt: *Dendrobium chrysotoxum, Eriopsis biloba*
Miguel Vieira: *Chloraea magellanica*

LITHOCASE IMAGES
Eric Hunt: *Acanthephippium mantinianum, Oncidium lutzi*
Jeremy Storey: *Pyrorchis forrestii*

CONTENTS

PREFACE

Orchids have given me an exceptional amount of pleasure over the years. For decades, it has been my mission to share that joy. Taking part in the creation of this book is the culmination of that desire to nurture and spread appreciation for what I believe to be the most extraordinary family of plants. Unquestionably lovely, orchids are far beyond being just beautiful. They are seemingly endless in their diversity, perpetually compelling, and astonishingly well adapted to a mind-boggling array of ecological niches and evolutionary partners. A geologically old family, members of the Orchidaceae have colonized the far reaches of our planet save those most inhospitable: extreme poles, high mountain peaks, the most desolate deserts, and, of course, the deep waters of our lakes, rivers, and oceans.

Having evolved to occur in such a wide variety of habitats, as well as perfecting the ability to interact with and exploit myriad creatures as symbionts, orchids are the ideal plant family to teach us about biodiversity and illustrate its importance. The remarkable structures and colors of each and every orchid species convey a story about their ecology, evolution, and survival strategy. Once analyzed and unlocked, these stories give us powerful insight into the processes that have shaped our world for millennia and, hopefully, inspire us to conserve that which took millennia to create.

Masters of deception and manipulation, orchids are famous for lying and cheating their way to their many evolutionary successes. Exploring the manner in which they co-opt pre-existing behaviors of a bewildering cohort of pollinators of lilliputian dimensions is not only outstandingly instructive, but is just plain fun to contemplate. Even the venerable Charles Darwin referred to orchids as "Splendid Sport" and maintained a passion

for them throughout his lifetime. It is undeniable that orchids have gripped the psyches of many humans. They have even, in recent years, become the most sold and cultivated type of ornamental plant. Their beauty alone does not explain this phenomenon.

Many theories exist as to why orchids are so alluring to us. It is thought that their zygomorphic (bilaterally symmetrical) flower structure influences us to see orchid flowers similarly to the way we see faces, attributing to them some "personality" in addition to their beauty. Some find the lip of certain orchids to be reminiscent of human anatomical parts that we normally keep covered, lending them a subliminal or feral attraction. Others simply find the combination of color, form, grace, and fragrance most appealing, yet not all orchids have traditionally attractive versions of these attributes. Some of the most compelling orchids are rank-smelling, muddy in coloration, and borne on clunky plants. Nothing adequately explains why people become so wildly obsessive about orchids. Ultimately, they are simply provocative creatures that manage to elicit strong reactions from pollinator and person alike.

In this ambitious book, we invite you to journey with us around the world and see orchids for the marvels of nature they truly are. It is our hope that the images and stories within will inspire appreciation and stewardship as well as give great pleasure to all, young and old, who choose to embark on the rewarding study of orchidology.

Tom Mirenda

INTRODUCTION

The orchid family, Orchidaceae, embraces 26,000 species in 749 genera and is one of the two largest families of flowering plants, or angiosperms— a broad group that includes herbs, trees, shrubs, and vines. The other large family is that of the daisies and lettuce, Asteraceae. Estimates of family size vary, depending on how the number of species is calculated, and which is the larger of the two is a hotly debated topic among botanists. Many people have a vague idea of what an orchid is, but it is likely that most would not recognize all the species included in this book as orchids. So, what is an orchid?

Orchids are divided into five subfamilies, Apostasioideae, Vanilloideae, Cypripedioideae, Epidendroideae, and Orchidoideae. This subdivision is based on DNA studies and morphology and reflects major differences in vegetative features and especially in the way orchid flowers are constructed. The five subfamilies have been recognized in the past as separate families by some botanists based on these distinctive characteristics, and the only characteristic they all share is that of how orchid embryos develop, from a structure called a protocorm, which is a small ball of cells without roots, stems, or leaves.

To develop into a mature orchid plant, a protocorm has to be successfully infected by a fungus, from which the developing orchid seedling obtains initially all the food (in the form of sugars) and minerals it needs to grow. As they start their life, orchids can be thought of as parasites on fungi. However, most but not all orchids as adults go on to develop roots and leaves, and produce their own food through photosynthesis. At a much later stage the continuing relationship of an orchid plant with the fungus can become mutually beneficial. In nature, the orchid exchanges sugars produced by its photosynthesis for minerals found more effectively by the fungus. In cultivation, the need of an orchid protocorm for a fungal partner can be replaced by manufactured sources of food and minerals, and many orchids are grown commercially using germination media with added sugars and minerals.

THE COLUMN

The other major trait that most botanists use to recognize an orchid is a structure called the gynostemium, or column, produced by the fusion of male (stamen) and female (stigma) parts in the flower. All but one of the five subfamilies share this feature. The exception is the subfamily Apostasioideae, consisting of only 14 species in two genera, *Apostasia* and *Neuwiedia*, which all lack complete fusion of the male and female parts.

In subfamily Cypripedioideae—which consists of five genera, *Cypripedium*, *Mexipedium*, *Paphiopedilum*, *Phragmipedium*, and *Selenipedium*, and 169 species, known as the slipper orchids—there are two stamens (the pollen-bearing structure of a flower), whereas only one occurs in the other three subfamilies—Vanilloideae (14 genera and 247 species), Orchidoideae (200 genera and around 3,630 species), and Epidendroideae (535 genera and around 22,000 species). Their single stamen is fused to three fused stigmas with a single female receptive region.

BELOW **Mexican species *Laelia gouldiana***, labeled to show the floral parts that make up a typical orchid flower.

THE PARTS OF AN ORCHID FLOWER

dorsal sepal

petal

petal

column

pollinia

lateral sepal

lateral sepal

lip
(or labellum)

ABOVE ***Epidendrum wallisii***, a species from Central and South America that is pollinated by butterflies searching for nectar.

The characteristically fused structure of the column, shared by 99.95 percent of all orchids, is responsible for the remarkable event where a pollinator, such as a bee, wasp, or moth, is maneuvered into doing exactly what the orchid wants. This allows the pollen, usually in the form of thousands of grains bound into a solid ball, or pollinium, to be placed on the animal in a precise manner and then, due to the close proximity of the stigma and anther (the part of the stamen holding the pollen), be precisely removed from that spot. Pollination in orchids is, therefore, a highly exact sequence of events, leading to fertilization of the thousands of developing orchid embryos in the carpel, or ovary, with just a single visit of a pollinator, provided that it has previously visited another flower of that same orchid species to pick up pollinia.

THE LIP

In most orchids the female receptive surface, or stigma, is a cavity on the side of the column that faces the other highly distinctive orchid structure: a modified petal (one of three) that is termed the labellum, or lip. This serves variously as a landing platform, a flag to attract the pollinator, or—playing an important part in various forms of deceit that orchids use to fool pollinators—a mimic of something the pollinator wants, such as nectar, pollen, a mate, or a place to lay its eggs.

There are many orchids that appear not to have a lip. A good example is the genus *Thelymitra* from Australia, where the member species are called sun orchids. Rather than a lip, the flowers of these plants have three sepals, which are initially a set of protective leaflike structures (that in many orchids also become colorful) and three similar petals (also colorful leaflike organs). Such similarity of all three petals, though, is the exception among orchids, most of which develop a highly modified lip.

Although it has long been known that orchids can control the appearance of the lip in isolation from the other showy parts of their flowers—the two remaining petals and three sepals—it was not clear until recently how the lip was controlled from a genetic or developmental perspective. In nearly all other plants that have been studied in this regard, the three petals are controlled by the same floral genes, and by and large they all three do the same thing and look the same. Think, for example, of a lily or a tulip, in which the three petals are identical. In orchids, there has been a duplication of the floral genes, and one of the duplicated copies is expressed just in the lip, making it possible for this petal—the lip—to look different and be involved in pollinator manipulation apart from the other two petals, in which the gene is not expressed. This more complicated set of genetic controls has made the flowers of orchids among the most complex in the plant world and undoubtedly is a major reason why their flowers are adapted for pollination by such a large range of animals.

DISTINGUISHING FEATURES

The combination of column, lip, and pollinia—the first unique to orchids, the others not unique but unusual among plants—makes it possible for botanists to recognize plants as orchids despite their capacity to look decidedly un-orchidlike. In biological terms, this amalgam of features has enabled orchids to become evolutionarily explosive, leading to the 26,000 species alive today.

BELOW *Epidendrum medusae* grows high in the Andes and is pollinated by moths attracted by its elaborate fringed lip.

Species numbers in the largest genera, *Epidendrum*, *Bulbophyllum*, *Dendrobium*, and *Lepanthes*, run into the thousands. No book could include all of them, so we have concentrated on illustrating 600 species, carefully chosen to display the wide range of orchid diversity and to cover all areas of the globe where the plants are found. They are presented in the five subfamilies, appearing alphabetically by Latin name within tribes (and subtribes where appropriate).

RIGHT: **Orchids can have** very limited geographical distribution; *Ceratocentron fesselii*, for example, occurs only in the mountains of Luzon Island, Philippines.

12

ORCHID EVOLUTION

Orchids evolved during the Late Cretaceous period, roughly 76 to 105 million years ago. This is much earlier than botanists once thought and makes Orchidaceae one of the 15 oldest angiosperm families, of which there are 416 in total. Few orchid fossils older than 20 to 30 million years have been found, and it was thought that orchids evolved relatively recently compared to many other groups of flowering plants. That they have a poor fossil record is not surprising because most orchids are herbs, which generally do not fossilize well, and their highly modified pollinia are difficult to recognize in the fossil record.

DINOSAUR DEPENDENCE

All five orchid subfamilies evolved before the end of the Cretaceous period, which means that orchids and dinosaurs overlapped. Considering the great diversity of orchid pollinators, we can only wonder if orchids managed to

Apostasioideae

Cypripedioideae

Vanilloideae

adapt to pollination by dinosaurs before the latter became extinct 65 million years ago. Vertebrates in general are uncommon orchid pollinators, and nearly all of those recorded are birds—direct descendants of the dinosaurs. There were many small species of dinosaurs, so it is possible that some visited flowers to collect nectar and, like many animals today, were deceived into pollinating orchids. Any orchids adapted to dinosaur pollination would have become extinct with their pollinator, and so are now lost to us.

DISTRIBUTION

The discovery that orchids were much older than previously thought was a result of the widespread sequencing of DNA that only became possible in the mid-1990s. This greater age makes a good deal of sense when it comes to understanding the geographic distribution of orchids. It was long assumed that orchids could have reached their current worldwide distribution relatively recently by long-distance dispersal of their small, almost microscopic seeds. Due to their dependence for food and minerals on the fungi with which they associate, orchids do not include food reserves or minerals in their seeds, unlike, for example, a bean in which the stored food and minerals make up the bulk of its much larger seed. Orchid seeds are, therefore, light and easily distributed by the wind, which theoretically could propel them over long distances. However, the longer an orchid seed remains aloft, the more the small embryo dries out, making most orchid seeds inviable before they can travel great distances. So, most orchid species have a limited distribution, even as constrained as a single mountain. Orchids have instead achieved their worldwide distribution by passively riding the continents, which at the time the plants evolved were much closer than they are today.

13

LEFT TO RIGHT **Species from each subfamily:** *Neuwiedia veratrifolia, Cypripedium kentuckiense, Vanilla aphylla, Platanthera ciliaris, Warczewiczella marginata.*

Orchidoideae

Epidendroideae

RIGHT **Bulbophyllum frostii** from Vietnam is pollinated by flies thinking food is present but then becoming trapped in its pouch-like lip. The only way out is past the reproductive organs of the orchid.

POLLINATION

ABOVE **Charles Darwin** was fascinated by orchids and at his home in Kent, England, studied tropical species in addition to native species.

Orchids are well known for elaborate pollination mechanisms that have evolved to achieve the mating of different plants, or cross-fertilization. Flowers of most plants, including orchids, contain organs of both sexes, but self-pollination is as generally undesirable in plants as it is in animals. Most plants, and orchids in particular, have evolved methods, often exceedingly complicated, to avoid self-pollination happening. This process has long fascinated scientists, including Charles Darwin, who studied pollination of orchids in detail and was so enthralled by the plants that his first book after publication of *On the Origin of Species* (1859) was entirely dedicated to orchids. The short title, *Fertilization of Orchids*, gave little hint of its main hypothesis, unlike its full and explanatory title, *On the Various Contrivances By Which British and Foreign Orchids Are Fertilized By Insects, and On the Good Effects of Intercrossing* (1862). Among the orchids studied by Darwin were a large number of tropical species provided by the then Director of the Royal Botanic Gardens, Kew, Sir Joseph D. Hooker.

POLLINATOR DECEPTION

Most orchids produce pollen in two to six tight bundles, called pollinia. These are often attached to ancillary structures that together are called a pollinarium, which attaches the pollinia to the pollinator's body, usually in a position that makes it difficult for the animal to remove them. Most orchids look as if they contain a reward for pollinators but few actually offer it. Some even produce long nectar spurs that are devoid of nectar. Rates of visitation by pollinating insects to such deceptive flowers are, understandably, low. Insects learn quickly to avoid these rewardless flowers, but they make the mistake often enough for it to be effective in a system in which a single visit can result in deposition of thousands of pollen grains, each fertilizing one of the thousands of orchid ovules produced by each flower. A rare mistake by a deceived pollinator is enough for the orchid to produce large numbers of seeds.

Darwin himself came to the conclusion that outcrossing, or pollination between unrelated plants, is so advantageous for most orchids that deceit and corresponding low rates of visitation are the general rule. Apparently, setting seeds in only a few flowers but guaranteeing that these are of high quality (due to cross-fertilization involving flowers on different plants) makes deceit a successful strategy. In this case, the cheating orchids have prospered, despite the fact that they so badly treat the insects upon which they depend. There is no mutual benefit for the orchid and its pollinators as there is in pollination systems with rewarding plants; the deceiving orchid could go extinct and the animal would only experience a slight improvement in its condition due to fewer floral visits without a reward. However, if the animal pollinating a deceitful orchid species becomes extinct, then the orchid also disappears or develops a method by which to self-pollinate its flowers, which has been known to evolve when an orchid species reaches an island without its pollinator accompanying it.

ABOVE *Zelenkoa onusta*, a species from Peru and Ecuador, is visited by bees that are fooled into thinking a reward is present.

15

BELOW *Dendrobium aphyllum*, an Asian species that produces a large inflorescence of non-rewarding flowers, only a few of which ever produce seeds.

SEED PRODUCTION

The combination of delivery of whole pollinaria on a single visit and fertilization of a correspondingly large number of ovules in the ovary means that from a single pollinator visit a massive number of seeds can be produced. That many orchids, such as some species of *Dendrobium*, *Epidendrum*, and *Oncidium*, bear large inflorescences with hundreds of flowers may seem like an extreme waste of energy, but production of mature ovules ready for fertilization is delayed until pollination takes place, thus reducing energy inputs associated with these large numbers of flowers.

MIMICRY AND DECEIT

Deceit involving mimicry of other local plants that produce a reward for their pollinator is another common habit for orchids. Although not offering a reward itself, the orchid benefits from pollinators that fail to distinguish between a cheating orchid and the rewarding species, and so the former obtains a degree of pollinator service that drops dramatically if the latter is not present. In other cases, a deceitful orchid species is not mimicking a single reward-offering species in the immediate neighborhood, but rather is using a suite of the traits associated by pollinators with the presence of a reward. These include fragrance, color, "nectar guides" to direct a pollinator to the center of the flower, and a nectarless cavity or spur of the correct shape and size to suggest that nectar is present. A quick look at the species illustrated on this page demonstrates many of these features in what is termed "general" or "non-specific" deceit.

In many groups of orchids, a much more specific type of deceit, involving sexual attraction, has evolved. Darwin was unaware of this phenomenon, although he speculated on what might be happening with native British bee and fly orchids (genus *Ophrys*). The details would probably have shocked him and many other botanists of that time. It is thought that mimicry of the female of a species of bee, wasp, or fly begins as some other more general type of deceit and subsequently becomes more complicated and specific. For example, the orchid *Anacamptis papilionacea* appears not to be mimicking any specific nectar-producing species in its

16

RIGHT **Cyrtochilum aureum**, a rewardless species from the Andes of Peru and Bolivia, attracts pollinators by appearing like several reward-offering flowers among which it grows.

LEFT **Central American** *Chysis tricostata* offers no reward but nonetheless attracts enough bees to achieve pollination and successful production of seeds.

17

habitat and is instead just a general reward-flower mimic. However, there are more males than females among the insects it attracts, so it appears that some sort of sexual attraction is operating, which could lead to further change on the part of the orchid to enhance this aspect of the deceit.

Many orchids using visual sexual mimicry also produce floral fragrances that are identical to the sex pheromones produced by the female of the insect species to attract a male. This at first sounds wholly preposterous: how can a flower evolve to produce something so alien to a plant as an animal sex pheromone? However, once it became known how the biochemical pathways operate by which such animal hormones are produced, it also became clear that plants share these same general pathways and often produce minor amounts of such compounds as part of their general bouquet of scents. Thus, the assembly of a highly specific sexual pheromone starts out with production of small amounts of similar compounds that become predominant when an increased presence in the mixture generates higher rates of male visitation, such as that observed in *A. papilionacea*. When combined with visual cues, such fragrance compounds reinforce the "message" being sent to male insects, and sexual mimicry is the result. Orchids in many distantly related groups have independently evolved this sexual mimicry syndrome, which, now that we know the genetic and biochemical details, is not as surprising as it first appeared.

BELOW **Anacamptis** *papilionacea*, a widespread southern European species, exhibits a mixed syndrome of deceit pollination and attracts more male than female bees.

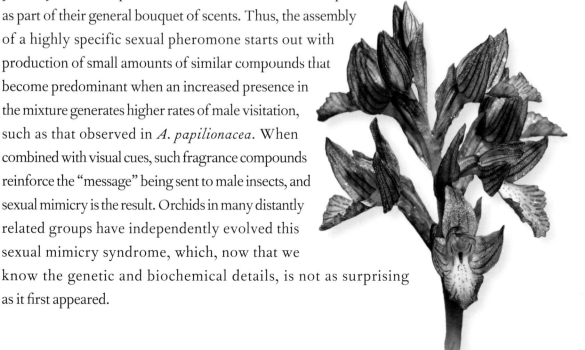

SYMBIOTIC RELATIONSHIPS

BELOW **A species from northern South America**, *Anguloa virginalis* offers floral fragrance compounds as rewards to its pollinating bees.

Orchids have a symbiotic relationship with soil fungi that enables germination of their seeds and sustains them in early phases of their development, when they are unable to be photosynthetic and make their own food. These fungi are so-called "wood-rot fungi" that break down dead wood in the soil and form masses of fungal tissue, known as pelotons, inside the cells of the orchid embryo. The exchange that occurs in the early stages of germination is entirely one-way in favor of the orchid, and it is not clear why the fungi participate in this process. There is no obvious benefit to the fungal partner; the embryonic orchid prospers, but there are only costs for the fungus. Once the orchid seedling forms its own leaves, then sugars that are produced by the orchid are exchanged for minerals from the fungus, which is much better at retrieving minerals from the soil than the plants. However, some orchids continue throughout their life to be a drain on the food reserves of their fungal partner.

FUNGAL PARTNER SWAP

Some ground orchids are known to switch fungal partners as they grow older and associate instead with "ectomychorrhizal" fungi, which regularly exchange minerals for sugars with forest trees. Orchids associating with ectomychorrhizal fungi have been found to contain sugars produced by the trees, the sugars recognized as distinct from those produced by the orchids as they leave a clear chemical fingerprint created by the fungus as they pass through it. These orchids abandon the wood-rot fungi that helped them germinate, without ever giving those fungi a reward for this service, and then switch to a fungal relationship that provides them with sugar produced by the trees in their habitat. We do not yet know how orchids manage these complicated relationships nor why the fungi involved should participate in such a decidedly one-sided relationship.

FUNGAL RELIANCE

A number of ground orchids carry the parasitic relationship one step further and forego ever carrying out photosynthesis—a phenomenon termed "holomycotrophy" or, more literally, "totally fungus eating." These orchids, such as the Bird's Nest Orchid (*Neottia nidus-avis*) of Eurasia, also switch to an ectomycorrhizal fungus as described above and obtain all their sugar indirectly from the neighboring trees. The underground orchids from Australia, genus *Rhizanthella*, not only produce none of their own food but also avoid raising their flowers above the soil surface. Unsurprisingly, the subterranean pollinator of *Rhizanthella* species is unknown.

Holomycotrophy is not confined among plants to orchids—for example, some members of the rhododendron and blueberry family, Ericaceae, form similar parasitic relationships with ectomycorrhizal fungi. All holomycotrophic plants, including orchids, that get their food entirely from fungi have in the past been classified as "saprophytes," meaning plants that live off decomposing material in the soil. This, however, is not an appropriate term because such plants are fungal parasites and not directly living off decaying material. Moreover, the food these orchids are stealing comes not from the fungi involved with rotting of wood in the soil but rather from fungi that are living in symbiosis with nearby forest trees.

19

TOP **The inflorescence** of the underground orchid from southwestern Australia, *Rhizanthella gardneri*, causes a crack in the soil by which its pollinator reaches it.

ABOVE ***Neottia nidus-avis*** steals its food and minerals from nearby trees via a fungal partner that is exchanging minerals for sugars with the trees.

THREATS TO WILD ORCHIDS

CONSERVATION

Plant conservation has long been the poor relation of animal conservation. It is much easier to get the public's attention if the plea for money involves the so-called "charismatic megafauna" such as elephants, tigers, pandas, rhinoceros, and cheetahs. Few plants have the same potential, but orchids come close. In a horticultural context, orchids grab the attention of the public, with many thousands drawn to orchid exhibitions.

Orchid conservation has had a few successes. For example, the Yellow Lady's Slipper, *Cypripedium calceolus*, has been the focus of a long-term restoration project funded by English Nature, the conservation arm of the United Kingdom Government. There are now flowering plants in several

RIGHT ***Cypripedium calceolus*** was reduced to a single plant in the wild in the UK and has been the subject of successful reintroduction efforts.

wild areas that were reintroduced as cultivated seedlings a decade ago. So far, none of these plants has produced seeds, but conservation is a slow process, even at its speediest. It takes a long time to overcome the problems created by our forebears, just as it will take a long time for our children to overcome the damage caused by the present generation.

MEETING HORTICULTURAL DEMAND

Other efforts to restore seedlings produced in cultivation to their natural habitats have failed dismally. Poachers removed a tropical Asian slipper orchid species, the spectacular *Paphiopedilum rothschildianum*, within months of it being replanted in forest reserves on Mount Kinabalu, in Sabah state, Malaysia—a UN World Heritage Site. For many showy orchid species, collection for the horticultural trade is a major threat, one that has proven almost impossible to surmount.

ABOVE **When small adult plants** of *Paphiopedilum rothschildianum* were reintroduced to its native Malaysian habitat they were quickly removed by orchid thieves.

The outcomes for the two lady's slipper species mentioned above were entirely different, but largely because the sites for *C. calceolus* were kept completely secret and guarded 24 hours a day while the plants were in flower. Also, horticultural demand for plants of *Cypripedium* is minimal as they have a partly justifiable reputation for being difficult to cultivate, in contrast to the high demand for the easily cultivated *P. rothschildianum*.

LEFT **Three of the four plants** on display at this orchid exhibition are species rather than hybrids, illustrating the horticultural appeal of orchid species.

RIGHT **Forest destruction** for agriculture, mining and human habitation is the major threat to orchid populations rather than collection for horticulture.

Relative to the number of orchid species in the world, those that are threatened by unsustainable collection for horticulture are a small percentage. For the great majority of orchid species, there is so little demand that concerns over their extinction for this reason can be discarded. There is legislation in place—the Convention on International Trade in Endangered Species of Fauna and Flora (CITES)—that has sought to control the unsustainable harvest of many species, mostly animals, but also of many plants, including all orchid species. For those species that are horticulturally desirable, the CITES provisions have not prevented the commercial exploitation of wild-collected plants, such as *P. rothschildianum*, and have to be considered a failure. The greatest threats to orchid species

RIGHT **Orchids such as *Dendrobium nobile*** are collected in huge numbers from several countries to supply the traditional medicine trade in China and India.

in many countries come, in fact, from conversion of their natural habitats for agriculture, mining, and human habitation, about which the CITES regulations can do nothing. The only orchids that have been formally and extensively assessed by the IUCN (International Union for Conservation of Nature) are the slipper orchids (Cypripedioideae).

HUMAN CONSUMPTION

Another major threat is posed by the use of many orchid species as food and medicine. Salep is a kind of starch made from the tubers of many terrestrial orchid species in the eastern Mediterranean and the Middle East, including members of the genera *Anacamptis*, *Orchis*, and *Ophrys*; it is used to make a dessert or a beverage. The tubers are collected unsustainably from the wild, and in many areas of Turkey all terrestrial orchid species are becoming rare as a result. Making a bad situation worse, orchid tubers are now being collected in many of the surrounding countries where salep is not consumed to supply the demand in those where it is.

In several East African countries, including Zambia, a bread called chikanda or African polony is made from peanuts and the pounded tubers of ground orchids—mostly, but not always, of the genera *Habenaria*, *Disa*, *Satyrium*, and *Brachycorythis*. Like salep, the bread is increasing in popularity, leading to many areas being stripped of all orchids and collection shifting to nearby countries.

In East Asia, use of *Dendrobium* species in traditional Chinese medicine is causing local extinction. As these wild populations collapse due to

23

FAR LEFT: ***Orchis italica* is harvested** to produce salep, a kind of starch.

LEFT: **The tubers of *Brachycorythis angolensis* are** ground and used to make chikanda, a kind of bread popular in East Africa.

RIGHT AND BELOW
Vanilla seed pods are fermented and then dried to produce the commercial flavoring vanilla.

due to unsustainable collection, plants are collected in neighboring countries to supply the burgeoning demand. In Europe, chemical extracts of several orchid species are added to shampoos and cosmetics. Although the containers claim that this is "sustainably harvested," there is no proof that any ground orchids are capable of being cultivated in quantities large enough to support this trade.

None of the above uses of orchids as food and medicine is being regulated by CITES, which was not designed to control such practices. Our advice is not to purchase any products that contain orchids, regardless of what the labels on these products might say. There is plenty of evidence that collection of these orchids is unsustainable and will result in at least local extinction of many species.

VANILLA

By far the most important orchid species economically is *Vanilla planifolia*, or vanilla, originally from Mexico and now widely cultivated in the tropics. Away from the plant's natural range and pollinators, the flowers are hand-pollinated to produce pods, with the nearly mature seed capsules, which contain the flavoring vanillin, fermented to produce vanilla on a commercial scale in areas such as Madagascar and Réunion. Tahitian Vanilla (a hybrid, *V.* × *tahitensis*) and West Indian Vanilla (*V. pompona*) are minor crops elsewhere. In Brazil and Paraguay, the orchid *Leptotes bicolor* is grown for its vanillin-rich seedpods. These orchids are propagated specifically for these purposes, and so this use is sustainable. We can continue eating ice cream made with real vanilla without feeling guilty.

25

OTHER USES OF ORCHIDS

- The perfume industry extracts scents from many orchids such as *Dendrobium moniliforme* and *Cattleya trianae*.

- Some orchid flowers are used to flavor drinks. In Réunion, the species *Jumellea fragrans*, found only on that island, flavors one of the rums known locally as *rhum arrangé*, which in turn now threatens the plant with extinction.

- In India, some species of *Dendrobium* are so abundant that they are used as cattle fodder.

- Species of *Gastrodia* (non-photosynthetic or holomycotrophic ground orchids) are widely employed in China and other Asian countries as traditional herbal medicine.

- Native Australian species of *Gastrodia* have been eaten as a source of starch by Aboriginal Australians.

BELOW **Hybrids of many orchid groups** are produced for sale in grocery and other stores, but those of *Phalaenopsis* are the most popular.

ORCHIDS IN THE HOME

Ultimately, the biggest use of orchids is in horticulture, as hugely popular ornamental houseplants and cut flowers, with hybrids of *Cattleya*, *Cymbidium*, *Oncidium*, *Phalaenopsis*, *Paphiopedilum*, and *Vanda* the most widely grown. There are currently more than 100,000 cultivars (mostly hybrids) in the trade, many of them not officially named. Hybrid seedlings are shipped in flasks containing thousands of plants from China and Japan to the Netherlands and the United States, where they are quickly grown to flowering size in greenhouses and sold to supermarkets and garden centers.

ORCHIDELIRIUM

The combination of column, lip, and pollinia have made it possible for orchid flowers to use pollinators from a wider range of animals than any other group of plants. The resulting diversity of form, size, shape, and color has provided the orchid hybridizer with an artist's palette full of amazing possibilities. The tropical orchid species that first started to appear in Europe, however, engendered great interest but no thought of hybridization, as no one at that time knew how to hybridize them or how to grow orchids from seed.

EARLY ARRIVALS IN THE WEST

The first recorded non-native "exotic" orchid in western Europe was *Bletia purpurea*, sent to England from the Bahamas in 1731. Soon thereafter, plants of the genus *Vanilla* were also introduced into English greenhouses, but these and other early introductions from the tropics were treated as heat-loving plants and kept in ridiculously hot conditions, where they soon perished. By 1794, 15 tropical orchid species were being grown more or less successfully at the Royal Botanic Gardens, Kew, England, almost all of them from the West Indies. It had taken nearly 65 years to figure out that orchids were not heat lovers, but, once this was realized, the era of their successful cultivation in Europe was underway.

EAST ASIAN TRADITIONS

Beyond Europe, the Chinese had been successfully cultivating orchids since at least the time of the Han Dynasty (206 BCE–CE 220), when Chinese nobility were growing plants collected from the wild in their private gardens. There are earlier references to "lan" (the modern Chinese word for orchid) in Chinese literature, but it may have been used simply for any fragrant plant, including species of the orchid genus *Cymbidium*. It was not until the Tang Dynasty (CE 618–907) that orchids gained popularity among the common people, and books started to appear that covered all aspects of their cultivation, including quality of plants, types of orchids, and care and watering.

Given the interest of the Chinese in cultural aspects of orchid growing, it is surprising that nothing was ever written by them about growing orchids from seed. For the Chinese, propagation was confined solely to the division of large plants into several smaller ones.

27

EARLIEST HYBRIDS

Not too long after tropical orchids began to appear in Europe, the first orchid hybrid emerged, raised from seed in 1853 and produced by crossing two species of *Calanthe—C. masuca* and *C. furcata*. The matter of how to get the peculiarly small seeds to germinate was not understood, but by sprinkling the seeds on the pots of the parent plants germination was achieved, presumably facilitated by an appropriate fungal species living in the potting medium of the mature orchid.

The first intergeneric (a cross between species in different genera) hybrid was produced in 1863, although this and the first trigeneric orchid hybrid, grown in 1892, are today all crosses between species considered to be members of the genus *Cattleya*. Records of every orchid hybrid produced, and its parentage, were started in 1906 and are maintained today by the Royal Horticultural Society in the United Kingdom.

TOP ***Cymbidium ensifolium*** has been cultivated for centuries in China, although no production from seed has been recorded there until the twentieth century.

ABOVE ***Calanthe masuca*** was one of the parent species of the first orchid hybrid known to have been produced in Europe.

GERMINATION DISCOVERIES

In 1922, American botanist Lewis Knudson (1884–1958) discovered that orchid seeds would germinate if spread over a nutrient-containing agar culture medium, so setting the stage for the mass production of orchid

28

ABOVE **Many terrestrial orchid species,** such as *Dactylorhiza fuchsii*, can be successfully grown in cultivation by inoculating them with the appropriate fungus.

species and hybrids by seed produced in cultivation. The dependence of orchids on fungi for natural germination was not realized, though, for a long time after orchid seeds were being germinated around the roots of the adult plants. That there were interactions between plant roots and fungi was reported frequently, without anyone understanding the nature of the interactions taking place.

In 1885 German botanist Albert Bernhard Frank (1839–1900) introduced the term "mycorrhiza" for the association between fungi and plant roots, but it was not until 1899 that orchid seeds were found by French mycologist Noël Bernard (1874–1911) to be infected by fungi. Bernard proved that this infection by an appropriate fungus induced orchid seed germination in 1903, when he infected orchid seeds with a pure culture of an orchid root fungus and followed their development into seedlings. He published a more general study of orchid germination in 1909.

Although nearly all epiphytic orchid species and many terrestrial species can be grown from seed on nutrient-enriched agar, most groups of orchids can be germinated by mycorrhizal fungi.

MASS PRODUCTION AND DIVERSITY

BELOW **Andean *Trichoceros antennifer*** is pollinated by sexually deceived male flies, but despite its fly-like appearance it is popular in horticulture.

Although orchids were initially only grown in Europe and North America by the well-to-do, as time has passed and mass artificial production of plants, both by seeds germinated on agar and with a fungus, has brought increased availability and lower prices, they have become much more widely cultivated. Today, mass-produced orchid hybrids are grown in the Netherlands, the United States, and East Asia so efficiently that flowering orchid plants are cheaply available in supermarkets and commercial nurseries in great quantities. Nevertheless, particular hybrids and rare species still command much higher prices and remain almost solely within the realm of specialist collections. Ultimately, the fascination with orchids is almost certainly due to the huge diversity of orchid flower types and their incongruous combination of relatively unattractive plants with spectacularly beautiful flowers. No one could describe a *Cattleya* plant as even vaguely interesting, but when this horrible thing bursts into flower it inspires such admiration that the ugly plant itself is forgotten.

If you attend an orchid show, then it is easy to think that all orchid flowers are big and showy. This, though, is far from true. We have illustrated here many of the smaller non-showy species so that a more balanced view of orchid diversity can be gained. All of the 600 orchids featured in these pages are species (not hybrids), and we have purposely focused on their history post-discovery and what little may be known about their biology and ecology, as well as shedding light on some uses other than in horticulture. Above all, our aim in this book is to convey a real sense of the astonishing species diversity that exists in nature within probably the most remarkable of all plant families.

29

ABOVE *Gavilea araucana* is an attractive species from Chile and Argentina but due to its problematic cultural requirements it is not seen in horticulture.

NOTES ON THE DESCRIPTIONS

Plant sizes are provided to give a general sense of how large or small these species are. The reader is likely to observe some plants that are larger or smaller than the size indicated. For some orchids, it is easy to estimate height and width because each season the plant dies back to an underground tuber, but for many others, especially the tropical epiphytic species, each stem (often with a pseudobulb, a swollen portion of the stem) is perennial. Each year a new stem is produced, and thus a plant increases in size over its lifetime. The height of these older plants is more or less constant but their width increases. The widths provided here are for the single growth produced annually, not for a clump of such growths, which will become larger as the plant ages. The height of such plants always includes the pseudobulb plus the leaf itself. In some cases, the leaf falls off at the end of the season, but the height provided always includes the leaf, even if at the time of flowering the leaf has fallen.

Flower size is measured from the top of the petal to the bottom of the lip or the two lateral sepals, if the latter are longer than the lip (they are often shorter). As with plant size, these sizes are given to provide a general impression of the flower size and smaller as well as larger examples will be encountered.

The area shaded in the distribution maps shows the native range of the species, and flowering times are given in months and/or seasons as appropriate to the region.

THE ORCHIDS

APOSTASIOIDEAE, VANILLOIDEAE & CYPRIPEDIOIDEAE

These three subfamilies account for few of the great number of orchid species, but they do contribute much variety in terms of vegetative and floral diversity. The smallest subfamily is Apostasioideae, its two genera and 14 species entirely confined to the tropics of Asia, where they are rarely recognized as orchids, lacking the usual fusion of the male and female parts of the flower. This has caused some botanists to consider them to be primitive orchids, although, in fact, other than their lack of fusion, Apostasioideae species are highly modified and unlike what we would imagine to be a primitive orchid. The 247 species in the 14 genera of Vanilloideae on the other hand have flowers that look like orchids but are vegetatively unlike an orchid, being tropical vines and small leafy and leafless plants, mostly herbs, of the temperate zones. The slipper orchids, subfamily Cypripedioideae (five genera, 169 species), are both tropical and north temperate species. They differ mostly in their retention of two anthers, although these are completely fused to the female parts, making them otherwise true orchids. Cypripedioideae species are mostly herbaceous plants, although a few resemble bamboos and can grow to a height of 20 feet (6 m).

SUBFAMILY	Apostasioideae
TRIBE AND SUBTRIBE	Not applicable
NATIVE RANGE	Tropical East Asia, from eastern Himalayas through Southeast Asia, north to Yakushima Island (Japan), and south to Queensland (Australia), at about 650–5,600 ft (200–1,700 m)
HABITAT	Tropical, broadleaf, evergreen humid forests, often in zones of mist and splash from rivers, cascades, or waterfalls
TYPE AND PLACEMENT	Terrestrial or on rocks
CONSERVATION STATUS	Locally common
FLOWERING TIME	June to September (wet season)

FLOWER SIZE
⅜ in (1 cm)

PLANT SIZE
Up to 16 × 14 in
(40 × 36 cm)

34

APOSTASIA WALLICHII
YELLOW GRASS ORCHID
R. BROWN, 1830

The Yellow Grass Orchid looks like a clump of grass, and its flowers are unlike other orchids. They are not resupinate (with the lip lowermost), and the anthers are only partially fused with the style. The flowers are almost regularly symmetrical and resemble those of the grasslike herb yellow star grass (*Hypoxis hirsuta*). Due to their unusual features, the genus *Apostasia* along with the genus *Neuwiedia* had been placed in a separate family, reflected in the name *Apostasia*, from the Greek for "separation" or "divorce."

The orchid's pollen is shed in response to vibration by pollinating insects. The roots have a strong smell of manure and are sometimes used medicinally to treat diarrhea and sore eyes. Nodules on the roots could be associated with a symbiotic relationship with mycorrhizal fungi that is typical of such orchids.

Actual size

The flower of the Yellow Grass Orchid is fragrant and has six slightly fleshy, boat-shaped, yellow tepals. There are two free stamens held parallel to the style, with their filaments partly fused to it, and the anthers clasp the style.

SUBFAMILY	Apostasioideae
TRIBE AND SUBTRIBE	Not applicable
NATIVE RANGE	Malesia to Melanesia, from Borneo and Java to the Philippines and Vanuatu, from sea level to 3,300 ft (1,000 m)
HABITAT	Evergreen dipterocarp forests on sandstone, limestone, ultramafic soil or shale, usually in deep shade under humid conditions
TYPE AND PLACEMENT	Terrestrial or on rocks
CONSERVATION STATUS	Locally abundant
FLOWERING TIME	June to September (wet season)

NEUWIEDIA VERATRIFOLIA

FALSE HELLEBORE ORCHID

BLUME, 1834

FLOWER SIZE
1⅜ in (3.5 cm)

PLANT SIZE
22 × 18 in
(56 × 46 cm)

35

Few people when they look at *Neuwiedia veratrifolia* think it is an orchid. Named for German naturalist, ethnologist, and explorer Prince Maximilian Alexander Philipp zu Wied-Neuwied (1782–1867), these large hairy plants produce up to ten plicate leaves that more closely resemble false hellebore (*Veratrum*, in the family Melanthiaceae). Like the genus *Apostasia*, *Neuwiedia* was placed in a separate family in the past because it has three free anthers instead of the single fused anther found in most other orchids. The two genera, however, share some unique traits with orchids and are now considered to be members of the Orchidaceae family.

Neuwiedia veratrifolia is self-compatible and mostly self-pollinating. In addition, stingless *Trigona* bees visit the flowers, vibrate the anthers, and are then dusted with the pollen released.

The flower of the False Hellebore Orchid has white crystals in its tissues. The upper sepals and petals are asymmetrical, whereas the lip is symmetrical and broader than the petals. Three stamens emerge from the column base, and the anthers are free from the style.

Actual size

SUBFAMILY	Vanilloideae
TRIBE	Pogonieae
NATIVE RANGE	Southeastern United States (New Jersey to Florida, eastern Tennessee, and Kentucky)
HABITAT	Pine barrens, bogs, wet meadows, stream courses
TYPE AND PLACEMENT	Terrestrial
CONSERVATION STATUS	Threatened or endangered
FLOWERING TIME	April to June (spring)

FLOWER SIZE
4½ in (11.4 cm)

PLANT SIZE
Stem up to 24 in (61 cm)

CLEISTESIOPSIS DIVARICATA
ROSEBUD ORCHID
(LINNAEUS) PANSARIN & F. BARROS, 2008

36

The fragrant, vanilla-scented Rosebud Orchid can be found in wetland areas of southeastern North America. The slender, long-stemmed plant typically bears one showy flower, subtended by a leafy bract that is usually longer than the ovary. Bees gather nectar from a pair of glands at the labellum base. Underground, the plant has a mass of thick roots attached to a rhizome and no tuber.

Cleistes, on which the genus name is based, comes from the Greek word for "closed," referring to the petals and lip, which form a tube, concealing the column. This makes the flower appear unopened, like a bud—hence its common name. The other part of the genus name, *-opsis*, refers to the plant's similarity to the large Neotropical genus *Cleistes*, in which it was previously included until DNA studies demonstrated that it should be segregated.

Actual size

The flower of the Rosebud Orchid has long, acuminate, usually maroon sepals and petals of soft rose pink, the latter never opening fully. The petals and long-keeled labellum, which is also pink with darker markings, form a long tunnel-like tube.

SUBFAMILY	Vanilloideae
TRIBE	Pogonieae
NATIVE RANGE	Michigan through Ontario to New England, south to Tennessee, Georgia, and South Carolina
HABITAT	Semi-open, mesic forests of eastern North America
TYPE AND PLACEMENT	Terrestrial
CONSERVATION STATUS	Threatened
FLOWERING TIME	April to May (spring)

ISOTRIA MEDEOLOIDES
SMALL WHORLED POGONIA
(PURSH) RAFINESQUE, 1838

FLOWER SIZE
1½ in (3.8 cm)

PLANT SIZE
Stem up to 12 in (30 cm),
with whorl of leaves just
below the flower

37

Considered to be the rarest orchid east of the Mississippi River, this species is found in temperate woodlands, where its ecology is tied deeply to the trees around it. The species name derives from the Small Whorled Pogonia's superficial resemblance to the plant *Medeola virginiana*, or Indian cucumber-root, which grows in similar habitats. The plant structure, unusual for an orchid, consists of a hollow stem with five or six bladelike leaves arranged in a whorl at its apex just below a single flower, although two flowers occasionally occur. Underground there is a mass of roots and no tuber.

Unlike its showier sister species, *Isotria verticillata*, *I. medeoloides* is sparse, often solitary, or found in small colonies. Like many woodland terrestrials, this species has been known to disappear or retreat underground for years at a time, making population studies difficult.

Actual size

The flower of the Small Whorled Pogonia has pale green sepals and petals and a whitish lip. The flowers do not open fully and are often short lived.

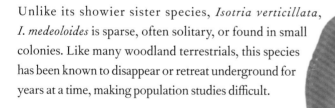

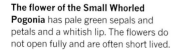

SUBFAMILY	Vanilloideae
TRIBE	Pogonieae
NATIVE RANGE	Eastern North America, from Canada to Florida and west to Minnesota
HABITAT	Wet meadows, bogs, stream sides, often occurring in poorly drained roadside ditches
TYPE AND PLACEMENT	Terrestrial
CONSERVATION STATUS	Threatened or endangered
FLOWERING TIME	Early spring in south to midsummer in northern part of range

FLOWER SIZE
1½ –2 in (3.8–5 cm)

PLANT SIZE
6–10 in (15–25 cm),
including inflorescence

POGONIA OPHIOGLOSSOIDES
ROSE POGONIA
(LINNAEUS) KER GAWLER, 1816

Actual size

A slender, semi-aquatic plant, often occurring in bogs and beside streams, the pretty Rose Pogonia can be locally abundant, often proliferating into lush, multi-growth colonies. Preferring to grow where there is easily available, pure water, this species is scarce in years with sparse rainfall but will rebound in wet periods. The short-lived, mostly pale pink flowers can vary in color and intensity and probably use their darker fringed labellum with yellowish filamentous crests to attract pollinators. This open-jawed appearance explains the plant's alternative common names, Adder's Mouth or Snake Mouth. Underground, there is a mass of roots but no tuber.

Pogonias grow in dappled light, usually in moist sphagnum moss, and can produce massive colonies. The genus name comes from the Greek word *pogon*, meaning beard, which refers to the hairy labellum.

The flower of the Rose Pogonia is usually pale pink with a darker lip, fringed with purplish striations, and a yellow crest. Flowers appear singly on a stem, though up to three have been reported on vigorous plants.

SUBFAMILY	Vanilloideae
TRIBE	Vanilleae
NATIVE RANGE	Southeastern China, Korea, Japan, and the Ryukyu Islands, at 3,300–4,300 ft (1,000–1,300 m)
HABITAT	Shaded woods
TYPE AND PLACEMENT	Terrestrial, mycoheterotrophic on wood-decaying fungi
CONSERVATION STATUS	Not assessed, but locally frequent
FLOWERING TIME	May to July (spring)

CYRTOSIA SEPTENTRIONALIS

NORTHERN BANANA ORCHID

(REICHENBACH FILS) GARAY, 1986

FLOWER SIZE
1⁹⁄₁₆ in (4 cm)

PLANT SIZE
Vegetative parts
underground, flowering
stems up to 36 in (91 cm)

39

The leafless Northern Banana Orchid lives underground until it flowers. Seedlings parasitize wood-decaying fungi (*Armillaria* species) and fulfill their carbon needs from this fungus, on which the plant is dependent for its entire life. The flowers do not produce nectar or scent, so it is difficult to imagine why insects or other animals might visit them. However, studies have shown that these flowers are actively self-pollinating, and every flower sets seed. Bright red, banana-like fruits grow from the pollinated flowers and are distributed by rodents and birds.

The genus name *Cyrtosia* is derived from the Greek *kyrtos*, meaning curved, which refers to the curved column, and *septentrionalis* is Latin for "northern." In Japan, the fruits have been used to treat urinary disease, gonorrhea, and dandruff.

The flower of the Northern Banana Orchid is orange brown and held in clusters. Sepals are warty outside and the petals are thinner and shorter. The lip is cup-shaped with a fringed edge. The column is strongly curved with two lateral, toothed wings and capped by two mealy pollinia, or pollen masses.

Actual size

SUBFAMILY	Vanilloideae
TRIBE	Vanilleae
NATIVE RANGE	Tropical South America, at 330–2,950 ft (100–900 m)
HABITAT	Open places in rain forests and savanna
TYPE AND PLACEMENT	Terrestrial
CONSERVATION STATUS	Not assessed
FLOWERING TIME	All year

FLOWER SIZE
4–5 in (10–12 cm)

PLANT SIZE
30–75 × 10–15 in
(76–190 cm),
76–191 × 25–38 in
(193–485 × 64–97 cm),
including inflorescence

40

EPISTEPHIUM SCLEROPHYLLUM
LEATHER-LEAFED CROWN ORCHID
LINDLEY, 1840

This large, ground-dwelling orchid produces erect stems covered with leathery, rigid, ovate leaves, while underground there is a branching horizontal rhizome with many tough roots. The inflorescences are terminal and have small floral bracts with many flowers that open successively, two to three at a time. At the top of the ovary the flowers are inserted into a scalloped ridge. This crownlike structure is the basis of the genus name (Greek, *epi-*, "upon," and *stephanos*, "crown"). The plant is a member of the same tribe as the genus *Vanilla*, to which it is closely related.

The showy flowers have a classical orchid shape (like species of the genus *Cattleya*), which indicates that they are probably pollinated by bees. In spite of their fantastically beautiful flowers, these orchids have never been successfully cultivated.

The flower of the Leather-leafed Crown Orchid has three relatively narrow, pink sepals and two broader petals. The massive pink lip is wrapped around the column and has yellow and white nectar guide markings with a cluster of long hairs near its middle.

Actual size

SUBFAMILY	Vanilloideae
TRIBE	Vanilleae
NATIVE RANGE	New Caledonia
HABITAT	Open, sunny savannas
TYPE AND PLACEMENT	Terrestrial
CONSERVATION STATUS	Not threatened
FLOWERING TIME	Spring

FLOWER SIZE
1³⁄₁₆–2 in (3–5 cm)

PLANT SIZE
2–3 ft (60–92 cm) tall,
including flowers

ERIAXIS RIGIDA
MAQUIS ORCHID
REICHENBACH FILS, 1876

41

The Maquis Orchid is endemic to the remote Pacific island of New Caledonia, which has a tropical climate. New Caledonia houses some of the most ancient flora on Earth. The orchid produces a tough wiry stem with leaves along its length and up to a dozen flowers at the top.

Growing in full sun, the buds and inflorescence are covered with minute white hairs. The plants have adapted to grow in the maquis, a vegetation on nutrient-poor soil laden with heavy metals that would be toxic to many other plants. The lip bears a row of sharp, hinged, inward-pointing scales that make it difficult for a pollinating insect to retreat, encouraging it to position itself to best carry the friable pollen.

The flower of the Maquis Orchid is fleshy and short-lived. They are usually borne two at a time on a successively flowering inflorescence, and are white or pale pink, with a tubular, reddish-purple lip.

Actual size

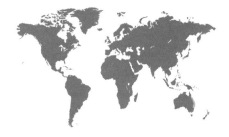

SUBFAMILY	Vanilloideae
TRIBE	Vanilleae
NATIVE RANGE	Northern and eastern Australia (Queensland, New South Wales)
HABITAT	Sclerophyll forests in full sun, scrambling over eucalyptus logs, tree stumps, and decaying wood, at 165–1,640 ft (50–500 m)
TYPE AND PLACEMENT	Terrestrial vine, mycoheterotrophic
CONSERVATION STATUS	Not assessed
FLOWERING TIME	September to December (spring to early summer)

FLOWER SIZE
½ in (1.3 cm)

PLANT SIZE
Up to 20 ft (6 m) long,
leafless

42

ERYTHRORCHIS CASSYTHOIDES
BLACK BOOTLACE ORCHID
(A. CUNNINGHAM EX LINDLEY) GARAY, 1986

The Black Bootlace Orchid is mycoheterotrophic, which means it lacks chlorophyll and instead takes all its nutrients from a fungus that it parasitizes. It has slender, brown, climbing branches, not unlike Devil's Twine, a parasitic vine of the genus *Cassytha*—hence the scientific name. The common name comes from the straplike branches, which resemble shoelaces.

The plant stems are held in place by thick, fleshy roots, and the highly branched inflorescence bears many scented flowers that attract small bees for pollination. The flowers closely resemble those of the genus *Vanilla*, to which the species is related, along with other chlorophyll-free genera, such as *Galeola*. Plants are reported to be short-lived and more vigorous and frequent in places where there is rotting timber.

The flower of the Black Bootlace Orchid has three creamy, spreading sepals and two similar spreading, creamy-white petals. The lip is also creamy and tubular, curved around the column with an irregularly notched or lobed margin.

Actual size

SUBFAMILY	Vanilloideae
TRIBE	Vanilleae
NATIVE RANGE	Tropical Southeast Asia and Malesia, from Hainan Island to New Guinea, from sea level to 5,600 ft (1,700 m)
HABITAT	Decaying tree stumps, often in treefall gaps or along streams in sunny but damp locations
TYPE AND PLACEMENT	Terrestrial and climbing, mycoheterotrophic
CONSERVATION STATUS	Not assessed, but locally frequent
FLOWERING TIME	April to June

FLOWER SIZE
1³⁄₁₆ in (3 cm)

PLANT SIZE
Climbing trees up to
66 ft (20 m) tall

GALEOLA NUDIFOLIA
LEAFLESS HELMET VINE
LOUREIRO, 1790

43

The tip of the column of the Leafless Helmet Vine has a pair of projecting appendages that resemble a small helmet, called *galeole* in Greek, which gives the genus its scientific name. This large vining orchid lacks chlorophyll and climbs with its reddish stems and roots over rotting tree trunks, where it parasitizes the wood-decaying fungus that provides its carbon and nutrients. The vibrant flowers are most likely pollinated by small bees, but no specific observations on pollination have been made.

The winged seeds of the *Galeola* species are much bigger than those of most orchids, which are generally dustlike. The wings help to disperse the seeds across dense rain forest habitats, where there is little wind to help them drift to new, suitable locations. A related Japanese species, *Galeola septentrionalis*, becomes a weed in log beds used to cultivate shiitake mushrooms.

The flower of the Leafless Helmet Vine is fleshy and has free, spreading sepals and thinner, similarly sized petals. The lip is fleshy, round, and cup-shaped with a margin turned inward. The column is strongly curved forward, bears a pair of projections, and has two pollinia.

Actual size

SUBFAMILY	Vanilloideae
TRIBE	Vanilleae
NATIVE RANGE	Southeast Asia, peninsular Malaysia, western Java, Taiwan, Philippines, New Guinea
HABITAT	Dark humid forests, at 985–5,250 ft (300–1,600 m)
TYPE AND PLACEMENT	Terrestrial
CONSERVATION STATUS	Unknown
FLOWERING TIME	March to June (spring)

FLOWER SIZE
⅝ in (1.6 cm)

PLANT SIZE
Vegetative parts underground, flowering stems up to 20 in (51 cm)

LECANORCHIS JAVANICA
BASIN ORCHID
BLUME, 1856

44

This leafless plant lives on fungi for all its life and only emerges from the soil when flowering. The flower is surrounded by a cup-shaped structure that persists on top of the ovary, hence the common name of Basin Orchid and the genus name derived from the Greek *lecane*, meaning a pot.

The flowers open only slightly, are not showy, and lack a scent. They may, therefore, be mostly self-pollinating, although the lips bear long hairs that could be an indication of fly pollination. However, very little is known about the ecological interactions of this plant.

The flower of the Basin Orchid has three greenish-yellow sepals and greenish-yellow, oblong petals. The three-lobed lip is white or light yellow, with a hairy middle lobe irregularly notched at the margin. The column is hairless and bears semicircular wings at the tip.

Actual size

SUBFAMILY	Vanilloideae
TRIBE	Vanilleae
NATIVE RANGE	Seychelles (islands of Mahé, Praslin, Silhouette, and Félicité)
HABITAT	Granite outcrops and other dry open areas, from sea level to 1,300 ft (400 m)
TYPE AND PLACEMENT	Climbing on rocks and trees
CONSERVATION STATUS	Least concern
FLOWERING TIME	December to February (wet season)

VANILLA PHALAENOPSIS
SEYCHELLES VANILLA
REICHENBACH FILS EX VAN HOUTTE, 1867

FLOWER SIZE
3⅛ in (8 cm)

PLANT SIZE
Stems grow to around
18 ft (5.5 m) long

45

One of the most beautiful species of the genus *Vanilla*, this leafless orchid has climbing green, succulent stems with which it scrambles over rocks and trees, supported by aerial roots. It flowers after heavy rains but only on plants that have grown taller than their support and have pendent branches on which the inflorescences are formed. Up to three fragrant flowers are open at the same time, giving an impressive display.

Even though its distribution is restricted to the granitic Seychelles, the species is relatively common there. It is similar to *Vanilla aphylla*, found in Southeast Asia and Madagascar, which has a lilac instead of orange lip. The green fruits do not produce the aromatic compounds typical of commercial *V. planifolia*.

The flower of the Seychelles Vanilla is pure white. The oblong sepals are spreading, and the two petals are of the same length, but with wavy margins. The lip is entire and folded back at the margin, with an apricot-colored center.

Actual size

SUBFAMILY	Vanilloideae
TRIBE	Vanilleae
NATIVE RANGE	Mexico (but widely cultivated and naturalized elsewhere in the tropics)
HABITAT	Lowland tropical forests
TYPE AND PLACEMENT	Terrestrial, but climbs trees
CONSERVATION STATUS	Not threatened
FLOWERING TIME	Throughout the year

FLOWER SIZE
2½ in (6.4 cm)

PLANT SIZE
20 ft (6 m) or more

VANILLA PLANIFOLIA

VANILLA ORCHID

JACKSON EX ANDREWS, 1808

46

The most commercially important orchid, Vanilla is cultivated in tropical places around the world for its "beans," which, when fermented and dried, produce the popular flavoring. The genus *Vanilla* occurs on five continents, with more than a hundred species, and is one of only five vining orchid genera, which need the support of trees to grow to their full potential. The vines can grow to great lengths.

Short-lived flowers with tubular lips are produced successively on axial racemes. They are pollinated by a wasp in their natural range in Mexico, but pollination has to be done by hand in plantations in places such as Madagascar, Réunion, and Tahiti, where the plants are cultivated in large numbers. This one species provides 95 percent of the world's commercially produced vanilla pods.

The flower of the Vanilla Orchid is usually yellow or greenish, with similarly colored sepals, petals, and tubular-keeled lip. Although flowers last a single day, the plants bloom frequently and successively over a long period.

Actual size

SUBFAMILY	Cypripedioideae
TRIBE AND SUBTRIBE	Not applicable
NATIVE RANGE	Temperate northern Europe and Asia, from the British Isles to Korea and Japan
HABITAT	Temperate woodlands and scrub, at up to 6,600 ft (2,000 m)
TYPE AND PLACEMENT	Terrestrial
CONSERVATION STATUS	Widespread, although endangered in places
FLOWERING TIME	April to June (spring)

CYPRIPEDIUM CALCEOLUS
YELLOW LADY'S SLIPPER
LINNAEUS, 1753

FLOWER SIZE
2–3 in (5–8 cm)

PLANT SIZE
15–30 in (38–76 cm),
including flowers

47

The Yellow Lady's Slipper has a vast native range over expansive areas of the Northern Hemisphere. Some have considered the North American species *Cypripedium parviflorum* to be the same or merely a varietal form of this beautiful orchid, but it is now known to be a distinct species. It thrives on damp substrate in limestone-rich areas, which may be why the species is amenable to cultivation. Despite the plant's widespread distribution, poaching and urban sprawl threaten some populations.

In the species name, *calceolus* (Latin) means "little shoe," and it is the slipperlike shape of the lip that has inspired both the scientific and common names. The pouch acts as an insect trap, waylaying hapless pollinators, usually bees, but with no reward for their services. They are released covered with the pollen.

The flower of the Yellow Lady's Slipper varies but generally has yellow to brown sepals and petals, with a brilliant-yellow, pouch-shaped lip. Usually solitary flowers appear at the apex of a pubescent stem and are subtended by a leaflike bract.

Actual size

SUBFAMILY	Cypripedioideae
TRIBE AND SUBTRIBE	Not applicable
NATIVE RANGE	Central mountains of Taiwan
HABITAT	Beside streams and in moist riparian woodlands
TYPE AND PLACEMENT	Terrestrial
CONSERVATION STATUS	Endangered
FLOWERING TIME	April to May

FLOWER SIZE
3½–4 in (9–10 cm)

PLANT SIZE
Basal leaves up to
8 in (20 cm) tall; inflorescence
up to 24 in (61 cm) tall,
including flower

48

CYPRIPEDIUM FORMOSANUM
TAIWANESE LADY'S SLIPPER
HAYATA, 1916

The flower of the Taiwanese Lady's Slipper
usually bears white to pale pink sepals and
petals and a deeper pink, pouch-shaped lip,
often with purple markings. The pouch interior
can be intensely purple.

With paired and pleated basal leaves that resemble a frilled, Elizabethan collar, this is one of the most distinctive of the lady's slipper orchids. Prettier and more diminutive than its coarser, more widely distributed sister species *Cypripedium japonicum*, the Taiwanese Lady's Slipper can form massive clumps of more than a hundred blooming stems on an individual plant. A hairy stem supports a single, remarkable flower (rarely two). White or pink-blushed segments are the background for the purplish speckled pouch lip, with an often heart-shaped front opening through which the pollinating bees force themselves. Underground it has a mass of thick roots with no tuber.

It is a hardy, temperate plant, with exceptional ornamental appeal, which is easy to propagate. The highly endemic and localized wild populations, however, are becoming increasingly rare due to collection pressure.

Actual size

SUBFAMILY	Cypripedioideae
TRIBE AND SUBTRIBE	Not applicable
NATIVE RANGE	Eastern and central North America
HABITAT	Bog edges and moist temperate woodlands
TYPE AND PLACEMENT	Terrestrial
CONSERVATION STATUS	Widespread but locally threatened by poaching
FLOWERING TIME	May to June

FLOWER SIZE
4–4¼ in (10–11 cm)

PLANT SIZE
15–40 in (38–102 cm) tall,
including flowers

CYPRIPEDIUM REGINAE
SHOWY LADY'S SLIPPER
WALTER, 1788

49

The Showy Lady's Slipper is revered and admired for its stately size and spectacular blooms. Taller and larger-flowered than many species of this genus, it is difficult to miss in the wild when in full show. Like all others in its subfamily, the plant has a colorful pouch that acts as a trap, forcing the bees to exit by one route, where they pick up the pollen. Bees approach the flower expecting the colorful blossoms to be sources of nectar, only to find no reward. Underground, these plants have a mass of thick roots but no tubers.

Because of its exceptional beauty and hardiness, the species is often poached from the wild for use in gardens. Such dug up plants rarely survive, but plants grown from seed in nurseries have a much improved success rate.

The flower of the Showy Lady's Slipper generally has white sepals and petals with a large, inflated, rose-pink (rarely white) lip. Flowers are usually subtended by a leafy bract.

Actual size

SUBFAMILY	Cypripedioideae
TRIBE AND SUBTRIBE	Not applicable
NATIVE RANGE	Oaxaca, Mexico
HABITAT	Seasonally arid limestone outcrops
TYPE AND PLACEMENT	Terrestrial or lithophyte
CONSERVATION STATUS	Critically endangered
FLOWERING TIME	Usually September (fall)

FLOWER SIZE
¾–1 in (2–2.5 cm)

PLANT SIZE
2–4 in (5–10 cm),
excluding inflorescence

MEXIPEDIUM XEROPHYTICUM
MEXICAN LADY'S SLIPPER
(SOTO ARENAS, SALAZAR & HÁGSATER) V. A. ALBERT & M. W. CHASE, 1992

50

Actual size

Slipper orchid enthusiasts were amazed at the discovery of this succulent miniature species in 1985. Shielded from direct sunlight by the rocky terrain, the original collection site in southern Mexico yielded a mere seven plants. This site was later badly burned during a dry year, and no further plants have been found there (although a second locality is now known). The Greek genus and species names refer to the orchid's country of origin and its xeric, or dry, habitat. There is, in fact, reasonably plentiful water during much of the year where *Mexipedium xerophyticum* grows but also a three-month period of extreme dryness in midwinter.

The species spreads easily due to production of horizontal shoots, called stolons. The plants can produce several blooms on their successive flowering inflorescences.

The flower of the Mexican Lady's Slipper
is tiny, white to pale pink, with some pink in its central staminode. The petals are scythe-shaped and the flower also bears a miniature, pinkish-white pouch.

SUBFAMILY	Cypripedioideae
TRIBE AND SUBTRIBE	Not applicable
NATIVE RANGE	Southern India (Kerala)
HABITAT	Exposed grassland in rock crevasses
TYPE AND PLACEMENT	Terrestrial on rocks
CONSERVATION STATUS	Critically endangered
FLOWERING TIME	March to April (spring)

FLOWER SIZE
3 in (7.5 cm)

PLANT SIZE
8–14 × 10–20 in
(20–36 × 25–51 cm),
excluding inflorescence

PAPHIOPEDILUM DRURYI
DRURY'S SLIPPER ORCHID
(BEDDOME) STEIN, 1892

51

Colonel Drury, a member of the British military in India, recorded this orchid in 1856, and it was subsequently named for him. From a fan of green, leathery leaves without any markings, its solitary flower emerges at the top of a purple, hairy stem with a green sheath subtended by a hairy bract. The plant is one of the few species of *Paphiopedilum* with a rhizome that creeps along or just below the surface of the substrate. There are no underground tubers in members of this subfamily.

Drury's Slipper Orchid has not recently been collected and is thought to be extinct in the wild. It persists in cultivation, where it is popular for its beautiful flowers. The pollination of the species is unknown, although, given its coloring, it might be pollinated by bees.

The flower of Drury's Slipper Orchid is mostly yellow and has two lateral sepals fused and placed behind the slipper-shaped lip. The dorsal sepal arches over the lip and has a brown stripe and a white margin. The two spreading petals have a brown stripe and spots.

Actual size

SUBFAMILY	Cypripedioideae
TRIBE AND SUBTRIBE	Not applicable
NATIVE RANGE	Southeastern Yunnan province (China), southeastern Laos, and northern Vietnam
HABITAT	Primary, broadleaved, permanently humid cloud forests, at 2,950–6,200 ft (900–1,900 m)
TYPE AND PLACEMENT	Terrestrial, usually in shady leaf-litter pockets on silicate soil or on granite cliffs
CONSERVATION STATUS	Critically endangered
FLOWERING TIME	September to December (fall to early winter)

FLOWER SIZE
3¼ in (8 cm)

PLANT SIZE
12–20 × 14–30 in
(31–51 × 36–76 cm),
including inflorescence

PAPHIOPEDILUM GRATRIXIANUM
GRATRIX'S SLIPPER ORCHID
ROLFE, 1905

52

First described in 1905 from a plant collected by Wilhelm Micholitz (1854–1932) in Laos and exhibited in the UK by Messrs. Sanders' nursery, Gratrix's Slipper Orchid was named for Manchester industrialist and hobbyist orchid grower Samuel Gratrix. The plants produce fans of leaves from which single-flowered, purple stems grow. The leaves are slightly purple-spotted below, near their base, and have a shallowly notched or three-toothed tip. There are no underground tubers.

Pollination of the species has not been studied, but its floral morphology, especially the spotting on the dorsal sepal, probably indicates pollination involving deceit by mimicking a brood site for flies (tricking them into laying their eggs and falling into the lip). The only way out of this trap is to climb up the rear side of the lip and out near its base, where the pollen and stigma are located.

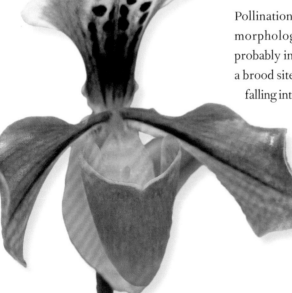

The flower of Gratrix's Slipper Orchid has the two lateral sepals fused behind the cup-shaped lip. The dorsal sepal is elaborately spotted, recurved at the base, and held above the lip. The petals are reddish, incurved, and slightly hairy along their margins.

Actual size

SUBFAMILY	Cypripedioideae
TRIBE AND SUBTRIBE	Not applicable
NATIVE RANGE	Northern Vietnam to southwestern China, including western and northern Guangxi, southeastern Yunnan, and western Guizhou provinces
HABITAT	Limestone cliffs and crevices, at 1,200–5,250 ft (360–1,600 m)
TYPE AND PLACEMENT	Terrestrial, or lithophyte growing on steep rocky slopes
CONSERVATION STATUS	Critically endangered due to overcollection
FLOWERING TIME	April to May (spring)

PAPHIOPEDILUM MICRANTHUM
TROPICAL PINK LADY'S SLIPPER
TANG & F. T. WANG, 1951

FLOWER SIZE
4–5½ in (10–14 cm)

PLANT SIZE
12–15 × 8–12 in
(30–40 × 20–30 cm),
including inflorescence

53

Collectors have coveted *Paphiopedilum micranthum*, one of the most ornamental of the slipper orchids, since its discovery in 1951. It is a member of the section *Parvisepalum*, a group notable for reduced sepals but large, colorful petals and lip. Thought to be described from an immature flowered specimen, the species name *micranthum* (Greek meaning "tiny flower") seems mismatched to the large, stunning bloom borne on upright stems, with its enormous bowl-shaped lip.

The tough, leathery foliage is exotically patterned with dark green and white, suffused with purple in blotches that cover the undersides of the leaves. Several color forms have been described, and its striking good looks have made it a popular parent for innumerable hybrids. Winter temperatures often approach freezing in its natural habitats, which induces spring flowering.

The flower of the Tropical Pink Lady's Slipper
is usually pink or white with reddish veins on the
petals, sometimes with a yellow or golden base.
The rounded, bowl-shaped lip is usually pink or
white, often with pale purple suffusion.

Actual size

SUBFAMILY	Cypripedioideae
TRIBE AND SUBTRIBE	Not applicable
NATIVE RANGE	Rain forests around Mount Kinabalu in northern Borneo, at 1,640–3,950 ft (500–1,200 m)
HABITAT	Steep serpentine cliffs near streams or seeps
TYPE AND PLACEMENT	Terrestrial
CONSERVATION STATUS	Critically endangered due to poaching
FLOWERING TIME	April to May (spring)

FLOWER SIZE
6–10 in (15–25 cm)

PLANT SIZE
10–15 × 12–20 in
(25–38 × 30–51 cm),
excluding inflorescence

54

PAPHIOPEDILUM ROTHSCHILDIANUM
ROTHSCHILD'S SLIPPER ORCHID
(REICHENBACH FILS) STEIN, 1892

Often referred to as the "king of orchids," this impressive multi-floral species was named for Ferdinand James von Rothschild (1839–98), a member of the Rothschild banking family and a supporter of horticultural science. Its large size and strong colors have made it a coveted collector's item and an outstanding parent of hybrids. Known from only a few sites on Mount Kinabalu, it has been close to extinction several times due to overzealous collectors.

The outstretched petals have an array of fine hairs and spots that lure flies to these flowers. They try to lay their eggs on the staminode, a sterile stamen, but instead fall into the traplike pouch and pick up a mass of pollen as they exit through the top part of the lip.

The flower of Rothschild's Slipper Orchid has cream-colored sepals and petals, overlaid with bold mahogany stripes and spots. The color of the pouch-shaped, forward-jutting lip varies from light reddish-brown to deep maroon red.

Actual size

SUBFAMILY	Cypripedioideae
TRIBE AND SUBTRIBE	Not applicable
NATIVE RANGE	Southern Mexico through Central America, Venezuela, Colombia, Ecuador, and Peru
HABITAT	Grassy rocky slopes near streams and seeps
TYPE AND PLACEMENT	Terrestrial on cliffs, steep embankments, and seeps; reported as occasionally epiphytic in Central America
CONSERVATION STATUS	Endangered
FLOWERING TIME	Throughout the year, but more likely winter to spring

PHRAGMIPEDIUM CAUDATUM
LONG-TAILED SLIPPER ORCHID
(LINDLEY) ROLFE, 1896

FLOWER SIZE
Up to 30 in (75 cm)

PLANT SIZE
12–28 × 20–36 in
(30–71 × 51–91 cm),
excluding inflorescence

55

Petals of exceptional length are what particularly distinguish the Long-tailed Slipper Orchid, one of about six or so slipper orchid species displaying this improbable yet fascinating floral trait. The petals continue elongating until they touch a hard surface, twisting and dangling in the breeze. They produce a bad-smelling scent that helps attract the plant's pollinators.

The flowers are pollinated by syrphid hoverflies searching for brood sites. The spots encircling the pouch rim are thought to resemble small aphids or other insects that might be consumed by newly hatched syrphid larvae. As on all other slipper orchids, the lip rim is slippery, and the insects fall inside the pouch and then exit the flowers at the top of the lip, where they contact the pollen (and stigma on subsequent visits).

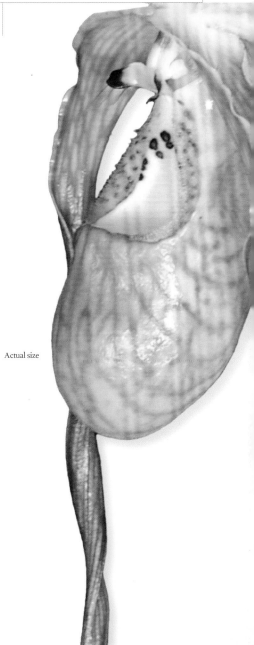

Actual size

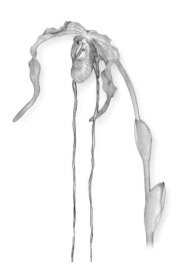

The flower of the Long-tailed Slipper Orchid is variable but generally displays pale shades of tawny brown, tan, and green. The pendent and twisting petals are often darker greenish-brown, and the lip often bears a reticulate pattern.

SUBFAMILY	Cypripedioideae
TRIBE AND SUBTRIBE	Not applicable
NATIVE RANGE	Northeastern Peruvian Andes
HABITAT	Cloud forests, at 5,900–7,200 ft (1,800–2,200 m), on steep rocky cliff faces
TYPE AND PLACEMENT	Terrestrial on steep embankments
CONSERVATION STATUS	Critically endangered
FLOWERING TIME	Spring

FLOWER SIZE
6 in (15 cm), sometimes
exceeding 7 in (18 cm)

PLANT SIZE
12–28 × 16–36 in
(30–71 × 41–91 cm),
excluding inflorescence

56

PHRAGMIPEDIUM KOVACHII
PERUVIAN GIANT
SLIPPER ORCHID

J. T. ATWOOD, DALSTRÖM & RICARDO FERNÁNDEZ GONZALES, 2002

The flower of the Peruvian Giant Slipper Orchid
is huge by slipper orchid standards, topping an
impressive 3 ft (1 m) successively flowering
inflorescence. Brilliant fuchsia to purple, it bears
large, rounded petals, an unusual shape for most
slippers, and an even more intensely colored pouch.

The spectacular, huge-flowered, brilliant fuchsia-colored
Phragmipedium kovachii, said to be the greatest orchid discovery
of the past century, is steeped in intrigue and scandal. Michael
Kovach, an American collector, bought the plant from a
roadside vendor in Peru. He transported the plant illegally to
Florida, where it was named and described, but the Peruvian
government intervened, and Kovach was fined and put on
probation for two years.

The Peruvian government developed a conservation and
propagation program that has supplied seed-grown plants to
horticulture. Thanks to that program, this extraordinary slipper
orchid is now enjoyed around the world, revolutionizing the
breeding of hybrids due to the color, size, and vigor
it contributes. In nature, the species is rare
and still subject to overcollection.

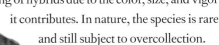

Actual size

SUBFAMILY	Cypripedioideae
TRIBE AND SUBTRIBE	Not applicable
NATIVE RANGE	Ecuador
HABITAT	Forest clearings in poor, sandy, and sometimes swampy conditions, at 1,800–3,300 ft (550–1,000 m)
TYPE AND PLACEMENT	Terrestrial
CONSERVATION STATUS	Endangered—found in rare localized populations, never in abundance
FLOWERING TIME	Spring to summer

SELENIPEDIUM AEQUINOCTIALE
EQUATORIAL MOON-SLIPPER
GARAY, 1978

FLOWER SIZE
1½–2 in (4–5 cm)

PLANT SIZE
6–10 ft (1.8–3 m),
including flowers

57

Though extremely rare in the wild, this is likely the most commonly encountered species in *Selenipedium*, the strangest of the slipper orchid genera. The Equatorial Moon-slipper is not well studied and next to impossible to cultivate. It is vegetatively a giant, with a terminal raceme of relatively insignificant flowers. Like all members of this subfamily, the flowers trap their pollinators, which are permitted to exit the pouch at its top, where they contact the pollen on their first visit to a flower and deposit the pollen on their second visit.

The plant has slender stems and soft, plicate leaves, resembling a bamboo or tall grass.

The flower of the Equatorial Moon-slipper is borne at the apex of a tall stem. Segments are generally buff yellow, with a rounded dorsal sepal and fused ventral sepal. The large pouch is blood-red centrally and yellow elsewhere.

Actual size

ORCHIDOIDEAE

The 200 genera and around 3,630 species of Orchidoideae make it the second-largest orchid subfamily and the dominant orchid element in a number of temperate regions, including Eurasia and North America (tribe Orchideae, subtribe Orchidinae), southern Africa (tribe Orchideae, subtribes Disinae and Satyriinae), southern South America (tribe Cranichideae, subtribe Chloraeinae), and Australia (tribe Diurideae). Its species are also diverse in the tropics, but their presence there is overshadowed by the huge diversity in the subfamily Epidendroideae. They share with Epidendroideae a single anther fully fused to the stigma, but, unlike the members of that much larger subfamily, are almost entirely composed of herbaceous plants with tubers or thick clusters of roots attached to a short stem, or rhizome. Their leaves also lack tough fibers that are typical of Epidendroideae. The type genus of Orchidoideae, which gives its name to the family, is Eurasian *Orchis*, referring to its testicle-like tubers, and the type species is *O. militaris*, named by Linnaeus.

SUBFAMILY	Orchidoideae
TRIBE	Codonorchideae
NATIVE RANGE	Central and southern Chile, southern Argentina, and the Falkland Islands
HABITAT	Forested areas with constant high humidity, at 65–6,200 ft (20–1,900 m)
TYPE AND PLACEMENT	Terrestrial
CONSERVATION STATUS	Locally common but not frequent
FLOWERING TIME	Late January to June (fall to winter)

FLOWER SIZE
1 in (2.5 cm)

PLANT SIZE
20 × 2 in (50 × 5 cm),
including inflorescence

60

CODONORCHIS LESSONII
LITTLE PIGEON ORCHID
(D'URVILLE) LINDLEY, 1840

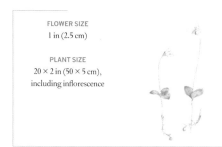

Actual size

The Little Pigeon Orchid, or *palomita* in Spanish, is a delicate plant with a whorl of three or four leaves partway along a stem topped by a single white flower. The species previously perplexed taxonomists because, although clearly in the subfamily Orchidoideae, its features were so unusual that its relation to the others was unclear. DNA studies confirmed its isolation in the subfamily, and so it became the sole member of its tribe. It grows terrestrially in low vegetation and has a single underground tuber from which a single collared stem arises.

The flowers are probably mostly self-fertilizing, but the floral structure suggests some sort of bee could be the natural pollinator. The orchid's name is derived from Greek *kodon*, meaning "bell," referring to the prominent appendages on the lip.

The flower of the Little Pigeon Orchid is white, with three spreading sepals. The two petals are purple-blotched and curve over the upright column. The lip is ornamented with purple or green warts, and the column is striped green and winged.

SUBFAMILY	Orchidoideae
TRIBE AND SUBTRIBE	Cranichideae, Chloraeinae
NATIVE RANGE	Central Chile
HABITAT	Low-elevation coastal areas, at 0–650 ft (0–200 m)
TYPE AND PLACEMENT	Terrestrial
CONSERVATION STATUS	Not assessed
FLOWERING TIME	July to November (winter to spring)

BIPINNULA FIMBRIATA

GREEN BEARD ORCHID

(POEPPIG) I. M. JOHNSTON, 1929

FLOWER SIZE
1½–2⅜ in (4–6 cm)

PLANT SIZE
7–10 × 10–28 in
(18–25 × 25–71 cm),
including flowers

61

Most species of *Bipinnula* exhibit one of the most bizarre floral structures of the orchid family. *Bipinnula fimbriata*, a Chilean endemic, bears the characteristic sepaline extensions, illustrating well the features that give the genus its name, meaning "two feathers" in Latin. This structure on the sepals may produce a scent to attract a pollinator, but no pollinator has ever been reported.

A resident of dry scrub in lowland coastal areas of central Chile, *B. fimbriata* blooms during late winter as many other plants are just coming out of dormancy. It can bear between 5 and 15 flowers per stem on an upright raceme, which emerges from a basal rosette of leaves. Underground, this species has a cluster of thick roots (but no tubers) and can form small colonies of closely spaced plants.

The flower of the Green Beard Orchid
has green sepals with extreme fimbriate extensions from the laterals at their terminus. Petals are white overlaid green, and the white lip has many dark green, wartlike structures on its surface.

Actual size

SUBFAMILY	Orchidoideae
TRIBE AND SUBTRIBE	Cranichideae, Chloraeinae
NATIVE RANGE	Central and southern Chile to southwestern Argentina
HABITAT	Shrubs or forests and along lakes and rivers, usually in humid environments
TYPE AND PLACEMENT	Terrestrial
CONSERVATION STATUS	Not assessed
FLOWERING TIME	Mid-November to early January (spring to summer)

FLOWER SIZE
1⅛ in (3 cm)

PLANT SIZE
Up to 30 × 4 in
(75 × 10 cm),
including inflorescence

62

GAVILEA ARAUCANA
ARAUCANIA ORCHID
(PHILIPPI). M. N. CORREA, 1956

Actual size

When the pollinator of the Araucania Orchid, most likely a wasp, tries to drink nectar produced at the base of the column, it lands on the hinged lip and moves inward, eventually passing the balance point and finding itself pressed into the column. The pollen of this species is not formed into pollinia, and the pollinator first gets smeared with stigmatic fluid and then dusted with dry pollen. However, in some cases the flowers may also self-pollinate. Underground is a short stem (rhizome) that has a cluster of thick hairy roots radiating outward.

Gavilea araucana is named for the region in Chile called Araucanía, a former independent kingdom ruled by the indigenous Mapuche people. It was here that botanists first encountered the species.

The flower of the Araucania Orchid has creamy white, ovate-lanceolate sepals with long green tips, and the petals have green dashed lines or spots. The lip is cupped with a red, narrow claw and three lobes, the middle lobe irregularly green-striped, the two lateral erect lobes yellow.

SUBFAMILY	Orchidoideae
TRIBE AND SUBTRIBE	Cranichideae, Chloraeinae
NATIVE RANGE	Montane regions of Chile and Argentina
HABITAT	Patagonian steppe, pre-Andean shrublands
TYPE AND PLACEMENT	Terrestrial
CONSERVATION STATUS	Not threatened
FLOWERING TIME	Early spring after snow melts

FLOWER SIZE
2–3 in (5–8 cm)

PLANT SIZE
10–20 basal leaves
up to 6 in (15 cm) long,
generally flat to the
ground in a basal rosette

CHLORAEA MAGELLANICA
PORCELAIN ORCHID
J. D. HOOKER, 1846

63

The Porcelain Orchid is one of the most widespread and exceptionally beautiful species of *Chloraea* found in the extreme southern latitudes of South America. The plant's common name comes from the porcelainlike quality of its relatively large, white flowers with bold, vivid green veins, carried above compact foliage. The species, like all members of *Chloraea*, manages to survive in the harsh and punishing Patagonian climate, blooming as the snow melts. The species name refers to its occurrence in the Straits of Magellan on the tip of South America.

It has been suggested that the lip has scent-producing glands on its surface. The floral morphology is compatible with pollination by bees, and there are images of pollination of some species of *Chloraea* by bumblebees. No reward is obvious, so this is likely to be a case of deceit. Underground, this species has a cluster of thick roots.

The flower of the Porcelain Orchid is greenish-white, highly textured, with dark green reticulations. While the sepals spread, the petals draw inward, providing a hood for the column and downturned, inverted spotted lip.

Actual size

SUBFAMILY	Orchidoideae
TRIBE AND SUBTRIBE	Cranichideae, Cranichidinae
NATIVE RANGE	Colombia and Ecuador
HABITAT	Steep, upper elevation embankments near the snow line
TYPE AND PLACEMENT	Terrestrial
CONSERVATION STATUS	May be endangered due to climate changes
FLOWERING TIME	Winter

FLOWER SIZE
¼–½ in (0.7–1 cm)

PLANT SIZE
12–24 × 8–12 in
(30–61 × 20–30 cm),
including inflorescence

AA COLOMBIANA
PAPER ORCHID
SCHLECHTER, 1920

64

The distinctive *Aa colombiana*, found on high montane steep banks and other open areas in forests, has flowers clustered near the top of a long, upright inflorescence. The non-resupinate (with lip uppermost) flowers are subtended by a translucent, often papery floral bract that sometimes covers the blooms and protects the buds from volatile weather in their stormy, windy, and rainy alpine habitat. Underground, these plants have a cluster of fat hairy roots.

With an equally stormy taxonomic history, members of the genus *Aa* have been tossed about by the German orchid expert Heinrich Gustav Reichenbach (1823–89) between *Aa* and the closely related *Altensteinia*. It is thought that he coined the name "*Aa*" so that one of his genera would always appear first in any alphabetical listing of plant genera. Pollination of this species and its relatives has not been studied.

The flower of the Paper Orchid is small and globular, clumped with others in a tight, cylindrical raceme at the apex of the inflorescence. The upright lip is usually whitish, hoodlike, and fringed at its margin. Sepals, petals, and bracts are often hyaline and brownish.

Actual size

SUBFAMILY	Orchidoideae
TRIBE AND SUBTRIBE	Cranichideae, Cranichidinae
NATIVE RANGE	Andean South America, from northwestern Venezuela to Bolivia
HABITAT	Open areas, humid meadows, or open cloud forests, at 5,900–14,100 ft (1,800–4,300 m)
TYPE AND PLACEMENT	Terrestrial, usually on granitic substrates
CONSERVATION STATUS	Locally common, not threatened
FLOWERING TIME	March to May

ALTENSTEINIA FIMBRIATA
FRINGED GROUND ORCHID
KUNTH, 1816

FLOWER SIZE
¾ in (2 cm)

PLANT SIZE
20 × 5 in (50 × 10 cm),
including inflorescence

65

The cold-climate Fringed Ground Orchid from the Andes forms a rosette of leaves on the ground and has an underground stem with a radiating system of hairy, fleshy white roots. In the flowering season, the plant produces a flower stem covered by small papery bracts from which greenish-white flowers protrude. The flowers are small and non-resupinate, unusual for an orchid in that the lip is on the upper side of the flower.

Altensteinia species in general are found in high-elevation meadows and on rock faces in the mountains, where they can be common, forming large stands. The genus is named for Baron Karl Sigmund Franz Freiherr von Stein zum Altenstein (1770–1840), a Prussian politician.

The flower of the Fringed Ground Orchid has three hairy, recurved sepals and two small, recurved petals. The greenish lip is much larger than the petals and held upright like a hood, with a fringed edge surrounding the hairy column.

Actual size

SUBFAMILY	Orchidoideae
TRIBE AND SUBTRIBE	Cranichideae, Cranichidinae
NATIVE RANGE	Tropical America, from southern Florida to Peru and Brazil
HABITAT	Wet forests and shady banks, at 650–9,850 ft (200–3,000 m)
TYPE AND PLACEMENT	Terrestrial
CONSERVATION STATUS	Not assessed, but widespread and locally common
FLOWERING TIME	Most of the year, except August to October

FLOWER SIZE
¼ in (0.6 cm)

PLANT SIZE
Up to 12 × 5–8 in
(30 × 13–20 cm),
including inflorescence

CRANICHIS MUSCOSA
MOSSY HELMET ORCHID
SWARTZ, 1788

Cranichis comes from the Greek word for helmet (*kranos*), a reference to the hoodlike upwardly pointing lip of the species. In this plant, a basal rosette of up to seven elliptic, totally green leaves with a distinct petiole, or stalk, emerges in fall from finger-like, fat, hairy roots. A terminal upright purple-green flower spike appears with leaflike bracts and many densely packed flowers that open in succession. The species makes seed capsules from nearly every flower, so it is likely that it self-pollinates.

The Mossy Helmet Orchid is one of the most common terrestrial orchids throughout its broad range. It occurs in both primary and secondary forests, even under plantations of pine and eucalyptus, both of which are known for producing compounds that are toxic to most tropical American terrestrial orchids.

The flower of the Mossy Helmet Orchid is non-resupinate, with two greenish-white, recurved sepals, one much smaller sepal, and two white petals that point downward. The upward-pointing lip is hooded and marked with green spots and stripes.

Actual size

SUBFAMILY	Orchidoideae
TRIBE AND SUBTRIBE	Cranichideae, Cranichidinae
NATIVE RANGE	Andean Ecuador and Peru
HABITAT	High elevation grassland and scrubland
TYPE AND PLACEMENT	Terrestrial
CONSERVATION STATUS	Not assessed
FLOWERING TIME	Spring

GOMPHICHIS MACBRIDEI
PÁRAMO HELMET ORCHID
C. SCHWEINFURTH, 1941

FLOWER SIZE
⅜ in (1 cm)

PLANT SIZE
39–55 × 3–6 in
(100–140 × 8–14 cm),
including inflorescence

67

This orchid grows high in the Andes mountains among grasses and short shrubs and forms a cluster of soft, linear leaves with a knot of thick hair-covered roots. Among these leaves, a bract-covered, woolly spike emerges with many non-resupinate (with the lip upwards) flowers. The flowers do not have a nectary and presumably are pollinated by flies or small bees, although this aspect has not yet been documented. There are some glandular hairs on the lip, and if these produce nectar then this might attract the pollinator.

Overall, the biology and ecology of the genus *Gomphichis* have had little study. The name is from the Greek for "nail" or "club" (*gomphos*), which could refer to the glands on the lip or the shape of the column. Páramo is a high-elevation area of grassland in the Andes, where this species is found.

The flower of the Páramo Helmet Orchid has three sepals that are white inside and hairy outside. The petals are enclosed in the lower, hooded sepal. The ribbed lip is yellow with greenish veins and curves upward.

Actual size

SUBFAMILY	Orchidoideae
TRIBE AND SUBTRIBE	Cranichideae, Cranichidinae
NATIVE RANGE	Northwestern South America, Galapagos Islands
HABITAT	Rain and cloud forests, at 1,640–9,850 ft (500–3,000 m)
TYPE AND PLACEMENT	Terrestrial
CONSERVATION STATUS	Not assessed
FLOWERING TIME	February to April (late winter to spring)

FLOWER SIZE
1¼ in (3 cm)

PLANT SIZE
12–18 × 6-12 in
(30–45 × 15–30 cm),
including inflorescence

PONTHIEVA MACULATA
SPOTTED SHADOW WITCH
LINDLEY, 1845

68

Actual size

Species of the genus *Ponthieva* prefer moist sites on the floor of cloud and other wet forests. One species, *P. racemosa*, reaches as far north as Virginia in eastern North America, which is unusual for a principally tropical group of orchids. The 15–20 long-stalked flowers of the Spotted Shadow Witch are carried on a small-bracted hairy inflorescence that emerges from a rosette of hairy, almost stemless leaves with a cluster of fat hairy roots. Oil, rather than nectar, is produced on glands on the lip and lateral sepals, so it is assumed (though not observed) that the pollinators are oil-gathering anthophorid bees.

The flowers are non-resupinate, and, unusually among orchids, it is the erect lateral sepals that are the most attractive part of these flowers. In Costa Rica, the roots are reportedly used as a substitute for ipecacuanha, an emetic.

The flower of the Spotted Shadow Witch is covered with hairs on the outside. The two heavily spotted lateral sepals point upward. The two yellow petals are clawed, with a bulging gland at their base. The lip has a cavity in its center.

SUBFAMILY	Orchidoideae
TRIBE AND SUBTRIBE	Cranichideae, Cranichidinae
NATIVE RANGE	Southeastern Brazil and Uruguay
HABITAT	Open areas, natural grasslands
TYPE AND PLACEMENT	Terrestrial in humid but well-drained sites
CONSERVATION STATUS	Not assessed, but widespread
FLOWERING TIME	October (spring)

PRESCOTTIA DENSIFLORA
SNOW-WHITE LADY'S TRESSES
(BRONGNIART) LINDLEY, 1840

FLOWER SIZE
¼ in (0.6 cm)

PLANT SIZE
10–18 × 6–10 in
(25–46 × 15–25 cm),
including inflorescence

69

The species name, *densiflora*, is particularly appropriate for this orchid—its flower stem is densely set with up to a hundred tiny, non-resupinate, white flowers, each subtended by a large green bract. The stem itself rises from a rosette of oval, hairless, stalkless leaves, which emerges from a short rhizome with a cluster of thick hairy roots. In the early nineteenth century, a member of the genus *Prescottia* was the first orchid species to be grown from seed in cultivation, a process that long mystified European nurserymen.

Prescottia densiflora is pollinated by halictid bees collecting nectar and attracted by the plant's sweet scent. Pollinia are fixed on the ventral surface of the insects' mouthparts. When visiting another flower, a bee brushes the stigmatic surface with the pollinia and leaves clumps of pollen, thus effecting pollination.

The flower of the Snow-white Lady's Tresses
has three recurved sepals, with the one normally in the uppermost position lowermost here. The petals are small, blunt, and recurved. The lip, which surrounds the short column, is fleshy, cup-shaped, and hairy inside.

Actual size

SUBFAMILY	Orchidoideae
TRIBE AND SUBTRIBE	Cranichideae, Cranichidinae
NATIVE RANGE	Western South America (Colombia, Ecuador, and Peru)
HABITAT	Páramo (montane grassland in the Andes), at 8,530–13,450 ft (2,600–4,100 m)
TYPE AND PLACEMENT	Terrestrial-lithophytic, in moss on rocks
CONSERVATION STATUS	Not assessed
FLOWERING TIME	August to October

FLOWER SIZE
⅛ in (0.5 cm)

PLANT SIZE
3½–21⅝ in (9–55 cm),
including inflorescence,
the leaf being held more
or less upright

PTERICHIS TRILOBA
SUNNY PÁRAMO ORCHID
(LINDLEY) SCHLECHTER, 1911

70

Actual size

Underground, the short stem of the Sunny Páramo Orchid produces a cluster of fat hairy roots, from which emerges a single upright leaf (sometimes two leaves). The leaf is glandular with a short stalk (petiole). From the base of the leaf (often after the leaf has withered), a lax inflorescence bearing 10–20 non-resupinate flowers grows. These plants are found well above the tree line in the mountains of South America and may hold the record for the orchid species growing at the highest elevation.

Although fairly common in these habitats, the orchid has been little studied. Nothing is known about its pollination, and the flowers do not appear to produce any sort of reward. There are marginal glandular spots on the lip, but their function has not been investigated.

The flower of the Sunny Páramo Orchid is not resupinate and has two partially fused, brownish upper sepals and a green-streaked lower sepal. The brownish petals point down and overlap with the lower sepal. The concave yellow lip is hooded and projects forward.

SUBFAMILY	Orchidoideae
TRIBE AND SUBTRIBE	Cranichideae, Cranichidinae
NATIVE RANGE	Peru (Puno) and Brazil (Mato Grosso)
HABITAT	Cloud forests at 9,850–12,100 ft (3,000–3,700 m)
TYPE AND PLACEMENT	Terrestrial
CONSERVATION STATUS	Not assessed
FLOWERING TIME	March (summer)

FLOWER SIZE
⅜ in (1 cm)

PLANT SIZE
24–39 × 8–12 in
(60–100 × 20–30 cm),
including inflorescence

STENOPTERA ACUTA
NARROW-WINGED ORCHID
LINDLEY, 1840

71

From the Narrow-winged Orchid's cluster of fleshy roots a rosette of oblanceolate, nearly stalkless leaves grows. In the middle of these leaves a tall stem with small green flowers emerges, subtended by green bracts. Flowers and leaves are present at the same time. Species of the genus *Stenoptera* share with other genera of this subtribe a non-resupinate lip, but they are much larger plants than many and do not grow at the extreme elevations of other Andean genera such as *Altensteinia*, with which they share many floral characters.

Pollination has not been observed, but the lack of an obvious nectary suggests some form of deceit pollination, perhaps by flies or small bees.

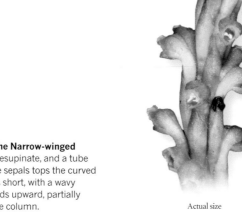

The flower of the Narrow-winged Orchid is non-resupinate, and a tube formed of three sepals tops the curved ovary. The lip is short, with a wavy margin, and folds upward, partially surrounding the column.

Actual size

SUBFAMILY	Orchidoideae
TRIBE AND SUBTRIBE	Cranichideae, Goodyerinae
NATIVE RANGE	Yunnan province (China), Burma, Thailand, Laos, and Peninsular Malaysia
HABITAT	Evergreen forests, at 1,300–4,600 ft (400–1,400 m)
TYPE AND PLACEMENT	Terrestrial in shady rich humus
CONSERVATION STATUS	Not assessed
FLOWERING TIME	October to December (fall to early winter)

FLOWER SIZE
⅝ in (1.5 cm)

PLANT SIZE
6–12 × 5–10 in
(15–31 × 13–25 cm),
including inflorescence

ANOECTOCHILUS BURMANNICUS
YELLOW JEWEL ORCHID
ROLFE, 1922

72

This orchid is often illustrated and sold in the horticultural trade under the name *Anoectochilus chapaensis*, but that species (from Vietnam and adjacent China) has a shorter, white lip and caudate sepals. The erect, somewhat thickened stem of the Yellow Jewel Orchid has three to six closely spaced petiolate, bronzy green leaves with copper-colored veins. It forms an upright, finely hairy, laxly flowered inflorescence, with each flower subtended by a small bract. The stem or rhizome creeps along the substrate in dark conditions in at least seasonally wet forests. The plants are evergeeen and there is no tuber or other storage organ.

Pollination of the species is unstudied, but it is likely to be pollinated by butterflies, given its floral morphology, which in related taxa has been associated with this mode of pollination. No nectar spur is present.

The flower of the Yellow Jewel Orchid is not regular in orientation and rarely resupinate. There are two hairy, glandular sepals and the dorsal sepal and petals make a cup around the short column. The long lip points outward, with an apex composed of two flaring lobes.

Actual size

SUBFAMILY	Orchidoideae
TRIBE AND SUBTRIBE	Cranichideae, Goodyerinae
NATIVE RANGE	Northern India, Nepal, Bhutan, Burma, China (Chekiang, Guangdong, and Hainan provinces), Thailand, Laos, and Vietnam
HABITAT	Lowland tropical forests often near streams and seeps
TYPE AND PLACEMENT	Terrestrial or on rocks
CONSERVATION STATUS	Threatened by collection for medicine
FLOWERING TIME	Summer and fall

ANOECTOCHILUS ROXBURGHII

ROXBURGH'S JEWEL ORCHID

(WALLICH) LINDLEY, 1839

FLOWER SIZE
Up to ¾ in (1.85 cm)

PLANT SIZE
9–20 × 10–18 in
(23–51 × 25–46 cm),
including flowers

73

This species, bearing some of the most attractive foliage in the plant kingdom, is known as a jewel orchid, so named for the astonishing crystalline texture and colorful, often glistening, gold veins that glow against its deep green to almost maroon leaves. The plant has an elongate rhizome with distantly spaced roots that creeps along the forest floor, and its leaves are clustered near the base of the upright flower stem.

Even though jewel orchids are appreciated more for their foliage than for their flowers, those of *Aneoctochilus roxburghii* are small and intricate. A related species from Taiwan, *A. formosanus*, is the source of an extract said to inhibit growth in malignant tumors. Pollination of species that grow in dense shade and produce flowers like this with a short spur is likely to be by a moth or butterfly.

The flower of Roxburgh's Jewel Orchid has greenish-purple sepals and petals and a complex lobed, whitish lip, with a deep fringe along the midrib that branches into two spatulate lobes at its terminus.

Actual size

SUBFAMILY	Orchidoideae
TRIBE AND SUBTRIBE	Cranichideae, Goodyerinae
NATIVE RANGE	Southeastern Brazil
HABITAT	Constantly wet shady places in forests
TYPE AND PLACEMENT	Terrestrial
CONSERVATION STATUS	Not assessed
FLOWERING TIME	September to March (spring to fall)

FLOWER SIZE
½ in (1.3 cm)

PLANT SIZE
6–8½ × 2⅜–4 in
(15–22 × 6–10 cm),
including inflorescence

ASPIDOGYNE FIMBRILLARIS
SILVER-STRIPED ORCHID
(B. S. WILLIAMS) GARAY, 1977

74

The Silver-striped Orchid often grows in dark places in wet forests. It produces nectar as a reward to its pollinators and is visited by hesperid butterflies and halictid bees. Only the latter are effective pollen vectors, with the pollinia becoming stuck to the bee's labrum, a difficult place for a bee to clean.

This attractive little plant has creeping rhizomes topped with a closely spaced set of green leaves with a broad silver streak down the middle. The flower stem is erect and terminates in a loose to densely clustered spike of delicate flowers. Underground, there are thick, hairy roots loosely spaced along the rambling rhizome. Stems sometimes branch, which results in a cluster of closely spaced plants.

Actual size

The flower of the Silver-striped Orchid is small and the three sepals are white with a greenish-brown streak down the middle. The petals are placed inside the hooded upper sepal. The lip is short and envelops the column.

SUBFAMILY	Orchidoideae
TRIBE AND SUBTRIBE	Cranichideae, Goodyerinae
NATIVE RANGE	Louisiana, Florida, and the West Indies to northern South America
HABITAT	Swamps, hardwood forests, and hammocks, up to 650 ft (200 m)
TYPE AND PLACEMENT	Terrestrial in humus
CONSERVATION STATUS	Not assessed
FLOWERING TIME	September to early May (fall to spring)

ASPIDOGYNE QUERCETICOLA
JUG ORCHID
(LINDLEY) MENEGUZZO, 2012

FLOWER SIZE
⅛ in (0.5 cm)

PLANT SIZE
5–12 × 3–5 in
(13–30 × 8–13 cm),
including inflorescence

75

The Jug Orchid grows in low elevation habitats, especially in shady wet sites. It has a creeping rhizome (stem) with two or three roots at each node, and forms upright shoots with up to six spirally arranged leaves that are often white-veined and have an inflated, sheathing base. Its relatives in Central and South America can have dark leaves with white to gold veins. The Jug Orchid sometimes branches and forms small colonies of plants. At almost any time of the year a flower stem can emerge, bearing many tiny flowers each with a bract beneath it.

No data exist about the pollination of these flowers. However, their broad lip, with a short, rounded spur containing nectar, resupinate flowers, and their flower color suggest pollination by bees or perhaps butterflies.

The flower of the Jug Orchid has three cupped, white sepals, two small petals that are tinged brown and held together to form a tube with the upper sepal, and a recurved, three-lobed, white lip, basally extended into a sacklike nectar spur.

Actual size

SUBFAMILY	Orchidoideae
TRIBE AND SUBTRIBE	Cranichideae, Goodyerinae
NATIVE RANGE	Himalayas to Yunnan (China), Burma, and Thailand
HABITAT	Damp places in forests, at 3,300–7,875 ft (1,000–2,400 m)
TYPE AND PLACEMENT	Terrestrial
CONSERVATION STATUS	Probably endangered, but not formally assessed
FLOWERING TIME	September (fall)

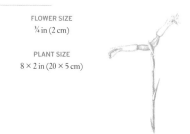

FLOWER SIZE
¾ in (2 cm)

PLANT SIZE
8 × 2 in (20 × 5 cm)

CHEIROSTYLIS GRIFFITHII
HIMALAYAN JEWEL ORCHID
LINDLEY, 1857

Actual size

The genus name *Cheirostylis* is derived from the Greek word *cheilos*, meaning a hand, which refers to the fingerlike lobes on the column of the species. *Cheirostylis griffithii* is named in honor of the English physician and naturalist William Griffith (1810–45). He was a prolific collector, and among his 12,000 specimens he discovered this delicate little jewel in the Himalayas. When in vegetative growth, the species and its relatives have a tightly clustered set of dark greenish-brown leaves with whitish silvery markings. It is these highly decorative leaves that are responsible for the common name, Jewel Orchid.

Between one and three flowers appear when the leaves have already withered during summer. The flowers are insect pollinated, perhaps by a small moth, although little is known about the pollinator interactions.

The flower of the Himalayan Jewel Orchid has partially fused sepals forming a tube covered by brown hairs with petals inside. The lip has two green spots at the base and two deeply split lobes, each of which has eight to ten fingerlike divisions. The column has four spoon-shaped fingers.

SUBFAMILY	Orchidoideae
TRIBE AND SUBTRIBE	Cranichideae, Goodyerinae
NATIVE RANGE	Temperate Northern Hemisphere
HABITAT	Shady places in most coniferous or mixed forests, rarely in bogs or cedar swamps, from sea level to 9,500 ft (2,900 m)
TYPE AND PLACEMENT	Terrestrial in humus
CONSERVATION STATUS	Not assessed, but widespread
FLOWERING TIME	July to early September (summer)

FLOWER SIZE
¼ in (0.6 cm)

PLANT SIZE
4–8 × 4–6 in
(10–20 × 10–15 cm),
including inflorescence

77

GOODYERA REPENS
DWARF RATTLESNAKE PLANTAIN
(LINNAEUS) R. BROWN, 1813

This lovely little orchid is abundant in coniferous forests throughout the Northern Hemisphere. It has a shallowly creeping rhizome with spirally arranged leaves that can be plain green or have various degrees of white mottling, spots, or stripes. The hairy flower spikes have white, externally hairy flowers, each subtended by a long, pointy green bract. The orchid's common name in the UK is Creeping Lady's Tresses, which refers to its resemblance to the lady's tresses species of the genus *Spiranthes*, which do not "creep."

Nectar is secreted in the cuplike lower part of the lip, and the flowers are sweetly scented. In North America this attracts bumblebees, which get pollen masses attached to their proboscis, although they may not be the most effective pollinators. In Europe, the bee *Lasioglossum morio* is of the right size and shows appropriate pollinating behavior.

The flower of the Dwarf Rattlesnake Plantain
has three equal, white, hairy sepals forming a
small bell. The petals and lip, enclosed by the
sepals, are also white.

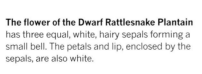

Actual size

SUBFAMILY	Orchidoideae
TRIBE AND SUBTRIBE	Cranichideae, Goodyerinae
NATIVE RANGE	Southern China, Hainan Island, northern Thailand, Vietnam
HABITAT	Seasonal evergreen forests and lower mixed deciduous forests, at around 390–2,950 ft (120–900 m)
TYPE AND PLACEMENT	Terrestrial
CONSERVATION STATUS	Not assessed
FLOWERING TIME	March to April (early spring)

FLOWER SIZE
⅛ in (0.5 cm)

PLANT SIZE
8–16 × 3–8 in
(20–40 × 7.5–20 cm),
including inflorescence

78

HETAERIA YOUNGSAYEI
YOUNG SAYE'S LADY'S TRESSES
ORMEROD, 2004

The genus name comes from *hetairos*, the Greek word for "companionship" or "brotherhood," and refers to the other genera, such as *Goodyera*, to which *Hetaeria* is closely related, and where it was formerly placed. Like *Goodyera*, its species make a creeping rhizome (underground stem) with one or two fat, hairy roots emerging from the joints on the rhizome and a spirally arranged cluster of leaves near the soil level. The leaves are broadly ovate and stalked and topped by a dense hairy inflorescence that bears small flowers with the lip uppermost. The flowers are also covered with short hairs.

This species is named for J. L. Young Saye, a Hong Kong orchid enthusiast, and was surprisingly only described in 2004.

The flower of Young Saye's Lady's Tresses has the lip pointing upward with three hooded, purplish-green, hairy sepals with pinkish-white tips. The petals are not hairy and white. The lip is yellowish-white and shorter than the rest of the flower parts.

Actual size

SUBFAMILY	Orchidoideae
TRIBE AND SUBTRIBE	Cranichideae, Goodyerinae
NATIVE RANGE	Southern China, through Indochina, Sumatra, Borneo, and the Philippines
HABITAT	Evergreen lowland forests, in shade and often near streams, at up to 4,300 ft (1,300 m)
TYPE AND PLACEMENT	Terrestrial, or lithophytic on rocky substrates
CONSERVATION STATUS	Not assessed
FLOWERING TIME	February to April (spring)

FLOWER SIZE
⅝ in (1.7 cm)

PLANT SIZE
10–18 × 6–10 in
(25–46 × 15–25 cm),
including inflorescence

LUDISIA DISCOLOR
COMMON JEWEL ORCHID
(KER GAWLER) A. RICHARD, 1825

79

The name "jewel orchid" refers to the reddish-brown leaves with pinkish-white venation, which also occur in the closely related genera *Aneoctochilus*, *Dossina*, *Macodes*, and *Cheirostylis*. The leaves are spirally arranged on stems that emerge from fleshy creeping rhizomes with one to three fuzzy roots at most horizontal nodes. From the middle of the leaves, a bract-covered hairy inflorescence is formed carrying 10–25 flowers with woolly ovaries, each subtended by a pinkish bract. The column is twisted to one side, with the anther and viscidium making a hook at the tip.

This jewel orchid grows on rocky substrates, and the white flowers attract a butterfly (*Pleisconeura asmara*), which drinks from a nectar cavity at the base of the lip. As the butterfly does this, the pollen masses attach to its legs. It is fairly commonly grown as a potplant.

The flower of the Common Jewel Orchid
has two recurved white sepals and one hooded sepal covering the yellow column and the small, curved, white lip. Two apical lip-lobes project to the side.

Actual size

SUBFAMILY	Orchidoideae
TRIBE AND SUBTRIBE	Cranichideae, Goodyerinae
NATIVE RANGE	Peninsular Thailand and Malaysia, Sumatra, Borneo, Java, Philippines, and Ryukyu Islands (Japan)
HABITAT	Lowland and lower montane forests, at up to 4,920 ft (1,500 m)
TYPE AND PLACEMENT	Terrestrial
CONSERVATION STATUS	Not assessed; common in cultivation
FLOWERING TIME	At any time

FLOWER SIZE
⅜ in (1 cm)

PLANT SIZE
8–12 × 6–8 in
(20–30 × 14–20 cm),
including inflorescence

MACODES PETOLA
SPARKLING JEWEL ORCHID
(BLUME) LINDLEY, 1840

80

Jewel orchids are known for their elaborate leaf markings rather than their flowers. This species has a loose rosette of beautiful, silvery veined leaves that sparkle in the light. Plants scramble more or less along the ground, with an elongate rhizome at or just below the soil surface. Roots emerge along the rhizome and may be clustered basally.

Because of the elaborate veins, resembling writing, it was once believed that the juice of the plant, when dropped into the eyes, could improve writing skills. Medicinal use has also been recorded, although the exact nature of the treatments is unknown. Pollination is most likely by moths or butterflies, given the morphology of the flower, but their approach must be from one side due to the twisted nature of the lip and column. Presence of nectar has not been reported.

The flower of the Sparkling Jewel Orchid is non-resupinate and has spreading sepals that are hairy outside and have brown-red markings inside. Petals are curved and partly fused with the lower sepal. The elaborate lip is fused with and surrounds the column.

Actual size

SUBFAMILY	Orchidoideae
TRIBE AND SUBTRIBE	Cranichideae, Goodyerinae
NATIVE RANGE	Caribbean islands and northern Venezuela
HABITAT	Montane rain forests and cloud forests
TYPE AND PLACEMENT	Terrestrial
CONSERVATION STATUS	Threatened by habitat loss
FLOWERING TIME	September to May (fall to spring)

FLOWER SIZE
⅛ in (0.5 cm)

PLANT SIZE
12–18 × 3–5 in
(30–46 × 8–13 cm),
including inflorescence

MICROCHILUS PLANTAGINEUS
CARIBBEAN FALSE HELMET ORCHID
(LINNAEUS) D. DIETRICH, 1852

81

This orchid is one of a genus with about 40 species. It is related to *Goodyera* and has a similar long, horizontal rhizome with one to two thick hairy roots arising from the joints. When the rhizome becomes erect it produces a series of spirally arranged leaves with a distinct petiole, or stalk, from which the flower stem emerges. Some individuals have white mottling on their leaves, which are hairless, unlike the often densely haired other parts of the plant.

The nectar spur implies pollination by bees or butterflies, although pollination has not been observed in the field. A bee carrying pollinia from a closely related extinct species (*Meliorchis caribea*) was found in Miocene amber from the Dominican Republic, suggesting that this group of orchids was already present in the Caribbean 20 million years ago.

The flower of the Caribbean False Helmet Orchid is white, with an upper sepal that forms a hood over the column and is flanked by two small petals. The lip points downward and has two small, mustachelike lobes at the apex. At the rear is a short nectar spur.

Actual size

SUBFAMILY	Orchidoideae
TRIBE AND SUBTRIBE	Cranichideae, Goodyerinae
NATIVE RANGE	Sikkim (India) to Taiwan and south to Indochina
HABITAT	Evergreen broadleaf forests and valleys, at 2,625–7,200 ft (800–2,200 m)
TYPE AND PLACEMENT	Terrestrial or on rocks in damp places
CONSERVATION STATUS	Not assessed
FLOWERING TIME	June to September (summer to early fall)

FLOWER SIZE
¾ in (2 cm)

PLANT SIZE
6–12 in (15–30 cm),
including terminal
inflorescence

82

ODONTOCHILUS LANCEOLATUS
YELLOW FISHBONE ORCHID
(LINDLEY) BLUME, 1859

The Yellow Fishbone Orchid grows in dark forests, where its brightly colored flowers make it easily seen and a ready attraction for pollinators. It produces a creeping stem that advances across the leafy forest floor and then grows upright to produce between two and eight white-veined leaves crowned with a stem holding up to 12 flowers. The genus name is derived from the Greek words *odontos* and *cheilos*, meaning "tooth lip," in reference to the fingerlike teeth along the sides of the lip, which also look like the bones of a fish—hence the common name. The species name refers to the lancelike shape of the leaves.

Little is known about the life history of this species, including any possible pollinators. The bright color, shape of the flowers and occurrence in the darker parts of the forest would, however, suggest a butterfly as pollinator.

Actual size

The flower of the Yellow Fishbone Orchid has greenish-yellow sepals and petals that project forward, forming a tube around the column. The yellow lip has a series of fingerlike teeth along its sides, and the apex is split into two large, flaglike lobes.

SUBFAMILY	Orchidoideae
TRIBE AND SUBTRIBE	Cranichideae, Goodyerinae
NATIVE RANGE	New Caledonia
HABITAT	Rain forests, at 330–3,300 ft (100–1,000 m)
TYPE AND PLACEMENT	Terrestrial in humid forests
CONSERVATION STATUS	Not assessed, but locally protected
FLOWERING TIME	September to October and April to May (spring and fall)

PACHYPLECTRON ARIFOLIUM
ARUM-LEAVED SPURLIP ORCHID
SCHLECHTER, 1906

FLOWER SIZE
⅜ in (1 cm), including
spur length

PLANT SIZE
7–12 × 6–10 in
(18–30 × 15–25 cm),
including inflorescence

83

Because of its brown leaves and flowers, *Pachyplectron arifolium* is often mistaken for a dead plant and is thus overlooked. It also grows in dark conditions where there are few other plants and few people would expect to find an orchid. It is common in places on the Pacific island of New Caledonia, but little is known about its ecology.

From a creeping rhizome, this unusual orchid produces a few arrowlike, bronze-colored leaves that lie flat on the ground. When the plant flowers, a wiry inflorescence bearing a few spurred flowers appears. The flowers have a short nectar spur, which makes pollination by flies appear to be a possibility. One species in this genus of three has no leaves (and depends entirely on its fungal partner for nutrition) and has rarely been collected.

Actual size

The flower of the Arum-leaved Spurlip Orchid has creamy-white and brown-spotted sepals and petals, of which two are recurved, the other held forward. The brown lip is short at the front, while a thick spur protrudes at the back of the flower.

SUBFAMILY	Orchidoideae
TRIBE AND SUBTRIBE	Cranichideae, Goodyerinae
NATIVE RANGE	Tropical and southern Africa
HABITAT	Swampy forests, at 100–4,360 ft (30–1,330 m)
TYPE AND PLACEMENT	Terrestrial in deep shade
CONSERVATION STATUS	Not assessed formally, but seemingly not under any threat
FLOWERING TIME	December to April (summer to fall)

FLOWER SIZE
⅜ in (1 cm)

PLANT SIZE
8–25 × 6–10 in
(20–63 × 15–25 cm),
including inflorescence

PLATYLEPIS GLANDULOSA
STICKY LADY'S TRESSES
(LINDLEY) REICHENBACH FILS, 1876

Damp, humid conditions contribute to the Sticky Lady's Tresses'
habit of growing a tuber-less stem that creeps near or at the soil
surface, with fleshy hairy roots arising singly from its nodes.
From these horizontal rhizomes, an erect stem is produced with
three to seven spirally arranged leaves showing three prominent
veins. A dense, glandular-hairy inflorescence tops the stem, with
a broad bract subtending each of the 15–45 flowers. The genus
name, from the Greek *platys* and *lepis*, meaning "flat scale,"
refers to these prominent floral bracts.

Lack of a nectar spur, and the overall flower shape and color,
suggest that small bees pollinate the plant, although this is
only speculation. The differences between the genera *Platylepis*
and *Goodyera* are mostly in the nature of the bracts subtending
the flowers and the lumpy to cristate thickenings on each side
of the lip base.

Actual size

The flower of the Sticky Lady's Tresses has
greenish-tan sepals, the laterals reflexed and the
dorsal forming a hood. The petals are small and lie
inside the dorsal sepal. The fleshy lip is fused to the
column for half its length and has two basal calli.

SUBFAMILY	Orchidoideae
TRIBE AND SUBTRIBE	Cranichideae, Goodyerinae
NATIVE RANGE	Northeastern India, southern Southeast Asia to New Guinea and southern Japan (Kyushu)
HABITAT	Forests in deep shade, at about 4,920 ft (1,500 m)
TYPE AND PLACEMENT	Terrestrial
CONSERVATION STATUS	Not assessed
FLOWERING TIME	October to November (fall)

FLOWER SIZE
⅟₁₆ in (0.3 cm)

PLANT SIZE
12–28 × 8–14 in
(31–71 × 20–36 cm),
including inflorescence

RHOMBODA LANCEOLATA
STRIPED JEWEL ORCHID
(LINDLEY) ORMEROD, 1995

85

The branching, creeping rhizomes of these plants produce one or two hairy roots on some of their nodes, which in turn form hairy, erect leafy stems with stalked, ovate-lanceolate leaves that are white-striped and sometimes dark red. When flowering, the stems carry large hairy bracts at each node and 10–14 small flowers. Like most orchids that inhabit moist dark forests, no tubers are formed and leaves are typically always present.

There are no reports of pollination of the genus *Rhomboda*, but the flowers have an orifice through which an insect could gain entry and find a reward. *Rhomboda lanceolata* has broad petals and a fleshy base on its lip, which would seem to guide visiting insects to the correct spot.

The flower of the Striped Jewel Orchid is minute and has three similar reddish-tan sepals, two upcurved, white petals, and a clawed, white lip with two large lobes that resemble a mustache.

Actual size

SUBFAMILY	Orchidoideae
TRIBE AND SUBTRIBE	Cranichideae, Goodyerinae
NATIVE RANGE	Tropical Southeast Asia, from Assam to New Guinea and the Philippines
HABITAT	Lower montane tropical rain forests, along stream banks in dark, damp places, at 820–6,000 ft (250–1,830 m)
TYPE AND PLACEMENT	Terrestrial
CONSERVATION STATUS	Not assessed
FLOWERING TIME	May to October (late spring to early fall)

FLOWER SIZE
⅜ in (1 cm)

PLANT SIZE
12–26 × 8–12 in
(30–66 × 20–30 cm),
including infloresence

VRYDAGZYNEA ALBIDA
TONSIL ORCHID
(BLUME) BLUME, 1858

86

Creeping stems, with one or two thick hairy roots at the nodes, give rise to evergreen branches with stalked ovate leaves, sometimes with a central white stripe, and clasping leaf bases. Inflorescences are terminal, dense racemes with a large bract subtending each flower. Like *Goodyera* and related genera, the Tonsil Orchid prefers lowland wet forests, where it grows in dark conditions. The rhizomes branch frequently, and the plants can form large colonies.

Because the flowers do not open much, the plant may be self-pollinating, though a moth could also be the pollinator, given the orchid's habitat and morphology, particularly the relatively long spur. The perplexing genus name was in honor of T. Vrydag Zynen, a Dutch friend of the author, Blume.

Actual size

The flower of the Tonsil Orchid does not open much. It has yellowish-green sepals that, together with the lip, form a tube enclosing the shorter petals and column. They turn darker as they mature and close.

SUBFAMILY	Orchidoideae
TRIBE AND SUBTRIBE	Cranichideae, Goodyerinae
NATIVE RANGE	Tropical and subtropical Asia, from Iran and Turkmenistan to Japan and New Guinea, naturalized in the American tropics and Hawaii
HABITAT	Open grassland, but often in disturbed sites such as lawns, roadsides, plant nurseries, gardens, fields, and sometimes pinelands
TYPE AND PLACEMENT	Terrestrial in moist soil
CONSERVATION STATUS	Common and weedy
FLOWERING TIME	October to February (fall to early spring), occasionally also later in spring

ZEUXINE STRATEUMATICA
SOLDIER'S LAWN ORCHID
(LINNAEUS) SCHLECHTER, 1911

FLOWER SIZE
½ in (1.2 cm)

PLANT SIZE
4–10 × 3–5 in
(10–25 × 8–12 cm),
including inflorescence

87

The Soldier's Lawn Orchid grows in full sun from an underground rhizome. Between one and three hairy roots emerge at each rhizome node, from which an erect stem grows, with spirally arranged, keeled narrow leaves. The inflorescence can hold up to around 50 flowers that are probably self-pollinating or apomictic. Many capsules are filled with seeds, and the plants set fruit even in insect-free environments.

The species is widespread in the Old World tropics but has also become invasive in southern North America and Brazil and is rapidly spreading. It has even been found growing in sidewalk cracks in Florida. The flowers lack a nectary, although the base of the lip forms a cavity. If insects visit the plant, a bee of some form is the most likely candidate.

The flower of the Soldier's Lawn Orchid often does not fully open. The three sepals and two petals are white and project forward, and the yellow lip protrudes slightly beyond the petals and sepals.

Actual size

SUBFAMILY	Orchidoideae
TRIBE AND SUBTRIBE	Cranichideae, Pterostylidinae
NATIVE RANGE	New Caledonia
HABITAT	Dense humid forests, sometimes in shaded places under shrubs in maquis
TYPE AND PLACEMENT	Terrestrial
CONSERVATION STATUS	Not assessed, but widespread on the island
FLOWERING TIME	September to March (summer to winter)

FLOWER SIZE
1¾ in (4.5 cm)

PLANT SIZE
20–36 × 10–16 in
(51–91 × 25–41 cm),
including inflorescence

88

ACHLYDOSA GLANDULOSA
GREEN FAIRY ORCHID
(SCHLECHTER) M. A. CLEMENTS & D. L. JONES, 2002

Actual size

The large terrestrial Green Fairy Orchid makes a cluster of fairly big, stiff elongate leaves with a long stem, from which tall, unbranched inflorescences emerge, covered with glandular hairs and several well-spaced flowers. Underground, the plant has a cluster of fat hairy roots.

The species was previously placed in the genus *Megastylis*, which contains large, relatively common terrestrial orchids on New Caledonia. *Achlydosa glandulosa*, however, is not related to that genus or even that tribe of orchids (Diurideae) and is instead the closest relative of the greenhoods, the genus *Pterostylis*, a member of tribe Cranichideae, and common throughout Australasia. Although the orchid is reported to produce nectar, there is no information about its pollinator. The green color might imply a moth as pollinator, but the shape of the flower suggests instead a bee.

The flower of the Green Fairy Orchid has three green sepals, of which the upper is held over the flower and forms a hood with the two green petals. The lip has a broad, green limb with a nectar spur at its base.

SUBFAMILY	Orchidoideae
TRIBE AND SUBTRIBE	Cranichideae, Pterostylidinae
NATIVE RANGE	Eastern and southeastern Australia, Lord Howe Island, New Caledonia
HABITAT	Open forests, near streams, at up to 4,600 ft (1,400 m)
TYPE AND PLACEMENT	Terrestrial
CONSERVATION STATUS	Not threatened
FLOWERING TIME	April to October (late fall to spring)

PTEROSTYLIS CURTA

BLUNT GREENHOOD

R. BROWN, 1810

FLOWER SIZE
1⅛ in (2.8 cm)

PLANT SIZE
6–10 × 2–3 in
(15–25 × 5–8 cm),
including flower

89

Vegetative reproduction allows this orchid to produce large colonies from a single seedling. It forms small globose tubers underground, from which a series of roots emanates, creating daughter tubers at a short distance from the mother plant. Each tuber forms a small rosette of up to six ovate to oblong leaves with wavy margins. A stem then emerges, bearing a single terminal flower.

Each green flower is lighter at the back, which entices the pollinator, in this case a male fungus gnat (genus *Mycomya*), to enter the bloom. During its investigation, the insect trips the "irritable" lip, pushing it into the pollen masses, where it is held for up to three hours. It has been suggested that the gnat enters the flowers to seek females, so there may be a scent (unnoticed by human noses) that attracts the male.

Actual size

The flower of the Blunt Greenhood carries a hood formed by the dorsal sepal and the petals. The lateral sepals are fused at the base and upturned. The lip is small and curved over the cleft of the sepals.

SUBFAMILY	Orchidoideae
TRIBE AND SUBTRIBE	Cranichideae, Pterostylidinae
NATIVE RANGE	Southwestern and southern Australia
HABITAT	Variable, from woodland to scrub and open grassland, common near granite outcrops in dry areas at up to 1,300 ft (400 m)
TYPE AND PLACEMENT	Terrestrial
CONSERVATION STATUS	Common and widespread
FLOWERING TIME	June to early September (winter to spring)

FLOWER SIZE
1 in (2.5 cm)

PLANT SIZE
3–12 × 3–5 in
(7–30 × 7–13 cm),
including inflorescence

PTEROSTYLIS SANGUINEA
DARK-BANDED GREENHOOD
D. L. JONES & M. A. CLEMENTS, 1989

90

Before the Dark-banded Greenhood reaches flowering size, a non-flowering soil-hugging rosette, with between three and ten leaves, is formed, followed in another season by a flowering shoot with no basal leaves but rather leaves spirally arranged up the flowering stem, which is topped by between two and eight hooded flowers. The bract subtending each flower is also leafy. Underground there is a tuber that forms a replacement tuber nearby and daughter tubers at the end of roots, resulting in a sparse cluster of plants.

Gnats and mosquitoes visit the flowers, becoming trapped on the column by an "irritable" lip that flips back when touched. The insects can escape along the winged column, where they pass first the stigmas and then pick up the pollinia.

The flower of the Dark-banded Greenhood
has two purplish, free sepals that project downward. The third sepal and both petals are fused into a green-striped hood over the prominently winged column. The lip is moveable, short, brown, and hairy.

Actual size

SUBFAMILY	Orchidoideae
TRIBE AND SUBTRIBE	Cranichideae, Spiranthinae
NATIVE RANGE	Southern North America (Mexico) south to Costa Rica in Central America
HABITAT	Ravines and grassy slopes, among rocks or in deep humus soils, in oak or oak-pine forests, at 4,920–7,200 ft (1,500–2,200 m), and also in disturbed sites, such as roads and fields
TYPE AND PLACEMENT	Terrestrial
CONSERVATION STATUS	Not assessed, but weedy and locally common
FLOWERING TIME	April to July (spring to summer)

AULOSEPALUM PYRAMIDALE
CONE ORCHID
(LINDLEY) M. A. DIX & M. W. DIX, 2000

FLOWER SIZE
¼ in (0.6 cm)

PLANT SIZE
18–30 × 8–15 in
(46–76 × 20–38 cm),
including inflorescence

91

In the fall and winter, the Cone Orchid produces a rosette of leaves, followed, as the leaves die and dry up, by a tall flower stem bearing many, densely packed, spirally arranged flowers subtended by somewhat papery bracts. Underground, the plant has a cluster of fat, almost tuberlike roots covered in dense hair.

The species does not mind disturbance and is often found in artificial environments, such as road verges, safety islands, and backyards. The shape of the flower and the stripe (a nectar guide) down its sepals, petals, and lip would seem to indicate that the pollinator is a bee. No one, however, has reported nectar in the basal cavity of the lip or observed any flower visitors.

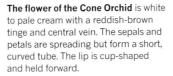

The flower of the Cone Orchid is white to pale cream with a reddish-brown tinge and central vein. The sepals and petals are spreading but form a short, curved tube. The lip is cup-shaped and held forward.

Actual size

SUBFAMILY	Orchidoideae
TRIBE AND SUBTRIBE	Cranichideae, Spiranthinae
NATIVE RANGE	Peru and Chile to southwestern Argentina
HABITAT	Open scrub and forests in dry to humid areas
TYPE AND PLACEMENT	Terrestrial
CONSERVATION STATUS	Uncommon but not assessed
FLOWERING TIME	September to November (spring)

FLOWER SIZE
⅛ in (0.5 cm)

PLANT SIZE
15 × 8 in (38 × 20 cm)

BRACHYSTELE UNILATERALIS
EMERALD LADY'S TRESSES
(POIRET) SCHLECHTER, 1920

An inconspicuous but not small species, this orchid has many small green flowers with white markings, forming rows that spiral up the floral stem. Bumblebees visit the flowers frequently to drink the nectar and in the process get the pollinia stuck to their proboscis, which they then insert into another flower and so effect pollination. The short column (*brachys* is Greek for "short," hence the genus name) is situated close to the nectar glands to allow the best chance for the pollen to be picked up by the bees.

When not in flower, the plants have a rosette of leaves that grows terrestrially. The species name refers to the flowers pointing in one direction in this orchid. Underground, a set of three to six thick, furry roots radiate from a short stem.

The flower of the Emerald Lady's Tresses has three white, green-veined, cup-shaped sepals and two petals that are shorter than the sepals. The downward-pointing and resupinate lip is green with a wavy edge. The column is placed deep inside the tube-shaped flower.

Actual size

SUBFAMILY	Orchidoideae
TRIBE AND SUBTRIBE	Cranichideae, Spiranthinae
NATIVE RANGE	Nicaragua to Ecuador
HABITAT	Forests, at 5,900–9,500 ft (1,800–2,900 m)
TYPE AND PLACEMENT	Terrestrial
CONSERVATION STATUS	Not assessed
FLOWERING TIME	March to June (fall and early winter)

COCCINEORCHIS BRACTEOSA
HUMMINGBIRD LADY'S TRESSES
(AMES & C. SCHWEINFURTH) GARAY, 1980

FLOWER SIZE
1½ in (4 cm)

PLANT SIZE
6–12 × 8–12 in
(15–30 × 20–30 cm),
including inflorescence

93

The genus name comes from the Greek word *kokkinos* (scarlet) and *orchis* (orchid), and the often bright-red or orange flowers of this species are showy and pollinated by hummingbirds. Previously, all red-flowered species related to the ladies tresses (*Spiranthes*) were placed in the genus *Stenorrhynchos*, but genetic (DNA) studies demonstrated that the similarity in shape and color of these orchids is due to independent adaptation to hummingbird pollination in distantly related species. The Hummingbird Lady's Tresses grows in mountain areas either on the ground or in the trees (mostly on mossy trunks or in branch crotches).

Underground, *Coccineorchis* species have a short rhizome with clustered, thick hairy roots. Their leaves have a distinct stalk (petiole) and an abruptly widening blade, and flowers are produced at the same time as the leaves.

The flowers of the Hummingbird Lady's Tresses grow in a cluster, in a congested inflorescence with colorful, prominent bracts. The orange or red sepals are similar and fused into an arching tube, in which the petals, lip, and column lie.

Actual size

SUBFAMILY	Orchidoideae
TRIBE AND SUBTRIBE	Cranichideae, Spiranthinae
NATIVE RANGE	Tropical America, from southern Florida to northern South America, and in southeastern South America
HABITAT	In humus on moist rocky hammocks, at up to 65 ft (20 m)
TYPE AND PLACEMENT	Terrestrial
CONSERVATION STATUS	Not assessed, but widespread and locally common
FLOWERING TIME	Throughout the year but mostly in spring

FLOWER SIZE
½ in (1.3 cm)

PLANT SIZE
15–28 × 8–12 in
(38–71 × 20–31 cm),
including inflorescence

CYCLOPOGON ELATUS
TALL LADY'S TRESSES
(SWARTZ) SCHLECHTER, 1919

94

How the genus name *Cyclopogon* relates to this species is obscure. It is derived from the Greek *cyclo*, meaning "circle," and *pogon*, the word for "beard," which could refer to the way the parts of the flower form a circle around the opening into the lip. A rosette of two to six leaves with a distinct petiole—and often a silvery stripe or other pattern—emerges from a short stem with many thick hairy roots. Up to 50, small, greenish-brown, white-lipped flowers appear at the same time as the leaves on a greenish-brown stem.

The morphological variation is substantial in this species. Variants can occur together and be locally distinct, with local variation perhaps maintained by self-pollination. If normal, insect-mediated pollination is taking place, there are no reports of what sort of insect is visiting these flowers.

The flower of the Tall Lady's Tresses has the white lip lowermost, with the other flower parts greenish-brown to creamy white. The lip and petals form a tube, which has a nearly circular opening.

Actual size

SUBFAMILY	Orchidoideae
TRIBE AND SUBTRIBE	Cranichideae, Spiranthinae
NATIVE RANGE	Central and southwestern Mexico
HABITAT	Grassland under pine forests, at about 6,600 ft (2,000 m)
TYPE AND PLACEMENT	Terrestrial
CONSERVATION STATUS	Not assessed
FLOWERING TIME	February (late winter)

DEIREGYNE CHARTACEA
NECK ORCHID
(L. O. WILLIAMS) GARAY, 1980

FLOWER SIZE
⅜ in (1 cm)

PLANT SIZE
12–18 × 6–11 in
(30–45 × 15–28 cm),
including inflorescence

95

The genus name *Deiregyne* is derived from the Greek *deire*, meaning "neck," and *gyne*, "woman" or "ovary," in reference to the ovary that is constricted into a neck—hence also the common name. This relatively common ground orchid grows in the mountains of central and southwestern Mexico, where it produces its leaves in the summer months and then flowers after the leaves have withered in the winter.

Plants produce a rosette of leaves with a cluster of fat hairy roots, and there are many closely spaced and spirally arranged flowers, partially covered by papery bracts. Flowers are diurnally fragrant and produce nectar at the base of the lip. The most likely pollinator is some type of small bee, although no direct observations of pollination have been made so far.

Actual size

The flower of the Neck Orchid has three brownish-red sepals forming a tube with recurved tips, the top one making a hood over the flower. The lip is white and pointed at the tip.

SUBFAMILY	Orchidoideae
TRIBE AND SUBTRIBE	Cranichideae, Spiranthinae
NATIVE RANGE	Central Mexico (states of Durango, Guerrero, Mexico D.F., Morelos, Nuevo Leon, Oaxaca, and Veracruz)
HABITAT	Grasslands and open areas in pine-oak forests, at 4,920–10,500 ft (1,500–3,200 m)
TYPE AND PLACEMENT	Terrestrial
CONSERVATION STATUS	Not assessed
FLOWERING TIME	January to March (winter to spring)

FLOWER SIZE
1⅛ in (3 cm)

PLANT SIZE
1½–2¾ in (4–7 cm) tall,
leaf rosette 4–7 in
(10–18 cm) tall,
3–6 in (8–14 cm) across

96

DEIREGYNE ERIOPHORA
WOOLLY ORCHID
(B. L. ROBINSON & GREENMAN) GARAY, 1980

Woolly Orchids flower after the basal rosette of petiolate leaves has withered. Underground, there is a short rhizome with a cluster of fat hairy roots. The inflorescence is enveloped in papery, nearly transparent bracts, and the stem is covered in woolly hairs. The species name comes from the Greek words for "wool-bearing" (*erio* and *phorein*), in reference to the plant's furry stem. This species has often been listed as a member of the genus *Schiedeella*, but recent genetic studies have indicated that it should instead be placed in *Deiregyne*.

The flowers are spirally arranged and scented during the daytime. Pollinators are not known, but given the shape and color of the flowers and the presence of a nectar cavity at the base of the lip, supplied with nectar by two small glands, the most likely candidates are bees, probably bumblebees.

The flower of the Woolly Orchid has narrowly lanceolate, white sepals, the upper one forming a tube with the falcate, white petals. The white lip has a thickened yellow callus in the throat.

Actual size

SUBFAMILY	Orchidoideae
TRIBE AND SUBTRIBE	Cranichideae, Spiranthinae
NATIVE RANGE	Mexico to Honduras
HABITAT	Seasonally wet forests, meadows, pastures, and fallow fields, at 2,625–9,850 ft (800–3,000 m)
TYPE AND PLACEMENT	Terrestrial
CONSERVATION STATUS	Not assessed, but a common weed in fields so no concern
FLOWERING TIME	June to October (summer to fall)

DICHROMANTHUS AURANTIACUS
ORANGE LADY'S TRESSES
(LEXARZA) SALAZAR & SOTO ARENAS, 2002

FLOWER SIZE
1 in (2.5 cm) long,
⅛ in (0.5 cm) across

PLANT SIZE
10–18 × 8–12 in
(25–46 × 20–31 cm),
including inflorescence

97

The tough orange flowers of *Dichromanthus aurantiacus* prevent nectar robbery by bees, which typically are pollinators of this group of orchids, the spiranthoids. This species is pollinated by hummingbirds and has to protect itself against bees that chew through the base of the flowers to remove nectar without pollinating flowers. Underground, there is a short stem with fat hairy roots, but there are no basal, rosette-forming leaves. Instead, the orchid produces a leafy stem during the rainy season and is dormant underground during the dry season.

This species, and other related red-flowered orchids, used to be considered members of the genus *Stenorrhynchos*, but DNA studies demonstrated that they were unrelated, being independently adapted to hummingbirds. An actual related species, *D. cinnabarinus*, also has orange flowers but with a yellow lip, resulting in the genus name *Dichromanthus*, meaning "two-toned flower."

The flower of Orange Lady's Tresses is tubular and emerges from behind a reddish-orange bract. The three orange sepals are hairy outside and flare at the tip. The two petals and the lip are similar to the sepals and are also bright orange.

Actual size

SUBFAMILY	Orchidoideae
TRIBE AND SUBTRIBE	Cranichideae, Spiranthinae
NATIVE RANGE	Southern Florida, the Caribbean, and tropical South America
HABITAT	Alongside streams in semi-deciduous forests
TYPE AND PLACEMENT	Terrestrial
CONSERVATION STATUS	Threatened in USA (southern Florida)
FLOWERING TIME	January to April

FLOWER SIZE
1–2 in (2.5–5 cm)

PLANT SIZE
20–32 × 3–9 in
(51–81 × 8–23 cm),
including inflorescence

98

ELTROPLECTRIS CALCARATA
LONGCLAW ORCHID
(SWARTZ) GARAY & H. R. SWEET, 1972

The large-flowered Longclaw Orchid grows near streams in the Caribbean and northern South America and as far north as southern Florida. It is often overlooked until it is flowering. It bears up to 15 winged and spurred flowers held high on a long stem approaching 3 ft (1 m) in height. Underground, the species has a cluster of thick hairy roots, and in some cases the leaf has withered by the time of flowering. It often bears a single, upright leaf, in some cases with white veins or diffuse spotting.

The blooms emit a sweet scent, and like other white flowers with fringed lips and strong fragrances *Eltroplectris calcarata* is probably attracting a moth as its pollinator. Nectar has been reported to be present in the spur.

The flower of the Longclaw Orchid has long, sharply pointed, white petals and sepals. The lip is fringed marginally. The petals are held at a 45-degree angle, giving it the appearance of a dancer with outstretched arms.

Actual size

SUBFAMILY	Orchidoideae
TRIBE AND SUBTRIBE	Cranichideae, Spiranthinae
NATIVE RANGE	Costa Rica
HABITAT	Cloud forests, at 4,600–4,920 ft (1,400–1,500 m)
TYPE AND PLACEMENT	Epiphytic in shaded and humid conditions
CONSERVATION STATUS	Not assessed
FLOWERING TIME	Throughout the year

EURYSTYLES STANDLEYI

CUSTARD ORCHID

AMES, 1925

FLOWER SIZE
³⁄₁₆ in (0.4 cm)

PLANT SIZE
1⅛ × 4 in (3 × 10 cm),
including inflorescence

99

The strange genus *Eurystyles* of dwarf rosette-forming epiphytic species was originally described as a member of the ginger family (Zingiberaceae), within which it would have been unusual in having such a thick style (the stem supporting the female receptive organ, the stigma). The genus name, meaning "thick style" in Greek, is logical, but this feature is not extraordinary in the orchid family, where all species have a comparatively large fused style and anther (column).

The Custard Orchid has a rosette of olive green, sometimes red-suffused waxy leaves. A pendent spike hangs from the rosette, with a short, hairy head of closely set flowers, each flower emerging from an axil of an ovoid, long-pointed bract. The roots are thick, spreading, and hairy. Together with the closely related *Lankesterella*, *Eurystyles* is one of only two truly epiphytic genera in subfamily Orchidoideae, in which the majority of species are ground-dwellers.

Actual size

The flower of the Custard Orchid has two cup-shaped sepals. The middle sepal points down, as do the green, flat petals. The lip is broad and short with a green stripe and a pale yellow edge, and its margins adhere to the column.

SUBFAMILY	Orchidoideae
TRIBE AND SUBTRIBE	Cranichideae, Spiranthinae
NATIVE RANGE	Mexico to Guatemala
HABITAT	Mossy banks in high elevation fir forests as well as alpine meadows, up to 13,100 ft (4,000 m)
TYPE AND PLACEMENT	Terrestrial
CONSERVATION STATUS	Not assessed
FLOWERING TIME	December to February (winter)

FLOWER SIZE
⅝ in (1.5 cm)

PLANT SIZE
6–14 × 4–8 in
(15–35 × 10–20 cm),
including inflorescence

FUNKIELLA HYEMALIS

MONARCH ORCHID

(A. RICHARD & GALEOTTI) SCHLECHTER, 1920

100

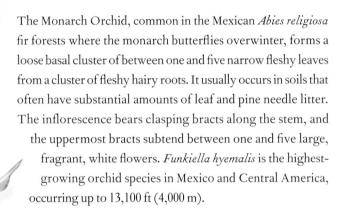

The Monarch Orchid, common in the Mexican *Abies religiosa* fir forests where the monarch butterflies overwinter, forms a loose basal cluster of between one and five narrow fleshy leaves from a cluster of fleshy hairy roots. It usually occurs in soils that often have substantial amounts of leaf and pine needle litter. The inflorescence bears clasping bracts along the stem, and the uppermost bracts subtend between one and five large, fragrant, white flowers. *Funkiella hyemalis* is the highest-growing orchid species in Mexico and Central America, occurring up to 13,100 ft (4,000 m).

Flowers are probably pollinated by bumblebees, which force their heads into the narrow chamber formed by lip, petals, and dorsal sepal. The genus is named in honor of the Belgian explorer and orchid collector, Nicolas Funck (1816–96).

The flower of the Monarch Orchid is white and emerges from a bract. The elongate, narrow lateral sepals flare outward. The upper (dorsal) sepal and petals form an upper lip. The lip, tinged red in the throat, is flared and clawed and has a V-shaped limb.

Actual size

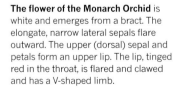

SUBFAMILY	Orchidoideae
TRIBE AND SUBTRIBE	Cranichideae, Spiranthinae
NATIVE RANGE	Venezuela, eastern and southeastern Brazil, Paraguay, and northeastern Argentina
HABITAT	Wet montane forests, at 1,300–4,920 ft (400–1,500 m)
TYPE AND PLACEMENT	Epiphytic in shade on mossy tree trunks
CONSERVATION STATUS	Not formally assessed, but under threat locally by deforestation and collectors
FLOWERING TIME	June to October (winter to spring)

FLOWER SIZE
⅝ in (1.5 cm)

PLANT SIZE
1–3 × 2–3 in
(2.5–7.5 × 5–7.5 cm),
including inflorescence

LANKESTERELLA CERACIFOLIA
WAX-LEAVED ORCHID
(BARBOSA RODRIGUES) MANSFELD, 1940

101

The epiphytic genus *Lankesterella* was named in honor of Charles Lankester, an English orchidophile who lived most of his life in Costa Rica and founded the botanical garden there that now bears his name. Its species have hairy but not particularly thick roots and form a cluster of succulent, keeled, ovate, glossy leaves with fine hairs along their margins.

The Wax-leaved Orchid produces an arching, hairy flower stem with large bracts subtending each of up to four white, spurred flowers. The genus is closely related to the other epiphytic genus in its subtribe, *Eurystyles*, which has different flowers. They may be mistaken for small bromeliads (looking much like Spanish moss and relatives) when not in flower. Some plants in this species may have more grayish leaves than others. The pollinator of this species is unknown, but its morphology and color suggest a small moth.

Actual size

The flower of the Wax-leaved Orchid is hairy and tubular, with three greenish-white sepals that are hairy outside. The two petals are creamy white and curve upward, opposite the broad, white lip with its green markings. A large nectar spur projects backward.

SUBFAMILY	Orchidoideae
TRIBE AND SUBTRIBE	Cranichideae, Spiranthinae
NATIVE RANGE	Tropical northeastern and eastern South America, southeastern Brazil, and on Trinidad
HABITAT	Open savannas
TYPE AND PLACEMENT	Terrestrial
CONSERVATION STATUS	Not assessed, but probably not threatened
FLOWERING TIME	June to December

FLOWER SIZE
½ in (1.25 cm)

PLANT SIZE
4–6 in (10–15 cm),
including inflorescence

102

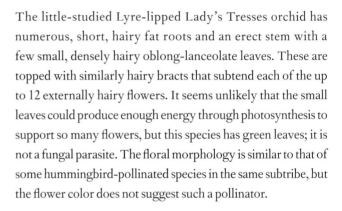

LYROGLOSSA GRISEBACHII
LYRE-LIPPED LADY'S TRESSES
(COGNIAUX) SCHLECHTER, 1921

The little-studied Lyre-lipped Lady's Tresses orchid has numerous, short, hairy fat roots and an erect stem with a few small, densely hairy oblong-lanceolate leaves. These are topped with similarly hairy bracts that subtend each of the up to 12 externally hairy flowers. It seems unlikely that the small leaves could produce enough energy through photosynthesis to support so many flowers, but this species has green leaves; it is not a fungal parasite. The floral morphology is similar to that of some hummingbird-pollinated species in the same subtribe, but the flower color does not suggest such a pollinator.

The species name honors German botanist August Grisebach (1814–79), who was a specialist on the flora of the West Indies. The genus name refers to the lip (*glossa* being Greek for "tongue," or in this case "lip"), which is lyre-shaped.

Actual size

The flower of the Lyre-lipped Lady's Tresses is bell shaped and has green sepals. The two white and green-striped petals are curved upward inside the sepal tube, enclosing the column. The lip is narrower at the base and broadly rounded at the tip with green lines.

SUBFAMILY	Orchidoideae
TRIBE AND SUBTRIBE	Cranichideae, Spiranthinae
NATIVE RANGE	Tropical America, from southern Mexico to northeastern Argentina
HABITAT	Rain forests, at up to 6,900 ft (2,100 m)
TYPE AND PLACEMENT	Terrestrial, or on rocks
CONSERVATION STATUS	Not assessed
FLOWERING TIME	Throughout the year

FLOWER SIZE
⅜ in (1 cm)

PLANT SIZE
8–14 × 6–8 in
(20–35 × 15–20 cm),
including inflorescence

PELEXIA LAXA
LAX HELMET ORCHID
(POEPPIG & ENDLICHER) LINDLEY, 1840

103

The Lax Helmet Orchid grows in a variety of habitats, but mostly wet forests. A rosette of between four and eight thick, mottled, petiolate leaves—sometimes with whitish cream spots or stripes—emerges from a cluster of thick, hairy roots. The inflorescence is densely hairy, and each flower is subtended by a bract. The genus name derives from the Greek word *pelex*, meaning "helmet," while *laxa* is Latin for "loose," both in reference to the shape of the dorsal sepal and petals.

The flowers are pollinated by bumblebees. These insects contact the pointed tip of the column while their mouthparts extend in an attempt to reach the nectar deeply buried in the long spur.

The flower of the Lax Helmet Orchid has a hood formed by the dorsal sepal and petals. The lateral sepals project forward, and there is a saccate spur at the back. The lip and sepals are white, and the rest of the flower is reddish-green to pale green.

Actual size

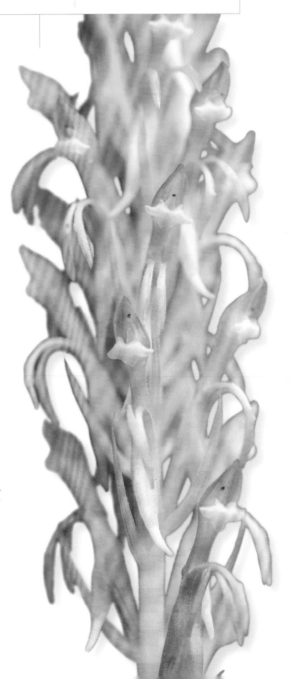

SUBFAMILY	Orchidoideae
TRIBE AND SUBTRIBE	Cranichideae, Spiranthinae
NATIVE RANGE	Tropical America, from southern Mexico to Paraguay and southern Brazil
HABITAT	Shady places in forest margins at middle elevations
TYPE AND PLACEMENT	Terrestrial in humus, gravel, or sand
CONSERVATION STATUS	Not assessed, but widespread
FLOWERING TIME	Throughout the year

FLOWER SIZE
⅞ in (2.3 cm)

PLANT SIZE
6–16 × 4–16 in
(15–41 × 10–41 cm),
including inflorescence

PTEROGLOSSA ROSEOALBA
SPOTTED JEWEL ORCHID
(REICHENBACH FILS) SALAZAR & M. W. CHASE, 2002

104

This species is perhaps better known under another genus, *Eltroplectris*, to which it is not closely related. An erect, few-to-many-flowered inflorescence is produced from a rosette of oblanceolate (elongate oval), yellow- or white-spotted leaves with a cluster of fat hairy roots. The bracts on the flower stem are tinged with purple and cover most of the stem and the emerging flowers. The leaves are seasonal, and after flowering the plant has a dormant period, usually during the dry season, when it disappears underground.

The flowers are flushed with pink in many cases and scentless, but there is abundant nectar in the long spur. The pollinators of this species are unknown, but the thick floral tube and the more darkly pigmented column suggest that hummingbird pollination is a possibility.

The flower of the Spotted Jewel Orchid is tubular and has three free, narrow sepals and two similar petals. The flowers are white but often tinged with pink, and the anther and stigma are more brightly colored pink or red.

Actual size

SUBFAMILY	Orchidoideae
TRIBE AND SUBTRIBE	Cranichideae, Spiranthinae
NATIVE RANGE	Tropical America from southern Florida and Mexico to Paraguay
HABITAT	Open and grassy places in tropical deciduous forests, at up to 5,600 ft (1,700 m)
TYPE AND PLACEMENT	Terrestrial
CONSERVATION STATUS	Not assessed, but locally common and also in disturbed habitats
FLOWERING TIME	July to November (summer to fall)

SACOILA LANCEOLATA
SCARLET LADY'S TRESSES
(AUBLET) GARAY, 1980

FLOWER SIZE
⅞ in (2.3 cm)

PLANT SIZE
12–36 in (30–90 cm),
including the inflorescence,
but not the leaves, which are
absent during flowering

105

From a mass of thick, hairy, fleshy roots connected to a short rhizome, the Scarlet Lady's Tresses produces a basal rosette of lanceolate leaves. After the leaves die, or as they are dying, a stout raceme is formed with 15–30 coral, pink, red, or orange flowers. *Sacoila lanceolata* is widespread in the tropics and subtropics of the New World, but there are populations that produce seeds without pollination (even without self-pollination) and others that are visited by hummingbirds. In some areas, the plant becomes weedy and occurs in high densities in disturbed sites.

The tubular flowers are odorless and are frequently visited by hummingbirds, which have no sense of smell. They pick up the pollinia, which, unlike those on most hummingbird-pollinated orchids, are yellow and not the usual gray matching the color of the bird's bill.

The flower of the Scarlet Lady's Tresses has sepals that are hairy outside. Two sepals are free and form a short basal spur, while the upper one is fused with the petals. The lip is recurved and, with the other floral parts, forms a tube that leads to the spur.

Actual size

SUBFAMILY	Orchidoideae
TRIBE AND SUBTRIBE	Cranichideae, Spiranthinae
NATIVE RANGE	Tropical South America, Trinidad, Tobago, and Grenada
HABITAT	Tropical wet lowland and montane forests, from sea level to 8,900 ft (2,700 m)
TYPE AND PLACEMENT	Terrestrial
CONSERVATION STATUS	Not assessed
FLOWERING TIME	February to October (summer to spring)

FLOWER SIZE
¾ in (2 cm)

PLANT SIZE
25–40 × 30–40 in
(63–102 × 76–102 cm),
including inflorescence

106

SARCOGLOTTIS ACAULIS
STEMLESS JEWEL ORCHID
(SMITH) SCHLECHTER, 1919

On the forest floor, the Stemless Jewel Orchid forms a rosette of obovate-elliptic, mottled or white-striped leaves, from which an erect hairy inflorescence emerges, with upright flowers subtended by lanceolate, pointed bracts. Underground is a large cluster of fat, hairy roots.

The flowers contain nectar, produced in the cavity on the base of the lip, and pollinaria of this species have been found on the mouth parts of an euglossine bee (*Eulaema cingulata*). The male bee is well known for the collection of orchid floral fragrances in other, unrelated groups of orchids. However, in other *Sarcoglottis* species, both male and female euglossine bees have been observed pollinating flowers. Female euglossine bees collect nectar, so this is clearly not a case of floral fragrance being the reward.

The flower of the Stemless Jewel Orchid is green and upright, with the tip curved outward. The sepals are velvety outside, with the upper sepal forming a hood with the green-striped petals. The lip, recurved between the two lower sepals, is pale green with darker green markings.

Actual size

SUBFAMILY	Orchidoideae
TRIBE AND SUBTRIBE	Cranichideae, Spiranthinae
NATIVE RANGE	Colombia and Ecuador, southern Brazil to northern Argentina, and Bolivia
HABITAT	Wet forests and grasslands at 5,900–8,200 ft (1,800–2,500 m)
TYPE AND PLACEMENT	Terrestrial
CONSERVATION STATUS	Not assessed
FLOWERING TIME	Throughout the year

SAUROGLOSSUM ELATUM

LIZARD-TONGUE ORCHID

LINDLEY, 1833

FLOWER SIZE
⅛ in (0.5 cm)

PLANT SIZE
12–32 × 10–20 in
(30–81 × 25–51 cm),
including inflorescence

107

The leaves and flowers of the Lizard-tongue Orchid are produced at the same time. A rosette of thick-veined, oblong-lanceolate leaves grows from a cluster of fat, hairy roots, with an erect inflorescence emerging, with bracts at each node.

The species is similar to *Sauroglossum nitidum* and is often confused with it. *Sauroglossum elatum*, however, has fewer, herbaceous, petiolate leaves, a less dense inflorescence, and a lip with more upturned sides. The genus name *Sauroglossum* comes from the Latin for "lizard" (*sauros*) and "tongue" (*glossa*), referring to the shape of the sepals, which are held upright. The flowers appear to produce nectar at the base of the lip, but pollination has not been observed, although the general floral morphology is similar to species of the genus *Pelexia*, which are pollinated by bumblebees.

The flower of the Lizard-tongue Orchid has green sepals. Two are curved upward, whereas the upper one curves over the flower and forms an elongate tube with the petals and lip. The lip is white and shortly recurved at the apex.

Actual size

SUBFAMILY	Orchidoideae
TRIBE AND SUBTRIBE	Cranichideae, Spiranthinae
NATIVE RANGE	Southern Brazil and Uruguay
HABITAT	Open grassland and scrubland, from sea level to 5,250 ft (1,600 m)
TYPE AND PLACEMENT	Terrestrial
CONSERVATION STATUS	Not assessed
FLOWERING TIME	October to December (late spring to early summer)

FLOWER SIZE
¼ in (0.75 cm)

PLANT SIZE
14–26 × 6–10 in
(36–66 × 15–25 cm),
including inflorescence

108

SKEPTROSTACHYS ARECHAVALETANII
SCEPTER ORCHID
(BARBOSA RODRIGUES) GARAY, 1980

The Scepter Orchid grows in open areas, including wet meadows and marshes that are subject to seasonal fires. It also occurs in rocky fields, known as *campos rupestres* in Brazil. From a short rhizome, with a cluster of fingerlike, fleshy hairy roots, a central stem is formed with a series of spirally arranged leaves that quickly become smaller and merge with the large pale bracts subtending each flower. The flower stem is topped with 25–50 densely packed flowers arranged in a spike.

There is no information available on pollination of this species. Its similarity in shape and color to *Sacoila lanceolata* (see page 105), however, might suggest that it is pollinated by hummingbirds.

The flower of the Scepter Orchid is coral to orange red. It is small, fleshy, and does not fully open. The sepals and petals are held forward, and the lip points down between the two lateral sepals, forming a channel into the nectar cavity at the lip base.

Actual size

SUBFAMILY	Orchidoideae
TRIBE AND SUBTRIBE	Cranichideae, Spiranthinae
NATIVE RANGE	Central and eastern North America (including southeastern Canada), absent from Florida
HABITAT	Open, moist, and sandy areas as well as bogs
TYPE AND PLACEMENT	Terrestrial
CONSERVATION STATUS	Not threatened
FLOWERING TIME	September and October, sometimes blooming well into November

FLOWER SIZE
⅜ in (1 cm)

PLANT SIZE
6–16 × 4–6 in
(15–41 × 10–15 cm),
including inflorescence

SPIRANTHES CERNUA
NODDING LADY'S TRESSES
(LINNAEUS) RICHARD, 1817

109

The Nodding Lady's Tresses is one of the most common orchids of North America, often forming large lush colonies in moist or marshy areas and grasslands that are difficult to spot until the small, brilliant white fragrant flowers are produced. It was a well-known species even before it was described by Carl Linnaeus in 1753. The "lady's tresses" of the common name (applied to all species in the genus) refers to the resemblance of the inflorescence to the braided hair worn by ladies.

Able to produce seed by both sexual and asexual means, and with a propensity to frantically produce new plantlets by trailing stolons in prime environmental conditions, *Spiranthes cernua* often colonizes disturbed areas, such as roadside ditches and even lawns, where its late-flowering allows it to survive mowing. Large drifts of pristine blooms often appear in older cemeteries in the fall. Pollination is by bees, especially bumblebees.

The flower of the Nodding Lady's Tresses
is pure crystalline white and deeply cupped, usually arranged in attractive spirals on the upper portion of its inflorescence.

Actual size

SUBFAMILY	Orchidoideae
TRIBE AND SUBTRIBE	Cranichideae, Spiranthinae
NATIVE RANGE	Japan, the Korean Peninsula, part of Russia, Iraq, Southeast Asia, Indonesia, Australia, New Zealand, and many Pacific islands
HABITAT	Open, moist grassy areas, bogs, and even lawns, and often in disturbed habitats such as the berms of rice paddies (naturalized in many parts of the world)
TYPE AND PLACEMENT	Terrestrial
CONSERVATION STATUS	Not threatened
FLOWERING TIME	July to August

FLOWER SIZE
⅜ in (1 cm)

PLANT SIZE
6–15 × 4–8 in
(15–38 × 10–20 cm),
including inflorescence

SPIRANTHES SINENSIS
PINK LADY'S TRESSES
(PERSOON) AMES, 1908

110

This pretty, ubiquitous weed of temperate and tropical Asia thrives, spreads, and seeds itself readily in almost any open sunny, grassy areas that have ample moisture. With an incredible latitudinal range, the species has adapted to a wide variety of climates, tolerating harsh frozen winters and torrid humidity. Happily colonizing disturbed and cultivated areas, it often appears in prepared flower beds.

Plants are perennial but short-lived—about five years—and are thought to replace themselves regularly by producing copious seed. The species is the only *Spiranthes* that strays from white or yellow in its color scheme. The striking flowers, variable in color, are usually pink or purple, although lavender, red, and white forms also exist. They arrange themselves gracefully in a spiral around the upright inflorescence. Pollination by megachilid (leaf-cutting) bees has been documented.

The flower of the Pink Lady's Tresses is small, with the blooms arranged in a spiral raceme on an upright stem. Sepals and petals are vibrant pink or purple often with a pure crystalline-white lip.

Actual size

SUBFAMILY	Orchidoideae
TRIBE AND SUBTRIBE	Cranichideae, Spiranthinae
NATIVE RANGE	Mexico, the Caribbean to Peru
HABITAT	Moister areas of seasonally dry semi-deciduous forests, often on steep embankments near seeps, at 3,950–9,850 ft (1,200–3,000 m)
TYPE AND PLACEMENT	Terrestrial, occasionally epiphytic
CONSERVATION STATUS	Not threatened
FLOWERING TIME	Winter

FLOWER SIZE
¾ in (1.8 cm)

PLANT SIZE
8–36 × 6–10 in
(20–91 × 15–25 cm),
including inflorescence

STENORRHYNCHOS SPECIOSUM
VERMILION LADY'S TRESSES
(JACQUIN) RICHARD, 1817

111

One of the more spectacular of the terrestrial orchids, also amenable to cultivation, the widespread Vermilion Lady's Tresses has a basal rosette of spirally arranged, variegated (striped or spotted) leaves reminiscent of the garden plant *Hosta*. True spectacle ensues when brilliant red spikes emerge from the center of the rosettes, bearing dazzling, torchlike racemes of up to 50 (usually 20–30) small red and white flowers, each subtended by a bright red bract. Pollinating hummingbirds are irresistibly drawn to these flowers, which are waxy and tough enough to stand up to the onslaught of a bird's beak.

The handsome plants have a cluster of thick, hairy roots that sustain them through dry seasons when leaves wither. This dormancy usually occurs shortly after blooming takes place. Some forms of this species have entirely green leaves without any variegation.

The flower of the Vermilion Lady's Tresses
is small, cupped, and white but infused with brilliant red and subtended by a dazzling red bract, which makes the blooms and inflorescence appear to be completely red.

Actual size

SUBFAMILY	Orchidoideae
TRIBE AND SUBTRIBE	Diurideae, Acianthinae
NATIVE RANGE	Coastal regions of New South Wales and Queensland in eastern Australia
HABITAT	Sand dunes and coastal scrub near gullies, often under shrubs
TYPE AND PLACEMENT	Terrestrial
CONSERVATION STATUS	Not threatened
FLOWERING TIME	April to May (fall)

FLOWER SIZE
⅜ in (1 cm)

PLANT SIZE
4–6 × 1½–2½ in
(10–15 × 4–6 cm),
including inflorescence

ACIANTHUS FORNICATUS
LARGE MOSQUITO ORCHID
ROBERT BROWN, 1810

112

Actual size

The extremely common Large Mosquito Orchid, found in the coastal scrub of eastern Australia, is said to be so plentiful and grow so densely that it is impossible not to step on one while exploring its habitat. A deciduous, colony-forming plant with small, subterranean tubers, *Acianthus fornicatus* blooms in the Australian fall after a two-month hot dry rest in the sweltering summer months.

With the return of the rains, a distinctive heart-shaped leaf, suffused with purple markings underneath, is first to emerge. From each leaf center, a single, slender inflorescence of up to 6 in (15 cm) arises, bearing small, hooded flowers arranged around the stem. The common name is from its small mosquito-like flowers. Pollinators are thought to be small flies attracted to the smell of rotting plant material or fungi—small amounts of nectar are also present.

The flower of the Large Mosquito Orchid is small and green with fused, pointed segments and a hooded dorsal sepal.

SUBFAMILY	Orchidoideae
TRIBE AND SUBTRIBE	Diurideae, Acianthinae
NATIVE RANGE	Southeastern Australia, from southern South Australia and Kangaroo Island, to Victoria, New South Wales, and Tasmania
HABITAT	Woodland and scrub with Mediterranean climate
TYPE AND PLACEMENT	Terrestrial
CONSERVATION STATUS	Not assessed, but locally abundant
FLOWERING TIME	June to October (winter to spring)

FLOWER SIZE
⅝ in (1.5 cm)

PLANT SIZE
1⅟₁₆ × ⅞₁₆ in
(4 × 1.5 cm),
including inflorescence

CORYBAS DIEMENICUS
PURPLE HELMET ORCHID
(LINDLEY) RUPP, 1928

113

The intriguing, intricate, and tiny Purple Helmet Orchid is believed to be a mushroom mimic. The flower has small glands on each side of the column, which, when punctured, produce a liquid of unknown function or purpose. A fungus gnat carrying pollinia fits neatly in the small space between the glands and pollinia. Pollination is, however, not very effective as only a small number of fertilized ovaries have been observed. Deep underground, there is a single small tuber with fine lateral roots spreading outward.

The genus name is derived from the Phrygian mythological male dancers, *korybantes*, who worshipped the goddess Cybele, mother of Corybas, and were depicted wearing crested helmets. The epithet, *diemenicus*, refers to Van Diemen's Land, the former name of Tasmania, the island where the species was first encountered.

Actual size

The flower of the Purple Helmet Orchid is glossy reddish-brown and sits directly on a heart-shaped leaf. The fringed upper sepal is curved over the rest of the flower, and all other parts are highly reduced and form a tube. The column is short and hidden, flanked by two glands.

SUBFAMILY	Orchidoideae
TRIBE AND SUBTRIBE	Diurideae, Acianthinae
NATIVE RANGE	Malaysia and western Indonesia (Java, Sumatra, Borneo)
HABITAT	Middle elevation forests
TYPE AND PLACEMENT	Terrestrial on mossy rocks and tree trunks, usually on steep slopes
CONSERVATION STATUS	Not assessed but rare
FLOWERING TIME	October to February

FLOWER SIZE
½ in (1.3 cm)

PLANT SIZE
1⅛ × ⅜ in (3 × 1 cm),
including inflorescence

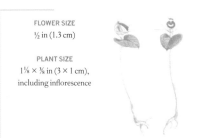

CORYBAS PICTUS
PAINTED HELMET ORCHID
(BLUME) REICHENBACH FILS, 1871

114

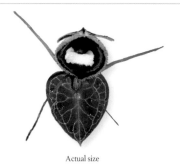

Actual size

This delicate, tiny orchid looks almost extraterrestrial with its hood and long tentacles. It is a tropical species that lives in mountain forests, where it needs dark and moist conditions and often grows on fibrous trunks of tree ferns or on mossy trunks of other forest trees. The name *pictus* derives from the Latin *pingere*, meaning "to decorate," referring to the attractive white venation on the tiny leaf.

Like most species of *Corybas*, the Painted Helmet Orchid is pollinated by fungus gnats. The long extensions on the sepals and petals occur only in the tropical species of this genus, and it is thought this may help lead flies to the center of the flower. The function of such floral specializations has never been experimentally tested, and such statements about potential roles are only speculative.

The flower of the Painted Helmet Orchid
sits atop a heart-shaped, silver-veined leaf.
The upper sepal (helmet) is dark brownish-
red with a white edge. The lip has a white
"cushion," and sepals and petals are
elongated into antennae.

SUBFAMILY	Orchidoideae
TRIBE AND SUBTRIBE	Diurideae, Acianthinae
NATIVE RANGE	East Asia, in southern Japan, China (southern Hunan and northern Fujian provinces), Taiwan, and northeastern Thailand
HABITAT	Humus-rich gullies and ditches in deep shade of dense forests
TYPE AND PLACEMENT	Terrestrial in humus and other decaying leaf litter
CONSERVATION STATUS	Not assessed, but threatened in Japan and likely uncommon elsewhere
FLOWERING TIME	August to September (late summer)

FLOWER SIZE
1 in (2.5 cm)

PLANT SIZE
4–8 × 2–3 in
(10–20 × 5–7 cm),
including flowers

STIGMATODACTYLUS SIKOKIANUS

ASIAN CRICKET ORCHID

MAXIMOWICZ EX MAKINO, 1891

115

Below ground, the Asian Cricket Orchid has a tuber ¾–1⅛ in (2–3 cm) long, from which a short scaly rhizome grows. Above ground, there are delicate stems, up to 8 in (20 cm) tall, bearing one or two small, round leaves, about 2 in (5 cm) across, topped with one to three flowers. Like many orchids growing in such dark sites, this species may derive some of the carbon it needs to sustain itself through its fungal symbiont, supplementing what it can gain through limited photosynthesis.

Stigmatodactylus sikokianus has an unusual distribution for a member of the tribe Diurideae, the species of which are largely confined to Australasia. The genus name derives from the Greek *stigmatos*, "a stigmalike structure," and *daktylos*, "finger," referring to the fingerlike structure on the column of the genus.

The flower of the Asian Cricket Orchid is small and delicate. It has threadlike, spreading sepals and petals and a broad lip with a pink, centrally horned callus. The column is green, winged, and protrudes forward over the lip.

Actual size

SUBFAMILY	Orchidoideae
TRIBE AND SUBTRIBE	Diurideae, Caladeniinae
NATIVE RANGE	New Zealand
HABITAT	High elevation southern beech (*Nothofagus*) forests, alpine herb fields, and alpine meadows, near the coast in the South Island
TYPE AND PLACEMENT	Terrestrial
CONSERVATION STATUS	Not threatened
FLOWERING TIME	December to January (summer)

FLOWER SIZE
¾ in (2 cm)

PLANT SIZE
4½ × 2½ in
(11.5 × 6 cm),
including inflorescence

APOROSTYLIS BIFOLIA
ODD-LEAF ORCHID
(J. D. HOOKER) RUPP & HATCH, 1946

116

The small Odd-leaf Orchid grows in the high southern mountains of New Zealand (the southern fjordlands), where it forms colonies of closely spaced plants, each with reddish-green to green stems that are covered with glandular hairs. The two hairy leaves are of different sizes—the upper one shorter—both often with purple blotches.

The orchid's attractive flowers have spots (nectar guides) to steer the approach of the pollinator. The species is apparently cross-pollinated, although the exact pollinator has not been observed. Below ground, small tubers form droppers (stems that make additional small tubers nearby), which explains why the species grows in clusters of closely spaced plants. Previously, *Aporostylis bifolia* was included in the species-rich and widespread genus *Caladenia*, to which it is related, although not closely.

Actual size

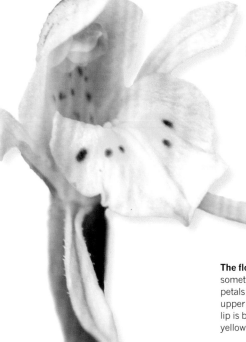

The flower of the Odd-leaf Orchid is mostly white, sometimes tinged pale pink. The lateral sepals and petals are spoon-shaped and of a similar size. The upper sepal forms a hood over the column. The lip is broad, with a smooth margin and variable yellow-to-brownish markings.

SUBFAMILY	Orchidoideae
TRIBE AND SUBTRIBE	Diurideae, Caladeniinae
NATIVE RANGE	Southwestern Australia
HABITAT	Forests and rocky outcrops, swamps, lake margins, and coastal scrub, at up to 985 ft (300 m)
TYPE AND PLACEMENT	Terrestrial on sand, clayey loam, laterite, or gravel
CONSERVATION STATUS	Not threatened
FLOWERING TIME	July to November (winter to spring)

CALADENIA LONGICAUDA
LARGE WHITE SPIDER ORCHID
LINDLEY, 1839

FLOWER SIZE
2⅜–3½ in (6–9 cm)

PLANT SIZE
12–18 × 2–3 in
(30–46 × 5–8 cm),
including inflorescence

117

Also called "daddy long legs," the Large White Spider Orchid, found close to Perth, has beautiful spidery flowers and is emblematic of the spring flora of Western Australia. It has an underground tuber with which it survives the dry season and fires. A single leaf and an inflorescence of one or six large flowers are produced, and both leaf and stem are hairy.

Spider orchid species are most likely pollinated by food deceit, and visits by a large variety of bees, wasps, and flies, all species that feed on nectar and pollen, have been reported. Some authors have suggested that these white-flowered species exhibit a mixed pollination syndrome and are also sexually attractive to male thynnine wasps, but this has yet to be demonstrated by detailed studies.

The flower of the Large White Spider Orchid
has long and spidery sepals and petals. Four are reflexed, but the dorsal sepal is held upright or over the column. The lip is white, recurving, and fringed with brown to purple-red hairs.

Actual size

SUBFAMILY	Orchidoideae
TRIBE AND SUBTRIBE	Diurideae, Caladeniinae
NATIVE RANGE	Southwestern Australia
HABITAT	*Casuarina* and *Eucalyptus* thickets, near granite outcrops
TYPE AND PLACEMENT	Terrestrial, on sides of rivers or creeks
CONSERVATION STATUS	Not threatened
FLOWERING TIME	September to October (spring)

FLOWER SIZE
1⅛–1⅝ in (3–4 cm)

PLANT SIZE
5–10 × 3–8 in
(13–25 × 8–20 cm),
including inflorescence

CALADENIA MULTICLAVIA
LAZY SPIDER ORCHID
REICHENBACH FILS, 1871

118

Actual size

This spider orchid usually has only a single flower held horizontally or in an inclined manner—hence the "lazy" part of the common name. Often up to six plants occur together in a clump, each with a single underground tuber and a single hairy leaf.

The species epitomizes pollination by sexual deceit. Its lip is covered in calli and trembles in the wind, making it look insect-like. The tips of the upturned sepals and petals produce sex pheromones, attracting male thynnine wasps that are forced to approach the lip from above. The male tries to grab the "female" mimic on the lip and fly off, but the hinged lip throws it downward into the column, where broad column wings catch the hapless insect. Pollen is then attached to the wasp's thorax.

The flower of the Lazy Spider Orchid has upcurved, reddish-brown petals and sepals that surround the motile, insect-like, striped lip. The column has broad wings and is surrounded by the petals and dorsal sepal.

SUBFAMILY	Orchidoideae
TRIBE AND SUBTRIBE	Diurideae, Caladeniinae
NATIVE RANGE	Southwestern Western Australia, between York and Bindoon
HABITAT	Open places in *Eucalyptus* (jarrah, wandoo) forests, usually only flowering after summer fires
TYPE AND PLACEMENT	Terrestrial on heavy lateritic soils
CONSERVATION STATUS	Not assessed
FLOWERING TIME	Late August to October (spring)

FLOWER SIZE
1⅝ in (4 cm)

PLANT SIZE
2–6 × 2–3 in
(5–15 × 5–8 cm),
including flower

CYANICULA IXIOIDES
YELLOW CHINA ORCHID
(LINDLEY) HOPPER & A. P. BROWN, 2000

119

The Yellow China Orchid is the only yellow-flowered *Cyanicula* species. Most of the others are blue, hence their genus name, from *cyano*, Greek for "blue," and *-icula*, a diminutive. The species has one or more small globular tubers underground and a single, densely hairy leaf held close to, but not against, the ground. Reproduction is almost entirely by seed, and daughter tubers are rarely formed. Flowering is much more frequent after a summer fire, and the orchid seems to prefer open sites without much competition from other plants.

Nectar-seeking scarab beetles are the pollinators for this species and its close relative, *C. gemmata*, which is blue. These two may turn out to be only color forms of a single species, given that they share the same pollinator. Genetic studies indicate few differences between populations of the two, but mixed populations are rare.

Actual size

The flower of the Yellow China Orchid has spreading, yellow sepals and petals, all of more or less the same size and shape. The lip is much shorter and recurved, with a brown callus in its center.

SUBFAMILY	Orchidoideae
TRIBE AND SUBTRIBE	Diurideae, Caladeniinae
NATIVE RANGE	Coastal areas of southwestern Western Australia
HABITAT	Sand, among open coastal scrub vegetation
TYPE AND PLACEMENT	Terrestrial
CONSERVATION STATUS	Not threatened
FLOWERING TIME	October or November (spring)

FLOWER SIZE
¾–1½ in (2–4 cm)

PLANT SIZE
6–12 × 1.5–2.5 in
(15–30 × 4–6 cm),
including inflorescence

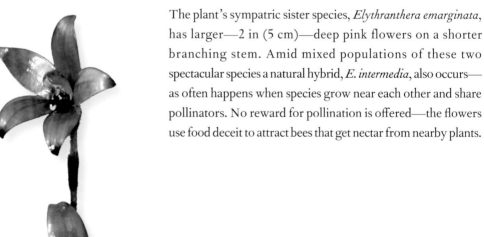

ELYTHRANTHERA BRUNONIS
PURPLE ENAMEL ORCHID
(ENDLICHER) A. S. GEORGE, 1963

120

One of the most extraordinary floral delights of coastal Western Australia, the small tuber-forming Purple Enamel Orchid is a common feature in the spring landscape. Emerging with the return of the spring rains on delicate stems from the base of a hairy elongate leaf, the glossy flowers appear to be hard-glazed like fine china but are actually fragile and easily bruised. Often holding up to three flowers per stem, their glory is short-lived, as they wither quickly if temperatures soar.

The plant's sympatric sister species, *Elythranthera emarginata*, has larger—2 in (5 cm)—deep pink flowers on a shorter branching stem. Amid mixed populations of these two spectacular species a natural hybrid, *E. intermedia*, also occurs— as often happens when species grow near each other and share pollinators. No reward for pollination is offered—the flowers use food deceit to attract bees that get nectar from nearby plants.

The flower of the Purple Enamel Orchid
is glossy purple with a pale, almost whitish reverse, bearing dark maroon spots. Petals and sepals are almost the same size, and extended, hoodlike column wings cover the reproductive parts and lip.

Actual size

SUBFAMILY	Orchidoideae
TRIBE AND SUBTRIBE	Diurideae, Caladeniinae
NATIVE RANGE	Western and southwestern Australia
HABITAT	Coastal scrub and *Eucalyptus* woodland (jarrah)
TYPE AND PLACEMENT	Terrestrial on sandy or rocky soil
CONSERVATION STATUS	Not assessed
FLOWERING TIME	March to June (fall to winter)

FLOWER SIZE
⅝ in (1.5 cm)

PLANT SIZE
4–14 × 3–4 in
(10–35 × 8–10 cm),
including flowers

ERIOCHILUS DILATATUS
EASTER BUNNY ORCHID
LINDLEY, 1840

121

The prominent hairs on the lip of the Easter Bunny Orchid and its related species explain the genus name, *Eriochilus*, which comes from the Greek, *erion*, meaning "wool," and *cheilos*, "lip." One or two long lanceolate leaves emerge from an underground tuber, and the stem eventually produces between three and eight (exceptionally up to 20) flowers. Flowering appears to be enhanced when summer fires have gone through the bush, although fire is not essential.

Small bees pollinate the orchids, probing with their mouthparts between the column and lip, where they contact the pollinia. To promote cross-pollination, anther flaps shield the stigma from contact when the bee enters but ensure contact with, and the removal of, pollinia when the bee withdraws. A strong scent attracts the bees, although this is false advertising, as the flowers seem to offer no reward.

The flower of the Easter Bunny Orchid has two white, downward-pointing sepals. The third, upper, sepal and the two petals are darkly colored (usually reddish-brown) and form a hood over the column. The lip is woolly, yellow, and recurved, and bears reddish-purple spots.

Actual size

SUBFAMILY	Orchidoideae
TRIBE AND SUBTRIBE	Diurideae, Caladeniinae
NATIVE RANGE	Southern Australia and Tasmania
HABITAT	Winter-wet areas, but flowers appear only after summer burns or disturbance, often around granite outcrops, creek margins, or near swamps
TYPE AND PLACEMENT	Terrestrial
CONSERVATION STATUS	Not threatened
FLOWERING TIME	September to October (spring)

FLOWER SIZE
⅝ in (1.5 cm)

PLANT SIZE
4–12 × 2⅜–4 in
(10–30 × 6–10 cm),
including inflorescence

LEPTOCERAS MENZIESII
RABBIT ORCHID
(R. BROWN) LINDLEY, 1840

122

The little Rabbit Orchid grows in large colonies and flowers especially well after fires. It has one to two tubers and produces daughter tubers at the ends of roots, allowing it to form closely spaced and abundant plants. From a single, basal, hairless leaf, a short stem arises bearing one to four flowers. The genus name is derived from the Greek for "slender" (*leptos*) and "horn" (*keras*), in reference to the erect petals' resemblance to horns.

Little is known about pollination of this sole species in the genus, but there are reports of small bees visiting the flowers, perhaps attracted initially to the stamenlike petals, but soon discovering droplets of nectar formed at the base of the column, where they may pick up or deposit pollinia. Some reports also have indicated that the flowers are sweetly fragrant.

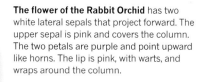

The flower of the Rabbit Orchid has two white lateral sepals that project forward. The upper sepal is pink and covers the column. The two petals are purple and point upward like horns. The lip is pink, with warts, and wraps around the column.

Actual size

SUBFAMILY	Orchidoideae
TRIBE AND SUBTRIBE	Diurideae, Caladeniinae
NATIVE RANGE	Southwestern Western Australia and southeastern Australia
HABITAT	Open, sandy coastal scrub vegetation
TYPE AND PLACEMENT	Terrestrial
CONSERVATION STATUS	Not threatened
FLOWERING TIME	August to September (late winter to spring)

PHELADENIA DEFORMIS

BLUE FAIRIES

(ROBERT BROWN) D. L. JONES & M. A. CLEMENTS, 2001

FLOWER SIZE
1½–2 in (4–5 cm)

PLANT SIZE
5–8 × 2–4 in
(13–20 × 5–10 cm),
including inflorescence

123

One of the earliest orchids to bloom in the botanical hot spots of Australia, the tiny Blue Fairies, with its vivid blue flowers, emerges vigorously through seemingly barren sands. Deeply submerged bulbs allow the plants to survive intense summer heat and fires. Historically, the small bulbs were dug and collected by aboriginals for food.

Currently a monospecific genus, this species has previously been considered a member of other genera, such as *Cyanicula* and *Caladenia*, to which it is clearly allied. The plant was christened *deformis* as it was thought originally to be an aberrantly formed *Caladenia*. Exhilarating to observe in the wild, Blue Fairies can form large, dense colonies, even on well-worn trails. They are pollinated by a halictid bee (sweat bee) that appears to be attracted to the glandular protuberances at the base of the lip.

Actual size

The flower of Blue Fairies is spidery, with narrow segments, pale to medium blue or violet, with a darker, vivid purple, upward-facing lip bearing small, purple and sometimes yellow, wartlike calli.

SUBFAMILY	Orchidoideae
TRIBE AND SUBTRIBE	Diurideae, Cryptostylidinae
NATIVE RANGE	Southwestern Australia
HABITAT	Low coastal scrub, dense woodland, and forests
TYPE AND PLACEMENT	Terrestrial on sand
CONSERVATION STATUS	Not threatened
FLOWERING TIME	November to April (summer)

FLOWER SIZE
1 in (2.5 cm)

PLANT SIZE
15–25 × 4–10 in
(38–64 × 10–25 cm),
excluding stem

124

CRYPTOSTYLIS OVATA
AUSTRALIAN
SLIPPER ORCHID
R. BROWN, 1810

In 1938, this was the first species of Australian orchid in which pollination by sexual deceit was documented. *Cryptostylis ovata* is a highly distinctive orchid that attracts male ichneumonid wasps. These parasitic wasps land on the lip of this upside-down flower, clasping it with thin legs. The wasp then probes the tube of the lip and picks up the pollinia on its abdomen.

The genus name refers to the column that is hidden (*crypto* in Greek) within the lip, because the flower is upside down and the column is short. The epithet refers to the ovate evergreen leaves, which, together with the orchid's long stem (petiole) and many-flowered inflorescence make it easy to recognize. Underground, it has a short stem with many thick smooth roots without any sort of tuber.

The flower of the Australian Slipper Orchid
is often reversed, with the middle sepal and lip pointed down and the two laterals pointing up. The green, threadlike petals are reflexed, and the reddish-brown lip is slipper shaped, forming a chamber around the column.

Actual size

SUBFAMILY	Orchidoideae
TRIBE AND SUBTRIBE	Diurideae, Diuridinae
NATIVE RANGE	Southwestern Australia
HABITAT	Winter-wet swamps
TYPE AND PLACEMENT	Terrestrial on sandy clay among sedges
CONSERVATION STATUS	Not threatened
FLOWERING TIME	October to November (spring)

FLOWER SIZE
1⅛ in (3 cm)

PLANT SIZE
20–30 × 8 in
(51–76 × 20 cm),
including inflorescence

DIURIS CARINATA
TALL BEE ORCHID
LINDLEY, 1840

125

In the past, this species was confused with the Australian Bee Orchid (*Diuris laxiflora*), but it differs from it in flowering period, taller inflorescences, and larger flowers. Underground, the Tall Bee Orchid has a tuber that may be relatively elongate. Daughter tubers are rarely formed, so the species tends to occur as single isolated plants. There are from one to five glossy grasslike leaves per plant.

Bee orchids (also called donkey orchids due to their upright petals that look like donkey ears) offer no reward to their pollinator and are believed to mimic spring-flowering species of the Fabaceae family, such as the genera *Daviesia* and *Pultenaea*, with which they often grow. They are pollinated by various species of bees, often halictid (sweat) bees, although other types of bees have also been reported, including honey bees (not native).

Actual size

The flower of the Tall Bee Orchid is yellow with reddish-brown spots and has two narrow, downcurved sepals and one broad, upward-facing sepal. The two petals are on long stalks and stand upright. The lip has two broad lateral wings that embrace the column.

SUBFAMILY	Orchidoideae
TRIBE AND SUBTRIBE	Diurideae, Drakaeinae
NATIVE RANGE	Coastal eastern Australia (Queensland, New South Wales)
HABITAT	Forested slopes and ridges between grassy tufts or shrubs
TYPE AND PLACEMENT	Terrestrial on sandy and clayey loams
CONSERVATION STATUS	Threatened
FLOWERING TIME	December to February (summer)

FLOWER SIZE
⅜–¾ in (1–1.8 cm)

PLANT SIZE
12⅝ × 5½ in (32 × 14 cm)
including inflorescence

126

ARTHROCHILUS IRRITABILIS
CLUBBED ELBOW ORCHID
F. MUELLER, 1858

Actual size

All elbow orchids are sexually deceptive. Male thynnid wasps of the genus *Rhagigaster* are attracted by their floral fragrance, which resembles the sexual pheromone released by the female wasp to attract a male, tricking them into thinking that they can copulate with the fluffy, modified part of the lip. The male wasp grasps the lip and tries to fly off with this "female" to mate. The lip, however, is hinged, which throws the wasp into the apex of the column, where the pollinia are located. He receives or deposits a packet of pollen (pollinium) and then repeats this process with another orchid flower.

The Clubbed Elbow Orchid produces a rosette of three to seven leaves and, from the side of this rosette, a separate leafless growth with the inflorescence. Underground, it produces tubers and stolons on which new tubers are formed.

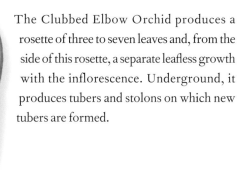

The flower of the Clubbed Elbow Orchid is pale green with purplish-red hairs. The sepals and petals are lanceolate and recurved. The lip is ornamented with a hairy horn and two-lobed tip. The column is curved and the wings are sickle-shaped.

SUBFAMILY	Orchidoideae
TRIBE AND SUBTRIBE	Diurideae, Drakaeinae
NATIVE RANGE	Eastern and southeastern Australia and Tasmania
HABITAT	*Eucalyptus* woodland, heath, and swampy scrub, usually on sand near the coast
TYPE AND PLACEMENT	Terrestrial
CONSERVATION STATUS	Not threatened in eastern Australia, but vulnerable in South Australia as a result of habitat loss
FLOWERING TIME	September to January (spring to summer)

FLOWER SIZE
¾–1 in (2–2.5 cm)

PLANT SIZE
6–20 in (15–51 cm),
including inflorescence

CALEANA MAJOR
FLYING DUCK ORCHID
R. BROWN, 1810

127

As the common name of this species suggests, its flowers resemble a duck in flight. Underground there is a single tuber, replaced each year, and daughter tubers can grow at the end of a short stolon, resulting in small colonies of closely spaced plants, each forming a single leaf and an inflorescence bearing two to four flowers.

The lip is moveable and actively traps a visiting insect in the bowl formed by the column wings, releasing it once the lip has reset itself. A volatile chemical is released from the flowers, attracting male sawflies of the genus *Pterygophorus*, which become trapped. In their struggle to escape, they spread pollen around on the stigma and remove the pollinia, which can be transferred to another flower.

Actual size

The flower of the Flying Duck Orchid has sepals forming the wings of the "duck," while the lip makes the neck and head. The body is composed of broad column wings, the tail is formed of the middle sepal, and the narrow, downward-pointing petals make the legs.

SUBFAMILY	Orchidoideae
TRIBE AND SUBTRIBE	Diurideae, Drakaeinae
NATIVE RANGE	Eastern Australia, Lord Howe Island, and the north island of New Zealand
HABITAT	Open forests and gullies, from sea level to 2,625 ft (800 m)
TYPE AND PLACEMENT	Terrestrial
CONSERVATION STATUS	Not uncommon in eastern Australia, but has not been seen in New Zealand since the early twentieth century; recently discovered on Lord Howe Island
FLOWERING TIME	August to November (late winter to spring)

FLOWER SIZE
⅝ in (1.5 cm)

PLANT SIZE
8 × ¾ in (20 × 2 cm)
including inflorescence

CHILOGLOTTIS FORMICIFERA
ANT ORCHID
FITZGERALD, 1877

128

Actual size

Raised warts on the lip of *Chiloglottis formicifera* mimic a flightless female of an ichneumonid wasp species. A floral scent is released that matches the female sex pheromone, to which male wasps are attracted. They attempt to grab and fly away with the lip, during which struggle one of the four pollinia becomes glued to the head of the wasp. Frustrated, the male flies off, only to pollinate another orchid flower during another visit and struggle.

The species is called Ant Orchid because the lip somewhat resembles an ant to the human eye, and it was hence named *formicifera*, meaning "ant-bearer." This association is spurious, but the plant was named long before its pollination syndrome was understood. Underground, there is a deeply buried globose tuber, from which a rosette of leaves is produced.

The flower of the Ant Orchid is greenish-brown, with the upper sepal bearing an osmophore (scent-producing gland). Lateral sepals and petals are narrow and reflexed. The lip has a narrow, black, glossy, antlike callus. The arching column hangs over the lip and is narrowly winged.

SUBFAMILY	Orchidoideae
TRIBE AND SUBTRIBE	Diurideae, Drakaeinae
NATIVE RANGE	Southwestern Australia
HABITAT	Open places on coastal plains, road verges, and swamp margins
TYPE AND PLACEMENT	Terrestrial on damp sandy soil
CONSERVATION STATUS	Locally common but threatened by human development and agriculture
FLOWERING TIME	August to October (spring)

DRAKAEA GLYPTODON
HAMMER ORCHID
FITZGERALD, 1882

FLOWER SIZE
1 in (2.5 cm)

PLANT SIZE
Up to 14 × 1.2 in
(35 × 3 cm),
including inflorescence

129

The heart-shaped leaf of the Hammer Orchid has a tiled pattern of pale and darker green. Above this, a thin stalk emerges, bearing one or two strange flowers. The lip has a narrow stalk, which is hinged and can move back toward the column, and bears an insect-like bulge at its tip. It produces a pheromone that specifically attracts a male wasp, *Zaspilothynnus trilobatus*, which mistakes the lip for a female sitting on a blade of grass and tries to carry it away to copulate in flight. The male is catapulted into the column, where pollinia get stuck to its thorax. Males are competitive and repeat this behavior many times and so transfer pollinia to the stigmas of other Hammer Orchid flowers.

The genus is named for Sarah Ann Drake (1803–57), a botanical illustrator and the governess for the botanist John Lindley's children.

Actual size

The flower of the Hammer Orchid is greenish-yellow with a red lip. Sepals and petals are tinged pink and reflexed, except for the upper sepal that supports the winged column. The lip bears a lobed structure covered in reddish hairs and black warts.

SUBFAMILY	Orchidoideae
TRIBE AND SUBTRIBE	Diurideae, Drakaeinae
NATIVE RANGE	Southwestern Australia, known only from a few places between Eneabba and Pingelly
HABITAT	Dense scrubland among *Banksia* species
TYPE AND PLACEMENT	Terrestrial in deep sandy soil
CONSERVATION STATUS	Listed as endangered in Australia
FLOWERING TIME	Late October to January (late spring to summer)

FLOWER SIZE
1 in (2.5 cm)

PLANT SIZE
4¾–7 in (12–18 cm)
tall; leaves are small
and close to the soil

PARACALEANA DIXONII
SANDPLAIN DUCK ORCHID
HOPPER & A. P. BROWN, 2006

130

Actual size

Duck orchids are pollinated by sexually deceived male thynnid wasps that attempt to snatch the hinged lip and fly off with it. This throws them against the column, where they pick up the pollina before flying away. Underground, the Sandplain Duck Orchid produces one or two tubers, and daughter tubers grow at the ends of roots. Most reproduction is by seeds, and large, though inconspicuous, colonies can be formed.

Paracaleana dixonii is named for Kingsley Dixon, a botanist and orchid scientist in Western Australia, who first recognized it as new. The plant is the largest of the duck orchids and flowers later than most species, when temperatures have risen to 98.6°F (37°C)—a temperature it can cope with by storing water and nutrients in its fleshy stem.

The flower of the Sandplain Duck Orchid
has red-suffused yellow sepals and petals,
which form the "body" of the duck. The
stalked and recurved lip is the "neck,"
a broadened limb the "head," and dark,
warty glands at the tip make up the "bill."

SUBFAMILY	Orchidoideae
TRIBE AND SUBTRIBE	Diurideae, Drakaeinae
NATIVE RANGE	Southwestern Australia
HABITAT	Granite outcrops
TYPE AND PLACEMENT	Terrestrial in shallow sandy soil overlaying granite, among mosses and cushion plants (especially genus *Borya*)
CONSERVATION STATUS	Not threatened
FLOWERING TIME	Late October to January (summer)

SPICULAEA CILIATA

ELBOW ORCHID

LINDLEY, 1840

FLOWER SIZE
¾ in (2 cm)

PLANT SIZE
3–7 × 5–6 in
(8–18 × 13–15 cm),
including inflorescence

131

Named for its loosely hinged lip, the Elbow Orchid grows during the summer, in temperatures that can exceed 98°F (37°C). In the wetter spring months, the plant forms a succulent fleshy stem, which provides moisture to the flowers in summer, when the leaf is shriveled and the tuber for the next year is already formed. Small colonies of daughter tubers are produced on short stolons, and the single leaf is held close to the substrate.

The anvil-shaped lip resembles a female thynnid wasp and gives off a pheromone to attract males. They attempt to fly off with the lip (in theory, to mate with it in flight) but instead are thrust onto the column and held by the hooked column wings, struggling to be released. If a male wasp repeats the process, cross-pollination of the flowers is achieved.

Actual size

The flower of the Elbow Orchid is straw-colored and elongate, with recurved lateral sepals and petals, and an erect dorsal sepal that flanks the winged column. The lip mimics a female wasp.

SUBFAMILY	Orchidoideae
TRIBE AND SUBTRIBE	Diurideae, Megastylidinae
NATIVE RANGE	Southeastern Australia and Tasmania
HABITAT	Swampy areas in scrubland
TYPE AND PLACEMENT	Terrestrial on wet peaty soil
CONSERVATION STATUS	Threatened
FLOWERING TIME	August to December (spring)

FLOWER SIZE
1 in (2.5 cm)

PLANT SIZE
Height of inflorescence
4–6 in (10–15 cm)

BURNETTIA CUNEATA
AUSTRALIAN LIZARD ORCHID
LINDLEY, 1840

132

The leafless, mycoheterotrophic Australian Lizard Orchid grows in swampy places, but only appears above ground after a fire, spending the rest of its life cycle underground. After germination, it forms small tubers that associate with fungi to obtain their food and minerals, waiting in the soil until the next fire passes through, after which the plant flowers and dies. A slender pale stem with reddish bracts forms from the tubers, topped with up to four flowers. No replacement tuber is formed—reproduction is strictly by seeds.

Burnettia cuneata grows in acid heaths, mostly under patches of scented paperbark (*Melaleuca squarrosa*, family Myrtaceae). Little is known about its life history, and pollination has never been observed, although the orchid appears not to set seed regularly, suggesting that it is pollinated by some animal.

The flower of the Australian Lizard Orchid has a hooded, purple-veined upper sepal and two flaring lateral sepals and petals that are white above and reddish below. The purple-streaked lip has the lateral lobes inside the hooded sepal. The three-lobed lip has short teeth and a basal callus.

Actual size

SUBFAMILY	Orchidoideae
TRIBE AND SUBTRIBE	Diurideae, Megastylidinae
NATIVE RANGE	Southern Australia and Tasmania
HABITAT	Woodland, scrubland, coastal heaths, and swamp margins
TYPE AND PLACEMENT	Terrestrial
CONSERVATION STATUS	Not threatened
FLOWERING TIME	March to June (fall to winter)

FLOWER SIZE
⅜ in (1 cm)

PLANT SIZE
4–10 × 3–5 in
(10–25 × 8–13 cm),
including inflorescence

LEPORELLA FIMBRIATA

HARE ORCHID

(LINDLEY) A. S. GEORGE, 1971

133

The flower of this species—the only one in its genus—resembles a hare, hence its common and genus names (*leporella* is Latin for "little hare"). The orchid has an ovoid fleshy tuber that forms new tubers away from the mother plant at the ends of some of its roots, which are produced from a bract-like structure just below the soil surface. One or two basal leaves lie more or less flat on the soil surface, and an erect inflorescence, with one or two flowers but no leaves, is produced.

Pollination is unusual because the flowers only attract flying male ants, which attempt to mate with the fringed lip (to which the species name *fimbriata* refers) and remove pollinia. This is the only plant known to have ants acting as regular pollen carriers.

Actual size

The flower of the Hare Orchid has two sepals pointing down and one sepal forming a hood over the column. Two petals point upward and resemble hare ears. The lip is fringed and brownish-red, with spots and a greenish-yellow center.

SUBFAMILY	Orchidoideae
TRIBE AND SUBTRIBE	Diurideae, Megastylidinae
NATIVE RANGE	Eastern Australia and Tasmania
HABITAT	Heath and acid woodland, in foothills on well-drained to moist soils
TYPE AND PLACEMENT	Terrestrial on peaty soils in partial shade
CONSERVATION STATUS	Not threatened and widespread
FLOWERING TIME	August to November (spring)

FLOWER SIZE
1⅛ in (3 cm)

PLANT SIZE
7–18 in
(18–46 cm) tall,
including inflorescence;
leaf is held erect

134

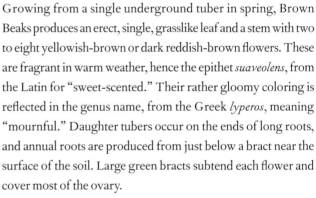

LYPERANTHUS SUAVEOLENS
BROWN BEAKS
R. BROWN, 1810

Growing from a single underground tuber in spring, Brown Beaks produces an erect, single, grasslike leaf and a stem with two to eight yellowish-brown or dark reddish-brown flowers. These are fragrant in warm weather, hence the epithet *suaveolens*, from the Latin for "sweet-scented." Their rather gloomy coloring is reflected in the genus name, from the Greek *lyperos*, meaning "mournful." Daughter tubers occur on the ends of long roots, and annual roots are produced from just below a bract near the surface of the soil. Large green bracts subtend each flower and cover most of the ovary.

The orchid's sweet fragrance and nectar suggest that it may be pollinated by bees. The labellum is brightly colored and covered with a crest and warty or glandular hairs, which might operate as pollen mimics.

The flower of Brown Beaks has narrow sepals and petals that are yellowish at the base and reddish-brown at the tip. The upper sepal is a hood. The lip is three-lobed and surrounds the column; the middle lobe is yellow and covered by a callus with fingerlike growths.

Actual size

SUBFAMILY	Orchidoideae
TRIBE AND SUBTRIBE	Diurideae, Megastylidinae
NATIVE RANGE	New Caledonia to southern Vanuatu
HABITAT	Open scrubland, at 330–650 ft (100–200 m)
TYPE AND PLACEMENT	Terrestrial, usually among ultrabasic rocks
CONSERVATION STATUS	Not assessed but locally frequent; habitat threatened by mining activity
FLOWERING TIME	September to December (spring to early summer)

FLOWER SIZE
3⅛ in (8 cm)

PLANT SIZE
30–40 × 12–15 in
(76–102 × 30–38 cm),
including inflorescence

MEGASTYLIS GIGAS
GIANT FAIRY ORCHID
(REICHENBACH FILS) SCHLECHTER, 1911

135

The large, mostly terrestrial but occasionally rock-inhabiting Giant Fairy Orchid produces an impressive display when in flower. It has long grassy leaves with no tuber underground, and its roots are fleshy and hairy. All reproduction is by seed. The fragrant flowers are generally white with a pink center, but there are variants with dark pink or entirely green flowers.

Nothing is known about the pollination of this orchid, but the flowers are similar to bee-pollinated species of the genus *Caladenia* from Australia. This means that bees may enter the tube formed by the dorsal sepal and lip to seek non-existent nectar at the base of the lip, which forms a perfect landing platform. The genus name is based on the Greek for "large" (*megas*) and "pillar" (*stylos*), referring to the column of this orchid.

The flower of the Giant Fairy Orchid has a middle sepal that forms a hood. The white lateral sepals and petals project down and forward. The pink to yellowish lip is covered by short stout hairs and, together with the sepal, helps to form a tube around the column.

Actual size

SUBFAMILY	Orchidoideae
TRIBE AND SUBTRIBE	Diurideae, Megastylidinae
NATIVE RANGE	Lower southwestern Western Australia between Augusta and Cheyne Beach
HABITAT	Winter-wet areas along creeks and swamps
TYPE AND PLACEMENT	Terrestrial in peaty soil
CONSERVATION STATUS	Not threatened
FLOWERING TIME	November to early December (summer)

FLOWER SIZE
¾ in (1.8 cm)

PLANT SIZE
4–12 × 3–4 in
(10–30 × 8–10 cm)

PYRORCHIS FORRESTII
PINK BEAKS
(F. MUELLER) D. L. JONES & M. A. CLEMENTS, 1994

136

Flowering almost exclusively after an intense summer fire, Pink Beaks carries two to seven fragrant blooms on a stem arising from an oval tuber and two to three leaves. Non-flowering individuals can live for years until a fire initiates flowering. They produce one or two annual ground-hugging leaves. Daughter tubers are formed near the mother plant, so colonies can appear in a mass of color in the spring following a fire. The genus name refers to this association with fire, from the Greek word *pyr*, for "fire."

Given its color, fragrance, presence of nectar, and shape, Pink Beaks is possibly pollinated by bees that force themselves between the column and the lip. Pollination has not been observed. Before the genus *Pyrorchis* was described in 1995, its two species were considered members of *Lyperanthus*.

The flower of Pink Beaks has pink or pink-tinged sepals and petals of more or less the same size and shape. The middle sepal is curved over the flower, forming a tube around the column with the curved, frilly-margined, dark pink or carmine-marked lip.

Actual size

SUBFAMILY	Orchidoideae
TRIBE AND SUBTRIBE	Diurideae, Prasophyllinae
NATIVE RANGE	Eastern Australia (southeastern Queensland and New South Wales, south to Nowra)
HABITAT	Dry or rocky *Eucalyptus* forests and heathland with frequent wildfires and a relatively high rainfall
TYPE AND PLACEMENT	Terrestrial, usually in moss gardens over sandstone sheets and in heath
CONSERVATION STATUS	Not assessed formally, but not under any current threat
FLOWERING TIME	January to May (fall)

FLOWER SIZE
⅜ in (1 cm)

PLANT SIZE
8–10 in (20–25 cm),
including inflorescence

GENOPLESIUM FIMBRIATUM
FRINGED MIDGE ORCHID
(R. BROWN) D. L. JONES & M. A. CLEMENTS, 1989

137

Starting its journey from a pair of lumpy underground tubers, the single inflorescence of the Fringed Midge Orchid bursts through the blade of a single hollow leaf near its tip. This raceme is composed of up to 30 loosely arranged flowers that hang upside down, with the lip filaments fluttering in any breeze.

Pollination is mainly carried out by small chloropoid flies, which may be attracted by the lemony fragrance and nectar of this orchid. The presence of nectar results in a high rate of self-pollination as the flies move around the open flowers before seeking out the next plant. The name *Genoplesium*, from the Greek for "race" or "kind" (*genos*) and "near" (*plesios*), refers to the resemblance of the genus to its close relative *Prasophyllum*.

Actual size

The flower of the Fringed Midge Orchid has a narrow, striped upper sepal with long cilia at the tip. Petals are striped with ciliate margins, and the motile lip has a recurved tip and margins fringed with coarse pink, red, or purple hairs.

SUBFAMILY	Orchidoideae
TRIBE AND SUBTRIBE	Diurideae, Prasophyllinae
NATIVE RANGE	Southwestern Australia (but has become a weed in many countries)
HABITAT	Variable, from granite outcrops to swamps, and from open bushland to woodland
TYPE AND PLACEMENT	Terrestrial
CONSERVATION STATUS	Not assessed, but locally common and not under threat
FLOWERING TIME	September to January (spring to early summer)

FLOWER SIZE
⅛ in (0.3 cm)

PLANT SIZE
8–24 in (20–60 cm),
including inflorescence
and upright leaf

138

MICROTIS MEDIA

MIGNONETTE ORCHID

R. BROWN, 1810

The Mignonette Orchid produces a single, hollow cylindrical stem that ends in a leaf enclosing the base of a densely flowered inflorescence. The genus is one of the most widespread in this tribe and grows as far north as Japan and east to Java. Its species seem to prefer wetter sites, some even growing in standing water. *Microtis media* has become a greenhouse and nursery weed.

Ants are rarely effective pollinators, but wingless ants of the genus *Iridomyrmex*, attracted by nectar, pollinate *Microtis* species. Its pollen is unaffected by the antifungal substance secreted on the insect's cuticle that usually prevents pollen from germinating. The genus name derives from the Greek words *mikros* and *otos*, meaning "small" and "ear," which refers to the pair of small arms on the sides of the column.

The flower of the Mignonette Orchid is small, with a relatively large ovary. There are two green, recurved sepals, one hooded dorsal sepal covering the column and the two tiny petals, and a small, ornate, downcurved lip with a rough surface and two arms.

Actual size

SUBFAMILY	Orchidoideae
TRIBE AND SUBTRIBE	Diurideae, Prasophyllinae
NATIVE RANGE	Coastal southwestern Australia, north to the Zuytdorp Cliffs near Kalbarri and east to Israelite Bay
HABITAT	Coastal heath and woodland often near winter-wet swamps
TYPE AND PLACEMENT	Terrestrial in deep sandy soil
CONSERVATION STATUS	Not threatened
FLOWERING TIME	September to November (spring)

PRASOPHYLLUM GIGANTEUM
BRONZE LEEK ORCHID
LINDLEY, 1840

FLOWER SIZE
¾ in (1.8 cm)

PLANT SIZE
16–48 in (41–122 cm) tall,
including inflorescence,
leaves are narrow
and upright

139

The genus name, *Prasophyllum*, refers to this orchid's onion-like habit, from the Greek *prason* and *phyllon*, meaning "leek leaf." A flower spike with up to 50 non-resupinate flowers emerges from a sheathing, hollow cylindrical leaf, but only after summer fires. Underground, there is usually a pair of globose tubers, but daughter tubers are rarely formed. Reproduction by seed is more frequent, and after a fire at the right time mass flowering occurs, which is often unnoticed due to heavy pigmentation of these plants.

The strongly scented flowers produce nectar and attract a variety of insects, such as flies, bees, wasps, and beetles. A similar selection of insects is attracted to the co-occurring grass trees, genus *Xanthorrhoea*, and it is believed that the Bronze Leek Orchid and other tall species of *Prasophyllum* may mimic grass tree inflorescences.

The flower of the Bronze Leek Orchid varies in color, from creamy to purplish and greenish-brown, and has fused sepals pointing upward. The petals are spreading and much narrower. The upward-pointing lip has a wavy margin, and the column has two conspicuous wings.

Actual size

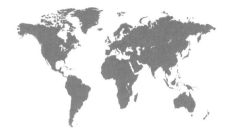

SUBFAMILY	Orchidoideae
TRIBE AND SUBTRIBE	Diurideae, Rhizanthellinae
NATIVE RANGE	Southwestern Australia
HABITAT	Broom bush shrublands
TYPE AND PLACEMENT	Terrestrial (underground) in sandy soil, low in nutrients
CONSERVATION STATUS	Vulnerable
FLOWERING TIME	May to July (winter, after good summer rains)

FLOWER SIZE
¼ in (0.6 cm)

PLANT SIZE
Subterranean,
flowering shoot about
10 in (25 cm) long

RHIZANTHELLA GARDNERI
UNDERGROUND ORCHID
R. S. ROGERS, 1928

140

This peculiar orchid lives its entire life underground. It makes rhizome-like tuberoids, has no leaves, and indirectly parasitizes the broom bush (*Melaleuca uncinata*, family Myrtaceae) through its mycorrhizal (fungal) link. The inflorescence is a head with the flowers facing inward, surrounded by large, fleshy overlapping bracts that expand near the soil surface and create fissures in the soil. Often these cracks remain covered by litter from the associated broom bush. The exact pollinators are not known, but gnats, small wasps, termites, and ants visit the flowers through the cracks in the soil. No reward is apparent.

The mode of seed dispersal is also unknown. Unlike other orchids, the Underground Orchid has large seeds enclosed in a fleshy fruit. They are not wind-dispersed, but probably pass undamaged through the digestive tract of small marsupials that may eat the fruits.

Actual size

The flower of the Underground Orchid is reddish, with three fleshy, upright sepals and two slightly smaller petals. The darker lip surrounds the upright column and is recurved. They are clustered together in a water-lily-like inflorescence.

SUBFAMILY	Orchidoideae
TRIBE AND SUBTRIBE	Diurideae, Thelymitrinae
NATIVE RANGE	Eastern and southern Australia, Tasmania, and North Island of New Zealand
HABITAT	Various habitats, from woodland and scrubland to heath and swamps
TYPE AND PLACEMENT	Terrestrial, often on poor clay soils
CONSERVATION STATUS	Not threatened in Australia, at risk in New Zealand
FLOWERING TIME	October to December (spring to early summer)

CALOCHILUS ROBERTSONII
PURPLE BEARD ORCHID
BENTHAM, 1873

FLOWER SIZE
¾–1 in (2–2.5 cm)

PLANT SIZE
12–18 × 2–3 in
(30–46 × 5–8 cm),
including inflorescence

141

Plants of the Purple Beard Orchid are relatively tall and have one to two narrow leaves held upright. They bear up to 15 flowers, each subtended by a large leaflike bract. Underground, there are one or two tubers, depending on the time of year, resulting in single, isolated plants, as daughter tubers are not formed. Vegetatively, this orchid resembles plants of the genus *Thelymitra* (the sun orchids), to which it is closely related.

Some populations appear to be composed of just self-pollinating individuals, but in others the beautiful, beardlike lip emits sex pheromones that attract male scollid wasps (*Campsomeris*). They try to mate with the lip, during which pollen is removed and then deposited on a subsequent visit. If a wasp fails to visit, the flowers self-pollinate, so every flower still makes seeds.

The flower of the Purple Beard Orchid has green sepals that are red-striped inside and petals that are shorter and striped on both sides. Together they enclose the short column. The lip is covered in long purple hairs, and its center has dark purple warts.

Actual size

SUBFAMILY	Orchidoideae
TRIBE AND SUBTRIBE	Diurideae, Thelymitrinae
NATIVE RANGE	Southwestern Western Australia, in coastal regions north to Kalbarri and east to Israelite Bay
HABITAT	*Banksia* woodland among low shrubs
TYPE AND PLACEMENT	Terrestrial on sandy soil
CONSERVATION STATUS	Not threatened
FLOWERING TIME	September to October (spring)

FLOWER SIZE
¾ in (2 cm)

PLANT SIZE
7–20 × 4–10 in
(18–51 × 10–25 cm),
including inflorescence

142

THELYMITRA CAMPANULATA
BELL SUN ORCHID
LINDLEY, 1840

Actual size

The flowers of the Bell Sun Orchid open up widely only when the weather is sunny and warm. The species is closely related to the Blue Sun Orchid (*Thelymitra canaliculata*) and Azure Sun Orchid (*T. azurea*), but the flowers are lighter blue with darker venation.

Some species of the genus are reported to be self-pollinating, but *T. campanulata* appears to be typical of those species that are pollinated by bees seeking nectar or pollen. No nectar is present, and it has been suggested that this species is mimicking blue-flowered species of the families Iridaceae, Asphodelaceae, and possibly Solanaceae. Buzz pollination (where the bee vibrates its wings, releasing pollen that is collected as food for its young) is likely, but the orchid does not provide pollen as a reward.

The flower of the Bell Sun Orchid has spreading sepals, petals, and lip, all of similar shape and size and pale blue with darker blue veins. The column is short and dark purple with a yellow anther cap and wings and a fuzzy white appendage.

SUBFAMILY	Orchidoideae
TRIBE AND SUBTRIBE	Diurideae, Thelymitrinae
NATIVE RANGE	Southwestern Western Australia
HABITAT	Open clearings among shrubs and grass tussocks
TYPE AND PLACEMENT	Terrestrial in sandy soil
CONSERVATION STATUS	Not assessed
FLOWERING TIME	June to September (late winter to spring)

THELYMITRA VARIEGATA
QUEEN OF SHEBA
(LINDLEY) F. MUELLER, 1865

FLOWER SIZE
1⅝ in (4 cm)

PLANT SIZE
6–14 × 2–4 in
(15–35 × 6–10 cm,
including inflorescence)

143

A spectacular orchid with a spectacular name, the Queen of Sheba has a peculiar spirally twisted leaf, emerging mid-season from a pair of subterranean tubers, and an inflorescence bearing up to five flowers. Its lip, unlike that of most orchids, is similar to the other floral parts, and the column, which is prominent and highly ornamented, seems to have taken on the role of attracting and positioning the pollinator.

The flowers of *Thelymitra variegata* mimic those of the Tinsel Flower (genus *Calectasia*, in the family Dasypogonaceae). These plants have reflective blue-lilac blooms and three yellow stamens, mimicked by the column lobes of this orchid.

Actual size

The flower of the Queen of Sheba has bluish-lavender, orange-striped, and dark maroon-spotted sepals, petals, and lip, all of the same shape and size. The column is cup-shaped and adorned with three yellow lobes.

SUBFAMILY	Orchidoideae
TRIBE AND SUBTRIBE	Orchideae, Brownleeinae
NATIVE RANGE	Eastern South Africa, Swaziland, and central Madagascar
HABITAT	Rock ledges in forest patches, in shade or in moss on trees, from sea level to 5,900 ft (1,800 m)
TYPE AND PLACEMENT	Terrestrial or occasionally epiphytic on moss, usually in sandy or gravelly soil that is dry in winter and wet in summer
CONSERVATION STATUS	Not assessed, but common locally
FLOWERING TIME	October to May (spring to early fall)

FLOWER SIZE
¾ in (2 cm)

PLANT SIZE
12–20 × 4–6 in
(30–50 × 10–15 cm),
including inflorescence

144

BROWNLEEA COERULEA
BALSAM ORCHID
HARVEY EX LINDLEY, 1842

This orchid produces a thin stem up to 20 in (50 cm) tall from two to four villous, oblong tubers with thick hairy roots. Along the stem are three ovate-lanceolate leaves, rounded and clasping at the base, and it is topped with an inflorescence of between five and ten large flowers, each subtended by a large lanceolate bract. The lilac flowers have a long spur at the base of the upper sepal, making them resemble those of balsam (genus *Impatiens*, family Balsaminaceae)—hence the common name.

Brownleea coerulea is pollinated by large tanglewing flies (*Prosoeca ganglbaueri*), which get the pollinia stuck on the underside of their proboscis. It has been suggested that this species mimics other plants flowering at the same time, perhaps of the genus *Cephalaria* (in the family Caprifoliaceae), and no nectar production has been noted.

The flower of the Balsam Orchid is bluish-pink, with more darkly spotted sepals that are fused—apart from at the tip—and form a hood with a long spur at the back. The two petals are free and point downward. The lip is small and spoon shaped.

Actual size

SUBFAMILY	Orchidoideae
TRIBE AND SUBTRIBE	Orchideae, Brownleeinae
NATIVE RANGE	Southwestern and southern Cape province of South Africa
HABITAT	Open grassland, in seepages, at 3,300–3,600 ft (1,000–1,100 m)
TYPE AND PLACEMENT	Terrestrial on damp sandstone
CONSERVATION STATUS	Populations are stable and local conservation status is "least concern"
FLOWERING TIME	July to September (winter to spring)

FLOWER SIZE
2–2½ in (5–6.4 cm)

PLANT SIZE
10–18 × 1 in
(25–46 × 3 cm),
including inflorescence

DISPERIS CAPENSIS
GRANNY'S BONNET
(LINNAEUS) SWARTZ, 1800

145

The species name refers to the orchid's origin in the Cape of South Africa. It may mimic the similarly colored and nectar-producing blooms of *Polygala bracteolata* (family Polygalaceae), which also occurs in the Cape and flowers at the same time. Carpenter bees of the genus *Xylocopa* were observed visiting *Polygala* flowers and carrying *Disperis capensis* pollinia on their thorax. Most *Disperis* species secrete oils near the base of the column, which are also collected by some bees, but this and nectar seem to be absent in *D. capensis*.

The orchid produces a small number of short, narrow leaves along the stem. Underground, it produces tubers, which are eaten locally and sometimes used as medicine. The common name refers to the flower's resemblance to an old-fashioned lady's head covering.

The flower of Granny's Bonnet has three dark red (or sometimes green) pouched sepals, of which two are elongated and recurved, and one hooded and spurred. The two petals are pink and curved, with a dark red edge. The lip has a narrow claw and a large, two-lobed appendage.

Actual size

SUBFAMILY	Orchidoideae
TRIBE AND SUBTRIBE	Orchideae, Coryciinae
NATIVE RANGE	Southern Cape province of South Africa
HABITAT	Seasonally wet coastal scrub, at up to 1,475 ft (450 m)
TYPE AND PLACEMENT	Terrestrial in sandy soil
CONSERVATION STATUS	Not assessed
FLOWERING TIME	November to January (summer)

FLOWER SIZE
1⅛ in (2.8 cm)

PLANT SIZE
20–36 × 6–10 in
(51–91 × 15–25 cm),
including inflorescence

CERATANDRA GRANDIFLORA
YELLOW HORNED ORCHID
LINDLEY, 1838

This robust orchid has erect stems with spirally arranged, stiff, narrow leaves. Underground is a tuber with a cluster of fat roots that occasionally produces a daughter tuber on the end of one of the longer roots. The plant is topped with a dense inflorescence of bright yellow flowers in which the buds are often red-tipped.

Monkey beetles (*Lepithrix hilarus* and *Heterochelus podagricus*), a type of scarab beetle, frequent the flowers. Males of especially the latter species set up shop on the inflorescences, where they defend their territory and wait for females to visit. The beetles show no interest in the oil secreted by the flowers, which seems to be an evolutionary holdover from a time when bees of the genus *Rediviva* visited this species to collect the oil.

The flower of the Yellow Horned Orchid has three spreading, cup-shaped sepals that are tinged red outside. The petals and lip are yellow, similar in shape and size, and clawed at the base. The column has two pronounced horns.

Actual size

SUBFAMILY	Orchidoideae
TRIBE AND SUBTRIBE	Orchideae, Coryciinae
NATIVE RANGE	Eastern Cape of South Africa and Lesotho
HABITAT	Plateau grassland
TYPE AND PLACEMENT	Terrestrial
CONSERVATION STATUS	Least concern
FLOWERING TIME	December to February (summer)

FLOWER SIZE
1 in (2.5 cm)

PLANT SIZE
15–23 × 5–9 in
(38–58 × 13–23 cm),
including inflorescence

CORYCIUM FLANAGANII

MONKSHOOD ORCHID

(BOLUS) KURZWEIL & H. P. LINDER, 1991

147

The name of the genus comes from the Greek for helmet
(*korys*), referring to the shape of the hooded resupinate flowers,
with a downward-pointing lip, in most member species.
The Monkshood Orchid, however, has non-resupinate and
unhooded flowers with an upward-pointing lip. It grows in
grasslands, occurring as distant individual plants with mostly
short leaves clothing the flower stems.

Underground, the plant has two tubers, one from the previous
season plus another that is formed during the current season.
There are also some longer thick roots. Conversion of grasslands
to agriculture is the main threat to the species, but the use of the
tubers in medicine and as a source of flour for production of
local dishes such as chikanda (a type of bread) and chinaka are
increasingly having a devastating effect.

The flower of the Monkshood Orchid
appears in a dense spike, tightly packed with
bracts and other blooms. They have three
reflexed sepals, two downcurved, inflated
lateral petals, and an upward-pointing lip
with swollen callus.

Actual size

SUBFAMILY	Orchidoideae
TRIBE AND SUBTRIBE	Orchideae, Coryciinae
NATIVE RANGE	Southwestern and southern Cape province of South Africa
HABITAT	Open scrub, at up to 5,250 ft (1,600 m)
TYPE AND PLACEMENT	Terrestrial in sandy wet soils
CONSERVATION STATUS	Least concern, with a stable population
FLOWERING TIME	September to October (spring)

FLOWER SIZE
1 in (2.5 cm)

PLANT SIZE
5–12 × 3–5 in
(13–30 × 8–13 cm),
including inflorescence

PTERYGODIUM CATHOLICUM
COWLED FRIAR
(LINNAEUS) SWARTZ, 1800

148

A cloaked appearance, like a friar with his cowl (hood) raised, gives this orchid its common name. The plant makes a tuber from which stems grow to form additional tubers, resulting in dense clonal colonies. A spike of flowers, which have a curious smell of crude oil, tops a single stem, with one larger and two to three smaller clasping leaves. Flowering is most frequent after a fire, which is atypical for most species in this genus.

Like several other *Pterygodium* species, this orchid is pollinated by female bees (*Rediviva peringueyi*, Melittidae) that collect oil from the lip. Even though the pollinator is the same bee, these orchid species reduce the chances of hybridization through differences in pollinarium sizes or the use of mutually exclusive pollinarium attachment sites on the body of the bee.

The flower of Cowled Friar is greenish-yellow. The sepals are hidden by a middle sepal that fuses with the broad petals to form a large hood over the column and lip, which are narrow and fit inside the hood, the former with two small arms.

Actual size

SUBFAMILY	Orchidoideae
TRIBE AND SUBTRIBE	Orchideae, Disinae
NATIVE RANGE	Southwestern and southern Cape province of South Africa,
HABITAT	Dense scrub on dry sandstone in full sun, at 985–3,300 ft (300–1,000 m)
TYPE AND PLACEMENT	Terrestrial
CONSERVATION STATUS	Not assessed, but locally common
FLOWERING TIME	January to March (summer)

DISA GRAMINIFOLIA
BLUE MOTHER'S CAP
KER GAWLER EX SPRENGEL, 1826

FLOWER SIZE
1½ in (3.8 cm)

PLANT SIZE
40 × 10 in
(102 × 25 cm),
including inflorescence

149

In Afrikaans, the orchid name is *bloumoederkappie*, meaning "blue mother's cap," which refers to the similarity of the flower to a traditional female headdress. The species bears up to ten magnificent blue flowers per inflorescence after the grasslike leaves have died off during the summer. Underground, there is a globose tuber, which is eaten locally and used to make flour for a type of bread.

The flowers have a sweet scent, but although they offer no nectar, carpenter bees frequent them in search of new nectar sources. The bees enter by pushing the robust column aside, and then pollinia become glued to their thorax. The floral differences from more typical species of the genus *Disa* resulted in *D. graminifolia* being widely known as *Herschelianthe graminifolia*, but most botanists today treat it as part of *Disa*.

The flower of Blue Mother's Cap has highly ornamented petals, often with purple and green spots, which flank the column under the hood. One sepal forms a club-shaped, hooded, upward-pointing spur. The lip is dark purple with downcurved margins, fading to white in the middle.

Actual size

SUBFAMILY	Orchidoideae
TRIBE AND SUBTRIBE	Orchideae, Disinae
NATIVE RANGE	Southwestern Cape province of South Africa
HABITAT	Near moist places such as streams and waterfalls
TYPE AND PLACEMENT	Terrestrial
CONSERVATION STATUS	Locally common in suitable habitats
FLOWERING TIME	December to March, peaking in February (late summer)

FLOWER SIZE
4 in (10 cm)

PLANT SIZE
9–20 × 5–9 in
(23–51 × 13–23 cm),
including inflorescence

DISA UNIFLORA
PRIDE OF TABLE MOUNTAIN
P. J. BERGIUS, 1767

150

This spectacular species, emblematic of South Africa's Table Mountain, is pollinated by the Mountain Pride butterfly (*Aeropetes tulbaghia*), which is strongly attracted to anything red, a color that many other insects cannot see. The orchid is the only *Disa* species that produces nectar. Pollinia are attached to the butterfly's legs and, while dangling there, easily come into contact with the stigma of another flower. A rare yellow mutant, popular in cultivation and used to make *Disa* hybrids, is ignored by the butterflies and rarely gets pollinated. In cultivation, numerous color forms have been bred.

Even though the species name is *uniflora* (one-flowered), flowering stems frequently bear more flowers, even up to eight. The leaves are evergreen, unlike those in other species of *Disa*, and underground there is a large globular tuber.

The flower of Pride of Table Mountain has three red sepals, of which the upper one is hooded and bears a short spur. The petals are reduced to two yellowish filaments flanking the column inside the hood. The lip is thin and goes almost unnoticed.

Actual size

SUBFAMILY	Orchidoideae
TRIBE AND SUBTRIBE	Orchideae, Disinae
NATIVE RANGE	Eastern Cape and KwaZulu-Natal provinces of South Africa
HABITAT	Temperate forests in deep shade, at 5,900–6,600 ft (1,800–2,000 m)
TYPE AND PLACEMENT	Terrestrial
CONSERVATION STATUS	Not assessed
FLOWERING TIME	October to December (late spring to summer)

FLOWER SIZE
½ in (1.25 cm)

PLANT SIZE
3–6 × 1–2 in
(8–15 × 3–5 cm),
including inflorescence

HUTTONAEA FIMBRIATA
FRINGED CAPE ORCHID
(HARVEY) REICHENBACH FILS, 1867

151

Although not tall, this beautiful orchid has between one and three cordate, petiolate leaves along the stem and may bear up to 20 flowers. There is no spur on these flowers, but they do have a rough patch of oil-producing glands in the middle of the petals near some reddish spots, which are thought to help the pollinator, an oil-collecting bee (*Rediviva colorata*), orientate and perhaps locate the oil-secreting glands. The bee scrapes both spots at the same time with its forelegs, to collect oil from the glands—and while this is taking place its abdomen contacts the column and removes the pollen masses.

This pollination strategy and unique morphology mark out *Huttonaea* as an unusual genus. It appears that oil-bee pollination evolved independently in the genus apart from other oil-flowers in this orchid subtribe. *Huttonaea* has no close relatives.

The flower of the Fringed Cape Orchid has two broadly lanceolate lateral sepals and a much smaller upper sepal. The two petals are stalked and cupped with heavily fringed margins and some reddish spots on their adjacent margins. The lip is broad and irregularly fringed along the lower margins.

Actual size

SUBFAMILY	Orchidoideae
TRIBE AND SUBTRIBE	Orchideae, Orchidinae
NATIVE RANGE	Widespread through southern Europe, Eurasia, and North Africa, bordering the Mediterranean
HABITAT	Calcareous meadows and fields, shrublands, and open woodlands
TYPE AND PLACEMENT	Terrestrial
CONSERVATION STATUS	Least concern
FLOWERING TIME	Late February through May (late winter to spring)

FLOWER SIZE
1 in (2.5 cm)

PLANT SIZE
5–12 × 4–6 in
(13–30 × 10–15 cm),
including inflorescence

152

ANACAMPTIS PAPILIONACEA
BUTTERFLY ORCHID
(LINNAEUS) R. M. BATEMAN, PRIDGEON & M. W. CHASE, 1997

Beloved throughout its Mediterranean haunts, the ubiquitous *Anacamptis papilionacea* is an uncommonly pretty orchid. Often occurring in grazed landscapes, the flower spikes pop up in early spring in easily accessible areas, making them one of the most readily recognizable European orchid species. Even though the plant has much floral diversity, and a good deal of regional variation, many botanists believe that this all fits neatly within one species, although others recognize several additional species.

The species exhibits a peculiar pollination system that appears to mix food deception (no nectar reward is offered and the flowers mimic other reward-offering flowers in their habitat) and sexual deception (only males appear to be attracted to the flowers). The bees that visit often stay in the flowers overnight, which also might be an additional attractant.

The flower of the Butterfly Orchid is usually pinkish with darker purple veins on the segments and especially the lip, though some forms vary. Stems and subtending bracts are also often highly colored.

Actual size

SUBFAMILY	Orchidoideae
TRIBE AND SUBTRIBE	Orchideae, Orchidinae
NATIVE RANGE	Western Cape to southwestern Namibia
HABITAT	Shrubby vegetation, from sea level to 5,900 ft (1,800 m)
TYPE AND PLACEMENT	Terrestrial on sand
CONSERVATION STATUS	Not assessed
FLOWERING TIME	September to December (spring to early summer)

BARTHOLINA ETHELIAE
CAPE SPIDER ORCHID
BOLUS, 1884

FLOWER SIZE
1 5/16 in (3.3 cm)

PLANT SIZE
4–7 × 1/2–1 in
(10–18 × 1.3–2.5 cm),
including flower

153

The single, spidery flower of the aptly named Cape Spider Orchid is carried on an upright, hairy inflorescence up to 7 in (18 cm) long. This emerges from a single, hairy, often red-tinged leaf that lies flat on the ground, having grown from a pair of underground tubers. The delicate orchid is found under the edges of shrubs, often more common after a fire, although its relatively small stature means that this is perhaps when the plant is more obvious.

The flower fits the moth-pollinated syndrome, but this is not yet confirmed by any studies. The species has what appear to be gland-tipped structures on the fingerlike lip appendages, which have caused some to call it the "fiber-optic orchid." There are no suggestions, as yet, on what their function might be.

Actual size

The flower of the Cape Spider Orchid has small, somewhat reflexed, green sepals and two white, upright petals. The lip resembles a white hydra, with its deeply split lobes bearing enlarged-tipped tentacles with a central opening leading to the nectar spur.

SUBFAMILY	Orchidoideae
TRIBE AND SUBTRIBE	Orchideae, Orchidinae
NATIVE RANGE	Coastal Mozambique to southern South Africa
HABITAT	Coastal bush, open deciduous woodlands, and forest edges at low elevation
TYPE AND PLACEMENT	Terrestrial
CONSERVATION STATUS	Not threatened
FLOWERING TIME	Spring and summer

FLOWER SIZE
2–2⅜ in (5–6 cm)

PLANT SIZE
15–45 × 8–20 in
(38–114 × 20–51 cm)

154

BONATEA SPECIOSA
BEAUTIFUL REIN ORCHID
(LINNAEUS FILS) WILLDENOW, 1805

As its common name suggests, the stately *Bonatea speciosa* has beautiful flowers, densely arranged on an upright raceme. Found in sandy, well-drained soils, the plant dies back to its large, woolly cylindrical tubers during the winter dry season. It is one of the most distinctive and plentiful South African orchid species, well known for more than 200 years, and, unlike many terrestrial orchids, is easily and commonly cultivated for its unusual, hooded blooms.

The plant is pollinated by a hawk moth, attracted to its stunning green and white, nocturnally fragrant spurred blossoms. This odd structure has been the subject of controversy since its first description. Thought originally to have a five-lobed lip, it instead has petals with two lobes, one of which fuses with the labellum to give the appearance of two additional lobes.

The flower of the Beautiful Rein Orchid is green and white with a hooded dorsal sepal. The bizarrely two-lobed petals form what appear to be more labellum lobes.

Actual size

SUBFAMILY	Orchidoideae
TRIBE AND SUBTRIBE	Orchideae, Orchidinae
NATIVE RANGE	Central Africa, from Congo and Angola to Zambia and Tanzania
HABITAT	Grassland and savanna
TYPE AND PLACEMENT	Terrestrial
CONSERVATION STATUS	Not assessed but threatened by overcollection
FLOWERING TIME	October to February

BRACHYCORYTHIS ANGOLENSIS
ANGEL ORCHID
(SCHLECHTER) SCHLECHTER, 1921

FLOWER SIZE
⅜ in (1 cm)

PLANT SIZE
11–23 × 4–8 in
(28–58 × 10–20 cm),
including inflorescence

155

Brachycorythis is derived from Greek *brachys*, meaning "short," and *korys*, "a helmet," referring to the hood formed by the petals and upper sepal. The genus is a member of the same subtribe that includes most European terrestrial orchids, and on morphological grounds it seems clear that the Angel Orchid is a member of this tribe. In terms of its general appearance, *B. angolensis* looks much like a species of *Dactylorhiza*, and it grows in the same wet grasslands and swamps as many of the species of that genus. It differs principally in the lack of a nectar spur and in having globose rather than fingerlike tubers.

This species is threatened due to its tubers, like those of other African terrestrial species, being used in folk medicines and to make a type of bread (chikanda). Their preferred habitats are subject to draining for agricultural purposes.

Actual size

The flower of the Angel Orchid has three white sepals, of which two spread outward, while the middle one curves over the flower. The lip is three-lobed, with the middle lobe the longest and covered in purple dots.

SUBFAMILY	Orchidoideae
TRIBE AND SUBTRIBE	Orchideae, Orchidinae
NATIVE RANGE	Central and southern Africa, from Angola and Tanzania to northern South Africa
HABITAT	Shallow wetlands and other marshy places, at 4,100–5,600 ft (1,250–1,700 m)
TYPE AND PLACEMENT	Terrestrial
CONSERVATION STATUS	Endangered
FLOWERING TIME	November to January (summer)

FLOWER SIZE
2 in (5 cm)

PLANT SIZE
30 × 6 in (76 × 15 cm)

CENTROSTIGMA OCCULTANS

HIDDEN SPUR ORCHID

(WELWITSCH EX REICHENBACH FILS) SCHLECHTER, 1915

The bases of the long spurs of this species are hidden by bracts, hence the common name and the scientific name (*occultans*, from the Latin for "concealing"). The flowers must have a pollinator with a long tongue that can reach the nectar, probably a kind of hawk moth, although the exact pollinator has not yet been observed in the wild. The masking of the spur among the bracts may prevent the robbing of nectar by other insects that would not effect pollination. Only when a moth of the correct size sticks its tongue inside the tube can its head touch the column in the hooded upper sepal and remove the pollinia.

The species has a leafy or bract-covered inflorescence but does not form a basal rosette of leaves. Underground, it has a round tuber and thin roots.

The flower of the Hidden Spur Orchid is pale yellowish-green, with the upper sepal hooded and lateral sepals smaller and deflexed. The petals are shorter and lie inside the upper sepal. The three-lobed lip has fringed lateral lobes with an entire linear midlobe. The 6-in-long (15 cm) slender spur points downward.

Actual size

SUBFAMILY	Orchidoideae
TRIBE AND SUBTRIBE	Orchideae, Orchidinae
NATIVE RANGE	Subarctic and subalpine Europe, from Alps and Carpathians to Scandinavia, northward reaching North Cape, Finnish Lapland, and the Kola Peninsula
HABITAT	Arctic and alpine meadows above the tree line, at 6,600–8,900 ft (2,000–2,700 m), but at lower elevations in the Arctic
TYPE AND PLACEMENT	Terrestrial in grassland on dry calcareous soil
CONSERVATION STATUS	Least concern; locally abundant, but not common
FLOWERING TIME	July to August (summer)

FLOWER SIZE
³⁄₁₆ in (0.4 cm)

PLANT SIZE
4 × ⅜ in (10 × 1 cm)

CHAMORCHIS ALPINA
ALPINE DWARF ORCHID
(LINNAEUS) RICHARD, 1817

157

Europe's smallest orchid (sometimes called the False Musk Orchid) can be difficult to spot, even when it grows in large colonies, which it often does. The greenish-brown flowers are so unremarkable that most people will not even recognize the plant as an orchid. Winter sport activity in the habitat, especially the use of ski lifts and slopes, in combination with warming temperatures are threats to the persistence of the species.

Tiny flies pollinate the orchid. When this is not successful, it multiplies vegetatively by its rhizomes. Underground, the species has a pair of tubers, one increasing in size as the season progresses to sustain the plant through the winter, and the other, which had been the overwintering tuber of the previous year, decreasing throughout the season.

Actual size

The flower of the Alpine Dwarf Orchid, which opens only slightly, is green, often with brownish-red streaks or spots. The upper sepal and the two petals form a small helmet that covers the flower, and the green, weakly three-lobed lip points downward.

SUBFAMILY	Orchidoideae
TRIBE AND SUBTRIBE	Orchideae, Orchidinae
NATIVE RANGE	Mascarene Islands, Comoro Islands, Seychelles, and Madagascar
HABITAT	Forest edges and stream valleys, evergreen forests or grasslands, often among rocks or at the bases of trees, at 330–6,600 ft (100–2,000 m), often in disturbed sites
TYPE AND PLACEMENT	Terrestrial or epiphytic on mossy tree trunks
CONSERVATION STATUS	Not assessed but locally common
FLOWERING TIME	Throughout the year

FLOWER SIZE
1⅛ in (3 cm)

PLANT SIZE
6–30 × 5–12 in
(15–76 × 13–30 cm),
including flowers

158

CYNORKIS FASTIGIATA
MASCARENE SWAN ORCHID
THOUARS, 1822

The genus name *Cynorkis* means "dog testicle" in Greek, in reference to the one or two small, globose to elongate tubers produced underground. From those tubers, with many smaller side roots, a rosette of one or two (sometimes up to four) strap-like leaves forms, which bears a flower stem of a few to many, successively opening flowers that are pink to white with pink or bluish markings. The flowers have a long backward-pointing nectar spur and depend on hawk moths for their pollination. Where the moths are absent (on most of the smaller islands), the plants are self-pollinating.

This and several other species in the genus can become greenhouse weeds, seeding into many types of compost. In its native habitat, *C. fastigiata* is a common orchid and seems to favor disturbed sites, such as road verges and abandoned fields.

The flower of the Mascarene Swan Orchid has two spreading sepals, with the upper sepal forming a hood along with the pinkish petals. The lip is deeply four-lobed and has a pink to bluish raised central area.

Actual size

SUBFAMILY	Orchidoideae
TRIBE AND SUBTRIBE	Orchideae, Orchidinae
NATIVE RANGE	Madagascar
HABITAT	Shady granitic rocks, steep banks, along streams, seepage areas, and forest edges, at 1,970–6,600 ft (600–2,000 m)
TYPE AND PLACEMENT	Terrestrial, often on rocks
CONSERVATION STATUS	Not assessed, but locally common
FLOWERING TIME	December to April (mid-spring to fall)

CYNORKIS GIBBOSA
LIPSTICK ORCHID
RIDLEY, 1883

FLOWER SIZE
1 in (2.5 cm),
excluding spur

PLANT SIZE
10–22 × 8–18 in
(25–60 × 20–46 cm),
including inflorescence

159

The Lipstick Orchid begins with between three and six hairy elongate tubers, from which one purple-spotted leaf emerges, its base clasping the stem. This gives rise to a pyramidal inflorescence covered with granular tubercles or bristly hairs. It can carry up to 40 flowers, with 10–15 open at the same time, each subtended by a narrow bract.

The pollinia of the species have unusually long stalks, and the flowers therefore appear to be adapted for hawk moth pollination, with a long-curving spur at the back of each flower. Nectar has not been reported. In some orchid nurseries, *Cynorkis gibbosa* has become a weed, spontaneously coming up in the pots of other plants. So, although the species is attractive to orchid growers, it should be prevented from setting seed, unless a large population is desired.

The flower of the Lipstick Orchid has two lateral sepals flared to the side, with the dorsal sepal curved over the column and forming a tube with the petals and the flared bases of the lip. The lip has three lobes, with the middle one in turn shallowly two-lobed.

Actual size

SUBFAMILY	Orchidoideae
TRIBE AND SUBTRIBE	Orchideae, Orchidinae
NATIVE RANGE	Europe, east to Mongolia and north to Siberia and southern Scandinavia
HABITAT	Forests, forest margins, or wet to dryish meadows, usually on calcareous soils, from sea level to 7,545 ft (2,300 m)
TYPE AND PLACEMENT	Terrestrial
CONSERVATION STATUS	Frequent
FLOWERING TIME	June to August (summer)

FLOWER SIZE
⅜ in (1 cm)

PLANT SIZE
Up to 24 × 4 in
(60 × 10 cm),
including inflorescence

DACTYLORHIZA FUCHSII

COMMON SPOTTED ORCHID

(DRUCE) SOÓ, 1962

160

The genus name *Dactylorhiza* refers to the finger-shaped roots, from the Greek word for finger, *daktulos*. The species is named in honor of the German botanist Leonhart Fuchs (1501–66). It is one of the most common of all Eurasian orchids, and is known to hybridize with other species.

In the spring, a loose cluster of usually purple-spotted leaves emerges and produces a leafy stem with up to 100 whitish to pink, purple-spotted flowers. Underground, a fingered tuber is produced each season. This root type is shared with its closest relatives in this subtribe, but it is otherwise unknown in orchids. Only one member of *Dactylorhiza*, *D. viridis*, produces nectar, so pollination of nearly all species of the genus happens by the deception of insects, including sawflies, wasps, and bees.

The flower of the Common Spotted Orchid is pink or white with a short spur. Sepals and petals form a hood over the column. The lip, usually decorated with purple loops or dots and dashes, has three lobes, with the midlobe half the size of the lateral ones.

Actual size

SUBFAMILY	Orchidoideae
TRIBE AND SUBTRIBE	Orchideae, Orchidinae
NATIVE RANGE	Europe, east to Ukraine
HABITAT	Dry meadows, alpine grasslands, open woods and clearings, on calcareous or siliceous soil, at 985–6,600 ft (300–2,000 m)
TYPE AND PLACEMENT	Terrestrial
CONSERVATION STATUS	Least concern, but decreasing
FLOWERING TIME	April to July (spring)

DACTYLORHIZA SAMBUCINA

ELDER-SCENTED ORCHID

(LINNAEUS) SOÓ, 1962

FLOWER SIZE
⅜ in (1 cm)

PLANT SIZE
Up to 16 × 4 in
(40 × 10 cm),
including inflorescence

161

This remarkable orchid has flowers that smell like elderflower (*Sambucus nigra*), hence the name *sambucina*. In Sweden, the plant is named "Adam and Eve," referring to the two color variants (yellow and purple). Flowers of this species are mostly pollinated by bumblebees, which are attracted by the scent but are offered no reward. The two color forms, however, make it more difficult for the bees to learn to avoid the plants and so increase the chances for more visits and higher pollination rates.

Dactylorhiza sambucina receives the royal treatment in that, unusually for an orchid, it is pollinated by naïve queen bumblebees just emerging from hibernation. Climate change can cause a mismatch in the timing of flowering and queen bumblebee emergence, which may be a factor in the recent decline observed in this species.

The flower of the Elder-scented Orchid comes in purple or yellow forms, typically in the same population without intermediate colors. It has a fat spur in which there is no nectar. The sepals are held above the flower like wings, with petals forming a hood over the column.

Actual size

SUBFAMILY	Orchidoideae
TRIBE AND SUBTRIBE	Orchideae, Orchidinae
NATIVE RANGE	Himalayas
HABITAT	Sandstone rocks in grasslands and lower mountain slopes, at 650–2,300 ft (200–700 m)
TYPE AND PLACEMENT	Terrestrial on mossy rocks
CONSERVATION STATUS	Not assessed, but probably vulnerable due to human activities, such as construction and overharvesting for medicinal purposes
FLOWERING TIME	Late July to early September (summer)

FLOWER SIZE
1⅛ in (3 cm)

PLANT SIZE
3 × 2 in (8 × 5 cm),
including inflorescence

DIPLOMERIS HIRSUTA
SNOW ORCHID
(LINDLEY) LINDLEY, 1835

162

With its pendent furry leaf and one or two flaring white flowers, the Snow Orchid, which grows in mosses on boulders, has a delicate beauty. The long, narrow spur and its preference for steep banks make pollination by night-flying moths likely. However, given the orchid's rarity, little is known about its specific biological interactions. Underground, there is a pair of globose tubers, from which between one and three leaves are produced during the monsoon season, loosely hugging the substrate.

The entire plant is covered with stiff silvery hairs, hence the species name, *hirsuta*. The genus name is Greek for "two-parted," referring to the divided stigma. With deforestation, this species is subjected to landslides, which, taken together with its use in local medicinal traditions, has resulted in decline and increased risk of extinction.

The flower of the Snow Orchid is hairy on the outside with a long green spur. Although sepals are small, the petals are larger and flaring. The white lip has a yellow central spot with two green glands at the base and two lobes.

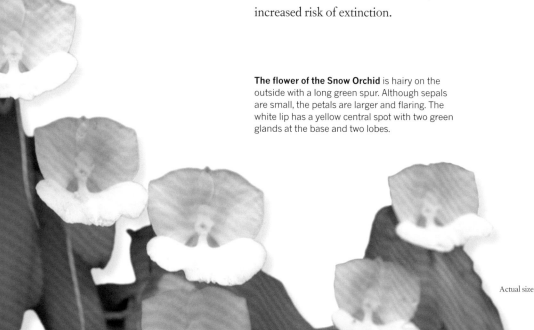

Actual size

SUBFAMILY	Orchidoideae
TRIBE AND SUBTRIBE	Orchideae, Orchidinae
NATIVE RANGE	Eastern North America, from Quebec to Alabama
HABITAT	Moist calcareous woodland, thickets, and old fields, at up to 4,300 ft (1,300 m)
TYPE AND PLACEMENT	Terrestrial
CONSERVATION STATUS	Not formally assessed
FLOWERING TIME	April to June (spring to early summer)

GALEARIS SPECTABILIS

SHOWY ORCHID

(LINNAEUS) RAFINESQUE, 1833

FLOWER SIZE
1 in (2.5 cm)

PLANT SIZE
10–14 × 10–20 in
(25–36 × 25–51 cm),
including flowers

163

In spring, from a cluster of fat roots and a short rhizome, the Showy Orchid produces a pair of broad leaves and then an inflorescence with leaflike green bracts subtending each flower. There are entirely white and entirely pink forms known in the wild, and a few plants have pale yellow lips with pink petals and sepals. The plants often produce additional shoots and can form tightly clustered clumps, although they do not like competition. They grow in shady places but the leaves emerge before the canopy leafs out, taking advantage of the sun hitting the forest floor.

The flowers attract long-tongued bees (*Bombus*) and especially bumblebee queens seeking the nectar produced in the end of a long, backward-projecting spur. Pollinia are deposited on the tongue or frons below the eyes of the bee.

The flower of the Showy Orchid has sepals and petals that form a pink hood over the column. The lip, which can be white, pink, or yellow, is rhomboid-shaped and has a spur at the back.

Actual size

SUBFAMILY	Orchidoideae
TRIBE AND SUBTRIBE	Orchideae, Orchidinae
NATIVE RANGE	Western Mediterranean (southern Iberian Peninsula, northern Maghreb, Balearic Islands, Sardinia, Elba), Madeira, and the Canary Islands
HABITAT	More or less shady places, slightly cool when flowering, dry and hot thereafter, often under pines (*Pinus halepensis, P. pinea*), in scrubland or laurel forests
TYPE AND PLACEMENT	Terrestrial on acidic or neutral soils
CONSERVATION STATUS	Endangered; locally abundant, but under threat due to urban development and deforestation in preferred coastal habitat
FLOWERING TIME	January to March (winter to early spring)

FLOWER SIZE
³⁄₁₆ in (0.4 cm)

PLANT SIZE
10–20 × 6–10 in
(25–51 × 15–25 cm),
including inflorescence

164

GENNARIA DIPHYLLA
TWO-LEAVED REIN ORCHID
(LINK) PARLATORE, 1860

From an underground tuber, the Two-leaved Rein Orchid produces a stem with two, clasping, heart-shaped leaves, the lower one larger than the upper. The stem is topped by an upright spike and downcurved green flowers that all face in the same direction. It is not a showy species, and many people completely overlook it. It is also rarely seen because of its early flowering period—in the late winter when few people venture out looking for orchids.

The flowers produce nectar and volatile compounds, which play important attraction roles for nocturnal pollinators. In Madeira, these are owlet moths (noctuids), mainly *Chrysodeixis chalcites*, *Phlogophora wollastoni*, and *Mythimna unipuncta*. The genus is named for an Italian artist, and the species name refers to the fact that it (mostly) has two leaves.

The flower of the Two-leaved Rein Orchid has a downcurved ovary topped with three dark green, slightly hooded sepals and two protruding, narrow, yellowish-green petals. The yellowish-green lip is deeply three-lobed, each lobe resembling a petal, with the two lateral lobes slightly narrower.

Actual size

SUBFAMILY	Orchidoideae
TRIBE AND SUBTRIBE	Orchideae, Orchidinae
NATIVE RANGE	Temperate Eurasia, from Ireland to Japan, north to Scandinavia and Siberia, and south to Spain, Turkey, and the Himalayas
HABITAT	Meadows, pastures, grassland, dunes, and cliffs, from sea level to 7,875 ft (2,400 m)
TYPE AND PLACEMENT	Terrestrial, usually on alkaline, poor soil
CONSERVATION STATUS	Least concern; widespread and locally abundant
FLOWERING TIME	June to July (summer)

GYMNADENIA CONOPSEA
FRAGRANT ORCHID
(LINNAEUS) R. BROWN, 1813

FLOWER SIZE
⅜ in (1 cm)

PLANT SIZE
24 × 4 in
(60 × 10 cm),
including inflorescence

165

The spur of the Fragrant Orchid produces a sweet clove-like scent, which attracts almost exclusively moths, especially the Elephant Hawk Moth (*Deilephila porcellus*) and Hummingbird Hawk Moth (*Macroglossum stellatarum*). Pollinators vary, however, because of the wide geographical range of the species. The genus *Gymnadenia* is closely related to *Dactylorhiza* and shares with it a fingerlike underground tuber. A loose cluster of narrow leaves forms in the spring, giving rise to the inflorescence. It is one of the latest species of orchid to flower in the summer.

The genus name is derived from the Greek *gymnos*, meaning "nude," and *aden*, a "gland," referring to the naked nectar glands. The epithet *conopsea* means "mosquito-like," probably because the thin spur is supposedly similar to the mouthparts of a mosquito.

The flower of the Fragrant Orchid is pink, rarely purplish or white, and arranged in a dense stalk. Each flower is subtended by a bract and bears a long spur. The three sepals are flaring. The two petals arch over the column, and the lip has three lobes.

Actual size

SUBFAMILY	Orchidoideae
TRIBE AND SUBTRIBE	Orchideae, Orchidinae
NATIVE RANGE	Indonesia
HABITAT	Monsoonal grassland
TYPE AND PLACEMENT	Terrestrial
CONSERVATION STATUS	Not threatened
FLOWERING TIME	Summer

FLOWER SIZE
3½ in (9 cm)

PLANT SIZE
8–18 × 5–10 in
(20–46 × 13–25 cm),
including flower stem

HABENARIA MEDUSA
MEDUSA'S REIN ORCHID
KRAENZLIN, 1892

166

Native to seasonally dry grasslands from Sumatra to Sulawesi, this extraordinary species bears one of the most spectacular and remarkable flowers. Sidelobes of its labellum are finely dissected to produce radiating filaments reminiscent of the mythological Medusa's head of snakes. Spurred and fringed flowers like this generally engage in a hawk moth pollination syndrome, seemingly related to their white color and cut edges, which gives them greater visibility at night and grabs the attention of a passing pollinator.

The plants have a prolonged, dry winter rest once the leaves die back to a small elongate tuber. Several daughter tubers are often formed, resulting in a closely spaced cluster of plants, each with its own flower stem. Growth resumes with spring rains.

The flower of Medusa's Rein Orchid has a white, three-lobed lip, with two sidelobes that are each highly dissected into up to 18 long, radiating filaments. The base of the lip is a dark red, and the rest of the inconspicuous floral parts are medium green.

Actual size

SUBFAMILY	Orchidoideae
TRIBE AND SUBTRIBE	Orchideae, Orchidinae
NATIVE RANGE	Southern China, southeast Asia, and the Philippines
HABITAT	Shady, seasonally dry forests, in mossy substrate over well-draining stones or boulders, dormant and leafless in the dry season
TYPE AND PLACEMENT	Terrestrial
CONSERVATION STATUS	Not threatened
FLOWERING TIME	July to September (summer)

HABENARIA RHODOCHEILA
RED REIN ORCHID
HANCE, 1866

FLOWER SIZE
1½ in (3.8 cm)

PLANT SIZE
15–25 × 6–8 in
(38–64 × 15–20 cm),
including inflorescence

167

With its often brilliant red, four-lobed lip, this species is probably the most colorful and showy of the megasized genus *Habenaria*. As is often the case with an orchid native to a vast range, it shows considerable variability and has some closely related species, such as *H. xanthocheila* and *H. erichmichaelii*, which some authors do not treat as separate. The plants begin to grow their pretty, often patterned lanceolate leaves rampantly with spring rains, bloom sumptuously in midsummer, and go dormant as the dry season begins.

The plants experience extreme drought in winter and die back to elongate tubers underground. Despite their striking colors, it is believed that the spurred flowers are still pollinated by crepuscular moths like most of the other members of this enormous, circumtropical, species-rich genus.

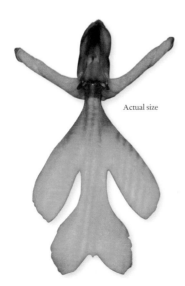

Actual size

The flower of the Red Rein Orchid is olive green suffused with salmon, with a four-lobed lip that is usually brilliant red but can be other colors, including yellow and mauve.

SUBFAMILY	Orchidoideae
TRIBE AND SUBTRIBE	Orchideae, Orchidinae
NATIVE RANGE	Tropical and subtropical America, from Mexico to Argentina
HABITAT	Mangrove swamps, wet grassland, and savannas up to 5,250 ft (1,600 m)
TYPE AND PLACEMENT	Terrestrial or nearly aquatic
CONSERVATION STATUS	Not assessed
FLOWERING TIME	Summer to fall, depending on the local seasons

FLOWER SIZE
¾ in (2 cm)

PLANT SIZE
10–24 × 4–8 in
(25–61 × 10–20 cm),
including inflorescence

HABENARIA TRIFIDA
THREE-LOBED BOG ORCHID
KUNTH, 1816

168

Actual size

From a pair of underground tubers (produced mid-season), the Three-lobed Bog Orchid forms a leafy stem with clasping leaves and large clasping leaflike bracts on the inflorescence, topped with up to four flowers. The plants often grow in wet habitats, sometimes with the roots and lower leaves submerged. In the dry season, these savanna-like habitats are subject to fires, which appear to benefit the orchids by removing competition and permitting easier access for pollinators to their flowers.

The flowers are most likely hawk moth pollinated, as are most other *Habenaria* species with this morphology and color, although no specific studies have been conducted. *Habenaria trifida* has a nectar spur up to 3 in (8 cm) long, with a two-lobed tip, again a trait associated with hawk moth pollination.

The flower of the Three-lobed Bog Orchid has green sepals, two spreading, and one forming a hood over the column, together with the petals, each of which also has a long, narrow lower lobe. The lip is creamy white, with three lobes and a long nectar spur.

SUBFAMILY	Orchidoideae
TRIBE AND SUBTRIBE	Orchideae, Orchidinae
NATIVE RANGE	Peninsular Malaysia
HABITAT	Mixed and deciduous forests, grassy pine forests, in bamboo thickets and near waterfalls, at 650–4,300 ft (200–1,300 m)
TYPE AND PLACEMENT	Terrestrial or on mossy rocks
CONSERVATION STATUS	Not assessed
FLOWERING TIME	April (spring)

HABENARIA XANTHOCHEILA
YELLOW REIN ORCHID
RIDLEY, 1896

FLOWER SIZE
1 in (2.5 cm),
excluding spur

PLANT SIZE
16–24 × 8–12 in
(40–60 × 20–30 cm),
including inflorescence

169

Often considered as a synonym of *Habenaria rhodocheila* (see page 167), this yellow-flowered form flowers at a different time and may in fact be a separate species. Underground, there are one or two globose tubers, which produce a rosette of green, lanceolate leaves from which a central spike emerges, clothed with between one and three leaflike lower bracts. In the upper parts of the stem, the flowers are produced in the axils of additional bracts that become less leaflike.

The flower somewhat resembles a person, with the body formed by the lip (with four lobes), and the hooded sepal and column looking like the head. A long, nearly straight nectar spur is at the back of the flower, and this, along with the long stalks on the pollinia and the flower color, probably indicate pollination by butterflies.

Actual size

The flower of the Yellow Rein Orchid has two crescent-shaped, spreading sepals, and the dorsal sepal and petals form a hood over the column. The lip is four-lobed, and the column is adorned with two horns (the pollinia stalks).

SUBFAMILY	Orchidoideae
TRIBE AND SUBTRIBE	Orchideae, Orchidinae
NATIVE RANGE	South-central China (northwestern Guizhou, southwestern Sichuan, central and northwestern Yunnan provinces)
HABITAT	Mossy and rocky places in open forests, at 5,250–10,500 ft (1,600–3,200 m)
TYPE AND PLACEMENT	Terrestrial on limestone
CONSERVATION STATUS	Not assessed
FLOWERING TIME	June to August (summer)

FLOWER SIZE
⅝ in (1.5 cm)

PLANT SIZE
6–11 × 4–6 in
(15–28 × 10–15 cm),
including inflorescence

HEMIPILIA FLABELLATA
FELT ORCHID
BUREAU & FRANCHET, 1891

170

The Felt Orchid has a pair of ellipsoid, underground tubers (produced in mid-season) from which a stem emerges with a single heart-shaped leaf that is purple below and green, often with purple spots or stripes, above. The inflorescence is a raceme with a few bracts and between four and eight (in some cases up to 15) flowers loosely spaced along the stem. The flower is characterized by a curved spur that is longer than the flower stalk. Flowers are variable in color, ranging from purplish-red to nearly pure white.

This orchid occurs in open sites among rocky outcrops with little competition from other plants. Although it grows in what are otherwise fairly warm areas, these are high enough in the mountains for the plant to be considered a cool-loving species.

Actual size

The flower of the Felt Orchid has pink and greenish, recurved lateral sepals and a hooded upper sepal, fused with the petals. The lip is generally also pinkish and has recurved margins, a long, curved spur, and a slightly fuzzy surface.

SUBFAMILY	Orchidoideae
TRIBE AND SUBTRIBE	Orchideae, Orchidinae
NATIVE RANGE	Temperate Eurasia, from Britain to Japan, south to Spain, Bulgaria, and the Himalayas to southern China
HABITAT	Chalk and limestone grassland and alpine meadows
TYPE AND PLACEMENT	Terrestrial on alkaline soil
CONSERVATION STATUS	Vulnerable, despite wide range and local abundance; changes in land use and overgrazing have caused loss of many populations
FLOWERING TIME	June to August (summer)

FLOWER SIZE
¼ in (0.7 cm)

PLANT SIZE
6 × 2 in (15 × 5 cm),
including inflorescence

HERMINIUM MONORCHIS
MUSK ORCHID
(LINNAEUS) R. BROWN, 1813

171

Most people overlook the tiny Musk Orchid because of its yellowish-green to whitish flowers, but it is worth kneeling down to smell the strongly honey-scented flowers. The scent attracts parasitic wasps that pollinate the flowers, but no reward is offered, resulting in the wasps learning to avoid the flowers, in spite of their lovely fragrance. This risky pollination strategy means seed production is low. The plant has another strategy, however, and reproduces vegetatively by underground stems that produce globose tubers some distance from the parent plant, allowing it to create large colonies. The tubers produce a cluster of narrow leaves in the spring, from which the inflorescence emerges.

The name *Herminium* is probably derived from the Greek *hermin*, meaning "bedpost." This refers to the two staminodia on either side of the anther, which resemble a beautifully carved bedpost.

Actual size

The flower of the Musk Orchid is pendent in the axil of a bract. The green sepals are shorter than the two green petals, all incurved. The lip is narrow, with two short lateral lobes, but is otherwise similar to the petals.

SUBFAMILY	Orchidoideae
TRIBE AND SUBTRIBE	Orchideae, Orchidinae
NATIVE RANGE	Central and southern Europe, northern Africa, Iraq
HABITAT	Rough ground, dunes, rocky places, scrub, road verges, hill slopes
TYPE AND PLACEMENT	Terrestrial in alkaline (limestone) soil
CONSERVATION STATUS	Least concern, but with a decreasing trend
FLOWERING TIME	April to June (late spring to early summer)

FLOWER SIZE
2 in (5 cm)

PLANT SIZE
20–28 × 8–12 in
(51–71 × 20–30 cm),
including inflorescence

HIMANTOGLOSSUM HIRCINUM
LIZARD ORCHID
(LINNAEUS) SPRENGEL, 1826

172

The Lizard Orchid is a species that fluctuates in numbers on the edges of its range in response to climate change and is currently expanding northward, including into England. The flowers smell like a goat (hence the epithet *hircinum*, meaning "goat-like") and are pollinated by leaf-cutter bees (Megachilidae). The common name refers to the flowers, which are said to resemble a lizard. The three long lobes of the lip are coiled like a clock spring when in bud and quickly uncoil as the flowers open.

Underground, there is a pair of globose tubers, one increasing in size (for the next season) and another decreasing (from the present season). The leaves are produced in the fall and remain green throughout the winter, turning yellow when the inflorescence emerges.

The flower of the Lizard Orchid has a hood and trilobed lip, the latter much larger than the rest of the flower, the middle lobe longest with a split tip. The callus and insides of the petals and sepals are marked with burgundy spots, blotches, and stripes.

Actual size

SUBFAMILY	Orchidoideae
TRIBE AND SUBTRIBE	Orchideae, Orchidinae
NATIVE RANGE	Southern tropical Africa, from Tanzania to Angola, Zambia, and Malawi
HABITAT	Open miombo (*Brachystegia*) woodland, in rock crevices, montane grassland overlying rock slabs, and on seepage slopes, at 5,400–6,200 ft (1,650–1,900 m)
TYPE AND PLACEMENT	Terrestrial
CONSERVATION STATUS	Not assessed
FLOWERING TIME	June to August (winter)

FLOWER SIZE
1 in (2.5 cm)

PLANT SIZE
9–21 × 4–8 in
(23–53 × 10–20 cm),
including inflorescence

HOLOTHRIX LONGIFLORA
AFRICAN SPIDER ORCHID
ROLFE, 1889

173

This terrestrial orchid grows from two hairy tubers about 1⅜ in (3.5 cm) long. They make two fleshy, heart-shaped leaves, one larger than the other, that press close against the ground. From these, a leafless stem emerges that bears between 6 and 20 scented, shortly spurred, spidery flowers. The plants often grow in seepy places among rocks and boulders. Leaves are still in good condition at the time of flowering, unlike some species of this genus in which flowers and leaves are produced at different times.

Pollination of the flower is unknown, but because of its spur and night scent, it is likely that moths visit the flowers, despite the shortness of the spur compared to other night-pollinated species. The genus name is derived from the Greek for "whole" (*holos*) and "hair" (*trichos*), referring to the hairy nature of some of its species.

The flower of the African Spider Orchid is white, with three short, green sepals and two much longer petals that are divided into nine threadlike lobes. The lip also bears between nine and eleven lobes. A conical, tightly curled spur is formed at the rear of the flower.

Actual size

SUBFAMILY	Orchidoideae
TRIBE AND SUBTRIBE	Orchideae, Orchidinae
NATIVE RANGE	Northwestern Europe to the Mediterranean, North Africa, the Levant, and Macaronesia
HABITAT	Grassland, garrigue, scrub, and conifer forests, at up to 6,600 ft (2,000 m)
TYPE AND PLACEMENT	Terrestrial in moist to dryish, primarily alkaline soils, but some plants also in slightly acidic soils
CONSERVATION STATUS	Not threatened in most of its distribution, but locally of conservation concern
FLOWERING TIME	March to June (spring)

FLOWER SIZE
⅛ in (0.5 cm)

PLANT SIZE
4–12 × 4–8 in
(10–30 × 10–20 cm),
including inflorescence

174

NEOTINEA MACULATA
DENSE-FLOWERED ORCHID
(DESFONTAINES) STEARN, 1974

In the spring, the Dense-flowered Orchid has a rosette of spotted leaves (produced during the winter), which begin to wither as the flower stem is produced. Underground, there are two globular tubers. It has a dense spike of small flowers, and two color forms—greenish-white and pink—have been recorded. In some areas, only one color form has been reported, whereas at other sites both occur.

The flowers are self-fertile, resulting in high levels of seed production. They are, however, also scented and have a nectar reward produced in a short spur at the base of the lip, possibly attracting small flies or wasps, although pollination has not been well documented. Other members of the genus are nectarless and deceive their pollinators, which have been recorded as beetles and bees.

Actual size

The flower of the Dense-flowered Orchid is greenish-white or pinkish and opens only slightly. The sepals and petals form a hood over the column. The lip has three or four lobes.

SUBFAMILY	Orchidoideae
TRIBE AND SUBTRIBE	Orchideae, Orchidinae
NATIVE RANGE	Europe, Mediterranean Basin, southwest Asia, and Caucasus
HABITAT	Grassland, scrub, meadows, and sand dunes
TYPE AND PLACEMENT	Terrestrial in alkaline soils
CONSERVATION STATUS	Least concern
FLOWERING TIME	Mid-April to July (spring)

OPHRYS APIFERA
BEE ORCHID
HUDSON, 1762

FLOWER SIZE
1 in (2.5 cm)

PLANT SIZE
6–20 × 4-8 in
(15–50 × 10-20 cm),
including inflorescence

175

Despite the resemblance of its flowers to bees, the Bee Orchid is the only species in the genus *Ophrys* that is regularly self-pollinating. The naturalist Charles Darwin (1809–82) studied the species and was perplexed by its relatively showy flowers and self-pollinating behavior. In the Mediterranean, flowers are visited by the solitary bees of the genera *Eucera* and *Tetralonia*, attracted by a pheromone-like scent mimicking that of a female bee. The lip functions as a decoy for a male bee to attempt to mate with it, with pollen transferred during this pseudocopulation.

Ophrys apifera produces a rosette of leaves in fall from a pair of globose tubers. In spring, as the leaves are dying, a stem emerges that can bear up to 12 flowers. Vegetative spread is rare, and population maintenance is dependent on seed dispersal.

Actual size

The flower of the Bee Orchid has sepals that are generally white to pink with a central green rib. Petals are short, pubescent, and yellowish-green. The lip has two hairy, erect side lobes, and the middle lobe is hairy and usually brownish-red with yellow markings.

SUBFAMILY	Orchidoideae
TRIBE AND SUBTRIBE	Orchideae, Orchidinae
NATIVE RANGE	Europe
HABITAT	Woodland, scrub, and grassland, and soil heaps near fens
TYPE AND PLACEMENT	Terrestrial in alkaline soil
CONSERVATION STATUS	Least concern, but with a declining trend in many countries
FLOWERING TIME	May to July (spring and early summer)

FLOWER SIZE
1 in (2.5 cm)

PLANT SIZE
6–24 × 3–7 in
(15–61 × 7.6–18 cm),
including inflorescence

OPHRYS INSECTIFERA
FLY ORCHID
LINNAEUS, 1753

176

Actual size

Like the Bee Orchid (see page 175), the Fly Orchid has a pair of globose tubers, from which leaves emerge in the spring, so unlike most species of the genus *Ophrys*, it is not wintergreen. Inflorescences resemble a stalk with flies sitting on them and are easily mistaken for such.

The flower of the Fly Orchid produces pheromones that mimic those of female wasps. Male digger wasps (genus *Argogorytes*) especially are attracted to this scent and attempt to mate with the flower, and in the process of this pseudocopulation transfer pollinia from one flower to the other. Tubers of many species of terrestrial Eurasian orchids, such as the members of *Ophrys*, are unsustainably harvested from the wild to produce a flour, called salep, which is used in drinks and desserts.

The flower of the Fly Orchid has three green sepals and short, threadlike, brown petals. The lip is dark brown, hairy, and three-lobed, the middle lobe split in two at the apex with a blue to gray, mirrorlike center marking.

SUBFAMILY	Orchidoideae
TRIBE AND SUBTRIBE	Orchideae, Orchidinae
NATIVE RANGE	Mediterranean Basin
HABITAT	Maquis, garrigue, open forests, scrubland, abandoned fields, and roadsides, from sea level to 3,950 ft (1,200 m)
TYPE AND PLACEMENT	Terrestrial in alkaline or neutral soils
CONSERVATION STATUS	Least concern, expanding in range due to climate warming in northwestern Europe
FLOWERING TIME	February to May (late winter to spring)

FLOWER SIZE
1 in (2.5 cm)

PLANT SIZE
6–12 × 4–8 in
(15–30 × 10–20 cm),
including inflorescence

OPHRYS TENTHREDINIFERA
SAWFLY ORCHID
WILLDENOW, 1805

177

Like the other species of the genus *Ophrys*, the Sawfly Orchid has flowers that are pollinated through pseudocopulation. The flower emits a pheromone-like scent that attracts male bees of the genus *Eucera*, especially *E. nigrilabris*, which attempt to mate with the lip and so remove or deposit the pollinia in the process. When conditions are right, this species can form dense colonies of seed-produced plants, and flowering can take place within one or two years following germination. The capsules of *Ophrys* species are large and can contain 10,000–15,000 seeds.

Ophrys is Greek for "eyebrow," in reference to the hairy parts of the plants in the genus. The name *tenthredinifera* means "wasp bearing" in Latin, because the lip resembles a wasp or a sawfly, which has also led to the common name.

Actual size

The flower of the Sawfly Orchid has three large, pale to dark pink sepals with a green rib and much smaller, velvety, pink petals. The hairy, three-lobed lip is usually yellow, with brown or purple and grayish markings in the center.

SUBFAMILY	Orchidoideae
TRIBE AND SUBTRIBE	Orchideae, Orchidinae
NATIVE RANGE	Europe, North Africa, western Asia to Iran, and on Madeira and the Canary Islands
HABITAT	Grassland, scrub, or open woods
TYPE AND PLACEMENT	Terrestrial in neutral or alkaline soil
CONSERVATION STATUS	Least concern, but threatened in its southeastern range
FLOWERING TIME	March to June (spring)

FLOWER SIZE
1 in (2.5 cm)

PLANT SIZE
10–20 × 8–14 in
(25–51 × 20–36 cm),
including inflorescence

ORCHIS MASCULA
EARLY-PURPLE ORCHID
(LINNAEUS) LINNAEUS, 1755

178

The scientific name *Orchis* derives from the Greek word for "testicle," which became synonymous with this plant and "orchid" is now used for the entire family. The name alludes to the two (waxing and waning) tubers, which the ancient Europeans compared to human testicles, resulting in the plant's reputation as an aphrodisiac. Like those of the Fly Orchid (see page 176), the tubers of the Early-purple Orchid are used to make salep, popular in the Middle East as a flavoring for a drink or dessert. Salep's undocumented benefits for men's health place orchid populations under threat of overexploitation, and plants are often collected while in flower, preventing seed release.

The pink to purple, spurred flowers smell strongly of urine, and the spur contains no nectar. The flowers are pollinated because they resemble other nectar-producing flowers.

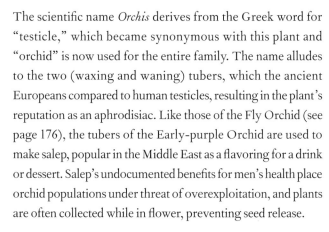

Actual size

The flower of the Early-purple Orchid has three sepals, two of which are spreading, and the middle sepal is curved down, forming a hood with the two petals. The lip is three-lobed and has an open throat at the center leading to an upturned spur.

SUBFAMILY	Orchidoideae
TRIBE AND SUBTRIBE	Orchideae, Orchidinae
NATIVE RANGE	Europe to Mongolia, north to the Åland Islands, and south to Iran and Afghanistan, most common in continental climates, rare in Atlantic and Mediterranean Europe
HABITAT	Chalk grassland, meadows, and forest edges, at up to 6,600 ft (2,000 m)
TYPE AND PLACEMENT	Terrestrial in alkaline soil
CONSERVATION STATUS	Least concern, but rare and protected in several countries
FLOWERING TIME	April to June (spring)

ORCHIS MILITARIS
MILITARY ORCHID
LINNAEUS, 1753

FLOWER SIZE
¾ in (2 cm)

PLANT SIZE
10–28 × 8–18 in
(25–71 × 20–46 cm),
including inflorescence

179

The common and scientific names of this taxonomic type upon which the family name, Orchidaceae, is based, refers to the resemblance of the flowers to a soldier wearing a silvery helmet, with the lip forming the body and the column the face. The Military Orchid produces a tall stem from a pair of tubers, with a few spirally arranged, sheathing, plain green basal leaves topped by a purplish-red inflorescence that has bracts and bears up to 30 blooms.

The flowers deceive various insects, especially bees, which act as pollinators. There is a scent but no nectar, and pollinators are probably attracted by the large display of brightly colored flowers, although in some cases mimicry of other flowers in its habitats has been suggested.

The flower of the Military Orchid has sepals and petals fused partially together to form a "helmet" over the column. The lip is four-lobed, with two lateral ("arms") and two basal lobes ("legs"). The pink lip has purple spots and purple tips to the lobes. The broad spur is short.

Actual size

SUBFAMILY	Orchidoideae
TRIBE AND SUBTRIBE	Orchideae, Orchidinae
NATIVE RANGE	Temperate areas of China, Japan, Korea, and eastern Russia
HABITAT	Seasonally boggy forest glade areas adjacent to montane forests
TYPE AND PLACEMENT	Terrestrial
CONSERVATION STATUS	Not threatened
FLOWERING TIME	Late spring to early summer

FLOWER SIZE
1⅛ in (3 cm)

PLANT SIZE
6–20 × 4–8 in
(15–51 × 10–20 cm),
including inflorescence

180

PECTEILIS RADIATA
EGRET FLOWER
(THUNBERG) RAFINESQUE, 1837

Actual size

This widely cultivated small species bears a flower with an exquisitely fringed, feathery lip and an uncanny resemblance to the graceful egret (a silver heron), which gives the plant its common name. The white color, fringed lip, nectar spur, and nocturnal fragrance of its flowers are classic attributes in the attraction of moths as pollinators

The flowers appear usually two or three at a time at the apex of a delicate stem, which rises from a small, pea-sized tuber. Massively propagated in Asia for the pot-plant market, this is a common, beginner's windowsill orchid, although few plants survive for more than a year. In nature, this species experiences a cold and moderately dry winter and may be naturally short-lived as individuals.

The flower of the Egret Flower has pale green sepals and white petals with an exceptional three-lobed lip. The center lobe is a narrow blade, whereas the sidelobes are elaborately fringed like the spread wings of an egret.

SUBFAMILY	Orchidoideae
TRIBE AND SUBTRIBE	Orchideae, Orchidinae
NATIVE RANGE	India and the Himalayas to Indochina and the Philippines
HABITAT	Scrubby slopes, at 4,920–9,200 ft (1,500–2,800 m)
TYPE AND PLACEMENT	Terrestrial
CONSERVATION STATUS	Not assessed
FLOWERING TIME	May to August (spring to summer)

FLOWER SIZE
⅝ in (1.5 cm)

PLANT SIZE
12–30 × 10–16 in
(30–76 × 25–41 cm),
including inflorescence

PERISTYLUS CONSTRICTUS
CONSTRICTED BUTTERFLY ORCHID
(LINDLEY) LINDLEY, 1835

181

This robust plant grows in the edges of forests among shrubby vegetation. Stems emerge from a pair of oblong tubers, with a few short clasping sheaths followed by four to six spirally arranged leaves. The inflorescence bears several sterile, leaflike bracts and is densely flowered, producing 30 blooms or more. The name *Peristylus* derives from the Greek words *peri*, meaning "around," and *stylos*, "column," referring to the arms that flank the column of the genus members. It now appears, though, that these species merely form the smaller end of a continuum with those of the larger genus *Habenaria*. Herbal formulations of both genera are used in Indian traditional medicine to treat malaria.

Pollination has not been studied. However, the flower color and nectar spur suggest either a butterfly or moth pollinator.

The flower of the Constricted Butterfly Orchid is open, with spreading sepals that are pale greenish-brown to white and white petals. The lip is also white and three-lobed. The column has a short arm on each side of its base.

Actual size

SUBFAMILY	Orchidoideae
TRIBE AND SUBTRIBE	Orchideae, Orchidinae
NATIVE RANGE	East slope of Pico da Esperança, on São Jorge Island in the Azores archipelago (Portugal)
HABITAT	Alpine grassland, at about 3,600 ft (1,100 m)
TYPE AND PLACEMENT	Terrestrial
CONSERVATION STATUS	Not formally assessed
FLOWERING TIME	Early June (spring)

FLOWER SIZE
⅜ in (1 cm)

PLANT SIZE
6–15 × 6–8 in
(15–38 × 15–20 cm),
including inflorescence

182

PLATANTHERA AZORICA
HOCHSTETTER'S
BUTTERFLY ORCHID
SCHLECHTER, 1920

Platanthera azorica is one of Europe's rarest orchids. It was originally thought to be synonymous with the superficially similar, but much smaller flowered, Azorean species *P. micrantha*. First discovered by German botanist Karl Hochstetter in 1838 on a single ridge of the Pico da Esperança volcano, the orchid had been considered extinct, only to be rediscovered in 2013. The references of the common and species names are obvious.

Like other members of its genus, the orchid has a pair of globose tubers with a long-tapering tip and mostly two, but sometimes more, leaves close to the ground. There are large bracts subtending each flower, which is green to greenish-yellow. Given the color and relatively long nectar spur, it is likely to be pollinated by moths.

The flower of Hochstetter's Butterfly Orchid is pale green with two reflexed sepals. The dorsal sepal and petals form a hood over the column. The lip is linear and slightly curved, with an opening at the base leading to the slender spur.

Actual size

SUBFAMILY	Orchidoideae
TRIBE AND SUBTRIBE	Orchideae, Orchidinae
NATIVE RANGE	Central and eastern North America, from Ontario to Florida and Texas
HABITAT	Wet meadows, marshes, prairies, open woodland, and pine barrens, also on roadsides, sphagnum bogs, and seepage slopes
TYPE AND PLACEMENT	Terrestrial in peat or sand
CONSERVATION STATUS	Not assessed
FLOWERING TIME	June to September (summer to early fall)

FLOWER SIZE
1 in (2.5 cm)

PLANT SIZE
12–40 × 6–10 in
(30–102 × 15–25 cm),
including flowers

PLATANTHERA CILIARIS
YELLOW FRINGED ORCHID
(LINNAEUS) LINDLEY, 1835

183

The Yellow Fringed Orchid often grows in wet conditions but is also found on slopes in the Appalachian Mountains of eastern North America. It can grow to more than 3 ft (1 m) tall and form spectacular colonies. Up to four oblanceolate leaves occur along the lower stem, which emerges from a cluster of slender tubers that extend into long root tips. These leaves decrease in size and become bracts in the densely flowered apex of the stem.

The pollinators are large butterflies, mainly the Spicebush Swallowtail (*Papilio troilus*) in the mountains and the Laurel Swallowtail (*P. palamedes*) in the lowlands. *Papilio troilus* has a ⅛ in (5 mm) shorter proboscis than *P. palamedes*, and so spur lengths of this orchid differ about ¹⁄₁₆ in (2 mm) between mountain and coastal plain populations.

The flower of the Yellow Fringed Orchid is orange, with reflexed lateral sepals and a cup-shaped upper sepal. The petals are also orange, and similar to sepals. The lip margin is wildly jagged and has a long spur at the back. The column has two prominent, upright wings.

Actual size

SUBFAMILY	Orchidoideae
TRIBE AND SUBTRIBE	Orchideae, Orchidinae
NATIVE RANGE	Western tropical Africa as far south as Zimbabwe and northwestern Madagascar
HABITAT	Wet grassland, meadows, valleys, and seepage slopes, up to 3,600 ft (1,100 m)
TYPE AND PLACEMENT	Terrestrial
CONSERVATION STATUS	Frequent
FLOWERING TIME	January to April (summer and early autumn)

FLOWER SIZE
1½ in (3.8 cm)

PLANT SIZE
8–24 × 3–5 in
(20–61 × 8–13 cm),
including flowers

PLATYCORYNE PERVILLEI
SCARLET REIN ORCHID
REICHENBACH FILS, 1855

184

Underground, the Scarlet Rein Orchid has fleshy round tubers with many thin roots. Above ground, it has several narrow leaves along the stem, which grade into a series of prominent bracts among the flowers, these numbering between three and twelve. The upright flowers are bright reddish-orange or yellowish with a long nectar spur that is often tucked into the subtending bracts. The plant's preferred habitats are all wet, at least seasonally.

This and other species in the genus are almost certainly pollinated by butterflies, with the actual mechanism similar to that of *Platanthera cilaris*, which inhabits wet sites in North America and displays a similar column and nectar spur. The overall morphology of these flowers is much like that of the large, widespread genus *Habenaria*, and these may be just species of that genus adapted for butterfly pollination.

The flower of the Scarlet Rein Orchid is bright reddish-orange and has recurved lateral sepals and a convex upper sepal that forms a hood with the petals. The lip is entire and curved down or forward and bears a spur at the back.

Actual size

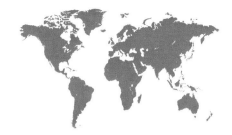

SUBFAMILY	Orchidoideae
TRIBE AND SUBTRIBE	Orchideae, Orchidinae
NATIVE RANGE	Southern Korea and southern Japan
HABITAT	Hanging from steep rocks in cracks and on ledges
TYPE AND PLACEMENT	Epilithic
CONSERVATION STATUS	Vulnerable and threatened due to overcollecting for horticulture
FLOWERING TIME	June to July (early summer)

PONERORCHIS GRAMINIFOLIA
SHOWY GRASS ORCHID
REICHENBACH, 1852

FLOWER SIZE
¾ in (2 cm)

PLANT SIZE
6–8 × 4–8 in
(15–20 × 10–20 cm),
including inflorescence

185

The Showy Grass Orchid is popular in horticulture, especially in Japan, and in many places populations have been extinguished or are sharply reduced. Considering the usually inaccessible places that the species grows, it is surprising how collectors manage to get close to them. Many color variants are known, from pure white to dark purple, with and without spots and stripes, and rare types can fetch high prices.

The plant forms underground tubers from which up to six grass-like narrow, glaucous leaves grow along the central stem, which can carry up to 20 flowers. There is a long nectar spur, although it has not been noted whether nectar is present, nor has pollination been observed. Pollination by butterflies, however, seems likely, given the long spur and the orchid's habitat preferences.

The flower of the Showy Grass Orchid has reflexed lateral sepals. A hood, formed by the middle sepal and petals, covers the column, and a spur and a deeply lobed lip are present. In cultivated plants, the lip can be variously shaped and colored.

Actual size

SUBFAMILY	Orchidoideae
TRIBE AND SUBTRIBE	Orchideae, Orchidinae
NATIVE RANGE	Central Honshu and Shikoku islands (Japan)
HABITAT	Rocks near mountain streams
TYPE AND PLACEMENT	Terrestrial in boggy places and wet rocky outcrops
CONSERVATION STATUS	Not assessed, but rare in the wild
FLOWERING TIME	April to June (spring)

FLOWER SIZE
¼ in (0.6 cm)

PLANT SIZE
3–6 × 4–6 in
(8–5 × 10–15 cm),
including inflorescence

PONERORCHIS KEISKEI
ROCK PLOVER ORCHID
(FINET) SCHLECHTER, 1919

The Rock Plover Orchid grows in wet mossy spots from underground tubers and has two to three, 1³⁄₁₆–2¾ in (3–7 cm) long, narrowly lanceolate leaves. An inflorescence can bear up to a dozen usually pink flowers. They have a large nectar spur at the base of their lip, but pollinators have not been reported, although butterflies might be expected to pollinate the species.

The flowers are variable in color patterning, and several unusual color selections are grown in Japanese horticulture. The species is named for the biologist and botanist Keisuke Ito (1803–1901), a Japanese student of the German physician and botanist Philipp Franz von Siebold (1796–1866) and one of the first Western-trained botanists in Japan. The common name is derived from the common name in Japanese, which seems to be more poetic than representational.

The flower of the Rock Plover Orchid has short, cup-shaped sepals and petals. Two of the sepals are spreading. The lip is large and three-lobed with the basal lobe split at the tip. A darker pink blotch or spots mark the base of the lip.

Actual size

SUBFAMILY	Orchidoideae
TRIBE AND SUBTRIBE	Orchideae, Orchidinae
NATIVE RANGE	Montane (including the Pyrenees and Apennines) and Arctic Europe, Siberia east to the Russian Far East, UK, Ireland, Greenland, and northeastern Canada (Newfoundland)
HABITAT	Upland or northern short grassland, open heath or woodland, or montane fens, at up to about 3,300 ft (1,000 m)
TYPE AND PLACEMENT	Terrestrial in alkaline or acidic soils, usually on the dryer or well-drained side
CONSERVATION STATUS	Least concern
FLOWERING TIME	June to July (late spring to early summer)

PSEUDORCHIS ALBIDA
SMALL-WHITE ORCHID
(LINNAEUS) A. LÖVE & D. LÖVE, 1969

FLOWER SIZE
⅛ in (0.5 cm)

PLANT SIZE
5–12 × 4–8 in
(13–30 × 10–20 cm),
including inflorescence

187

The only *Pseudorchis* species, it owes its genus name to its resemblance to a member of the genus *Orchis*. From a few fat fingerlike roots, and several thin ones, the stem emerges with up to seven clasping leaves at the base, terminating in a many-flowered raceme. The off-white to greenish blooms are typically turned to one side, which makes the species distinctive.

The flowers have a strong scent of vanilla and produce nectar, so have a high frequency of visitation by insects. Many of these do not remove pollinia, except for a small tineid moth. Nevertheless, the plants have a high fruit set that is probably due to spontaneous self-pollination, which is certainly likely in Greenland, where insects are scarce. It has been suggested that the Vikings accidentally introduced the orchid into Canada, but it could have reached the country of its own accord.

The flower of the Small-white Orchid is white to greenish-yellow and cup-shaped. Most are held horizontally (with the lip pointing to one side). All sepals and petals are partially fused to form a hood, which together with the three-lobed lip encloses the column.

Actual size

SUBFAMILY	Orchidoideae
TRIBE AND SUBTRIBE	Orchideae, Orchidinae
NATIVE RANGE	Western and southwestern Cape province of South Africa
HABITAT	Dry inland valleys in karroo and dry fynbos vegetation, at 820–4,920 ft (250–1,500 m)
TYPE AND PLACEMENT	Terrestrial in sandy soil
CONSERVATION STATUS	Not assessed
FLOWERING TIME	July to September (winter to early spring)

FLOWER SIZE
1 in (2.5 cm)

PLANT SIZE
11–23 × 8–16 in
(28–58 × 20–41 cm),
including inflorescence

SATYRIUM ERECTUM
PINK GROUND ORCHID
SWARTZ, 1800

188

The flower of the Pink Ground Orchid has petals and sepals that are all similar, small, and linear. The flower is non-resupinate, so the lip is uppermost and forms two spurs with entrances on either side of the column. Darker spots or stripes direct the pollinator to the spurs.

In fall, the fleshy tuber of the Pink Ground Orchid sprouts a pair (usually) of soil-hugging leaves, from which a bract-covered stalk emerges carrying between 25 and 60 doubly spurred flowers. The genus name comes from the Greek *satyros*, or satyr—the two-horned, part man-part horse or goat companion of the god Dionysus.

The species produces a sweet-pungent scent that attracts solitary bees, which probe the two spurs for nectar and pick up pollinaria. The morphology and coloration of this orchid are similar to bee-pollinated species of the genus *Orchis* in Europe. However, the two spurs are produced by the lip, which is held uppermost, so the effective landing platform is formed by the petals and middle sepal, leaving the lip itself to form a hood over the column.

Actual size

SUBFAMILY	Orchidoideae
TRIBE AND SUBTRIBE	Orchideae, Orchidinae
NATIVE RANGE	Mediterranean Basin
HABITAT	Damp meadows and open fields, dune slacks and marshes, guarrigue, scrubland, and heath, at up to 6,200 ft (1,900 m)
TYPE AND PLACEMENT	Terrestrial in alkaline to neutral soils
CONSERVATION STATUS	Least concern, but decreasing in its eastern range
FLOWERING TIME	April to June (spring)

SERAPIAS LINGUA
TONGUE ORCHID
LINNAEUS, 1753

FLOWER SIZE
1–1½ in (2.5–3.8 cm)

PLANT SIZE
10–20 × 6–12 in
(25–51 × 15–30 cm),
including inflorescence

189

Species of the genus *Serapias* prefer damper sites than many other Eurasian orchids and can occur in huge numbers. The Tongue Orchid has a pair of tubers, producing several linear leaves from which a lax inflorescence of 2–15 flowers is formed with large bracts, often colored similarly to the sepals, enclosing at least the base of the flowers. The large lip protrudes, and the entire flower forms a sleeping site for small bees, which remain there overnight and so effect pollination.

The genus name is from a Greco-Egyptian god, Serapis—a symbol of fertility. The Greeks applied the name to an orchid, probably one of the *Orchis* species, which was known as an aphrodisiac. *Serapias* tubers are unsustainably harvested from the wild to make a drink or dessert called salep (see pages 176 and 178), which is also considered an aphrodisiac.

The flower of the Tongue Orchid has three cream sepals with pinkish venation and shorter, similar petals, all usually enveloped in a pale, striped bract. The side lobes of the three-lobed, protruding lip are a much darker maroon and upturned to form a tube.

Actual size

SUBFAMILY	Orchidoideae
TRIBE AND SUBTRIBE	Orchideae, Orchidinae
NATIVE RANGE	KwaZulu-Natal, South Africa
HABITAT	Shady forests at varying elevations, in mossy humus over well-draining stones or boulders
TYPE AND PLACEMENT	Epilithic, sometimes epiphytic near the bases of moss-encrusted trees
CONSERVATION STATUS	Not threatened
FLOWERING TIME	Spring to fall

FLOWER SIZE
⅝–¾ in (1.5–2 cm)

PLANT SIZE
10–25 × 6–8 in
(25–64 × 15–20 cm),
including inflorescence

STENOGLOTTIS LONGIFOLIA
PLUME ORCHID
J. D. HOOKER , 1891

190

The largest and most robust of the exquisite and widely cultivated species of southern African *Stenoglottis*, this clump-forming species can produce more than a hundred small flowers over several months. The sturdy, cylindrical inflorescences emerge spear-like from a central rosette of spirally arranged leaves. The genus derives its name from the Greek for "narrow tongue," for the thin, tonguelike projections on its lip, which also afford the pink or lavender blooms a delicate and lacy character.

Variable in its foliar dimensions, the attractive lanceolate foliage is often undulate at the leaf margins and sometimes bears purplish markings, although this is more common for its more diminutive sister species *S. fimbriata*. Plants go dormant during their brief winter dry season. No pollinators are reported for any species of this genus, but floral morphology and color would suggest butterflies.

The flower of the Plume Orchid is white, pale pink, or lavender with pink or purple spots. The fringed lips give the flower stems a feathery appearance, hence the popular common name.

Actual size

SUBFAMILY	Orchidoideae
TRIBE AND SUBTRIBE	Orchideae, Orchidinae
NATIVE RANGE	Mountains in Europe and Caucasus
HABITAT	Nitrogen-poor, moist meadows, mountain pastures, marshes, and open areas in conifer forests, at 3,300–8,900 ft (1,000–2,700 m)
TYPE AND PLACEMENT	Terrestrial in limestone, shale, and slightly acidic granite-derived soils
CONSERVATION STATUS	Least concern
FLOWERING TIME	May to August (late spring to midsummer)

FLOWER SIZE
⅜ in (1 cm)

PLANT SIZE
8–20 × 4–6 in
(20–51 × 10–15 cm),
including inflorescence

TRAUNSTEINERA GLOBOSA
ALPINE GLOBE ORCHID
(LINNAEUS) REICHENBACH, 1842

191

The Alpine Globe Orchid has two ovoid tubers that give rise to a central stem producing several spirally arranged linear leaves, topped with a globose, dense raceme of flowers. A short spur produces no nectar reward for the pollinators. This globe of blooms mimics the inflorescences of other co-flowering species, such as Red Clover (*Trifolium pratense*) and Small Scabius (*Scabiosa columbaria*), and it has been shown that larger populations of these species were positively correlated with greater reproductive success of the Alpine Globe Orchid.

Pollinators appear to be diverse, but no definitive study has been carried out as yet. Many of the observed insects are probably too small to be effective at pollinia removal or deposition. *Traunsteinera* is closely related to the tiny *Chamorchis alpina* (see page 157), which together make a strongly contrasting pair.

The flower of the Alpine Globe Orchid is pale pink to white and cup-shaped, with protruding sepals and petals. The dorsal sepals, two petals, and the lip are tipped with a thickened, club-like stalk that mimics anthers. The three-lobed lip is purple spotted.

Actual size

EPIDENDROIDEAE

With 535 genera and about 22,000 species, Epidendroideae are by far the largest of the five orchid subfamilies. The greatest species diversity occurs in the wet tropics of South America and Asia, in which Orchidaceae is the major plant family. It has been suggested that this diversity is the result of a number of orchid traits, especially those associated with epiphytism—that is, roots covered with highly absorbent tissues (velamen) and pseudobulbs—and pollinator specificity—that is, features associated with complex pollinia. Another diversity factor is the long-distance dispersal of tribes Cymbidieae and Epidendreae to the New World, where they have formed some of the largest subtribes of the orchid family. Although some species of Cypripedioideae and Orchidoideae are epiphytic, epiphytism is a more predominant trait in Epidendroideae. The shift to epiphytism from terrestrial habitats is not a one-way street, however, and based on phylogenetic studies it can be seen that transitions in both directions are frequent. For example, subtribe Oncidiinae is mostly epiphytic, but a few species in several of its genera, such as *Cyrtochilum*, *Gomesa*, and *Oncidium*, grow in soil. Epidendroideae species are also diverse in terms of vegetative habits, including herbs (epiphytic and terrestrial), vines, and shrubs.

SUBFAMILY	Epidendroideae
TRIBE AND SUBTRIBE	Arethuseae, Arethusinae
NATIVE RANGE	Assam and Nepal to southern China and Indochina
HABITAT	Partially deciduous, dry forests and savanna-like woodland at 3,950–7,545 ft (1,200–2,300 m)
TYPE AND PLACEMENT	Terrestrial
CONSERVATION STATUS	Undetermined
FLOWERING TIME	September to October (late summer and fall)

FLOWER SIZE
1 in (2.5 cm)
from spur to lip

PLANT SIZE
7–15 × 10–12 in
(18–30 × 25–30 cm),
excluding inflorescence

194

ANTHOGONIUM GRACILE
FUMITORY ORCHID
WALLICH EX LINDLEY, 1836

The ovoid pseudobulbs of the Fumitory Orchid, partially buried in soil and leaf litter, produce up to five, narrowly lanceolate, deciduous leaves. The flower stalk carries three to twelve short-spurred flowers that resemble those of fumitory (*Corydalis*, Papaveraceae), hence the common name. Color is variable, ranging from white with pinkish-purple lip margins to pinkish-purple with white lip margins. It was thought that *Anthogonium* was closely related to *Bletia*, but the similarities may instead have more to do with adaptations to tropical grassland habitats. *Anthogonium* is actually more closely related to *Arundinia*, *Arethusa*, and *Calopogon*.

The orchid's long floral tube and color probably indicate a butterfly pollinator, but this is unstudied. The genus name comes from the Greek for "angled flower" (*anthos*, "flower," and *gonia*, "angle"), in reference to the distinct elbow on the flower.

Actual size

The flower of the Fumitory Orchid is non-resupinate and forms a long tube with a short spur. The mostly fused sepals are split on the upper side and have recurved tips. The petals and lip surround the long column, which curves upward into the opening of the floral tube.

SUBFAMILY	Epidendroideae
TRIBE AND SUBTRIBE	Arethuseae, Arethusinae
NATIVE RANGE	Temperate eastern North America, Canada to South Carolina
HABITAT	Acidic bogs, mostly in mossy substrates
TYPE AND PLACEMENT	Terrestrial
CONSERVATION STATUS	Not threatened
FLOWERING TIME	Late spring

FLOWER SIZE
2 in (5 cm)

PLANT SIZE
6–10 in (15–25 cm),
effectively just a leafless stem
at the time of flowering

ARETHUSA BULBOSA
AMERICAN DRAGON MOUTH
LINNAEUS, 1753

195

Found in sphagnum bogs and other wet areas, the lovely American Dragon Mouth, while not technically endangered, does not enjoy the range of populations it once had. Its southern locations are high in the mountains, where populations are smaller than they are in the colder northern reaches of the orchid's range. The genus name derives from the sea nymph Arethusa, who was transformed into a fountain in ancient Greek mythology.

Underground, the stem, which is nearly leafless at flowering, springs from a corm (a bulb-like swollen stem—hence the species name). Early in the season, the fragrant blooms attract bumblebee queens of the species *Bombus ternarius* and *B. terricola*, before they learn to avoid this non-reward-offering plant. Corms of *Arethusa bulbosa* have been used as a remedy for boils, toothaches, tumors, and various degenerative diseases.

The flower of the American Dragon Mouth is usually pinkish-purple with erect sepals and petals, the latter covering the column. The sharply recurving lip is pale pink to white with amethyst stripes and blotches, and the yellow keels are covered with fingerlike projections.

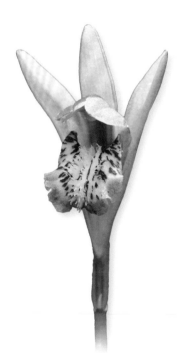

Actual size

SUBFAMILY	Epidendroideae
TRIBE AND SUBTRIBE	Arethuseae, Arethusinae
NATIVE RANGE	Tropical and subtropical Asia; also naturalized in the American and African tropics
HABITAT	Exposed situations, rocks, lava flows, and in meadows, up to 3,950 ft (1,200 m)
TYPE AND PLACEMENT	Terrestrial
CONSERVATION STATUS	A common species in its native and non-native areas
FLOWERING TIME	January to December (but also throughout the year)

FLOWER SIZE
2½ in (6.5 cm)

PLANT SIZE
50–100 × 12–30 in
(127–254 × 31–76 cm),
including inflorescence,
which occurs at the tips
of the tall plants

ARUNDINA GRAMINIFOLIA
BAMBOO ORCHID
(D. DON) HOCHREUTINER, 1910

The Bamboo Orchid is one of the most commonly encountered through the tropical regions of the world. It has escaped from gardens into the wild on many occasions, often becoming part of island floras, in places such as Puerto Rico, Jamaica, Guadeloupe, Hawaii, and Reunion, where it colonizes fresh lava flows. Alternating lanceolate leaves produced by a tall stem make the plant look like a bamboo. Each stem bears a bract-bearing inflorescence that carries up to six fragrant flowers, produced one at a time.

The flowers lack nectar but have extrafloral nectaries on the inflorescence, which are visited by many insects, especially ants. As its fruit set is often good, the species is probably self-pollinating. Otherwise, in many places where the orchid grows naturally or has been introduced, pollination is by carpenter bees of the genus *Xylocopa*.

The flower of the Bamboo Orchid is bright purple with narrow sepals—one upright, the others folded behind the trumpet-shaped lip that envelops the column, which often has a yellow, central spot. Petals are spreading, broad, and showy.

Actual size

SUBFAMILY	Epidendroideae
TRIBE AND SUBTRIBE	Arethuseae, Arethusinae
NATIVE RANGE	Eastern North America, from Canada to Florida, the Bahamas, and Cuba
HABITAT	Bogs, meadows, savannas, swamps, and seepage slopes
TYPE AND PLACEMENT	Terrestrial
CONSERVATION STATUS	Generally secure but endangered in Illinois, Kentucky, and Maryland, and vulnerable in New York State
FLOWERING TIME	April to July (spring)

FLOWER SIZE
1–1⅜ in (2.5–3.5 cm)

PLANT SIZE
20–35 × 6–10 in
(51–89 × 15–25 cm),
including apical erect
inflorescence, which is much
longer than the leaves

197

CALOPOGON TUBEROSUS
GRASS PINK
(LINNAEUS) BRITTON, STERNS & POGGENBURG, 1888

Grass Pinks are slender plants with narrow, folded leaves, sheathing at the base around the tip of an underground corm. The genus name is Greek for "beautiful beard" (*kalos* and *pogon*, respectively), in reference to the bright yellow hairs on the lip, which are thought to mimic pollen and attract pollinators.

The species is pollinated by bumblebees, which also visit the similar *Arethusa bulbosa* (American Dragon Mouth); of the wide variety that visit, however, only those of a suitable size and weight can be effective pollinators. The two orchid species can co-occur and have overlapping flowering times, but the pollen is transferred to the back of the thorax of the bumblebee in American Dragon Mouths and to the abdomen in Grass Pinks, so hybridization is prevented. Their floral similarity is due to convergence on pollen mimicry.

The flower of the Grass Pink is bright pinkish-purple and has the lip uppermost, with upcurved, spreading, lateral sepals and petals, and a hooded, slightly clawed, hinged lip covered with golden hairs. The column projects downward and is winged.

Actual size

SUBFAMILY	Epidendroideae
TRIBE AND SUBTRIBE	Arethuseae, Arethusinae
NATIVE RANGE	Northern Japan to the southern Kuril Islands
HABITAT	Bogs with sphagnum moss (acidic)
TYPE AND PLACEMENT	Terrestrial
CONSERVATION STATUS	Vulnerable
FLOWERING TIME	July (summer)

FLOWER SIZE
1 in (2.5 cm)

PLANT SIZE
5–10 × 3–6 in
(13–25 × 8–15 cm), including
apical inflorescence, which is
borne on a slender stem

ELEORCHIS JAPONICA
JAPANESE DRAGON MOUTH
(A. GRAY) MAEKAWA, 1935

198

The underground corm of the Japanese Dragon Mouth produces a stem enveloped basally in two sheathing bracts, followed by one sheathing leaf and topped with a single nodding flower, rarely two. The plant is closely related to the American Dragon Mouth, *Arethusa bulbosa*. The genus name comes from the Greek *helos*, "marsh," and *orchis*, "orchid," a reference to its sunny bog habitat. The common name alludes to the flower's resemblance to the mouth of a dragon with a protruding fiery tongue (the lip crest).

Pollination has not been studied, but it is likely to be similar to that discovered for its close North American relative *A. bulbosa*, which grows in similar bog habitats and is pollinated by bumblebees. Like its near relation, *Eleorchis japonica* has soft mealy pollen masses, which would adhere well to a hairy bee.

The flower of the Japanese Dragon Mouth has slender, pink, forward-pointing or slightly reflexed sepals. The pink petals and lip form a tube, beyond which the lip extends slightly and forms a tube around the column. The lip also has a ruffled, yellow to white central crest.

Actual size

SUBFAMILY	Epidendroideae
TRIBE AND SUBTRIBE	Arethuseae, Coelogyninae
NATIVE RANGE	Island of New Guinea
HABITAT	Forests, at 3,600–5,600 ft (1,100–1,700 m)
TYPE AND PLACEMENT	Epiphytic, rarely terrestrial
CONSERVATION STATUS	Not formally assessed
FLOWERING TIME	December to January

FLOWER SIZE
1 in (2.5 cm)

PLANT SIZE
10–35 × 8–20 in
(25–90 × 20–51 cm),
including inflorescence

AGLOSSORRHYNCHA LUCIDA
GLOSSY OIL-LAMP ORCHID
SCHLECHTER, 1912

199

The apex of the slightly compressed stem of the Glossy Oil-lamp Orchid is completely enveloped by the sheathing leaf bases. From it, a short inflorescence bearing a single flower (rarely two) emerges from a clasping floral bract. The glossy leaves, which bear a ligule at the base, fall off at the lower end as the stem extends upward. The base of the stems is often thickly covered by mosses, especially if the plants are on the ground.

The column is shaped like an old-fashioned oil lamp, hence the common name for the plant. The genus name means that it does not have a beak or snout on the lip, in contrast to the related *Glossorhyncha* (from the Greek for "tongue-snouted"), which has such outgrowths. Nothing is known about pollination, but given its floral morphology it is likely to be a bee.

The flower of the Glossy Oil-lamp Orchid is yellow-green and has spreading petals and a spreading upper sepal. The lateral sepals are curved around the lip at the base, and the column has a large anther cap.

Actual size

SUBFAMILY	Epidendroideae
TRIBE AND SUBTRIBE	Arethuseae, Coelogyninae
NATIVE RANGE	Northern Myanmar, throughout China and Korea to Japan
HABITAT	Evergreen broadleaved or coniferous forests, grassy meadows, or rock crevices, at 330–10,500 ft (100–3,200 m)
TYPE AND PLACEMENT	Terrestrial
CONSERVATION STATUS	Endangered in the wild, due to overcollecting, but commonly and easily cultivated
FLOWERING TIME	April to June (spring)

FLOWER SIZE
1½ in (4 cm)

PLANT SIZE
10–30 × 8–20 in
(25–76 × 20–51 cm),
excluding inflorescence, which
is 4–8 in (10–20 cm) longer
than the leaves

200

BLETILLA STRIATA
ASIAN HYACINTH ORCHID
(THUNBERG) REICHENBACH FILS, 1878

Large plicate leaves grow from the corm-like pseudobulbs of the Asian Hyacinth Orchid in spring. Among these leaves, a raceme emerges bearing a few, fragrant, showy, nodding flowers. These are usually pinkish-purple, though forms with white flowers or variegated leaves are known in horticulture. In Japan, pollination has been reported to be by male and female longhorn bees, *Tetralonia nipponensis*.

In Chinese medicine, the astringent bitter pseudobulbs of *Bletilla* are commonly used in a mix of medicinal compounds to reduce swelling and stop bleeding associated with lung, stomach, and liver disorders, and to promote tissue healing. The pseudobulbs also produce a mucilage, used in the production of ceramics. *Bletilla* root is sold in many Asian markets and specialty shops, but can also be cultivated, as the plant is hardy and easy to grow.

The flower of the Asian Hyacinth Orchid has spreading, dark-pink sepals and petals, which are slightly lobed. The three-lobed lip forms a tube and is hinged to the column base, the middle lip having lamellae.

Actual size

SUBFAMILY	Epidendroideae
TRIBE AND SUBTRIBE	Arethuseae, Coelogyninae
NATIVE RANGE	Sulawesi island (Indonesia)
HABITAT	Mossy montane forests, at about 5,250–6,600 ft (1,600–2,000 m)
TYPE AND PLACEMENT	Epiphytic on mossy trunks and branches
CONSERVATION STATUS	Not evaluated
FLOWERING TIME	September to November (fall)

FLOWER SIZE
1¾₆ in (3 cm)

PLANT SIZE
Each growth
8–15 × 4–6 in
(20–38 × 10–15 cm),
excluding stem

BRACISEPALUM DENSIFLORUM
TROUSER ORCHID
DE VOGEL, 1983

201

The Trouser Orchid grows on mossy, cool sites in the middle portions of trees, rather than in the canopy. It produces a cluster of egg-shaped, smooth pseudobulbs that each has a single, upright, elliptic leaf. From the top of an immature pseudobulb, an initially upright inflorescence becomes arching and pendent and has up to 30 densely packed, lily-scented flowers.

The scientific name *Bracisepalum* is derived from the Latin *braca*, meaning "trousers" and *sepalum*, "sepal," referring to the appearance of the lateral sepals that are fused into a two-lobed sack at the base of the lip, making it appear to be covered. Nothing is known about the orchid's pollination, but the floral morphology and sweet scent suggest a night-flying moth or a long-tongued bee.

The flower of the Trouser Orchid is yellowish-orange with laterally spreading petals and basally fused sepals. The lip is saccate at the base, and the heart-shaped blade envelops the hooded column.

Actual size

SUBFAMILY	Epidendroideae
TRIBE AND SUBTRIBE	Arethuseae, Coelogyninae
NATIVE RANGE	Island of Borneo
HABITAT	Elfin woodland and secondary forests, at 2,625–9,850 ft (800–3,000 m)
TYPE AND PLACEMENT	Epiphytic on moss-covered trunks and roots
CONSERVATION STATUS	Not assessed
FLOWERING TIME	Spring

FLOWER SIZE
1³⁄₁₆ in (3 cm)

PLANT SIZE
15–25 × 10–20 in
(38–64 × 25–51 cm),
excluding inflorescence
10–20 in (25–51 cm) long

202

CHELONISTELE LURIDA
SALLOW TURTLE ORCHID
PFITZER, 1907

When the Sallow Turtle Orchid is about to flower, a single, stiff plicate leaf grows out of a swollen pseudobulb, which is grooved lengthwise. From a newly emerging growth, an inflorescence appears that bears up to a dozen flowers, which are a striking yellow, tinged with bright red. Both the common name and the generic name, *Chelonistele* (from the Greek for "turtle-like"), refer to the broad, flat wing on the column, which resembles the carapace of a turtle.

No pollination reports for the species exist, but some sort of bee is thought to be effective. The lip has a pair of narrow side lobes that must function to position the pollinator on the lip so as to remove the pollinia. *Chelonistele* is properly a member of *Coelogyne*, but the combinations of names in that genus have not been made, so here it is treated as distinct.

Actual size

The flower of the Sallow Turtle Orchid has broad sepals and narrow petals. The lip is saddle-shaped and has a red and white fringe. The column is broad and forms a shell-shaped hood over the lip.

SUBFAMILY	Epidendroideae
TRIBE AND SUBTRIBE	Coelogyneae, Coelogyninae
NATIVE RANGE	Eastern Himalayas and Vietnam
HABITAT	Cool mountainous mossy areas, at 3,300–6,600 ft (1,000–2,000 m)
TYPE AND PLACEMENT	Epiphytic and lithophytic
CONSERVATION STATUS	Not threatened
FLOWERING TIME	Late winter to early spring

COELOGYNE CRISTATA
CRESTED SNOW ORCHID
LINDLEY, 1824

FLOWER SIZE
4 in (10 cm)

PLANT SIZE
6–10 × 4–8 in
(15–25 × 10–20 cm),
excluding the usually
arching-pendent inflorescence
5–13 in (20–33 cm) long

203

An exhilarating sight when in full bloom, this cool-growing Himalayan orchid is one of the most spectacular of the species in this genus, bearing numerous large, glistening white flowers with orange crests on their lips, from which the scientific name (*cristata*) is derived. Its beauty and heady fragrance have made it a favorite for cultivation in cool climates around the world.

In their native mountain setting, the flowers emerge from the bases of dormant, walnut-sized pseudobulbs as the snows melt in early spring. When well tended, plants can grow into enormous specimens with hundreds of blooms, often too heavy to move. A town in northern India near the epicenter of the species range was given the name Kurseong, which is derived from the language of ancient Lepcha settlers and means "land of the white orchids."

The flower of the Crested Snow Orchid has crystalline white sepals and petals and a broad, flaring lip bearing crests of orange, comblike papillae. Completely alba forms exist. It is one of the largest flowers in this genus.

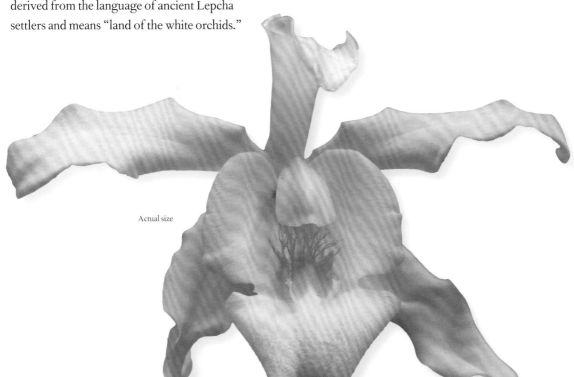

Actual size

SUBFAMILY	Epidendroideae
TRIBE AND SUBTRIBE	Arethuseae, Coelogyninae
NATIVE RANGE	Himalayas to China (Yunnan province), and Assam through Indochina
HABITAT	Evergreen wet forests in shaded valleys, at 1,970–6,600 ft (600–2,000 m)
TYPE AND PLACEMENT	Epiphytic
CONSERVATION STATUS	Not threatened
FLOWERING TIME	Summer to winter

FLOWER SIZE
1½ in (3.8 cm)

PLANT SIZE
6–10 × 3–6 in
(15–25 × 8–15 cm),
including arching, successively
flowering inflorescence
4–6 in (10–15 cm) long

204

COELOGYNE OVALIS
FALLING-SCALE ORCHID
LINDLEY, 1838

Actual size

The Falling-scale Orchid has elongate-oval pseudobulbs, a pair of lanceolate apical leaves, and a large deciduous floral bract that covers the apex of the inflorescence and gives this species its common name. Over time, a plant branches into often-massive specimens, with long, interwoven wiry rhizomes connecting each growth. When a flower fades, another replaces it, with up to five blooms in total. The genus name derives from the Greek *koilos,* "hollow," and *gyne,* "female," referring to the hollow spot (a cavity) where the stigma is located (a feature of nearly all orchids, not just species of this genus).

With its light fragrance, tubular lip, and strongly marked nectar guides, *Coelogyne ovalis* is probably pollinated by bees, although this has yet to be documented. Pollinators of other species of *Coelogyne* with similar flower shape and markings include bees and wasps, making this likely.

The flower of the Falling-scale Orchid bears lanceolate, creamy green to yellow sepals and narrow, threadlike, spreading petals. The relatively large lip is trilobed—the midlobe much larger and forward projecting—with a fringed margin and dark brown nectar guides. The sidelobes curl up around the column.

SUBFAMILY	Epidendroideae
TRIBE AND SUBTRIBE	Arethuseae, Coelogyninae
NATIVE RANGE	Malaysia, Sumatra, Borneo, and the Philippines
HABITAT	Tropical wet forests, often near streams , at 4,950–6,600 ft (1,500–2,000 m)
TYPE AND PLACEMENT	Epiphytic, climber
CONSERVATION STATUS	Not threatened
FLOWERING TIME	Mostly in summer

COELOGYNE PANDURATA

BLACK FIDDLE ORCHID

LINDLEY, 1853

FLOWER SIZE
3 in (8 cm)

PLANT SIZE
15–28 × 8–12 in
(38–71 × 20–30 cm),
excluding arching to pendent
10–30 in (25–76 cm) long
inflorescence, which can be
longer than the leaves

205

The Black Fiddle Orchid is renowned for its unusual black lip, shaped like a fiddle or lute, hence both the common name and scientific name, from the Latin for "fiddle-like." The large flattened pseudobulbs are widely spaced on a stem that climbs around large tree trunks. The impressive blooms, up to 15 at a time, open all at once and smell strongly of honey, but last for only a few days.

Although *Coelogyne pandurata* blooms almost year-round, many of the other species in the genus come from seasonally dry habitats where their pseudobulbs often shrivel alarmingly just prior to blooming. Pollination of these fantastically beautiful flowers has not been observed in nature, but the lack of a spur and the sidelobes of lip surrounding the column would most likely be adapted for a bee or wasp. There is no reward offered, in spite of the honey scent.

The flower of the Black Fiddle Orchid has chartreuse sepals and petals. The large, greenish lip is overlaid with black spots and stripes ornamented with a series of ridges, knobs, and keels, and its margin is frilly.

Actual size

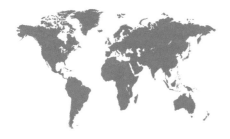

SUBFAMILY	Epidendroideae
TRIBE AND SUBTRIBE	Arethuseae, Coelogyninae
NATIVE RANGE	Philippines
HABITAT	Mid elevation outcrops and forests, often near streams
TYPE AND PLACEMENT	Mostly lithophytic, sometimes terrestrial or epiphytic
CONSERVATION STATUS	Not threatened
FLOWERING TIME	September to October (fall)

FLOWER SIZE
¾ in (2 cm)

PLANT SIZE
10–16 × 3–4 in
(25–41 × 8–10 cm),
excluding 12–20 in
(30–51 cm) tall
inflorescence with a
sharply pendent apex

DENDROCHILUM COBBIANUM
DANGLING CHAIN ORCHID
REICHENBACH FILS, 1880

The Dangling Chain Orchid is the best known of an ever-expanding genus (with now about 250 species) closely related to *Coelogyne*. These vigorous plants form large clumps and have been observed thriving under varied circumstances: epiphytic under shady tree canopies to sunny rocky outcrops.

Flower stems resembling upright fishing poles with a dangling line emerge with new growth and can carry up to 50 flowers neatly arranged in two ranks. Their fragrance varies from sweet to rotten fruit, which supports the contention that the small flowers are lures for small flies, such as fruit flies (*Drosophila*), as the likely pollinators. The genus name comes from the Greek *dendron*, "tree," and *chilos*, "food"—a reference to these plants gaining their food from living on trees, although orchids do not harm their host trees.

The flower of the Dangling Chain Orchid has pale yellow to whitish sepals and petals and a slightly broader, two-lobed lip that projects forward out of the plane of the other flower parts. A bead of nectar often collects in the middle of the grooved lip.

Actual size

SUBFAMILY	Epidendroideae
TRIBE AND SUBTRIBE	Arethuseae, Coelogyninae
NATIVE RANGE	Malesia, from Peninsular Thailand and Malaysia to New Guinea
HABITAT	Forest edges, usually on exposed rocky banks in full sun, at 490–6,600 ft (150–2,000 m)
TYPE AND PLACEMENT	Terrestrial on rocky banks or epiphytic
CONSERVATION STATUS	Not assessed, but due to its broad distribution not likely to be of conservation concern
FLOWERING TIME	March to April (spring)

FLOWER SIZE
1 in (2.5 cm)

PLANT SIZE
30–50 × 10–16 in
(76–127 × 25–41 cm),
excluding terminal, often
branched inflorescence
8–16 in (20–41 cm) tall

DILOCHIA WALLICHII
WALLICH'S BELL ORCHID
LINDLEY, 1830

207

Wallich's Bell Orchid forms an elongate stem that has elliptic leaves on opposite sides. An apical, poorly branched spike bears 10–20 somewhat pendent flowers, which are bell-like (hence the common name) and subtended by large bracts. The genus name comes from the Greek *di*, "two," and *lochos*, "rows," in reference to the leaf arrangement.

The shape of the flowers and lack of a nectar spur suggest that bees visit them, attracted by deceit. In some populations, however, they usually self-pollinate. The five species in this genus have been considered by some to be members of *Arundina*, a common genus throughout the Asian tropics. The species name is in honor of Nathaniel Wallich (1786–1854), who was a Danish surgeon later employed as the superintendent of the East India Company's botanical garden at Calcutta (now Kolkata).

The flower of Wallich's Bell Orchid has yellow to pinkish-yellow, cupped sepals and petals that surround the column. The lip has three lobes, the middle one curling back and exceeding the sepals, the two laterals upright and partially surrounding the column.

Actual size

SUBFAMILY	Epidendroideae
TRIBE AND SUBTRIBE	Arethuseae, Coelogyninae
NATIVE RANGE	Southwestern Pacific islands, from New Guinea to Fiji and Samoa
HABITAT	Rain forests, at 820–3,950 ft (250–1,200 m)
TYPE AND PLACEMENT	Epiphytic
CONSERVATION STATUS	Not assessed
FLOWERING TIME	September

FLOWER SIZE
½ in (1.4 cm)

PLANT SIZE
30–50 × 7–11 in
(76–127 × 18–28),
including terminal
pendent inflorescence
1–2 in long (2.5–5 cm)

GLOMERA MONTANA
PACIFIC GLOBE ORCHID
REICHENBACH FILS, 1876

208

The woody stems of the Pacific Globe Orchid are completely covered in basally sheathing, rigid, linear leaves with an unequal, obtusely bilobed tip. Like many tall orchids, they often become trailing or pendent with age. The globose head of the flowers is the basis of the genus name and common name, from the Latin *glomero*, which means "formed into a ball."

The densely packed, fleshy flowers and short nectar cavity, with a wide mouth to the cavity, would be suitable for pollination by birds. While feeding, they need a place to perch, which is provided by the woody stems and tough leaves of *Glomera montana*. The species also has darkly colored pollinia, which is thought to make them less noticeable to the birds after they have been deposited on their beaks.

The flower of the Pacific Globe Orchid is white and cup-shaped, with three slightly recurved sepals, forward-pointing petals, a column with a dark anther cap, and a short lip with a rounded, red-pink apex and a short nectar spur.

Actual size

SUBFAMILY	Epidendroideae
TRIBE AND SUBTRIBE	Arethuseae, Coelogyninae
NATIVE RANGE	Tropical Asia, from Nepal, southwestern China, and northeast India to Vietnam
HABITAT	Humid tropical forests and subtropical valleys, at 820–7,545 ft (250–2,300 m)
TYPE AND PLACEMENT	Epiphytic or lithophytic
CONSERVATION STATUS	Not assessed
FLOWERING TIME	October to March (fall to spring)

FLOWER SIZE
1 in (2.5 cm)

PLANT SIZE
4–8 × 3–5 in
(10–20 × 8–13 cm),
including single-flowered
basal inflorescence, which
is shorter than the leaves

PANISEA UNIFLORA
ORANGE-SPOTTED FLASK ORCHID
(LINDLEY) LINDLEY, 1854

209

The small Orange-spotted Flask Orchid has a fleshy rhizome covered with brown sheaths and clusters of egg- or flask-shaped pseudobulbs, each carrying two apical, linear, folded leaves. The common name refers to the peculiar shape of its pseudobulbs. From the base of these, an upright, usually single-flowered inflorescence appears, subtended by brown floral bracts.

Panisea is derived from Greek, meaning "all equal," in reference to the petals and sepals that are of the same shape and size. The species name refers to the single-flowered inflorescences. The shape of the flower (with no nectar spur and the proximity of column and lip) suggests bee pollination, as observed in the closely related genera *Coelogyne* and *Pleione*, which have a similar floral structure. However, this has not been verified by observation in nature.

The flower of the Orange-spotted Flask Orchid has three similar, forward-pointing, yellow to pale orange sepals and two smaller, basally clawed petals. The lip is darker and orange-spotted, with three ridges on the central lobe and two small lateral lobes that flank the column.

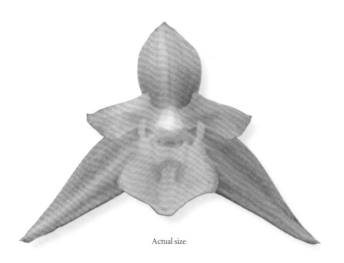

Actual size

SUBFAMILY	Epidendroideae
TRIBE AND SUBTRIBE	Arethuseae, Coelogyninae
NATIVE RANGE	Tropical and subtropical Asia to the islands of the southwestern Pacific
HABITAT	Evergreen tropical forests, at 1,640–3,300 ft (500–1,000 m)
TYPE AND PLACEMENT	Epiphytic, lithophytic
CONSERVATION STATUS	Not threatened
FLOWERING TIME	Spring to summer

FLOWER SIZE
⅜ in (1 cm)

PLANT SIZE
8–14 × 3–5 in
(20–36 × 8–13 cm),
excluding initially erect
inflorescence, which when
pendent gives a total length of
16–24 in (41–61 cm)

210

PHOLIDOTA IMBRICATA
NECKLACE ORCHID
HOOKER, 1825

A common and widely distributed species, the robust and sturdy Necklace Orchid gets its common name from the appearance of its developing inflorescence, which is flexible and looks remarkably like a necklace with many sparkling jewels. On a pendent terminal portion of the inflorescence, up to 60 cup-shaped flowers appear in two rows. Each flower is subtended by a scaly bract, from which the Latin species name is derived due to the overlapping, roof-tile-like appearance of the bracts before the flowers open. The genus name, from the Greek *pholidotos*, "clad in scales," refers to this same feature.

The plants have fleshy and succulent, tightly clustered pseudobulbs and inhabit open and seasonally dry environments. In China, an unidentified species of wasp has been observed as pollinator, but further study is required. On the basis of close-up photographs, nectar appears to be present in the lip cavity.

The flower of the Necklace Orchid is cream with a pinkish to brownish cast. It has three cupped sepals that cradle the other flower parts. The petals are parallel to the dorsal sepal, and the lip has a deep cavity with yellow veins and a pair of recurving apical lobes.

Actual size

SUBFAMILY	Epidendroideae
TRIBE AND SUBTRIBE	Arethuseae, Coelogyninae
NATIVE RANGE	Southeastern China and Taiwan
HABITAT	Seasonally dry, cold forests, at 4,920–8,200 ft (1,500–2,500 m)
TYPE AND PLACEMENT	Lithophytic, terrestrial in moss, epiphytic on tree bases
CONSERVATION STATUS	Not threatened
FLOWERING TIME	Spring

PLEIONE FORMOSANA
FORMOSAN ROCK ORCHID
HAYATA, 1911

FLOWER SIZE
3 in (8 cm)

PLANT SIZE
6–12 × 3–5 in
(15–30 × 8–13 cm),
including mostly single-
flowered, erect inflorescence,
which at the time of
flowering is leafless

211

The beautiful Formosan Rock Orchid bears showy blooms that emerge from leafless, cone-shaped, ridged pseudobulbs. Plants occur near the frost zone and often grow on vertical cliff faces, moss-encrusted rocks, and around bases of trees. A new pseudobulb and leaf appear on a newly emerging growth after the separate inflorescence, produced by the pseudobulb from the previous year, has withered. The genus name is the Greek word for "annual"—a reference to the annual cycle of leaf production and loss.

Delightfully fragrant and displaying attractive nectar guides on its lip, *Pleione formosana* attracts bumblebees (*Bombus eximius*, *B. flavescens*, and *B. trifasciatus*) as its pollinators, but there is no nectar reward for the attention of both queen and worker bees. This and other species in the genus are used in Chinese traditional medicine to treat tumors.

The flower of the Formosan Rock Orchid usually comes in shades of lavender purple with lanceolate sepals and petals that are generally similar in shape and color. The heavily fringed lip has reddish markings and yellow keels in the throat, and its sides wrap around the column, forming a tube.

Actual size

SUBFAMILY	Epidendroideae
TRIBE AND SUBTRIBE	Arethuseae, Coelogyninae
NATIVE RANGE	Southeast Asia, from the Himalayas to southern China and Peninsular Malaysia
HABITAT	Forests or shaded rocky places, at 3,300–7,545 ft (1,000–2,300 m)
TYPE AND PLACEMENT	Epiphytic on lower tree branches or lithophytic on rocks
CONSERVATION STATUS	Not assessed, but due to its broad distribution not likely to be of conservation concern
FLOWERING TIME	June (summer)

FLOWER SIZE
3½ in (9 cm)

PLANT SIZE
24–40 × 12–20 in
(61–102 × 30–51 cm),
including nodding
terminal inflorescence
8–12 in (20–30 cm) long

THUNIA ALBA
WHITE BAMBOO ORCHID
(LINDLEY) REICHENBACH FILS, 1852

212

The flower of the White Bamboo Orchid has spreading, white, lanceolate sepals and petals. The lip surrounds the column, is hairy inside, and bears a yellow blotch, which in some color forms may have darker orange to purple-red veins.

The White Bamboo Orchid is a big plant, with upright to pendent stems (in sunny as opposed to shady sites) bearing up to ten narrow leaves that have a sheathing base and are deciduous in winter—like a bamboo in habit, hence the common name. It produces large, strongly scented, short-lived flowers.

Pollination has not been observed in *Thunia alba*, but the flower shape, with the lip surrounding the column, and the presence of a spur and nectar guides suggest that bees are the most likely visitors. There are two color forms, one with yellow in the throat of the lip and the other with darker veins covering the series of ridges on the lip.

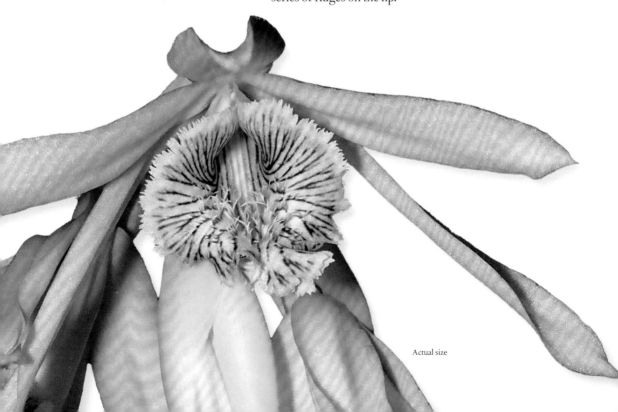

Actual size

SUBFAMILY	Epidendroideae
TRIBE	Collabieae
NATIVE RANGE	Philippines
HABITAT	Shady forests, at 1,640–4,950 ft (500–1,500 m)
TYPE AND PLACEMENT	Terrestrial
CONSERVATION STATUS	Not formally assessed, but having a restricted distribution is likely to be under threat
FLOWERING TIME	Summer

ACANTHEPHIPPIUM MANTINIANUM

MANTIN'S SADDLE ORCHID

L. LINDEN & COGNIAUX, 1896

FLOWER SIZE
2 in (5 cm)

PLANT SIZE
12–20 × 8–12 in
(30–51 × 20–30 cm),
excluding inflorescence,
8–12 in (20–30 cm) long

213

Mantin's Saddle Orchid has dark purple-green, elongate-oblong pseudobulbs, which produce three (rarely more), plicate, fleshy leaves that wither during the dry season. Like many epiphytes, it has pseudobulbs and roots covered in velamen (a spongy layer thought to aid in rapid uptake of water and nutrients) but is typically found growing on the ground in deep shade. A newly emerging vegetative growth produces a separate stalk bearing five or more fragrant, tightly clustered, eye-catching flowers.

The name *Acanthephippium* refers to the saddlelike lip with a thornlike growth on each side—from Greek *acanthos*, meaning "thorn," *epi*, "upon," and "*hippos*," a horse. Nothing is known about the plant's pollination, but the floral morphology and color suggest a bee. The pseudobulbs can reportedly be used to make glue.

The flower of Mantin's Saddle Orchid is bright yellow with red stripes and has a sack-shaped cavity at the base. The lateral sepals are recurved over the upper sepal, and the petals and ridged lip form a tube surrounding the column.

Actual size

SUBFAMILY	Epidendroideae
TRIBE	Collabieae
NATIVE RANGE	Guinea and Sierra Leone to Uganda
HABITAT	Rain forests with a brief dry season, at 1,300–2,950 ft (400–900 m)
TYPE AND PLACEMENT	Epiphytic
CONSERVATION STATUS	Threatened by deforestation
FLOWERING TIME	Mostly winter and spring

FLOWER SIZE
Up to 2½ in (6.5 cm)

PLANT SIZE
6–10 × 2–4 in
(15–25 × 5–10 cm),
excluding 6–12 in (15–30 cm)
long inflorescence, which is
as long or slightly longer
than the leaves

ANCISTROCHILUS ROTHSCHILDIANUS
FISHHOOK ORCHID
O'BRIEN, 1907

214

The more broadly distributed of the two species that comprise this genus, the Fishhook Orchid bears proportionally large, showy flowers. Between two and five of these remarkable blooms are held on erect stems that emerge from the bottom of squat, angular pseudobulbs that bear one to two broadly oblanceolate apical leaves, which are usually deciduous during the drier period of the year. The genus name comes from the Greek word *ankistron*, and this and the common name refer to the fishhook-like apex of the lip. The common name is in honor of Sir Nathan Rothschild, the 1st Baron Rothschild (1840–1915).

Pollination of the species has not been studied. The color and shape (open, without a nectar spur), however, suggest pollination by a bee that is deceived into thinking that the colorful orchid contains a reward.

The flower of the Fishhook Orchid has lavender to pale purple, ovate-elliptic sepals and petals and a lip with darker purple keels, a hooked tip, and upwardly curled sidelobes that partly surround the column.

Actual size

SUBFAMILY	Epidendroideae
TRIBE	Collabieae
NATIVE RANGE	Temperate East Asia
HABITAT	Lowland and lower montane forests
TYPE AND PLACEMENT	Terrestrial
CONSERVATION STATUS	Not formally assessed
FLOWERING TIME	February to June (late winter through spring)

FLOWER SIZE
2 in (5 cm)

PLANT SIZE
10–20 × 12–22 in
(25–51 × 30–59 cm),
excluding inflorescence
8–20 in (20–51 cm) long

215

CALANTHE STRIATA

LEMON SHRIMP-ROOT ORCHID

R. BROWN EX SPRENGEL, 1826

The small ovoid pseudobulb of the Lemon Shrimp-root Orchid is enveloped by leaf blades that give rise to broadly elliptic, plicate leaves. However, before these develop, an upright, lax raceme is formed that carries lemon-scented flowers. The shape of the pseudobulb with its roots gives rise to the common name of Shrimp-root Orchid in Japan. *Calanthe* is derived from Greek, meaning "beautiful flower."

Pollination of the nectar-less flowers by naïve, newly emerged carpenter bees, *Xylocopa appendiculata circumvolans*, has been reported. Currently, numerous cultivars and selections from this hardy orchid are offered in the trade under the name *C. sieboldii*, a nomenclatural synonym. Many medicinal uses, such as treatments for internal bleeding, aching bones, and diarrhea, have been reported for other species of *Calanthe*.

The flower of the Lemon Shrimp-root Orchid is yellow, with spreading, equal sepals and petals. The lip is three-lobed, with the middle lobe lamellate, and the column projects forward and is winged.

Actual size

SUBFAMILY	Epidendroideae
TRIBE	Collabieae
NATIVE RANGE	Tropical and southern Africa, from Tanzania to South Africa, Madagascar, and Mascarene Islands
HABITAT	Stream sides or wet, shady forests
TYPE AND PLACEMENT	Terrestrial
CONSERVATION STATUS	Not threatened
FLOWERING TIME	Late winter to early spring

FLOWER SIZE
1½–2⅜ in (4–6 cm)

PLANT SIZE
12–22 × 12–18 in
(33–56 × 33–46 cm),
excluding the inflorescence,
which is erect and taller
than the leaves,
18–30 in (46–76 cm) long

216

CALANTHE SYLVATICA
BROAD-LEAVED FOREST ORCHID
(THOUARS) LINDLEY, 1833

With a name derived from the Greek words for "beautiful flower" (*calos*, meaning "beautiful," and *anthos*, "flower") *Calanthe* is a genus of mostly terrestrial orchids, widely admired for their stately spikes of colorful blooms, set against striking broad, plicate foliage typical of shade-loving forest understory plants. Possibly one of the earliest tropical orchids planted in a garden setting, the very first orchid hybrid, *C. dominyi*, was bred from this lovely species.

The colorful well-spaced blossoms grow from sturdy upright spikes up to 16 in (40 cm) long and bloom over a long period. Usually mauve or pale purple, the flowers can display richer, vibrant purples and other colors, and often take on an orangey hue as they fade. Although many related species are deciduous in temperate zones, this tropical member of the genus is evergreen and not able to take a freeze.

The flower of the Broad-leaved Forest Orchid is deep or pale purple to violet and white and held on tall spikes above the foliage. Sepals are falcate or winglike, whereas the petals are narrower. The lip is trilobed with an indented midlobe and bears a central yellow to orange crest.

Actual size

SUBFAMILY	Epidendroideae
TRIBE	Collabieae
NATIVE RANGE	Tropical Asia from India and Sri Lanka to Taiwan, throughout Southeast Asia and Malesia to New Guinea, Vanuatu, New Caledonia, Samoa, and Fiji
HABITAT	Shaded and humid places in forests, at 2,300–5,600 ft (700–1,700 m), often on ridge tops
TYPE AND PLACEMENT	Terrestrial
CONSERVATION STATUS	Unassessed
FLOWERING TIME	April to June (spring)

FLOWER SIZE
¾ in (2 cm)

PLANT SIZE
Each growth
8–14 × 2–4 in
(20–36 × 5–10 cm),
excluding inflorescence

CHRYSOGLOSSUM ORNATUM
GOLDEN CICADA ORCHID
BLUME, 1825

217

The Golden Cicada Orchid grows in the shaded forest understory, often on or near rotting logs. It forms conical or cylindrical pseudobulbs, up to 2¾ in (7 cm) long, on an underground rhizome, each a short distance from the others and bearing a single elliptic, plicate (ridged) leaf. On a 20 in (50 cm) tall separate leafless pseudobulb, a laxly flowered inflorescence grows with 10–15 blooms.

The plant may be self-pollinating, but further study is needed. No nectar or fragrance has been reported, and little is really known about this species and its relatives. The genus name refers to the usually golden yellow, tongue-shaped lip of the orchid (from the Greek, *chrysos*, meaning "gold," and *glossa*, "tongue"). The common name is derived from the Chinese name for this species.

Actual size

The flower of the Golden Cicada Orchid is green with reddish-brown spots on the similar sepals and petals. The white to yellow lip has purple spots and two small auricles and is trilobed. The column is shortly winged.

SUBFAMILY	Epidendroideae
TRIBE	Collabieae
NATIVE RANGE	Peninsular Malaysia and western Indonesia, Sumatra to Sulawesi
HABITAT	Hills and lower mountain forests, at 1,640–6,200 ft (500–1,900 m)
TYPE AND PLACEMENT	Terrestrial in shade
CONSERVATION STATUS	Not assessed
FLOWERING TIME	January to February (summer)

FLOWER SIZE
¾ in (1.9 cm)

PLANT SIZE
6–10 × 3–5 in
(15–25 × 8–12 cm),
excluding inflorescence
25–35 in (64–89 cm) tall

COLLABIUM SIMPLEX
MODEST JEWEL ORCHID
REICHENBACH FILS, 1881

218

The Modest Jewel Orchid has a short, creeping stem with four-sided pseudobulbs, each carrying a single blue green to pale green, purple-blotched, broadly lanceolate leaf with a stem. Older pseudobulbs are leafless, often resulting in a single leaf being present at a given time. From the base of the pseudobulb, a usually erect inflorescence emerges with many brightly colored, but not large, flowers.

A nectar spur is present at the back of the flower, and although pollination has not been reported, a moth might be expected to pollinate these plants, as bees are not frequent in the forest understory where these plants grow. The genus name refers to the lateral sepals being joined to the base of the lip and nectar spur, from the Latin for "with" (*col*) and "lip" (*labium*).

The flower of the Modest Jewel Orchid has spreading sepals and petals that are all similar and greenish-yellow, tinged with pink along their edges. The lateral petals are attached broadly to the lip base. The white lip is broadly three-lobed to triangular.

Actual size

SUBFAMILY	Epidendroideae
TRIBE	Collabieae
NATIVE RANGE	Sri Lanka and southern India
HABITAT	Open patana grassland, above 3,950 ft (1,200 m)
TYPE AND PLACEMENT	Terrestrial
CONSERVATION STATUS	Not assessed, but rare and possibly endangered
FLOWERING TIME	February to May (spring)

IPSEA SPECIOSA
DAFFODIL ORCHID
LINDLEY, 1831

FLOWER SIZE
1 in (2.5 cm)

PLANT SIZE
8–15 × 3–4 in
(20–38 × 8–10 cm),
excluding erect,
unbranched inflorescence
9–18 in (23–46 cm) tall, which
is slightly taller than the leaf

219

A grasslike leaf emerges from the Daffodil Orchid's underground pseudobulb. From the apex of that pseudobulb, a shoot forms with one to three large, sweetly scented, daffodil-like flowers—hence the common name. The genus name comes from the Greek *ipse*, meaning "self," which refers to original single-species status in the genus *Ipsea* (three species are now recognized). *Speciosa* is Latin for "beautiful."

A Sinhalese legend tells of a princess who walked with her elder brother along a jungle path when she suddenly attempted to make love to him. The prince became angry and killed her, after which he found out that her unusual behavior must have been caused by a strange yam she ate, supposedly the pseudobulb of *Ipsea speciosa*. The plant is now often referred to as the "yam that killed the younger sister," and the roots are still often used in local medicine for love-potions.

Actual size

The flower of the Daffodil Orchid has spreading, bright yellow petals and sepals, shortly spurred at the back, and carrying a three-lobed lip with the sidelobes surrounding the column. The midlobe is ruffled and forward projecting.

SUBFAMILY	Epidendroideae
TRIBE	Collabieae
NATIVE RANGE	Eastern Himalayas to southern China and Indochina, the Philippines, Malaysia, and western Indonesia
HABITAT	Lower montane oak-laurel or mixed forests, at 985–6,600 ft (300–2,000 m)
TYPE AND PLACEMENT	Terrestrial in shady places
CONSERVATION STATUS	Not formally assessed, but widespread and not threatened
FLOWERING TIME	May to July (spring to summer), sometimes again later in the year

FLOWER SIZE
1 in (2.5 cm)

PLANT SIZE
5–8 × 3–6 in
(13–20 × 8–15 cm),
excluding inflorescence, which
is slightly taller than the leaves,
6–10 in (15–25 cm) tall

NEPHELAPHYLLUM PULCHRUM
SILVER JEWEL ORCHID
BLUME, 1825

Actual size

The Silver Jewel Orchid produces cylindrical pseudobulbs on a fleshy rhizome, topped with a single, ovate-triangular or cordate, dark and silver-green, often purple-tinged variegated leaf. It carries a densely packed inflorescence with between three and seven small flowers. The species is variable in flower form, and a number of different regional variants can be recognized, particularly var. *sikkimensis* from the Himalayas. The genus name refers to the cloud-like coloration of the leaf (from the Greek words *nephos*, for "cloud," and *phyllon*, "leaf"), and the species name means "beautiful" in Latin.

Pollination has not been studied, but flower shape, particularly the nectar spur, indicates pollination by moths. The roots are used in herbal medicine to make a paste for treating itching sores or as a tea to act as a diuretic.

The flower of the Silver Jewel Orchid is non-resupinate, with the arrow-shaped, yellow to pink-striped white lip pointing upward and the greenish-brown, strap-like petals and sepals curling downward.

SUBFAMILY	Epidendroideae
TRIBE	Collabieae
NATIVE RANGE	Southern China, Southeast Asia, and northern Australia to the Pacific islands; escaped from cultivation and becoming invasive in Hawaii, Florida, and other suitably tropical areas
HABITAT	Shady to bright wet-forest margins, from sea level to 2,625 ft (800 m)
TYPE AND PLACEMENT	Terrestrial
CONSERVATION STATUS	Not threatened
FLOWERING TIME	Late spring

PHAIUS TANKERVILLEAE

GREATER SWAMP ORCHID

(BANKS) BLUME, 1856

FLOWER SIZE
4¼ in (12 cm)

PLANT SIZE
35–60 × 20–35 in
(89–152 × 51–89 cm),
excluding erect
inflorescence 40–70 in
(102–178 cm) tall, which
is longer than the leaves

221

The striking Greater Swamp Orchid is popular in horticulture, both under glass and outside in tropical regions, where it has escaped repeatedly into nature. The genus name derives from the Greek word *phios*, "gray," referring to the peculiar gray color of the flowers as they die. This is due to the presence of indigo, which in the past was extracted and used in New Guinea to dye clothing. The plants are also employed in folk medicine, as a poultice for infected sores in Java and as an aid to conception in New Guinea. The species name honors Lady Emma Colebrooke, Countess of Tankerville (1752–1836), in whose collection at Walton-on-Thames, near London, it first flowered.

The stately upright stems hold 10–35 flowers. Pollination is by carpenter bees (*Xylocopa*), although no reward is offered. After flowering, the stems produce plantlets, which permits them to form large colonies.

The flower of the Greater Swamp Orchid has tan to pinkish-tan lanceolate sepals and petals with a white or pale yellowish reverse. The lip is pink to purple, basally fading to near white, and the sidelobes wrap around the column. White and yellow forms occur.

Actual size

SUBFAMILY	Epidendroideae
TRIBE	Collabieae
NATIVE RANGE	Malaysia, Indonesia, and the Philippines
HABITAT	Shady tropical forests near streams, at 650–3,300 ft (200–1,000 m)
TYPE AND PLACEMENT	Terrestrial
CONSERVATION STATUS	Not threatened
FLOWERING TIME	July to October

FLOWER SIZE
⅝ in (1.5 cm)

PLANT SIZE
10–20 × 5–8 in
(25–51 × 13–20 cm),
excluding erect
inflorescence 15–30 in
(38–76 cm), which is taller
than the leaves

222

PLOCOGLOTTIS PLICATA
YELLOW-SPOTTED FOREST ORCHID
(ROXBURG) OMEROD, 2001

Actual size

The Yellow-spotted Forest Orchid is a lover of shade and moist conditions. The yellow-spotted plicate (folded like a fan) leaves draw attention and are the basis for the species name—*plicata* meaning "folded" in Latin. The genus name focuses on the lip of the species, from the Greek words *ploke*, "twining," and *glotta*, "tongue," referring to its bent or twisted shape.

Plocoglottis plicata and its relatives in the genus attract flies that lay their eggs on rotting fruit—a probable example of brood-site deceit. A fly, drawn by the plant's rotten-fruit fragrance, lands on the spring-loaded lip, which lifts up and pushes the insect against the column. During struggles to extricate itself, the fly picks up pollinia, which are subsequently deposited when the mistake is repeated.

The flower of the Yellow-spotted Forest Orchid is usually yellow, overlaid with red to reddish-brown spots on its elongate lanceolate sepals and petals. The lip is without markings, but the column has red stripes and a pair of white basal lobes with purple splashes.

SUBFAMILY	Epidendroideae
TRIBE	Collabieae
NATIVE RANGE	Widespread throughout Southeast Asia, Indonesia, Taiwan, Philippines, Pacific islands, and northern Australia; invasive in Hawaii and the Caribbean
HABITAT	Rocky grassland and mixed dry lowland forests, usually on well-draining hillsides
TYPE AND PLACEMENT	Terrestrial
CONSERVATION STATUS	Not threatened
FLOWERING TIME	Throughout the year

SPATHOGLOTTIS PLICATA

LARGE PURPLE ORCHID

BLUME, 1825

FLOWER SIZE
1½ in (3.8 cm)

PLANT SIZE
20–35 × 20–30 in
(51–89 × 51–76 cm), excluding
inflorescence, which is erect
and longer than the leaves,
25–40 in (63–102 cm) long

223

A plant clearly bent on world domination, this practically ever-blooming species, one of about 40 in its genus, is thought to have originated in Southeast Asia but is so widespread that its true origin is hard to determine. Currently naturalized in tropical areas around the world, including Hawaii, Florida, Costa Rica, and Puerto Rico, the plant is often considered to be an invasive species. Its weedy nature is exacerbated by its cleistogamy (ability to self-pollinate), allowing rampant reproduction without a pollinator. The genus name refers to the broad midlobe of the lip (from the Greek words *spathe*, "spathe," and *glotta*, "tongue").

The large plicate leaves and the propensity to bloom continuously make this species, its hybrids, and other members of the genus, popular subjects for warm, humid, and rainy tropical gardens. A rainbow of color forms has been bred and can be found in many tropical plant nurseries.

The flower of the Large Purple Orchid is generally pinkish-purple, but other colors are known. It has similarly shaped sepals and petals and a distinctive three-lobed lip; the midlobe is shaped like an anchor and bears a prominent central orange crest.

Actual size

SUBFAMILY	Epidendroideae
TRIBE	Collabieae
NATIVE RANGE	Tropical Asia, eastern India to Taiwan and Peninsular Malaysia
HABITAT	Tropical evergreen forests and stream banks, at 1,970–4,600 ft (600–1,400 m)
TYPE AND PLACEMENT	Terrestrial
CONSERVATION STATUS	Not assessed, but due to its broad distribution not likely to be of conservation concern
FLOWERING TIME	February to March (late winter)

FLOWER SIZE
¾ in (2 cm)

PLANT SIZE
12–20 × 6–9 in
(30–51 × 15–23 cm),
excluding erect inflorescence
15–30 in (38–76 cm) tall,
produced from the base
of the pseudobulb

TAINIA PENANGIANA
GREEN-STRIPED ORCHID
HOOKER FILS, 1890

Actual size

This ground orchid often grows in densely shaded sites where its large ovoid pseudobulbs are topped by a single, plicate, deciduous leaf with a long petiole. From the base, an inflorescence with large bracts is formed, bearing 5–15, simultaneously opening, highly fragrant flowers. The genus name comes from the Greek word *tainia*, "ribbon," which refers to the long, narrow, ribbonlike shape of the petals and sepals in the member species.

The attractive scent and shape of the flowers of *Tainia penangiana* (especially the spur, with its broad opening) indicate that bees would be likely pollinators, but this has so far not been observed in nature. There are reports of self-pollination for some populations. Genetic (DNA) studies have indicated that the species, including this one, that were formerly referred to the genus *Ania* should be recognized again as a distinct genus.

The flower of the Green-striped Orchid has spreading, lanceolate sepals and petals with green or greenish-brown stripes. The three-lobed, white to cream lip has some darker spotting and lateral lobes that enfold the column, whereas the middle lobe recurves at its apex.

SUBFAMILY	Epidendroideae
TRIBE AND SUBTRIBE	Cymbidieae, Catasetinae
NATIVE RANGE	Northeastern Ecuador
HABITAT	Dry forests, at 65–4,920 ft (20–1,500 m)
TYPE AND PLACEMENT	Epiphyte
CONSERVATION STATUS	No concern
FLOWERING TIME	May to December (late spring to early winter)

CATASETUM EXPANSUM
BROAD-LIPPED TRIGGER ORCHID
REICHENBACH FILS, 1878

FLOWER SIZE
3–5 in (8–13 cm)

PLANT SIZE
Each growth 8–18 × 2–5 in
(20–46 × 5–13 cm),
with apical fan of leaves,
each leaf up to 14 in (36 cm)
long, excluding inflorescence

225

The Broad-lipped Trigger Orchid has ovoid, spindle-shaped pseudobulbs bearing a fan of several plicate, lanceolate, deciduous leaves, often absent when flowering begins. On a newly forming pseudobulb, the plant makes an inflorescence with half a dozen flowers that are either male or female (rarely bisexual) and variable in flower color—pure yellow, green, or white, and sometimes red-spotted.

Flowers are pollinated by fragrance-collecting male euglossine bees. The bees touch a trigger, connected to the anther, which violently slaps the pollinarium down on the thorax of the unsuspecting insect. When the bee then visits the female flower, it deposits the pollinia on the stigma. The male and female flowers are so different in structure that they were originally placed in different genera. Enlightenment came only when one plant produced male, female, and hermaphroditic flowers.

The flower of the Broad-lipped Trigger Orchid has similar spreading sepals and petals, but the lip is either spreading (male) or cupped (female), and the former usually has a central, blood-red callus. The winged column curves upward and, in the male, has one trigger-like appendage.

Actual size

SUBFAMILY	Epidendroideae
TRIBE AND SUBTRIBE	Cymbidieae, Catasetinae
NATIVE RANGE	Southern Mexico to Guatemala
HABITAT	Seasonally dry deciduous forests, at 1,650–4,950 ft (500–1,500 m)
TYPE AND PLACEMENT	Epiphytic
CONSERVATION STATUS	Not threatened
FLOWERING TIME	Late winter to early spring

FLOWER SIZE
1½–2⅜ in (4–6 cm)

PLANT SIZE
10–18 in
(25–46 cm),
excluding pendent
inflorescence
4–8 in (10–20 cm) long

CLOWESIA ROSEA
ROSE-COLORED
BASKET ORCHID
LINDLEY, 1843

The seasonally dry Pacific slopes of southern Mexico are home to some extraordinary species, including the Rose-colored Basket Orchid, which, with its pendent racemes of attractive flowers, has been a favorite of orchid connoisseurs for well over a century. The flowers emit a strong, delicious fragrance of cinnamon and citrus that is collected by male euglossine bees and used to produce a sex pheromone. Plants of genus *Clowesia* differ from the closely related genus *Catasetum* by having bisexual flowers rather than separate male or female flowers with different shapes.

The roots of the species first grow downward and then branch, the branches growing upward to form a leaf litter-trapping basket—giving this species its common name. This trait is rare in orchids and known to occur only in a small number of distantly related genera.

The flower of the Rose-colored Basket Orchid is pale pink, overlaid on white or cream segments. The petals are fringed on their margin, and the gullet-shaped lip is deeply fringed on its margin. The crest and column are pale yellow.

Actual size

SUBFAMILY	Epidendroideae
TRIBE AND SUBTRIBE	Cymbidieae, Catasetinae
NATIVE RANGE	Central and southeastern Brazil, Paraguay, and northeastern Argentina
HABITAT	Seasonally flooded sloping meadows, at around 3,300 ft (1,000 m)
TYPE AND PLACEMENT	Terrestrial
CONSERVATION STATUS	Not assessed, but probably threatened by habitat loss
FLOWERING TIME	November (late spring)

FLOWER SIZE
2⅛ in (5.5 cm)

PLANT SIZE
14–24 × 3–5 in
(36–61 × 8–13 cm),
including terminal,
erect inflorescence

CYANAEORCHIS ARUNDINAE
WATERNYMPH ORCHID
(REICHENBACH FILS) BARBOSA RODRIGUES, 1877

227

The short-lived flowers of the Waternymph Orchid are enveloped in bracts produced on the tip of erect, reedlike stems, with two to three apical stem-clasping, linear leaves. Between two and eight flowers are borne on the inflorescence. After flowering and setting fruit, the orchid disappears below ground, only to reappear when the rainy season returns. Its habitat may occasionally burn during the dry season. The genus name comes from the Greek words *Kyane*, "water nymph," and *orchis*, "orchid," and both common and genus names refer to the standing water present in its habitat during the wet season.

Nothing has been published on the pollination of the Waternymph Orchid. Its floral morphology, however, is similar to the genus *Eulophia*, for which pollination by various types of bees has been reported.

The flower of the Waternymph Orchid has three yellow to yellow green, spreading sepals and two smaller, forward-projecting petals. The lateral lobes of the lip curve around the column, forming a short tube. The lip curves downward and has a brighter yellow callus made up of short thick hairs.

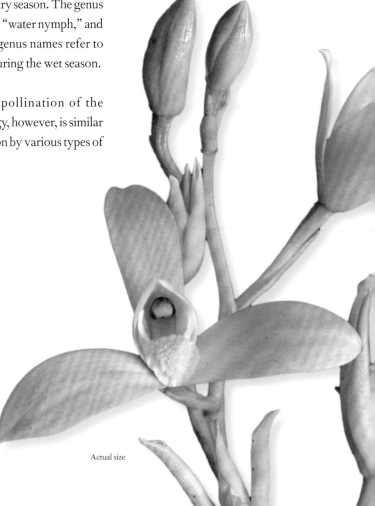

Actual size

SUBFAMILY	Epidendroideae
TRIBE AND SUBTRIBE	Cymbidieae, Catasetinae
NATIVE RANGE	Costa Rica and Nicaragua
HABITAT	Edges of wet forests near streams, at 1,870–2,950 ft (570–900 m)
TYPE AND PLACEMENT	Epiphytic on low trees
CONSERVATION STATUS	Not assessed
FLOWERING TIME	September to October (fall)

FLOWER SIZE
1 in (2.5 cm)

PLANT SIZE
8–15 × 10–25 in
(20–38 × 25–64 cm),
including erect to arching
inflorescence 7–12 in
(18–30 cm) long

228

DRESSLERIA EBURNEA
DRESSLER'S IVORY ORCHID
(ROLFE) DODSON, 1975

Actual size

Plants of Dressler's Ivory Orchid frequently grow on branches overhanging streams. They have elongate pseudobulbs enveloped by sheaths that in the upper part bear five or six plicate leaves. The species produces a raceme from the base of the pseudobulb with up to eight, highly fragrant flowers. The genus was named for the American botanist Robert L. Dressler (1927–), one of the world authorities on the orchid family, especially Neotropical orchids.

Males of the euglossine bee *Eulaema cingulata* land on any open part of the flower and try to collect fragrance compounds from the lip cavity where they are produced. The pollinarium, which is under tension, is released by this activity and is attached by its sticky disc (the viscidium) to the insect's legs. The bee flies off and tries to remove the pollen masses before visiting another flower, usually failing to do so.

The flower of Dressler's Ivory Orchid has creamy white sepals and petals that are folded back. The column is swollen and surrounded by the fleshy, cap-shaped lip. There is a narrow cavity in the center of the lip, within which floral fragrances are released.

SUBFAMILY	Epidendroideae
TRIBE AND SUBTRIBE	Cymbidieae, Catasetinae
NATIVE RANGE	Ecuador, Venezuela, and Brazil
HABITAT	Seasonally dry forests, at 330–1,650 ft (100–500 m)
TYPE AND PLACEMENT	Epiphytic
CONSERVATION STATUS	Not threatened
FLOWERING TIME	Summer

GALEANDRA LACUSTRIS
FLOODED-PALM ORCHID
BARBOSA RODRIGUES, 1877

FLOWER SIZE
2 in (5 cm)

PLANT SIZE
10–20 × 6–12 in
(25–51 × 15–30 cm),
excluding apical,
arching-pendent inflorescence
3–6 in (8–15 cm) long

229

The Flooded-palm Orchid has thin pseudobulbs and soft, grassy deciduous leaves with sheathing, lightly spotted bases. It grows on the trunks of palm trees that inhabit seasonally inundated areas, hence the common name and species name (derived from the Latin *lacus*, "lake"). The genus name comes from Greek words *galea*, "helmet," and andro, "anther," referring to the shape of the anther cap in some *Galeandra* species.

The remarkable flowers are suspended in air by a stem attached to the top of the flower (rather than below, as in most orchids). Although a long spur is present, there is no nectar reward for pollinators. *Galeandra lacustris* instead attracts fragrance-collecting male euglossine bees as its pollinators, which use the chemically complex fragrances as the precursors for sex pheromones, with which they attract females.

The flower of the Flooded-palm Orchid has lanceolate, dark olive to tan sepals and petals that set off a large, flaring, internally keeled, white lip with a purple pink patch near the apex. The lip is shaped like a scoop with a long, upward-pointing spur behind.

Actual size

SUBFAMILY	Epidendroideae
TRIBE AND SUBTRIBE	Cymbidieae, Catasetinae
NATIVE RANGE	Southeastern and southern Brazil
HABITAT	Atlantic Forest, wet coastal forests, mostly without a dry season
TYPE AND PLACEMENT	Epiphytic, rarely epilithic
CONSERVATION STATUS	Not assessed
FLOWERING TIME	April to May (summer)

FLOWER SIZE
¾ in (1.8 cm)

PLANT SIZE
8–12 × 5–8 in
(20–30 × 13–20 cm),
including arching to
pendent inflorescence
6–10 in (15–25 cm) long

230

GROBYA FASCIFERA
ATLANTIC GRASS ORCHID
REICHENBACH FILS, 1886

Actual size

The spherical pseudobulbs of the Atlantic Grass Orchid are clustered closely together, each topped with seven or eight grasslike, narrowly linear leaves up to 12 in (30 cm) long. The plant produces a dense inflorescence (with up to 20 flowers) from the base of a pseudobulb. The genus is named in honor of Lord Grey of Groby (d. 1836), a British orchid grower and enthusiast. The species name is Greek for "carrying a bundle," in reference to the bundle of flowers it produces, and the common name refers to its grassy leaves and restriction to the Atlantic Forest of Brazil.

The flowers produce oil in hairs on the lip, which is collected by anthophorid bees that mix it with pollen to feed to their larvae. While gathering the oil, the bees are thrown against the column by the hinged lip, where they pick up the pollinia.

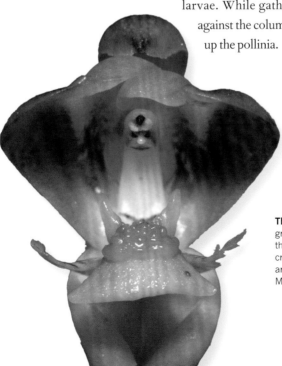

The flower of the Atlantic Grass Orchid is green, with the lower lateral sepals—larger than the upper sepal—fused at the base and crescent shaped. The petals are broader and arch over the lip, which is complexly lobed. Most parts are spotted or barred with red.

SUBFAMILY	Epidendroideae
TRIBE AND SUBTRIBE	Cymbidieae, Catasetinae
NATIVE RANGE	Costa Rica and Panama
HABITAT	Cloud forests, at 3,950–6,900 ft (1,200–2,100 m)
TYPE AND PLACEMENT	Epiphytic
CONSERVATION STATUS	Has a restricted distribution but occurs in reserves, therefore unlikely to be under threat of extinction
FLOWERING TIME	Spring

FLOWER SIZE
Up to 4¾ in (12 cm)

PLANT SIZE
24–32 × 15–24 in
(61–81 × 38–61 cm)
excluding the erect
inflorescence,
18–28 in (46–71 cm) tall

231

MORMODES COLOSSUS
FLYING BIRD ORCHID
REICHENBACH FILS, 1852

The large Flying Bird Orchid has cylindrical, tapering pseudobulbs with several (usually two to five) plicate leaves, which are deciduous in the dry season. At the base of the pseudobulbs, a long arching raceme emerges carrying 6–15 fragrant flowers that vary remarkably in size and color. The genus name comes from *Mormo*, a malevolent female spirit in Greek mythology, and *–oides*, "resembling," referring to the strange appearance of the flowers. The plants are often found on rotting stumps or fallen branches and have a mycorrhizal connection to wood-rot fungi.

Flowers of *Mormodes* are functionally male until the pollinia are removed. A long slender appendage on the anther cap is brushed by visiting male euglossine bees, causing the pollinarium to become attached to their back. The coiled stipe of the pollinarium takes 30 minutes to straighten and assume the correct position for pollination.

Actual size

The flower of the Flying Bird Orchid has spreading to recurved, lanceolate sepals and petals. The lip is shaped like a boat, with one end stalked and fixed to the flower, the other end pointing upward, overarching the column.

SUBFAMILY	Epidendroideae
TRIBE AND SUBTRIBE	Cymbidieae, Cymbidiinae
NATIVE RANGE	Indochina, Malay Peninsula, Indonesia, Philippines, New Guinea, Solomon Islands, and northern Australia
HABITAT	Swamps and semi-deciduous to dry forests up to 4,920 ft (1,500 m)
TYPE AND PLACEMENT	Epiphytic, rarely lithophytic
CONSERVATION STATUS	Not threatened
FLOWERING TIME	March to May (spring)

FLOWER SIZE
½ in (1.25 cm)

PLANT SIZE
8–18 × 6–10 in
(20–46 × 15–25 cm),
excluding the inflorescence,
which is arching and
longer than the leaves,
12–20 in (30–51 cm) long

232

ACRIOPSIS LILIFOLIA
GRASSHOPPER ORCHID
(J. KOENIG) SEIDENFADEN, 1995

Actual size

Though an easily grown small epiphytic orchid, the wide-ranging Grasshopper Orchid is barely known by many orchidists. A floriferous species, up to 200 pretty and intricate flowers have been observed on branching wiry racemes. This would normally be a spectacle, but the attractive flowers are so small—even in massive quantities—that most people fail to notice it.

Acriopsis in Greek means "like a grasshopper" (and hence the common name), referring to the strange insect-like lips. The lateral sepals are fused into a synsepal that curves and extends below the trilobed lip. A pair of rounded arms protrudes adjacent to the column, which also bears well-developed wings. Reportedly pollinated by small bees, the flowers are short-lived, staying fresh for only about three or four days.

The flower of the Grasshopper Orchid is small, intricate, and insect-like, with a cream background and purple central splotches. The lip is white with pinkish-purple markings in the center and trilobed.

SUBFAMILY	Epidendroideae
TRIBE AND SUBTRIBE	Cymbidieae, Cymbidiinae
NATIVE RANGE	Himalayas to temperate East Asia
HABITAT	Moist woodlands, conifer plantations, and pine forests on coastal sand dunes, usually on extremely steep rocky slopes in a thin humus layer over bedrock
TYPE AND PLACEMENT	Terrestrial or on rocks
CONSERVATION STATUS	Not formally assessed, but in China highly endangered due to forest clearance and mass collecting from the wild; better protected in other parts of its range
FLOWERING TIME	January to May (winter to spring) in southern part of range; March to April (spring) in Japan

CYMBIDIUM GOERINGII

NOBLE ORCHID

(REICHENBACH FILS) REICHENBACH FILS, 1852

FLOWER SIZE
2 in (5 cm)

PLANT SIZE
14–20 × 10–20 in
(35–51 × 25–51 cm),
excluding inflorescence
generally 4–12 in
(10–30 cm) long

233

The clustered, egg-shaped pseudobulbs of the Noble Orchid have linear, grasslike leaves. Among these, a short inflorescence with a large sheathing bract carries one (rarely two or three) strongly musk-scented flowers, which are used to flavor ceremonial tea in Japan. Chinese cultivars have a stronger fragrance than the Japanese.

Cymbidium goeringii is a symbol of refinement, elegance, and nobility in China, and because of its cold tolerance it has been popular in Japanese and Chinese horticulture since at least the fourteenth century. Several cultivars are known, and *C. goeringii* has been used in hybridization with other *Cymbidium* species. Cultivars often differ from the species as found in the wild, some having more rounded flower parts and others a rusty red color. The Noble Orchid is the state flower of Sikkim, India.

The flower of the Noble Orchid is inconspicuous, with spreading, green sepals and smaller green petals, sometimes with red markings basally, curved over the column and lip. The white lip is delicately marked with red and yellow, and the tip is recurved with wavy margins.

Actual size

SUBFAMILY	Epidendroideae
TRIBE AND SUBTRIBE	Cymbidieae, Cymbidiinae
NATIVE RANGE	Himalayas to central China
HABITAT	Wet forests, at 3,300–9,200 ft (1,000–2,800 m)
TYPE AND PLACEMENT	Epiphytic on mossy trees, mossy rocks, and limestone cliff banks
CONSERVATION STATUS	Not assessed
FLOWERING TIME	September to November (fall)

FLOWER SIZE
3–4 in (7.5–10 cm)

PLANT SIZE
20–40 × 14–30 in
(51–102 × 36–76 cm),
excluding inflorescence
20–36 in (51–91 cm) long

CYMBIDIUM IRIDIOIDES
TIGER-STRIPE ORCHID
D. DON, 1825

234

The Tiger-stripe Orchid has clusters of bilaterally compressed pseudobulbs, enveloped by persistent, overlapping leaf bases that become deciduous with age. Each pseudobulb has up to seven grasslike leaf blades articulated to a clasping, yellowish base. A horizontal-to-arching raceme arises among the leaves, producing between four and seven long-lasting flowers, which are slightly fragrant.

Several *Cymbidium* species with similar flowers are pollinated by *Trigona* bees. The labellum with its erect sidelobes forms a channel that forces the bees to make contact with the stigma and pollinia. The insects collect a sticky, waxlike compound from the center of the lip, which they use to close up cracks in their nest. *Cymbidium iridioides* is popular in horticulture, with several varieties in cultivation that vary in the degree of patterning of the sepals and petals. It has also been used in hybridization.

The flower of the Tiger-stripe Orchid is yellow with brown-striped, spreading petals and sepals. The brighter yellow lip has lateral wings that surround the column and a frilly margin with large, irregular, red spots.

Actual size

SUBFAMILY	Epidendroideae
TRIBE AND SUBTRIBE	Cymbidieae, Cymbidiinae
NATIVE RANGE	Myanmar, Thailand, Laos, Vietnam, Malaysia, and Indonesia
HABITAT	Hot wet forests, near streams and rivers, at 330–2,625 ft (100–800 m)
TYPE AND PLACEMENT	Epiphytic, occasional lithophytic
CONSERVATION STATUS	Threatened by poaching
FLOWERING TIME	July to October

GRAMMATOPHYLLUM SPECIOSUM
SHOWY TIGER ORCHID
BLUME, 1825

FLOWER SIZE
5 in (12.5 cm)

PLANT SIZE
40–400 × 20–38 in
(102–1,016 × 51–97 cm),
excluding erect-arching
inflorescence 100–500 in
(254–1270 cm) tall, which is
taller than the leafy stems

235

Considered by many to be the world's largest orchid (but not the tallest), the Showy Tiger Orchid is unquestionably the heaviest of all orchids. Its many long, lanceolate leaves borne on impressive cane-like pseudobulbs give the plants, at least superficially, the look of a palm. The genus name comes from the Greek words *gramma*, "letter," and *phyllon*, "leaf," a reference to the bold marking on the sepals and petals. The species name refers to the massive display of flowers, from the Latin for "showy"—also reflected in the common name.

Pollination of this species has been reported to be by large carpenter bees (*Xylocopa*) seeking nectar, although the flowers produce none. Low on the inflorescence are several sterile flowers (no column or lip are present), which produce only fragrance.

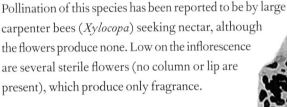

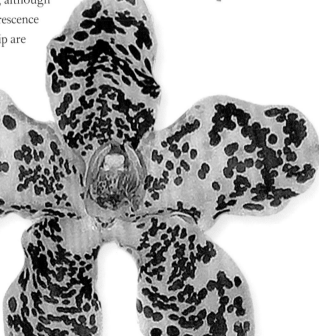

The flower of the Showy Tiger Orchid
generally has large yellow sepals and petals with an overlay of chestnut-brown spots and blotches. The lip is much smaller, usually pale yellow with some red marking on the callus and sidelobes, which surround the column.

Actual size

SUBFAMILY	Epidendroideae
TRIBE AND SUBTRIBE	Cymbidieae, Cymbidiinae
NATIVE RANGE	Widespread in Southeast Asia and Philippines
HABITAT	Lowland tropical wet forests
TYPE AND PLACEMENT	Epiphytic
CONSERVATION STATUS	Not threatened
FLOWERING TIME	Late winter to early spring

FLOWER SIZE
⅝ in (1.5 cm)

PLANT SIZE
4–9 × 3–6 in
(10–23 × 7–15 cm),
excluding the inflorescence,
which is pendent,
10–22 in (25–56 cm) long

236

THECOSTELE ALATA
JUMPING FROG ORCHID
(ROXBURGH) C. S. P. PARISH & REICHENBACH FILS, 1874

Actual size

The poorly known *Theocostele* is closely related to a similarly named genus, *Thecopus*. Rarely cultivated due to its diminutive flower size, the species produces long, pendent racemes from the base of its comparatively large pseudobulbs, each holding a dozen or more colorful flowers speckled with vibrant maroon markings and arranged spirally on their inflorescences. The common name refers to the similarity in appearance of the column apex to a jumping frog with its front legs outstretched.

An epiphyte from lowland tropical forests and widespread throughout much of Southeast Asia, the Jumping Frog Orchid grows year-round in warm, humid conditions. In some parts of Indonesia, the leaves and pseudobulbs of this species were boiled and the liquid then combined with arsenic to make a rat poison.

The flower of the Jumping Frog Orchid is small and arranged spirally on pendent racemes. The incurved sepals are cream with maroon-purple blotches, and the petals are narrow. The lip is outstretched but folded down and forms an interior box shape. The column has two prominent wings.

SUBFAMILY	Epidendroideae
TRIBE AND SUBTRIBE	Cymbidieae, Cyrtopodiinae
NATIVE RANGE	Southern Florida, the Caribbean to northern Venezuela
HABITAT	Swamps and grasslands, often low on trees, up to 3,950 ft (1,200 m)
TYPE AND PLACEMENT	Epiphytic, lithophytic, or terrestrial
CONSERVATION STATUS	Almost extirpated in Florida
FLOWERING TIME	Spring

CYRTOPODIUM PUNCTATUM
COWHORN ORCHID
(LINNAEUS) LINDLEY, 1833

FLOWER SIZE
1¼ in (3.25 cm)

PLANT SIZE
28–40 × 12–30 in
(71–102 × 30–76 cm),
excluding erect
inflorescence 50–72 in
(127–183 cm) long

237

Once one of the most common and impressive orchid sights of south Florida, the Cowhorn Orchid was aggressively removed from the swamps it inhabits by orchid enthusiasts. Even so, since its natural habitat still remains, the plant has become one of the most successful of reintroduced species. Elsewhere in its range, it has always been and remains common.

Growing low down on dying trees and stumps or on rocks and soil, the conspicuous plants grow quickly to prodigious sizes. The large, deliciously fragrant, spotted flowers appear on upright stems in big numbers. Pollination is by male euglossine bees (*Euglosssa*) seeking floral fragrance compounds in some parts of its range, although this genus of bee is not native to Florida, so the orchid presence in that region is considered unexplained. Native peoples use the mucilage from crushed pseudobulbs as glue.

The flower of the Cowhorn Orchid has curly, bright yellow segments and red-brown to purple-brown spots and blotches. The yellow and red-brown lip appears brilliant orange from a distance.

Actual size

SUBFAMILY	Epidendroideae
TRIBE AND SUBTRIBE	Cymbidieae, Eriopsidinae
NATIVE RANGE	Belize (Central America) to northern Brazil and the Guianas (South America)
HABITAT	Wet forests, at 1,640–6,600 ft (500–2,000 m)
TYPE AND PLACEMENT	Epiphytic or terrestrial on steep rocky or clayey slopes or exposed rocks
CONSERVATION STATUS	Not assessed, but due to its broad distribution not likely to be of conservation concern
FLOWERING TIME	July–August (summer)

FLOWER SIZE
1 in (2.5 cm)

PLANT SIZE
10–20 × 10–16 in
(25–51 × 25–41 cm),
excluding erect to
arching inflorescence
15–36 in (38–91 cm) long,
which is longer than the leaves

238

ERIOPSIS BILOBA
SUNSHINE ORCHID
LINDLEY, 1847

The stout Sunshine Orchid usually grows on rocks or in soil and has elongate-conical, dark green pseudobulbs with two or three weakly stalked, elliptic leaves. It produces long, densely flowered racemes with purple flower stems and up to 35 waxy, fragrant flowers that open simultaneously. The genus name refers to the similarity in habit of this species to that of the Asian genus *Eria* (*–opsis* is Greek for "likeness to"). The lip is in fact trilobed, not bilobed as the species name suggests.

Pollination has not been observed, but the open flower shape and lack of a nectar spur are consistent with pollination by bees attracted by deceit. Although widespread and moderately common in nature, these plants are difficult to cultivate successfully outside their natural habitat.

The flower of the Sunshine Orchid has spreading petals and sepals that are yellow with bronze markings. The column points downward, and the lip is curved around it and capped with a short, white, purple-spotted midlobe that is sometimes bilobed.

Actual size

SUBFAMILY	Epidendroideae
TRIBE AND SUBTRIBE	Cymbidieae, Eulophiinae
NATIVE RANGE	Southern South Africa, from the Cape northeast to the province of KwaZulu-Natal
HABITAT	Coastal scrub, from sea level to 2,300 ft (700 m)
TYPE AND PLACEMENT	Terrestrial on sand
CONSERVATION STATUS	Not formally assessed, but not uncommon where it grows
FLOWERING TIME	September to October (spring)

FLOWER SIZE
⅜ in (1 cm)

PLANT SIZE
24–36 × 16–24 in
(60–90 cm × 40–60 cm),
including inflorescence

ACROLOPHIA COCHLEARIS
IMPIMPI ORCHID
(LINDLEY) SCHLECHTER & BOLUS, 1894

239

The robust Impimpi Orchid, the only member of its subfamily found so far south in Africa, grows from a cluster of fleshy roots and produces a clump of grasslike leaves. An upright, laxly branched inflorescence bearing up to 100 flowers is formed after the cool rainy season that is winter in the Southern Hemisphere. Flowering is apparently stimulated by fire. The common name is based on a local name in South Africa.

Even though *Acrolophia* is closely allied to *Eulophia*, a genus with deceptive flowers that offer no reward to their pollinator, *A. cochlearis* does produce some nectar for the pollinating solitary bee species *Colletes claripes*—the only insect known to effectively pollinate the flowers. However, up to 90 percent of fruits may be the result of self-pollination in this species—that is, the bees move between flowers on the same stem.

The flower of the Impimpi Orchid is non-resupinate, with the white lip uppermost and pointing outward, covering the rest of the flower, whereas the brown sepals and petals are spreading, and the column is shortly winged.

Actual size

SUBFAMILY	Epidendroideae
TRIBE AND SUBTRIBE	Cymbidieae, Eulophiinae
NATIVE RANGE	Most of Sub-Saharan Africa
HABITAT	Lowland seasonally dry areas among scrubby vegetation, although often high in trees
TYPE AND PLACEMENT	Epiphytic or lithophytic
CONSERVATION STATUS	Not threatened
FLOWERING TIME	August to October in Southern Hemisphere, spring in Northern Hemisphere

FLOWER SIZE
1¼ in (3.25 cm)

PLANT SIZE
25–35 × 12–25 in
(64–89 × 30–64 cm),
excluding inflorescence,
which is erect and produced
from the top of a pseudobulb,
10–20 in (25–51 cm) long

240

ANSELLIA AFRICANA
LEOPARD ORCHID
LINDLEY, 1844

The genus *Ansellia* has only one highly variable species well distributed throughout tropical and southern Africa. The flowers vary in color and shape. The mostly epiphytic plants vary in size as well and can grow into spectacular clumps.

The elongate, upright canes bear just a few leaves toward their apex and large paniculate inflorescences holding often up to 100 spotted yellow blooms. *Ansellia* produces two types of roots: thick fibrous support roots that wrap around or, in some cases, burrow deeply into the substrate, and curious upright, more filamentous roots that form an outstretched network called a "trash basket." Its purpose is thought to be to collect leaf litter and other detritus from the canopy, which is then decomposed by the mycorrhizal symbionts. Pollinators of this species are likely to be large bees.

The flower of the Leopard Orchid varies in color but not so much in structure. Starry subequal sepals and petals of yellow or pale green background are often, though not always, spotted with varying degrees and shades of brownish or tawny spots. The trilobed lip often has brilliant yellow and brown markings.

Actual size

SUBFAMILY	Epidendroideae
TRIBE AND SUBTRIBE	Cymbidieae, Eulophiinae
NATIVE RANGE	Peninsular Thailand and Malaysia, northern Indonesia
HABITAT	Open primary forests on sandstone, up to 2,625 ft (800 m)
TYPE AND PLACEMENT	Terrestrial on poor soil, becoming epiphytic, sometimes on rocks
CONSERVATION STATUS	Least concern—widespread, common, adaptable species
FLOWERING TIME	Spring and early summer

FLOWER SIZE
2 in (5 cm)

PLANT SIZE
8–15 × 10–18 in
(20–38 × 25–46 cm),
excluding inflorescence,
which can be about twice
the height of the plant

CLADERIA VIRIDIFLORA
GREEN CLIMBING ORCHID
HOOK FILS, 1890

241

The Green Climbing Orchid produces a thick, creeping rhizome from which leafy growths emerge. Eventually (sometimes after up to ten years), these stems reach a tree or rocks and clamber over the surfaces, reaching brighter areas where they flower. Each growth has up to six lanceolate, ribbed leaves that cover the stem at the base. A many-flowered inflorescence tops this leafy stem, and stiff bracts support the flowers, of which only one or two are open at any given time.

Nothing is known about pollination, and no uses have been reported. For many years, the position of this species in the orchid family was debated, and authors have suggested different ideas. Finally, DNA was employed to assess its position, and now it is considered to belong to genera related to the genus *Eulophia*, although the morphology of this species does not match well with theirs.

The flower of the Green Climbing Orchid is greenish-yellow and has three equal-spreading sepals and two forward-pointing petals that make a cup around the column together with the bilobed, green-veined lip.

Actual size

SUBFAMILY	Epidendroideae
TRIBE AND SUBTRIBE	Cymbidieae, Eulophiinae
NATIVE RANGE	Eastern forests of Madagascar, Perinet region
HABITAT	Mid-elevation forests on *Platycerium* (Staghorm Fern)
TYPE AND PLACEMENT	Epiphytic
CONSERVATION STATUS	Critically endangered
FLOWERING TIME	March to April

FLOWER SIZE
4 in (10 cm)

PLANT SIZE
15–38 × 18–30 in
(38–97 × 46–76 cm),
excluding inflorescence,
which is arching-erect,
15–24 in (38–61 cm) long

CYMBIDIELLA PARDALINA
BAT-EARED ORCHID
(REICHENBACH FILS) GARAY, 1976

242

The flower of the Bat-eared Orchid has green
sepals and petals mottled with purplish-black
spots. In striking contrast, the vivid red lip has
a central yellow band.

Like many large, mountainous islands, Madagascar is rich in
endemic species, including the Bat-eared Orchid, which, in
the wild, grows exclusively on the Staghorn Fern (*Platycerium
madagascariense*). This fern in turn is an epiphyte that grows
only on tall *Albizia fastigiata* trees, making the orchid extremely
specific in its natural requirements and therefore an extremely
rare plant in nature. This highly specific relationship, however,
has not prevented successful cultivation beyond the island.

Because of its fantastic and unusually colored flowers—green
with purplish-black spots and a vivid vermilion-red lip—this
species is popular in horticulture. The genus name is due to its
similarity to *Cymbidium* (in which it was once placed), and the
species name refers to its spotted petals. The common name is
based on the shape of the petals, which resemble the large ears
of a bat.

Actual size

SUBFAMILY	Epidendroideae
TRIBE AND SUBTRIBE	Cymbidieae, Eulophiinae
NATIVE RANGE	Australia, New Caledonia, and Vanuatu
HABITAT	Protected, shady positions in dry woodland and forests
TYPE AND PLACEMENT	Terrestrial mycoheterotrophic
CONSERVATION STATUS	Locally the species may be rare or threatened, but it is common in other places, so not of conservation concern
FLOWERING TIME	November to March (summer)

DIPODIUM SQUAMATUM

AUSTRALIAN HYACINTH ORCHID

(G. FORSTER) R. BROWN, 1810

FLOWER SIZE
1 in (2.5 cm)

PLANT SIZE
16–39 in
(40–99 cm) tall,
with no leaves

243

The Australian Hyacinth Orchid is leafless and takes all its nutrients and carbon from soil fungi that it parasitizes. The plant produces a dark red inflorescence in late spring, bearing many flowers in various shades of pink and white. In Australia, the species is usually known under its synonym *Dipodium punctatum*, while in Tasmania the closely related species *D. roseum* occurs. The orchid appears to be variable, and its taxonomy is not yet well explored.

The flower falsely advertises for bees, which are attracted by glossy yellow green areas on the column. However, because nectar is wanting, the visiting bee (a female megachilid bee has been observed to carry pollinia between its antennae) is forced to push further under the column and so pick up the pollinia. The blooms are protected from flower-eating insects by ants, thanks to extrafloral nectaries produced along the scape.

Actual size

The flower of the Australian Hyacinth Orchid has linear, recurved sepals and petals that are sometimes spotted pink. The lip forms a tube with the winged column, and the limb is narrow and tongue-shaped.

SUBFAMILY	Epidendroideae
TRIBE AND SUBTRIBE	Cymbidieae, Eulophiinae
NATIVE RANGE	Southern Africa northeast to the Arabian Peninsula
HABITAT	Grasslands, deciduous forests, and sand dunes, often in poor soils, at 650–6,600 ft (200–2,000 m)
TYPE AND PLACEMENT	Terrestrial
CONSERVATION STATUS	Not threatened
FLOWERING TIME	Late spring to summer

FLOWER SIZE
2 in (5 cm)

PLANT SIZE
12–20 × 6–10 in
(30–51 × 15–25 cm),
excluding basal erect
inflorescence 25–40 in
(64–102 cm) tall

244

EULOPHIA SPECIOSA
BEAUTIFUL PLUME
(R. BROWN) BOLUS, 1889

The stately Beautiful Plume occurs sporadically over a wide range of the African continent as well as the Arabian Peninsula and bears stems of up to 30, showy, long-lasting, and sweetly fragrant flowers. Arising from pseudobulbs just below or at the soil surface, the fleshy, slightly curled leaves are well adapted to survive dry conditions. Due to its tall stems and bright color, the orchid has become a common sight in gardens, particularly in southern Africa. The genus name comes from the Greek words *eu–*, "true," and lophos, "plume," referring to the large inflorescence. The common and species names refer to the showy flowers (*speciosus* is Latin for "showy" or "beautiful").

Eulophia speciosa is pollinated by carpenter bees (genus *Xylocopa*), which forage for nonexistent nectar. Preparations made from its pseudobulbs have been used in African folk medicine to treat many ailments.

The flower of the Beautiful Plume has a little variation in color over its extensive range, usually with greenish-yellow recurved sepals and broad, brilliant-yellow petals. The lip is also bright yellow, trilobed, and often marked centrally with bold red to purple striping.

Actual size

SUBFAMILY	Epidendroideae
TRIBE AND SUBTRIBE	Cymbidieae, Eulophiinae
NATIVE RANGE	Tropical and subtropical Asia, Australia, and western Pacific islands
HABITAT	Moist grassland, coastal dune valleys, in wet deciduous or evergreen forests and savanna woodland, at up to 5,900 ft (1,800 m)
TYPE AND PLACEMENT	Terrestrial
CONSERVATION STATUS	Not globally assessed, but endangered in Australia
FLOWERING TIME	Spring or fall, depending on locality and hemisphere

FLOWER SIZE
⅜ in (1 cm)

PLANT SIZE
10–30 × 10–20 in
(25–76 × 25–51 cm),
excluding inflorescence
8–25 in (20–64 cm) long

GEODORUM DENSIFLORUM
NODDING SWAMP ORCHID
(LAMARCK) SCHLECHTER, 1919

245

The underground, spherical pseudobulbs of the Nodding Swamp Orchid produce up to five broadly lanceolate, plicate leaves with a stalk. From their base, a many-flowered inflorescence with a nodding apex emerges, bearing waxy flowers that often do not fully open. The downward-pointing inflorescence straightens out after fertilization of the flowers and can be upright in fruit. The genus name derives from the pendent tip of the inflorescence (from the Greek *ge*, meaning "earth," and *doron*, meaning "gift"). The species name refers to this same characteristic.

Pollination by small native bees in Australia has been reported, but no specific bee species have been mentioned. The plant is sometimes used medicinally, as in the Maluku Islands, where the leaves are pounded into a paste applied to running sores.

The flower of the Nodding Swamp Orchid has white to pink, lanceolate, spreading sepals and petals with several rows of pink dots at their bases. The darker lip is cupped, with a central warty, yellow callus that has dark pink markings.

Actual size

SUBFAMILY	Epidendroideae
TRIBE AND SUBTRIBE	Cymbidieae, Eulophiinae
NATIVE RANGE	Eastern forests of Madagascar, reportedly on Sainte Marie Island (may be extirpated there)
HABITAT	Forests, up to 4,920 ft (1,500 m)
TYPE AND PLACEMENT	Epiphytic
CONSERVATION STATUS	Critically endangered
FLOWERING TIME	January

FLOWER SIZE
2¾–3 in (7–8 cm)

PLANT SIZE
25–40 × 15–30 in
(64–102 × 38–76 cm),
excluding basal
inflorescence, which is
arching-erect, 28–45 in
(71–114 cm) long

246

GRAMMANGIS ELLISII
BRONZE-BANDED
ORCHID
(LINDLEY) REICHENBACH FILS, 1860

The genus *Grammangis* is exclusive to the eastern forests of Madagascar and consists of two species—both spectacular plants similar in habit to their gigantic Asian counterpart, *Grammatophyllum*. The pseudobulbs of the Bronze-banded Orchid are oddly quadrangular, and the stunning, gracefully arching inflorescences hold up to 20 large tawny and bronze blooms. These are triangular due to the dark sepals spreading widely, while the petals form a tube around the column.

The species grows epiphytically from sea level to about 4,920 ft (1,500 m), often overhanging streams in woodlands. The robust habit, unusual angular bulbs, and fleshy leaves combine to make the orchid an impressive sight even when not in bloom. The genus name, from the Greek words *gramma*, meaning "mark," and *angos*, "vessel," alludes to the impressive markings on these flowers. The common name refers to the banded pattern of the prominent sepals.

The flower of the Bronze-banded Orchid can vary, but sepals are generally a combination of bronze-green and yellow, often with a yellow, transverse stripe near their tips. The much smaller, yellow petals and white to rose, keeled lip surround the column.

Actual size

SUBFAMILY	Epidendroideae
TRIBE AND SUBTRIBE	Cymbidieae, Eulophiinae
NATIVE RANGE	Indian Ocean islands (Madagascar, Comoros, Seychelles, Réunion, Mauritius)
HABITAT	Humid and littoral forests, from sea level to 985 ft (300 m)
TYPE AND PLACEMENT	Epiphytic
CONSERVATION STATUS	Not assessed, but fairly rare in nature
FLOWERING TIME	March to May (spring)

FLOWER SIZE
1 in (2.5 cm)

PLANT SIZE
10–18 × 8–16 in
(25–46 × 20–41 cm),
excluding erect to
arching inflorescence
16–40 in (41–102 cm) tall,
which is longer than the leaves

247

GRAPHORKIS CONCOLOR
UNPAINTED PAINTED ORCHID
(THOUARS) KUNTZE, 1891

A cluster of conical pseudobulbs forms along the short stem of the Unpainted Painted Orchid, with a new pseudobulb added each year. The bulbs are topped with linear-lanceolate deciduous leaves. A panicle with numerous flowers emerges at the base of a mature pseudobulb before or just as the plant initiates new vegetative growth. The genus name is derived from the Greek words *graphos*, "mark," and *orchis*, "orchid," a reference to the typical form of this species, which has darker script-like markings. The species name implies that the flowers are of a single color.

Pollination has not been observed in nature, but the open flower shape with a nectar spur would be suitable for bee pollination. This is one of the orchid species that produces upwardly growing roots that form a litter basket and collect detritus, which provides nutrition to the plant as decomposition takes place.

The flower of the Unpainted Painted Orchid has yellow to greenish-yellow spreading sepals and petals. The lip is brighter yellow and three-lobed, with the lateral lobes curved up around the column. The midlobe has two raised ridges and a papillate base.

Actual size

SUBFAMILY	Epidendroideae
TRIBE AND SUBTRIBE	Cymbidieae, Eulophiinae
NATIVE RANGE	Tropical central Africa, but has invaded the Neotropics, especially Florida and the Caribbean
HABITAT	Often in disturbed habitats in deep shade, from sea level to 3,950 ft (1,200 m)
TYPE AND PLACEMENT	Terrestrial
CONSERVATION STATUS	Not threatened
FLOWERING TIME	Winter to spring

FLOWER SIZE
¾ in (2 cm)

PLANT SIZE
8–14 × 2–4 in
(20–36 × 5–10 cm),
excluding erect inflorescence
10–16 in (25–41 cm) tall, which
is longer than the leaves

248

OECEOCLADES MACULATA
COW'S TONGUE
(LINDLEY) LINDLEY, 1833

The tough, adaptable Cow's Tongue has made its way across much of the tropical African and American regions of the world, readily colonizing disturbed environments. The succulent leaves are referred to as *lengua de vaca* ("cow's tongue") in Cuba, giving this orchid its English common name. The genus name derives from the Greek words *oikeios*, "private," and *klados*, "branch," apparently referring to Lindley's view that this genus has no close relatives.

In its native Africa, *Oeceoclades maculata* is pollinated by small bees, but in the areas in which it is an invasive it is self-pollinating. A peculiar mechanism requires a raindrop to dislodge the anther cap, permitting the pollinia to dry out and the pollinia stalk to curl, placing the pollen masses above the stigma. At this point a second bout of rain is required to cause the pollinia to fall into the stigma.

The flower of the Cow's Tongue has yellow green, lanceolate sepals and petals that arch up or forward from the center, and a three-lobed, white lip with pink markings in which the midlobe projects forward and the two side lobes flank the column.

Actual size

SUBFAMILY	Epidendroideae
TRIBE AND SUBTRIBE	Cymbidieae, Maxillariinae
NATIVE RANGE	Colombia, Venezuela, Ecuador, Peru, and Bolivia
HABITAT	Wet, cool forests, at 4,600–8,200 ft (1,400–2,500 m)
TYPE AND PLACEMENT	Terrestrial
CONSERVATION STATUS	Not threatened
FLOWERING TIME	Spring

ANGULOA VIRGINALIS
SWADDLED BABY IN CRADLE
LINDEN EX B. S. WILLIAMS, 1862

FLOWER SIZE
3 in (8 cm)

PLANT SIZE
24–30 × 15–24 in
(61–76 × 38–61 cm),
including erect, single-
flowered inflorescence
8–15 in (20–38 cm) tall,
which is shorter than the leaves

249

The Swaddled Baby in Cradle loves cool moist conditions on the forest floor. Deliciously and potently fragrant flowers appear on between two and eight sturdy upright stems, each emerging from the base of the newest pseudobulb, usually as the leaves develop. The stunning, fleshy flowers are borne on bract-covered stems and have one larger bract that covers their base. The genus was discovered during a ten-year expedition to Peru and Chile by botanists Hipòlito Ruiz and José Antonio Pavón (1777–88) and named in honor of Don Francisco de Angulo, Director-General of Mines in Peru. The common name refers to the appearance of the lip and column of this species (baby in the cradle) among the large fleshy sepals and petals (swaddling).

The complex floral scents of Swaddled Baby in Cradle are largely composed of p-dimethoxybenzene, which is a known attractant for several species of fragrance-collecting male euglossine bees.

The flower of the Swaddled Baby in Cradle
is fleshy and waxy pink to white overlaid with purplish-red spots. The dorsal sepal and petals form a hood, and the more elongate, pointed, lateral sepals project downward. The complex trilobed lip is hinged and hangs just below the column.

Actual size

SUBFAMILY	Epidendroideae
TRIBE AND SUBTRIBE	Cymbidieae, Maxillariinae
NATIVE RANGE	Mountainous regions of southern and southeastern Brazil
HABITAT	Rocky outcrops, at about 3,300 ft (1,000 m)
TYPE AND PLACEMENT	Lithophytic, sometimes epiphytic
CONSERVATION STATUS	Threatened
FLOWERING TIME	May to June

FLOWER SIZE
2¾ –3⅛ in (7–8 cm)

PLANT SIZE
10–15 × 4–8 in
(25–38 × 10–20 cm),
excluding erect inflorescences,
which are much shorter than
leaves, 3–6 in (8–15 cm) long

BIFRENARIA HARRISONIAE
FRAGRANT ROCK ORCHID
(HOOKER) REICHENBACH FILS, 1855

250

The exquisite, highly variable Fragrant Rock Orchid, which embeds its tenacious roots into crevices of stone outcrops, is one of the most beautiful and sought-after orchids. Growing in full sun and often harsh conditions, the large, deliciously perfumed, crystalline textured flowers arise from the bases of tough pseudobulbs and are shaded by broad, leathery foliage.

The heady fragrance is an attractant to the euglossine bees and queen bumblebees that pollinate them in the wild. The spurred flowers do not produce nectar, and this makes this species fit the deceit pollination syndrome. The origin of the common name is obvious from the above description, and the genus name comes from the Greek *bi-*, for "two," and *frenum*, for "strap," referring to the forked nature of the pollinarium of this genus.

The flower of the Fragrant Rock Orchid is variable but most commonly a clean crystalline white with a plum-purple, three-lobed lip, covered internally with glistening trichomes, leading to an empty spur at the back. Purple, yellow, bicolor, and blue forms exist but are rare.

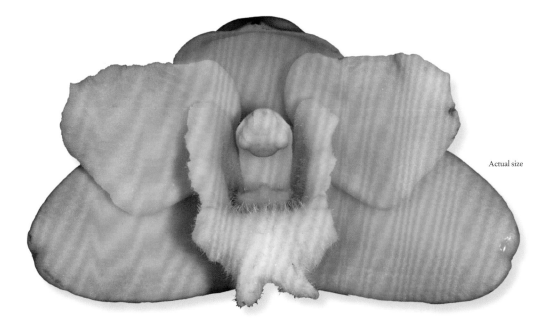

Actual size

SUBFAMILY	Epidendroideae
TRIBE AND SUBTRIBE	Cymbidieae, Maxillariinae
NATIVE RANGE	Mexico, Guatemala, Nicaragua, Belize, Honduras, and El Salvador
HABITAT	Seasonally dry forests, at 1,640–6,600 ft (500–2,000 m)
TYPE AND PLACEMENT	Epiphytic, terrestrial, and lithophytic
CONSERVATION STATUS	Threatened by overcollection
FLOWERING TIME	Late spring to summer

LYCASTE AROMATICA
CINNAMON ORCHID
(GRAHAM) LINDLEY, 1843

FLOWER SIZE
3 in (8 cm)

PLANT SIZE
12–20 × 12–18 in
(30–51 × 30–46 cm),
including erect, single-
flowered 4–6 in (10–15 cm)
long inflorescences

251

The Cinnamon Orchid produces up to ten vivid yellow, deliciously fragrant blossoms on short individual inflorescences from each leafless pseudobulb as the next season's growth is starting to emerge. Extremely adaptable, it can be found growing epiphytically on mossy branches as well as on steep seasonally moist limestone outcrops. Plants are deciduous during the dry season, with the bulbs bearing sharp spines at their apex. The genus name derives from Lycaste, the beautiful daughter of King Priam of Troy.

The sweet cinnamon aroma emitted by the flowers is a lure for male euglossine bees, which use the fragrance compounds to produce a sex pheromone that attracts females. There are several related species that look similar to *Lycaste aromatica*, but these display slightly different blooming seasons and fragrances.

The flower of the Cinnamon Orchid is greenish-yellow with ovate, acuminate sepals arranged in a triangle and brilliant yellow, incurved petals. The also bright yellow, three-lobed lip has a pubescent disc and a large, ridged callus that extends over the midlobe.

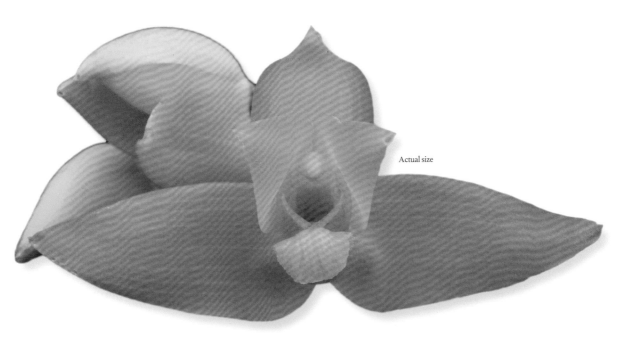

Actual size

SUBFAMILY	Epidendroideae
TRIBE AND SUBTRIBE	Cymbidieae, Maxillariinae
NATIVE RANGE	Mexico, Guatemala, Honduras, and El Salvador
HABITAT	Wet forests, at 3,950–6,600 ft (1,200–2,000 m)
TYPE AND PLACEMENT	Epiphytic, terrestrial, and lithophytic
CONSERVATION STATUS	Threatened by overcollection
FLOWERING TIME	Late spring to summer

FLOWER SIZE
6 in (15 cm)

PLANT SIZE
12–20 × 12–18 in
(30–51 × 30–46 cm),
including erect, single-
flowered 6–12 in (15–30 cm)
long inflorescences, which are
shorter than the leaves

252

LYCASTE VIRGINALIS
NUN ORCHID
(SCHEIDWEILER) LINDEN, 1888

With its exceptionally showy, large triangular flowers and an enchanting fragrance, the Nun Orchid has been cherished and passionately cultivated since it was first discovered, when hundreds of thousands were exported. As a result, it is still exceedingly rare in the wild.

The species varies in color, size, and form. The most coveted plants may be the pure white forms, once referred to as *Lycaste skinneri* (now considered a synonym of *L. virginalis*), which are the national flower of Guatemala, where it is known as *Monja Blanca* (the White Nun). Growing epiphytically but also sending their roots into decomposing leaf mulch, the plants are adaptable and vigorous, although they prefer the cooler temperatures of their high-elevation habitat. Pollination is, as with all species of *Lycaste*, by fragrance-collecting male euglossine bees.

The flower of the Nun Orchid is variable in color with wide, strap-like sepals, usually recurved at their tips, and shorter petals covering the column. Colors range from purest white to pale purple, rarely in apricot tones. The lip callus is yellow and tongue-shaped.

Actual size

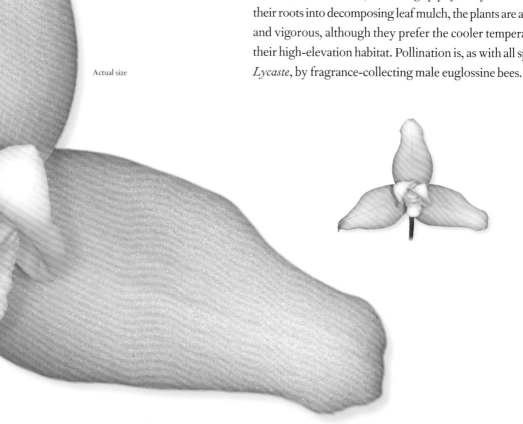

SUBFAMILY	Epidendroideae
TRIBE AND SUBTRIBE	Cymbidieae, Maxillariinae
NATIVE RANGE	Western South America (Colombia, Ecuador, and Peru)
HABITAT	Wet and cloud forests, at 2,950–9,850 ft (900–3,000 m)
TYPE AND PLACEMENT	Epiphytic or lithophytic
CONSERVATION STATUS	Not assessed
FLOWERING TIME	Throughout the year

FLOWER SIZE
⅜ in (1 cm)

PLANT SIZE
6–30 × 4–10
(15–76 × 10–25 cm),
including single-flowered
lateral inflorescences
1–2 in long (2.5–5 cm)

253

MAXILLARIA ALPESTRIS
MOUNTAIN JAW ORCHID
LINDLEY, 1845

The Mountain Jaw Orchid is a clump-forming species producing an upright rhizome with flattened, wrinkly pseudobulbs subtended by papery bracts and topped by a single, narrowly elliptic, leathery leaf. Older plants can topple over, making them pendent. The genus name comes from the Latin for jawbone, *maxilla*, referring to the prominent chin formed by the bases of the column and lip (termed a "column foot"). The common name combines this reference and the high-elevation nature of the plant's home.

The species is sometimes placed in the genus *Sauvetrea* but is now included in an expanded *Maxillaria*. Pollination has not been studied, but members of the genus with this same morphology and color that have been studied were pollinated by stingless meliponine bees, probably seeking nectar, which is never present in *Maxillaria*.

Actual size

The flower of the Mountain Jaw Orchid has cream to yellowish, recurved lateral sepals and a hooded upper sepal. The similarly colored petals project forward. The sidelobes of the yellow to orange lip surround the column and have a raised, elongate basal callus.

SUBFAMILY	Epidendroideae
TRIBE AND SUBTRIBE	Cymbidieae, Maxillariinae
NATIVE RANGE	Costa Rica, Panama, Colombia, and Ecuador
HABITAT	Continuously moist tropical forests, at 1,640–4,950 ft (500–1,500 m)
TYPE AND PLACEMENT	Epiphytic
CONSERVATION STATUS	Not threatened
FLOWERING TIME	Spring to fall

FLOWER SIZE
1 in (2.5 cm)

PLANT SIZE
6–10 × 6–12 in
(15–25 × 15–30 cm),
including single-flowered
arching inflorescences
4–10 in (10–25 cm) long

MAXILLARIA AMPLIFOLIATA
FOREST FAIRY STARS
MOLINARI, 2015

Actual size

The flower of Forest Fairy Stars is bright green to tawny brown, and the sepals are much longer than the petals, and both are narrow and pointed. The smaller but also pointed lip hugs the column with a long nectar spur behind, hidden by a bract.

Formerly considered a member of the genus *Cryptocentrum* (from the Greek *cryptos* and *kentron*, meaning "hidden spur"), the Forest Fairy Stars is atypical among the species of *Maxillaria* in exhibiting a fan-shaped, pseudobulb-less growth in which new leaves are added at the apex (rather than at the side of the previous growth or pseudobulb). Like most species without pseudobulbs, the plant occurs in perpetually wet areas in moderately shady sites. Many single-flowered inflorescences per plant bear starry flowers that are sweetly fragrant and produce nectar in long spurs, the last obscured by large sheathing bracts.

Maxillaria amplifoliata and its close relatives have all the features associated with nocturnal moth pollination. This partially explains the clear differences between these species formerly treated as members of genus *Cryptocentrum* and those considered to be more typical of *Maxillaria*, which are bee pollinated.

SUBFAMILY	Epidendroideae
TRIBE AND SUBTRIBE	Cymbidieae, Maxillariinae
NATIVE RANGE	Northwestern tropical America, from Costa Rica to Peru
HABITAT	Wet forests, at 165–4,600 ft (50–1,400 m)
TYPE AND PLACEMENT	Epiphytic
CONSERVATION STATUS	Least concern
FLOWERING TIME	December to March (summer to fall)

MAXILLARIA CHARTACIFOLIA
SUN-GOD ORCHID
AMES & C. SCHWEINFURTH, 1930

FLOWER SIZE
1 in (2.5 cm)

PLANT SIZE
15–30 × 15–30 in
(38–76 × 38–76 cm),
including axillary, single-
flowered inflorescences
3–5 in (8–13 cm) long

255

The rhizome of this elegant orchid terminates in an upright stem that forms a flat fan of basally sheathing, narrowly lanceolate, prominently veined leaves. Several foul-smelling flowers (described by some as "rancid cheese") are produced on individual stems and probably pollinated by flies attracted to a place to lay their eggs. This is a type of deceit pollination termed "brood-site deception," where females are lured to the source of the smell and, in their maneuvers around the flowers, land on the lip, contacting and removing the pollen.

The Sun-god Orchid is sometimes placed in the genus *Inti*, but it is now included in the enlarged genus *Maxillaria*. Inti is an Incan sun god, so the choice of genus name reflected the fan of leaves resembling the rays of the rising sun.

Actual size

The flower of the Sun-god Orchid has lanceolate, spreading, yellowish sepals and shorter, elliptic, forward-projecting yellow petals. The spade-shaped, reddish-brown lip has upcurved sidelobes, a yellow edge, and a heart-shaped, thickened callus with a mealy surface.

SUBFAMILY	Epidendroideae
TRIBE AND SUBTRIBE	Cymbidieae, Maxillariinae
NATIVE RANGE	Hispaniola, Puerto Rico, Lesser Antilles, and Trinidad
HABITAT	Primary rain and cloud forests, at 1,640–3,300 ft (500–1,000 m)
TYPE AND PLACEMENT	Epiphytic on mossy trunks, sometimes on rocks and mossy banks
CONSERVATION STATUS	Not formally assessed, but locally abundant
FLOWERING TIME	December to May (winter to spring)

FLOWER SIZE
1 in (2.5 cm)

PLANT SIZE
10–24 × 8–14 in
(25–61 × 20–36 cm),
including short, single-
flowered inflorescences
2–3 in (5–8 cm) long

MAXILLARIA COCCINEA
FLAME ORCHID
(JACQUIN) L. O. WILLIAMS, 1954

The elongate, leafy pseudobulbs of the Flame Orchid are arranged along a rhizome covered in papery sheaths. As the plants age and each pseudobulb climbs upon those before it, the stem grows longer and may become lax and almost pendent. From the base of the pseudobulb, a tight cluster of stems emerges, each with a single flower on a wiry stem. These flowers, like those of most *Maxillaria* species, have a short chin (hence the genus name from the Latin *maxilla*, "jawbone") and no nectar spur.

The species was formerly placed in the genus *Ornithidium*, which has recently been transferred to *Maxillaria*. The flowers produce copious amounts of nectar from finger-shaped glands on the base of their column, and the shape and color support pollination by hummingbirds, although field observations showing transferal of pollen by birds are wanting.

The flower of the Flame Orchid has forward-pointing, bright red sepals and petals, the latter forming a tube with the column around the lip. The column base produces nectar that is stored in the saccate base of the orange-red to yellow lip, which has upturned margins.

Actual size

SUBFAMILY	Epidendroideae
TRIBE AND SUBTRIBE	Cymbidieae, Maxillariinae
NATIVE RANGE	North America (Mexico) through Central America to Panama
HABITAT	Wet forests, at 2,300–10,825 ft (700–3,300 m)
TYPE AND PLACEMENT	Epiphytic
CONSERVATION STATUS	Not threatened
FLOWERING TIME	October to January (fall to winter)

MAXILLARIA CUCULLATA
BROWN FRIARS
LINDLEY, 1840

FLOWER SIZE
2 in (5 cm)

PLANT SIZE
8–14 × 2–3 in
(20–36 × 5–8 cm),
including erect, single-
flowered inflorescence
4–10 in (10–25 cm) tall

257

The widely distributed Brown Friars is variable in form, color, and habitat preferences. Consistent are its small, oblong, and moderately flattened pseudobulbs, with large sheathing basal bracts and a single, narrow, folded apical leaf. The plants form between one and five inflorescences per pseudobulb, each clothed with a series of sheathing bracts. The common name refers to the generally dull color of the flowers, usually in shades of brown to nearly black, and their hooded shape, like the cowl of a Franciscan brother. The species name, from the Latin for "hoodlike," also points to the clothing of a friar.

The fragrant flowers are likely targeting bees for pollination, but they also appear to be engaging in deception. The shiny callus on the lip may suggest nectar, but there are no glands on the lips of this species, so no reward is being provided to potential bee visitors.

The flower of Brown Friars has long, greenish-yellow to reddish-brown striped sepals and shorter petals, the latter often covering the column along with the dorsal sepal. The lip is usually almost black or dark brown with a shiny callus near its base.

Actual size

SUBFAMILY	Epidendroideae
TRIBE AND SUBTRIBE	Cymbidieae, Maxillariinae
NATIVE RANGE	Mesoamerica and northwestern South America, from southern Mexico to Ecuador and Guyana
HABITAT	Lowland and other forests, at up to 3,300 ft (1,000 m)
TYPE AND PLACEMENT	Epiphytic
CONSERVATION STATUS	Not assessed, but widespread and unlikely to be threatened
FLOWERING TIME	Throughout the year

FLOWER SIZE
1½ in (4 cm)

PLANT SIZE
10–18 × 3–6 in
(25–46 × 8–15 cm),
excluding single-flowered
inflorescence, which is upright
and about as long as the leaves

MAXILLARIA EGERTONIANUM
DRAGON'S MOUTH ORCHID
BATEMAN EX LINDLEY, 1838

258

The pseudobulbs of the Dragon's Mouth Orchid are clustered, ribbed, and topped with a pair of glossy linear-lanceolate folded leaves. From the base of each leaf, several stems arise, each with a single tubular flower. The flaring sepals give the flower a lily-like appearance. The much shorter petals are tipped with neon blue and resemble two eyes. Its former genus name, *Trigonidium*, comes from the Greek word *trigonos*, meaning "triangular," which refers to the three-sided nature of the flowers. It is now considered part of *Maxillaria*.

This is one of the few Neotropical orchids pollinated by sexual deception, whereby male bees are tricked into thinking the flower is a female. While trying to dislodge the "female" to mate with it, the male falls into the central floral cavity, where it struggles to get out and picks up or deposits pollinia.

The flower of the Dragon's Mouth Orchid has three prominent, recurved, brownish-yellow sepals that have darker veins and form a triangular cup. The shorter petals are tipped with dark neon-blue "eyes." The lip is hidden within the floral cup.

Actual size

SUBFAMILY	Epidendroideae
TRIBE AND SUBTRIBE	Cymbidieae, Maxillariinae
NATIVE RANGE	Colombia
HABITAT	Steep slopes and in trees of cloud forests, at 2,625–9,850 ft (800–3,000 m)
TYPE AND PLACEMENT	Epiphytic or lithophytic
CONSERVATION STATUS	Not assessed
FLOWERING TIME	Late spring to early summer

MAXILLARIA GRANDIFLORA
ELEGANT JAW ORCHID
(KUNTH) LINDLEY, 1832

FLOWER SIZE
Up to 4 in (10 cm)

PLANT SIZE
12–20 × 3–6 in
(30–51 × 8–15 cm),
including single-flowered
erect inflorescence
4–8 in (10–20 cm) tall,
which is shorter than the leaves

259

The Elegant Jaw Orchid produces a cluster of ovoid, compressed, pseudobulbs that are enveloped in dry sheaths and topped with a single, thin folded leaf. From the base of a mature pseudobulb a single flowered inflorescence arises. The flower stem is covered with loose, overlapping bracts, and the long-lasting flower is scented during the afternoon.

Female *Eulaema cingulata* bees in search of food visit this orchid, landing on the lip, which causes the flower to open. After collecting wax from the callus, the insect backs out and in doing so the pollinarium is stuck to its head or eye. The bee flies off but is unbalanced at first due to the changed distribution of weight. After resting, it visits another flower and so effects pollination.

The flower of the Elegant Jaw Orchid has fused, recurved sepals, the laterals fused to the column foot at the base, the upper sepal hooded. The petals are smaller and point forward. The lip forms a tube for the column and has a ruffled, recurved limb.

Actual size

SUBFAMILY	Epidendroideae
TRIBE AND SUBTRIBE	Cymbidieae, Maxillariinae
NATIVE RANGE	Tropical South America, from Colombia to southern Brazil
HABITAT	Wet forests, at 3,300–6,600 ft (1,000–2,000 m)
TYPE AND PLACEMENT	Epiphytic on mossy trunks
CONSERVATION STATUS	Not assessed, but the species is widely distributed and unlikely to be of conservation concern
FLOWERING TIME	March (spring)

FLOWER SIZE
¾ in (2 cm)

PLANT SIZE
6–10 × 4–5 in
(15–25 × 10–13 cm),
including inflorescence
4–6 in (10–15 cm) long, which
is shorter than the leaves

MAXILLARIA NOTYLIOGLOSSA
RESIN ORCHID
REICHENBACH FILS, 1854

Actual size

The flower of the Resin Orchid has spreading, yellowish-green to green lateral sepals and an arching upper sepal that forms a hood over the forward-pointing petals and column. The lip is spade-shaped, with two lateral ears and a V-shaped, resinous callus.

A small epiphyte, the Resin Orchid has a scrambling rhizome covered in bracts, on which ovoid, flattened pseudobulbs are carried, well separated from each other. At the tip of the pseudobulbs there are two strap-shaped leaves with a notched tip. An upright, single-flowered inflorescence forms from the base of a mature pseudobulb, with the flower stalk covered in overlapping, compressed bracts.

The flowers are fragrant and have a white V-shaped area on their lip that produces resin, hence the genus name in which this species was previously placed, *Rhetinantha* (from the Greek words *rhetinos*, "resin," and *anthos*, "flower"). Female bees collect this resin and use it for nest construction. They receive the pollinarium on their head when they back out of the flower and deposit it in the stigma of the next flower visited.

260

SUBFAMILY	Epidendroideae
TRIBE AND SUBTRIBE	Cymbidieae, Maxillariinae
NATIVE RANGE	Colombia and northwestern Venezuela
HABITAT	Cloud forests, at 6,600–8,200 ft (2,000–2,500 m)
TYPE AND PLACEMENT	Epiphytic
CONSERVATION STATUS	Not assessed
FLOWERING TIME	September to October (fall)

MAXILLARIA OAKES-AMESIANA
AMES' FIR ORCHID
SCHUITEMAN & M. W. CHASE, 2015

FLOWER SIZE
⅛ in (0.5 cm)

PLANT SIZE
4–10 × 1 in
(10–25 × 2.5 cm),
including short,
single-flowered inflorescence
⅛ in (0.5 cm) long

261

The small but delightful Ames' Fir Orchid does not look like an orchid. It has pendulous, simple or branched rhizomes that resemble little chains beset with conical, grooved, silver-spotted, glossy dark green pseudobulbs topped with a tuft of needlelike leaves, resembling those of a fir or larch. A single flower emerges on a bract-covered stalk. Pollination has not been studied, but most species of *Maxillaria* are bee-pollinated. The species name is in honor of the American orchidologist Oakes Ames (1874–1950), who built up the orchid herbarium at Harvard University.

Until recently, *M. oakes-amesiana* was placed in the genus *Pityphyllum* (Greek pitys, for "pine-like," and *phyllo*, "leaf"). *Maxillaria* has been expanded to include this closely related genus that has flowers like a small *Maxillaria* but diverges in its highly unusual habit.

Actual size

The flower of Ames' Fir Orchid has forward-pointing, white to cream flower parts. The sepals recurve at the tip, and the lip and petals form a tube around the column. There is no nectar spur, and sometimes the lip has pink spots.

SUBFAMILY	Epidendroideae
TRIBE AND SUBTRIBE	Cymbidieae, Maxillariinae
NATIVE RANGE	Southeastern and southern Brazil to Argentina (Misiones province) and Paraguay
HABITAT	Tropical forests, at 1,640–3,300 ft (500–1,000 m)
TYPE AND PLACEMENT	Epiphytic and terrestrial
CONSERVATION STATUS	Not threatened
FLOWERING TIME	Winter and spring

FLOWER SIZE
2 in (5 cm)

PLANT SIZE
10–18 × 8–12 in
(25–46 × 20–30 cm),
including single-flowered
inflorescences
6–12 in (15–30 cm) long, which
are shorter than the leaves

262

MAXILLARIA PICTA
INSIDE-OUT
SPOTTED ORCHID
HOOKER, 1832

A vigorous clump-forming species, the Inside-out Spotted Orchid has globose, often longitudinally ridged pseudobulbs, each with a pair of terminal leaves. Plants are often festooned with these showy blossoms, which strangely bear their spots on the outside of the petals and sepals rather than on the inside like most orchid flowers. This gives rise to the orchid's common name and species name, the Latin *picta*, "painted," referring to the spotting.

The species is pollinated by stingless trigonid bees (genus *Trigona*)—called *abelha-cachorro*, "dog bees," in Brazilian Portuguese. The bees are looking for honey, but, sadly, the wonderfully fragrant flowers produce no reward for their pollinators, which soon learn to avoid the attractive blooms. However, enough visits occur to guarantee a supply of seeds to produce the next generation of *Maxillaria picta*.

The flower of the Inside-out Spotted Orchid
has yellow to orange sepals and petals, with a white backside blotched heavily with cinnamon brown. These incurved segments encircle the white lip, which has maroon stripes/spots. The column is dark maroon red.

Actual size

SUBFAMILY	Epidendroideae
TRIBE AND SUBTRIBE	Cymbideae, Maxillariinae
NATIVE RANGE	Mexico, Belize, Guatemala, El Salvador, Honduras, Nicaragua, Costa Rica, and Panama
HABITAT	Wet forests, at 650–3,950 ft (200–1,200 m)
TYPE AND PLACEMENT	Epiphytic
CONSERVATION STATUS	Not threatened
FLOWERING TIME	Throughout year, but more frequently in summer to fall

FLOWER SIZE
2 in (5 cm)

PLANT SIZE
10–18 × 3–5 in
(25–46 × 8–13 cm),
including short,
single-flowered erect
2–4 in (5–10 cm) tall
inflorescences

MAXILLARIA RINGENS
BLACK-TONGUE ORCHID
(LINDLEY) GENTIL, 1907

263

The Black-tongue Orchid forms dense clusters of pseudobulbs, each bearing one grasslike leaf. A long wiry stem emerges from along the stem underneath the pseudobulb, carrying a single bizarre flower with an almost black lip—hence the common name. The former genus name, *Mormolyca*, derives from the Greek *mormo*, "goblin," and *lykos*, "wolf," again alluding to the unusual appearance of the flowers. The species are closely related to *Maxillaria*, in which genus they are now placed.

The flowers of *M. ringens* engage in sexual deceit, their moveable lips appearing to be females to naïve male bees that attempt copulation with the strange, seductive blooms. In this case the pollinators are males of *Nannotrigona testaceicornis* and a species of *Scaptotrigona*—both stingless meliponine bees. The males are first attracted by the scent, which excites them even before they land on the lip.

The flower of the Black-tongue Orchid has tawny to yellowish petals and sepals with dark red-brown stripes. The darkly colored lip, lying just underneath the prominent column, is insect-like, complete with small, winglike sidelobes and hairs.

Actual size

SUBFAMILY	Epidendroideae
TRIBE AND SUBTRIBE	Cymbidieae, Maxillariinae
NATIVE RANGE	Mexico, Guatemala, El Salvador, Honduras, Nicaragua, and Costa Rica
HABITAT	Wet tropical forests, at 1,970–4,920 ft (600–1,500 m)
TYPE AND PLACEMENT	Epiphytic, rarely terrestrial on slopes or embankments
CONSERVATION STATUS	Not threatened
FLOWERING TIME	Mostly in spring and summer

FLOWER SIZE
2 in (5 cm)

PLANT SIZE
10–18 × 3–5 in
(25–46 × 8–13 cm),
including single-flowered,
erect 2–4 in (5–10 cm) tall
inflorescences

264

MAXILLARIA TENUIFOLIA
COCONUT ORCHID
LINDLEY, 1837

Due to its delicious coconut fragrance and hardy nature, the Coconut Orchid is widely cultivated. It is a vigorous species, native to a wide range and tolerant of climatic conditions through much of Central America. It forms large clumps on tree trunks and major branches, with many small bulbs bearing one long, narrow apical leaf. The genus name is from the Latin word for "jaw," *maxilla*, a reference to the "chin" formed by the bases of the lip and column. The species name, from the Latin for "slender-leaved," refers to the narrow, grasslike leaves.

The fragrant blooms arise on short stems from the bases of the newest pseudobulbs. *Maxillaria tenuifolia* uses fragrance to lure bees, which in this case are meliponine bees (stingless bees). The pollinaria are attached to the scutellum (forward portion of the thorax), and in this case there is no reward for the pollinator.

Actual size

The flower of the Coconut Orchid is pale yellow, heavily overlaid with brown to maroon red blotches that coalesce into a solid red color on the tips. The lip is buff yellow with maroon red spotting. The column is creamy white.

SUBFAMILY	Epidendroideae
TRIBE AND SUBTRIBE	Cymbidieae, Maxillariinae
NATIVE RANGE	Widespread through tropical South America
HABITAT	Rain forests, from sea level to 4,920 ft (1,500 m)
TYPE AND PLACEMENT	Epiphytic
CONSERVATION STATUS	Not threatened
FLOWERING TIME	Throughout the year

MAXILLARIA UNCATA
CANDY-STRIPE ORCHID
LINDLEY, 1837

FLOWER SIZE
¼ in (2 cm)

PLANT SIZE
2–6 × 2–3 in
(5–15 × 5–8 cm),
including single-flowered
inflorescence
½–1 in (1.3–2.5 cm) long

265

This miniature orchid, despite less than remarkable flowers, is nontheless attractive, with compact and glossy clumps of foliage and ever-blooming habit. It was considered a member of *Christensonella*, a group of cluster-forming plants often referred to as "pincushions" due to their narrow and often needlelike leaves, but was recently re-included in the large, highly diverse genus *Maxillaria*. *Maxillaria uncata* has fleshy succulent leaves.

This orchid occurs in ever-wet forests, and there are two forms of the species, which can occur even in a single population. One has short stubby leaves, and the other is much more vigorous, with longer and thicker foliage. The common name refers to the white and red-striped flowers, although totally white to cream forms are also known. The genus *Christensonella* was named for the orchidologist Eric Christenson (1956–2011), who worked extensively on this group of orchids.

Actual size

The flower of the Candy-stripe Orchid is translucent white to cream, usually with red stripes. It has a contrasting stark white, hook-shaped lip.

SUBFAMILY	Epidendroideae
TRIBE AND SUBTRIBE	Cymbidieae, Maxillariinae
NATIVE RANGE	Panama, Colombia, and Ecuador
HABITAT	Wet forests, at 1,640–3,300 ft (500–1,000 m)
TYPE AND PLACEMENT	Terrestrial, occasionally epiphytic
CONSERVATION STATUS	Rare in the wild
FLOWERING TIME	April to May (spring)

FLOWER SIZE
2½ in (6.5 cm)

PLANT SIZE
30–50 × 14–25 in
(76–127 × 36–64 cm),
including erect
inflorescence
28–40 in (71–102 cm)

266

NEOMOOREA WALLISII
SHARP-TONGUED CHESTNUT ORCHID
(REICHENBACH FILS) SCHLECHTER, 1924

The Sharp-tongued Chestnut Orchid, the only species in its genus, has large round, lightly grooved pseudobulbs that bear stalked broad leaves with prominent veins and several large subtending bracts. The chestnut brown flowers, 8–20 or more per inflorescence, are held on upright racemes that emerge from the base of the pseudobulbs. The genus was named for Sir Frederick William Moore (1857–1949), who was superintendent of the National Botanic Gardens at Glasnevin, Dublin, and had bought and cultivated this orchid. An earlier genus name, *Moorea*, already existed, so the Greek *neos*, "new," was added.

Deliciously fragrant of jasmine and lemon, the flowers are likely visited by male euglossine bees that collect the complex aroma. No formal studies of its pollination, however, have been undertaken in the wild.

The flower of the Sharp-tongued Chestnut Orchid has dusky orange outspread segments, pure white bases, and a three-lobed lip. The winglike sidelobes are streaked with chocolate, while the central lobe is yellow and pointed, with a large yellow and red-spotted callus.

Actual size

SUBFAMILY	Epidendroideae
TRIBE AND SUBTRIBE	Cymbidieae, Maxillariinae
NATIVE RANGE	Northern South America to the Guianas and Trinidad
HABITAT	Tropical wet forests, at 330–1,970 ft (100–600 m)
TYPE AND PLACEMENT	Epiphytic on tree trunks and large branches
CONSERVATION STATUS	Not assessed but widespread, so not likely endangered
FLOWERING TIME	September to October (fall)

FLOWER SIZE
1⅜ in (3.4 cm)

PLANT SIZE
10–18 × 4–5 in
(25–46 × 10–13 cm),
excluding arching
inflorescence, which is much
longer than the leaves,
at 6–10 in (15–25 cm) long

267

RUDOLFIELLA AURANTIACA
ORANGE SCHLECHTER ORCHID
(LINDLEY) HOEHNE, 1949

The pseudobulbs of the Orange Schlechter Orchid are almost round but slightly compressed with purple-brown spots. Each pseudobulb carries a single, plicate leaf that is occasionally also spotted. From the base of the pseudobulb an inflorescence with two dark-gray bracts emerges, holding up to 15 flowers above the leaves. It has been suggested that pollinators (probably bees) feed from the putatively nutritious cells on the lip callus. There is no nectar reward in these flowers.

The species was first described by the pioneer orchidologist John Lindley (1799–1865), after which it was moved to *Lindleyella* in honor of him. It is now placed in *Rudolfiella*, a genus named for the German botanist Rudolf Schlechter (1872–1925), perhaps the leading orchidologist of the early twentieth century. Both species and common names also reference the orchid's golden, orange-yellow color.

The flowers of the Orange Schlechter Orchid have spreading yellow, red-dotted to blotched sepals and petals, the former fused to a column foot, to which the hinged lip is attached. The column has an irregularly lobed edge, and the lip has two lateral lobes.

Actual size

SUBFAMILY	Epidendroideae
TRIBE AND SUBTRIBE	Cymbidieae, Maxillariinae
NATIVE RANGE	Minas Gerais (Brazil)
HABITAT	Rocky outcrops, at 5,740–6,430 ft (1,750–1,960 m)
TYPE AND PLACEMENT	Lithophytic
CONSERVATION STATUS	Not assessed
FLOWERING TIME	December to January (summer)

FLOWER SIZE
3 in (7.5 cm)

PLANT SIZE
8–12 × 1 in
(20–30 × 2.5 cm),
excluding single-flowered
inflorescence,
8–15 in (20–38 cm) tall

268

SCUTICARIA IRWINIANA
MORAY EEL ORCHID

PABST, 1973

The creeping stem of the Moray Eel Orchid is attached to exposed rocks by short fleshy roots. The stem is covered with upright, short bracts that each envelop the base of a thick, terete, succulent leaf. A lateral, upright inflorescence emerges in the summer carrying a single, long-lived flower that is most fragrant in the morning. The lip has nutritive hairs that are collected by male and female euglossine bees.

Scuticaria irwiniana hosts a fungus, *Fusarium oxysporum*, in its stems and leaves. The fungus is antimicrobial, which protects the orchids against harmful infections, although some other strains of *Fusarium* are highly detrimental to some crops, such as banana plantations. This is the only *Scuticaria* species that grows on rocks, and the only one with erect leaves. Other species have pendent, whiplike leaves, similar in form to the moray eel referred to in the common name.

The flower of the Moray Eel Orchid has glossy-red, spreading sepals and similar but smaller, forward-pointing petals. The lip is white with red veins and upcurved at the edges to form a tube with the winged column.

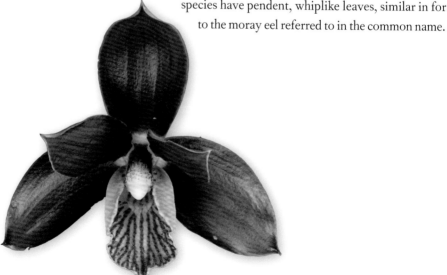

Actual size

SUBFAMILY	Epidendroideae
TRIBE AND SUBTRIBE	Cymbidieae, Maxillariinae
NATIVE RANGE	Southeastern Ecuador to northern Peru
HABITAT	Rain forests, at 3,120–5,900 ft (950–1,800 m)
TYPE AND PLACEMENT	Epiphytic, often on the undersides of branches
CONSERVATION STATUS	Not assessed
FLOWERING TIME	July to October (winter to spring)

FLOWER SIZE
3 in (7.5 cm)

PLANT SIZE
8–14 × 8–12 in
(20–36 × 20–30 cm),
including single-flowered
pendent inflorescences, which
are shorter than the leaves

SUDAMERLYCASTE DYERIANA
GREEN LANTERN
(SANDER EX MASTERS) ARCHILA, 2002

269

The Green Lantern often hangs under branches, where its bluish-green, almost circular, flattened pseudobulbs produce conduplicate, narrowly lanceolate leaves. From the base of the pseudobulbs, several inflorescences emerge, largely covered by inflated bracts. The flowers are fragrant and long-lived and face downward. The genus name comes from *Sudamerica*, Spanish for South America, and *Lycaste*, the name of the genus from which these species were segregated, referring to the fact that, unlike *Lycaste*, this set of species is found exclusively in South America.

Night-flying bees are speculated to be the pollinators of *Sudamerlycaste dyeriana*. Green and white flowers are nearly always associated with insects active at night, and the flowers do not have the long, narrow nectar spur that is typical of flowers pollinated by moths.

The flower of the Green Lantern has leathery, green to bluish-green petals and sepals forming a cup, the petals somewhat shorter than the sepals. The lip, also green, is fimbriate along the edge and has two upright sidelobes and a saddle-shaped callus.

Actual size

SUBFAMILY	Epidendroideae
TRIBE AND SUBTRIBE	Cymbidieae, Maxillariinae
NATIVE RANGE	Ecuador
HABITAT	Cloud forests, at 3,300–4,920 ft (1,000–1,500 m)
TYPE AND PLACEMENT	Epiphytic
CONSERVATION STATUS	Not assessed
FLOWERING TIME	March to May (spring)

FLOWER SIZE
½ in (1.3 cm)

PLANT SIZE
8–14 × 2–3 in
(20–36 × 5–8 cm),
including single-flowered
inflorescences 2–4 in
(5–10 cm) long, which are
much shorter than the leaves

270

TEUSCHERIA CORNUCOPIA
ECUADOREAN CORNUCOPIA ORCHID
GARAY, 1958

Actual size

The Ecuadorean Cornucopia Orchid has a long rhizome with widely spaced, rounded pseudobulbs, each carrying a single, narrowly lanceolate leaf basally enveloped by papery sheaths. It has an inflorescence with a single, upside-down, spurred flower. The genus is named for the orchidologist and landscape architect Henry Teuscher (1891–1984), who was one of the founders of the Montreal Botanical Garden and its first curator. The species and common names refer to the resemblance of the long tubular flower to a horn of plenty or cornucopia—in Greek mythology a broken goat's horn filled with whatever food its owner desired.

Pollination of this species has not been studied. However, its floral morphology, with a broadly opening nectar spur, is consistent with pollination by a bee that receives the pollinia on its scutellum.

The flower of the Ecuadorean Cornucopia Orchid has three forward-pointing, pale pink to white sepals. The similarly colored and sized petals are recurved, but the trilobed lip is shorter than the sepals and the midlobe is shorter than the lateral two, which often have some pink striping.

SUBFAMILY	Epidendroideae
TRIBE AND SUBTRIBE	Cymbidieae, Maxillariinae
NATIVE RANGE	Northwestern Venezuela to northern Peru
HABITAT	Cloud forests, at 6,600–9,850 ft (2,000–3,000 m)
TYPE AND PLACEMENT	Epiphytic on mossy trunks; sometimes terrestrial on mossy banks
CONSERVATION STATUS	Not assessed
FLOWERING TIME	December to February (winter)

XYLOBIUM LEONTOGLOSSUM
LION'S MOUTH
(REICHENBACH FILS) BENTHAM EX ROLFE, 1889

FLOWER SIZE
1¼ in (3 cm)

PLANT SIZE
10–18 × 4–6 in
(25–46 × 10–15 cm),
excluding inflorescence,
which is arching-erect and
shorter than the leaves,
6–12 in (15–30 cm) long

271

The Lion's Mouth forms a cluster of ovoid pseudobulbs, each carrying a single, plicate, long-stemmed, elliptic to lanceolate leaf at its apex. A lateral arching-erect, bracteate raceme is produced with 10–30 loosely carried flowers. The genus name is derived from the Greek *xylon* for "wood," and *bios*, "living," referring to the epiphytic nature of the member species, whereas the common name is based on the species name, *leo*, meaning "lion," and *glossum*, "tongue."

It has been observed that some *Xylobium* species are pollinated by stingless bees of the genus *Trigona*, but there are no observations for *X. leontoglossum*. No obvious reward is offered, and there is no spur, although the bases of the lateral sepals and lip do form an empty cavity. This species is also highly fragrant.

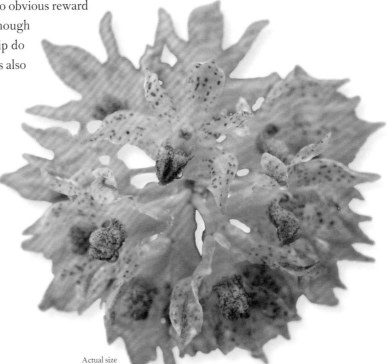

The flower of the Lion's Mouth has red-spotted cream sepals that are spreading at their tips. Petals are strap-shaped and widely spread. The lip has a central ridged callus and is trilobed, with the two lateral lobes clasping the short column.

Actual size

SUBFAMILY	Epidendroideae
TRIBE AND SUBTRIBE	Cymbidieae, Oncidiinae
NATIVE RANGE	Southeastern Brazil
HABITAT	Transitional areas of dense tropical forests to more open areas, at 650–2,300 ft (200–700 m)
TYPE AND PLACEMENT	Epiphytic
CONSERVATION STATUS	Not threatened
FLOWERING TIME	Spring

FLOWER SIZE
2¾ in (7 cm)

PLANT SIZE
5–10 × 3–5 in
(13–25 × 8–13 cm),
including erect to arching
single-flowered inflorescence
3–8 in (8–20 cm) long

272

ASPASIA LUNATA
CRESCENT MOON
LINDLEY, 1836

The Crescent Moon, a member of a genus closely related to *Brassia* and *Miltonia*, can form large localized populations and yet remain fairly rarely encountered. Primarily epiphytic, the species never receives full sunlight in nature. The genus is probably named for Aspasia, the lover and partner of the Athenian statesman Pericles, although Lindley was not clear about this when he originally described *Aspasia* in 1836. Also unclear are the reasons for the moon references in the species and common names.

Although pollination has not so far been observed for this species, another species in the genus, *A. principissa*, from Panama, was recorded as being pollinated by euglossine bees. Both male and female bees are attracted to the colorful fragrant flowers, seeking nectar in the (empty) cavity formed by the base of the lip and column.

The flower of the Crescent Moon has lanceolate, olive-green sepals and petals that are overlaid with mahogany spots and blotches. The flat, broad, white lip is partly fused to the column. The midlobe is stained purple to nearly blue and has two white keels leading up to the basal cavity.

Actual size

SUBFAMILY	Epidendroideae
TRIBE AND SUBTRIBE	Cymbidieae, Oncidiinae
NATIVE RANGE	Nicaragua, Costa Rica, Panama, Venezuela, Colombia, Ecuador, and Peru on both sides of the Andes
HABITAT	Tropical rain forests, at 985–4,920 ft (300–1,500 m)
TYPE AND PLACEMENT	Epiphytic, rarely on steep embankments
CONSERVATION STATUS	Not threatened
FLOWERING TIME	Anytime, but more often in summer

BRASSIA ARCUIGERA
ARCHING SPIDER ORCHID
REICHENBACH FILS, 1869

FLOWER SIZE
10 in (25 cm), but in some
plants the long sepals can reach
nearly twice this size

PLANT SIZE
12–20 × 2–3 in
(30–51 × 5–8 cm), excluding
arching inflorescence
20–28 in (51–71 cm) long

273

The flowers of the genus *Brassia* – and especially this species – have extremely long narrow sepals and petals that are certainly evocative of large spiders, hence the common name. Sturdy and vigorous, with stout pseudobulbs and large lanceolate leaves, these orchids can produce prodigious inflorescences bearing 15 or more huge blooms arranged alternately on large arching racemes (the species name reference). The genus was named for William Brass, an eighteenth-century illustrator and collector of African plants.

Female wasps, said to be attempting to sting the lip callus, have been observed as pollinators of *Brassia* flowers. These wasps paralyze spiders by first stinging and then laying eggs on them, leaving the larvae to consume the spider as their first meal. To human eyes, however, the lip of *B. arcuigera* does not appear particularly spiderlike.

The flower of the Arching Spider Orchid has long, spidery segments of chartreuse to buff yellow, spotted irregularly with brown. The lip is spade-shaped and paler with an acuminate tip and brown spotting near the basal yellow callus.

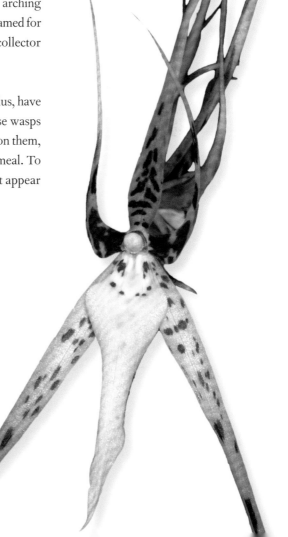

Actual size

SUBFAMILY	Epidendroideae
TRIBE AND SUBTRIBE	Cymbidieae, Oncidiinae
NATIVE RANGE	Venezuela, Colombia, and Ecuador
HABITAT	Cold wet tropical forests, at 6,600–8,250 ft (2,000–2,500 m)
TYPE AND PLACEMENT	Epiphytic
CONSERVATION STATUS	Not threatened
FLOWERING TIME	Late winter to early spring

FLOWER SIZE
1¼ in (3 cm)

PLANT SIZE
8–14 × 8–12 in
(20–36 × 20–30 cm),
excluding arching
inflorescence
12–20 in (30–51 cm) long,
which is longer than the leaves

274

BRASSIA AURANTIACA
ORANGE CLAW
(LINDLEY) M. W. CHASE, 2011

The flowers of the Orange Claw, unlike those of most *Brassia* species, do not open fully and are tightly bunched, with up to 18 flowers per stem. What the plants lack in individual flower form, they more than make up for in their displays of many brightly colored flowers. They grow in high-elevation, wet, cold forests, where the brilliant flowers stand out against the generally dark green background. The plants form clusters of many elongate spindle-shaped pseudobulbs, which can each produce between one and three inflorescences.

The Orange Claw is different from its closest relatives, the spider orchids, which make up the majority of *Brassia* members, and is the only *Brassia* adapted to pollination by hummingbirds. If the flower parts are forced open, then the resulting shape is similar to that of the typical spider orchid, with elongate sepals and petals.

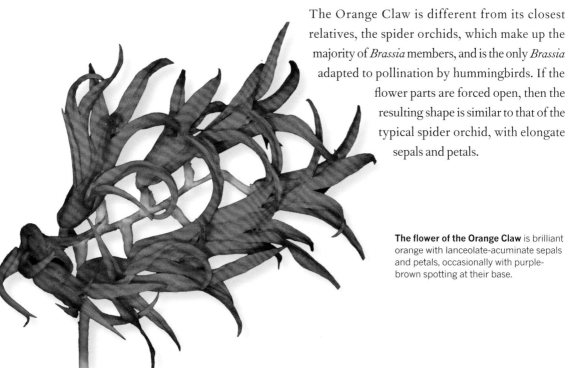

The flower of the Orange Claw is brilliant orange with lanceolate-acuminate sepals and petals, occasionally with purple-brown spotting at their base.

Actual size

SUBFAMILY	Epidendroideae
TRIBE AND SUBTRIBE	Cymbidieae, Oncidiinae
NATIVE RANGE	Napo province in Ecuador, and French Guiana
HABITAT	Low elevation rain forests in deep shade
TYPE AND PLACEMENT	Epiphytic on mossy twigs, often in the canopy
CONSERVATION STATUS	Not assessed, but may be more common than assumed as often overlooked and rarely collected
FLOWERING TIME	Spring

CALUERA VULPINA
PURPLE-SPOTTED LUER ORCHID
DODSON & DETERMANN, 1983

FLOWER SIZE
¼ in (0.6 cm)

PLANT SIZE
1 in (2.5 cm) in diameter,
excluding inflorescence,
which is about as long
as the leaves

275

The Purple-spotted Luer Orchid, and other members of the small genus *Caluera*, have an umbel (cluster) of four to five flowers. Arranged in this way, the flowers, probably pollinated by oil-collecting bees, help distinguish the orchids from the larger, related genus *Ornithocephalus*, widespread across Central and South America. These plants are fan-shaped, with leaves that have their inner surfaces fused, meaning the upper and lower surfaces are identical.

The species is probably more common than the few collections in herbaria might suggest as it is often overlooked, due to its small size. The genus is named for orchid specialist, Carlyle A. Luer (1922–), who was the founder of *Selbyana*, the journal of the Marie Selby Botanical Gardens in Sarasota, Florida, dedicated to the taxonomy of epiphytes, especially orchids.

Actual size

The flower of the Purple-spotted Luer Orchid has spreading, linear sepals and petals, and a lip that is heart-shaped with a cavity in the middle. The column is T-shaped and protrudes forward.

SUBFAMILY	Epidendroideae
TRIBE AND SUBTRIBE	Cymbidieae, Oncidiinae
NATIVE RANGE	Southern Brazil (Espiritu Santo state) to northern Argentina
HABITAT	Smaller branches of trees, among lichens and mosses, from sea level to 1,640 ft (500 m)
TYPE AND PLACEMENT	Epiphytic
CONSERVATION STATUS	Not threatened
FLOWERING TIME	Spring

FLOWER SIZE
⅜ in (1 cm)

PLANT SIZE
1–3 × 1 in
(2.5–8 × 2.5 cm),
excluding pendent
inflorescences
2–4 in (5–10 cm) long

276

CAPANEMIA SUPERFLUA
FLORIFEROUS NEEDLE ORCHID
(REICHENBACH FILS) GARAY, 1967

The Floriferous Needle Orchid is a miniature species with short needlelike leaves that can blanket itself with lovely little flowers. Approximately a dozen or more blooms appear on each of the densely packed, pendent racemes. Its terete leaves are subtended by basal papery bracts, and plants resemble a small species of Spanish Moss (Tillandsiaceae, bromeliads) when out of bloom.

Growing mostly on lichen and moss-encrusted thin branches, this orchid is an epiphyte confined to the outer parts of the canopy. In their native habitats, such plants tend to be short-lived, as the host twigs change their characteristics as they mature, staying wetter longer, making them unsuitable for species of this kind. Pollination of one species in the genus has been recorded. It was visited by a species of *Polybia*, a social wasp that collects nectar, which is produced in a cavity on the base of the lip.

The flower of the Floriferous Needle Orchid is small, usually white or pale pink with a yellow callus on the lip just in front of the nectar cavity. Sepals and petals are similar and incurved.

Actual size

SUBFAMILY	Epidendroideae
TRIBE AND SUBTRIBE	Cymbidieae, Oncidiinae
NATIVE RANGE	Colombia, Ecuador
HABITAT	Wet forests, at 8,900–9,200 ft (2,700–2,800 m)
TYPE AND PLACEMENT	Epiphyte
CONSERVATION STATUS	Restricted distribution and under threat due to overcollection and forest cutting
FLOWERING TIME	Throughout the year

CAUCAEA PHALAENOPSIS

ANDEAN MOTH ORCHID

(LINDEN & REICHENBACH FILS) N. H. WILLIAMS & M. W. CHASE, 2001

FLOWER SIZE
1½ in (3.8 cm)

PLANT SIZE
5–18 × 3–6 in
(13–46 × 8–15 cm),
excluding inflorescence

277

The Andean Moth Orchid has ovoid-conical pseudobulbs from which two linear-oblanceolate leaves grow. The slender, arching raceme can be up to 24 in (60 cm) long and carries up to 15 fragrant flowers. This cool-loving orchid and others of its genus suffer from exposure to warm conditions, resulting in flowers with a different shape and color when grown in cultivation. Some species described from cultivation have never been observed in the wild.

The genus is named after the Cauca River Valley in Colombia, but its species occur in the Andes from Venezuela to Ecuador. Pollination has never been documented, but flower color and shape suggest bees. Previously, these species were included in *Oncidium*, to which they are only distantly related.

Actual size

The flower of the Andean Moth Orchid has white, finely pink-spotted or blotched sepals and petals with slightly wavy edges. The lip is trilobed, with purple-spotted sidelobes and a cleft terminal lobe. The yellow callus is three-horned, and the column is hooded.

SUBFAMILY	Epidendroideae
TRIBE AND SUBTRIBE	Cymbidieae, Oncidiinae
NATIVE RANGE	Southeastern Brazil (Minas Gerais, Rio de Janeiro, São Paulo states)
HABITAT	Mossy forests
TYPE AND PLACEMENT	Epiphyte
CONSERVATION STATUS	Unassessed, but rare and most likely endangered
FLOWERING TIME	April to June (fall)

FLOWER SIZE
⅜ in (0.8 cm)

PLANT SIZE
2–3½ × 1–1½ in
(5–9 × 2.5–3.8 cm),
excluding inflorescence

278

CENTROGLOSSA TRIPOLLINICA
FLYING ANGEL ORCHID

BARBOSA RODRIGUES, 1882

Actual size

This tiny orchid has round, slightly compressed pseudobulbs that are enveloped at the base by one to three leaf-bearing sheaths, with a single leathery, linear leaf at the tip. Roots are thin but covered with short hairs. A thin inflorescence with a few flowers is about as long as the leaves.

The area in which the orchid grows, the Brazilian *Mata Atlantica* (Atlantic Forest), is famous for its biological diversity, and a number of other micro-orchid species grow there. Pollination for the species is unstudied, but most of the genera in this group—the former subtribe Ornithocephalinae—are visited by oil-collecting bees. The long spur of this species is empty, and pollination must be based on deceit of some form.

The flower of the Flying Angel Orchid has three green, spreading sepals and two white, rounded, spreading petals. The lip is white and funnel-shaped. The column is unusual due to the two long arms attached to its base that are almost as long as the column itself.

SUBFAMILY	Epidendroideae
TRIBE AND SUBTRIBE	Cymbidieae, Oncidiinae
NATIVE RANGE	Southeastern Brazil
HABITAT	Atlantic rain forests
TYPE AND PLACEMENT	Epiphytic
CONSERVATION STATUS	Not assessed
FLOWERING TIME	July to August (winter)

CHYTROGLOSSA AURATA
GOLDEN CAVE-LIP ORCHID
REICHENBACH FILS, 1863

FLOWER SIZE
½ in (1.3 cm)

PLANT SIZE
2–4 × 4–6 in
(5–10 × 10–15 cm),
excluding inflorescence up to
6 in (15 cm) long

279

The miniature Golden Cave-lip Orchid has a short stem with a cluster of several slender, keeled leaves. One or two leaves are attached to each small pseudobulb. The pendent, wiry raceme bears up to ten flowers that are yellow with splashes and spots of red. These are supported by amplexicaul (stem-clasping), heart-shaped bracts. Very often, *Chytroglossa aurata* produces more flowers by area than it has leaf tissue. Unless they are in flower, plants of this tiny species go unnoticed by most people.

Nothing is known about pollination, but like most members of this group of genera (formerly in their own subtribe, Ornithocephalinae) it is most likely pollinated by a species of oil-collecting bee. On the base of the lip is a shallow cavity covered with oil-secreting hairs, but generally there is no obvious oil present.

The flower of the Golden Cave-lip Orchid has similar greenish, spreading sepals and petals, all with an erose (irregularly notched) edge. The lip is broadly heart-shaped with a shallow hollow or "cave" in the middle.

Actual size

SUBFAMILY	Epidendroideae
TRIBE AND SUBTRIBE	Cymbidieae, Oncidiinae
NATIVE RANGE	Costa Rica, Panama, and Colombia
HABITAT	Tropical wet forests, at 1,640–3,300 ft (500–1,000 m)
TYPE AND PLACEMENT	Epiphytic
CONSERVATION STATUS	Not threatened
FLOWERING TIME	Spring

FLOWER SIZE
1 in (2.5 cm)

PLANT SIZE
5–8 × 3–4 in
(13–20 × 8–10 cm),
including arching-pendent
inflorescence
2–4 in (5–10 cm) long

280

CISCHWEINFIA PUSILLA
SCHWEINFURTH'S ORCHID
(C. SCHWEINFURTH) DRESSLER & N. H. WILLIAMS, 1970

Actual size

The flower of Schweinfurth's Orchid has
chestnut-brown tepals, usually tipped with
butter yellow. The lip is white with a yellow
to brown-blotched throat and yellow to
brown lip callus. The column has a hood
covering its apex.

This species forms dense clusters of pseudobulbs, each topped
by one leaf (rarely two leaves) with perhaps between one and
three leaf-bearing bracts at the base of each flattened pseudobulb.
From the base of a pseudobulb, one or two inflorescences arch
gracefully among the leaves. The genus was named in honor
of Charles I. Schweinfurth (1890–1970), the orchid curator at
the Oakes Ames Orchid Herbarium (Harvard University). The
genus *Schweinfurthia* already had been named in the unrelated
family Plantaginaceae, so a hybrid of his initials (CI) and part
of his surname was used instead.

Though the blossoms of Schweinfurth's Orchid emit no
discernable fragrance and produce no nectar, male and
female euglossine bees have been observed pollinating
them. These bees are searching for nectar, so the orchid
is deceiving them by appearing to have the sort of
flower that offers a reward for pollination.

SUBFAMILY	Epidendroideae
TRIBE AND SUBTRIBE	Cymbidieae, Oncidiinae
NATIVE RANGE	Colombia
HABITAT	Shrubs near watercourses, often on cultivated guava or citrus, at 3,950–5,900 ft (1,200–1,800 m)
TYPE AND PLACEMENT	Epiphytic
CONSERVATION STATUS	Not threatened
FLOWERING TIME	Winter

FLOWER SIZE
2–2⅜ in (5–6 cm)

PLANT SIZE
4–6 × 1 in (10–15 × 2.5 cm),
excluding arching-pendent,
sometimes sparsely
branched inflorescence
8–15 in (20–38 cm) long

281

COMPARETTIA MACROPLECTRON
SPOTTED BUTTERFLY ORCHID
REICHENBACH FILS & TRIANA, 1878

Even though most of the species of the genus *Comparettia* are miniature epiphytes, these bird and butterfly-pollinated orchids have brilliant, eye-catching coloration. The Spotted Butterfly Orchid has long, graceful racemes of showy pink flowers with lovely purple-pink spots. It is named for its lengthy nectar spur, from the Greek words for "long" (macro) and "spur" (*plectron*).

Comparettia species offer a nectar reward to their pollinators, butterflies. The nectar, though, is not of good quality, with low sugar content, and therefore butterflies tend to visit these flowers sparingly and sporadically since this takes more energy than they receive in return. Now found mostly on twigs of cultivated (often abandoned) fruit trees, the natural habitat of *Comparettia* species may actually be in tall forest trees. If, however, these trees are removed, the orchids move on to the cultivated trees that replace them.

The flower of the Spotted Butterfly Orchid is generally various shades of pink, with a peppering of purple to dark pink spots. The petals and sepals are somewhat reduced, with a broad and round labellum and a long, backward-projecting spur.

Actual size

SUBFAMILY	Epidendroideae
TRIBE AND SUBTRIBE	Cymbidieae, Oncidiinae
NATIVE RANGE	Venezuela and Colombia
HABITAT	Moderately wet forests, at 3,300–8,200 ft (1,000–2,500 m)
TYPE AND PLACEMENT	Epiphytic
CONSERVATION STATUS	Not threatened
FLOWERING TIME	April to May (spring)

FLOWER SIZE
⅛ in (0.5 cm)

PLANT SIZE
4–6 × 2–3 in
(10–15 × 5–8 cm),
excluding lateral, arching
to pendent inflorescence
3–6 in (8–15 cm) long

282

COMPARETTIA OTTONIS
YELLOW SCOOP ORCHID

(KLOTZSCH) M. W. CHASE & N. H. WILLIAMS, 2008

The Yellow Scoop Orchid produces small, cylindrical pseudobulbs, almost covered by large bracts, with a single, broadly lanceolate leaf on top. The inflorescence can bear up to 15 flowers. The common name refers to the shape of the blooms, which resemble a scoop used for flour or sugar. The species was formerly placed in the genus *Scelochilus*, which is named from the Greek words for *skelos*, "leg," and *kheilos*, "lip," referring to the two long, leg-like appendages on the base of the lip.

The color, sweet fragrance, and shape of the flowers, with their short, backward-projecting spur, are compatible with pollination by a small bee (although this has not been observed, so far). The blooms do not open widely, but they do open enough to admit a small insect seeking the nectar produced by the pair of short horns in the spur.

The flower of the Yellow Scoop Orchid has narrowly lanceolate, bright yellow sepals and petals that form a narrow tube surrounding the lip and are marked by reddish-purple stripes inside. The lip is also yellow, with reddish marking, and is slightly longer than the tube.

Actual size

SUBFAMILY	Epidendroideae
TRIBE AND SUBTRIBE	Cymbidieae, Oncidiinae
NATIVE RANGE	West and central Mexico, in the states of Jalisco, Michoacan, and Sinaloa
HABITAT	Pine-oak forests, at 4,600–7,200 ft (1,400–2,200 m)
TYPE AND PLACEMENT	Epiphytic
CONSERVATION STATUS	Endangered as a result of overcollecting and habitat loss
FLOWERING TIME	May to October (late spring to fall)

CUITLAUZINA PENDULA

AZTEC KING

LEXARZA, 1825

FLOWER SIZE
2 in (5 cm)

PLANT SIZE
8–12 × 5–10 in
(20–30 × 13–25 cm),
excluding pendent
inflorescences
10–30 in (25–76 cm) long

283

The Aztec King has tightly clustered ovoid, laterally flattened pseudobulbs, each topped with two broad, leathery leaves. In the axils of a new growth, before the pseudobulb itself emerges, a pendent raceme carries 6–20, waxy, long-lived, lemon-scented flowers. In its natural habitat, the species can form large clumps that make an impressively beautiful display of pendent inflorescences.

The handsome genus is named for the Aztec king Cuitlahuatzin, or Cuitlahuac, (1476–1520), brother of Montezuma and a noted designer of early public gardens in Mexico; the common name similarly references the king. Although nothing has been published on the plant's pollination, the floral morphology—despite the white to pink flowers—is the same as many yellow-flowered *Oncidium* species common in Mexico. These are pollinated by bees seeking floral oil to mix with pollen as food for their young, which may also occur in the Aztec King.

The flower of the Aztec King has white, broad petals and sepals that are sometimes pink or pink-suffused. The petals are shortly clawed, and the usually pink lip is narrow at its yellow and red-spotted base with two broad, apical lobes. The column has wings and a hood.

Actual size

SUBFAMILY	Epidendroideae
TRIBE AND SUBTRIBE	Cymbidieae, Oncidiinae
NATIVE RANGE	Northwestern South America (Colombia, Ecuador, northern Peru)
HABITAT	Cloud forests, up to 9,850 ft (3,000 m)
TYPE AND PLACEMENT	Epiphytic
CONSERVATION STATUS	Not assessed
FLOWERING TIME	Throughout the year, but mainly from fall to spring

284

FLOWER SIZE
4 in (10 cm)

PLANT SIZE
12–20 × 10–20 in
(30–51 × 25–51 cm),
excluding lateral, usually
vinelike inflorescence,
which can be
30–140 in (76–356 cm) long

CYRTOCHILUM MACRANTHUM
GOLDEN CLOUD ORCHID
(LINDLEY) KRAENZLIN, 1917

The spectacular Golden Cloud Orchid produces clumps of conical pseudobulbs that are enveloped in several leaf-bearing sheaths and topped with two linear-oblong leaves. A lateral, stout, branched inflorescence rambles through the surrounding vegetation, each side branch carrying up to five large flowers. Although the genus name comes from the Greek words for "curved lip," *cyrtos* and *cheilos*, this characteristic is not true of this particular species.

The orchid is pollinated by *Centris* bees, which normally collect floral oils that are mixed with pollen from other plant species (not orchids) and fed to their young. However, the Golden Cloud Orchid does not produce floral oils, so this is a case of deceit. The 140 or so species of *Cyrtochilum* are distinguished from *Oncidium*, where they used to be placed, by their often elongate rhizomes, pseudobulbs that are round in cross section rather than flattened, and their rambling or twining inflorescences.

The flower of the Golden Cloud Orchid has large, yellowish-brown or yellow, spreading, clawed sepals and petals. The yellow lip is triangular, with its lateral tips blood-red to purple, and bears an intricate series of horns and protuberances. The column has two wings.

Actual size

SUBFAMILY	Epidendroideae
TRIBE AND SUBTRIBE	Cymbidieae, Oncidiinae
NATIVE RANGE	Northwestern South America, Lesser Antilles, Puerto Rico
HABITAT	Humid forests, at 1,970–6,600 ft (600–2,000 m)
TYPE AND PLACEMENT	Epiphytic
CONSERVATION STATUS	Not assessed
FLOWERING TIME	December to March (late winter to early spring)

CYRTOCHILUM MEIRAX
ENCHANTED DANCING LADY
(REICHENBACH FILS) DALSTRÖM, 2000

FLOWER SIZE
⅝ in (1.5 cm)

PLANT SIZE
4–10 × 3–6 in
(10–25 × 8–15 cm),
excluding erect
inflorescence, which is usually
taller than the leaves,
6–15 in (15–38 cm) long

285

A single lanceolate or linear leaf tops the bright green, ovoid-compressed pseudobulbs of the Enchanted Dancing Lady. The inflorescence has a zigzag, flattened stem, which is triangular in cross section. This arises from the leaf sheath beside a mature pseudobulb.

As in many related species, the several, long-lasting flowers produce oil, and this is a putative reward for oil-collecting *Centris* and related genera of bees that pollinate the flowers. There may be enough oil to attract the bees but not enough to reward them, in which case this is a form of deceit pollination. The common name of the orchid refers to the common name of the genus *Oncidium* (dancing ladies), in which this species was formerly included. The epithet comes from the Greek *meirax*, meaning "enchanting."

Actual size

The flower of the Enchanted Dancing Lady has yellow, red-brown spotted, clawed sepals and slightly shorter and broader spreading petals. The yellow lip is broadly spade-shaped with a prominent callus and many reddish-brown spots.

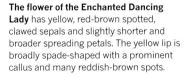

SUBFAMILY	Epidendroideae
TRIBE AND SUBTRIBE	Cymbidieae, Oncidiinae
NATIVE RANGE	Western South America, from Ecuador to northern Peru
HABITAT	Wet cloud and elfin (dwarf) forests, at 6,600–9,850 ft (2,000–3,000 m)
TYPE AND PLACEMENT	Epiphytic on mossy trees or terrestrial on steep slopes
CONSERVATION STATUS	Not assessed
FLOWERING TIME	Summer to fall

FLOWER SIZE
⅝ in (1.5 cm)

PLANT SIZE
20–35 × 20–30 in
(51–89 × 51–76 cm),
excluding erect-arching
branched inflorescence
30–55 in (76–140 cm) long,
which is longer than the leaves

CYRTOCHILUM TRICOSTATUM
YELLOW SPADE ORCHID
KRAENZLIN, 1922

Actual size

The large, conical pseudobulbs of the Yellow Spade Orchid are partially covered by several overlapping leaf-bearing sheaths and topped by two apical, narrowly elongate leaves. From a leaf axil of a mature pseudobulb, a much-branched inflorescence with zigzag side branches emerges and carries many fleshy flowers with a spade-shaped lip—hence the common name. The small size of the flowers makes a stark contrast to the overall substantial proportions of the plant. The genus name comes from the Greek *cyrtos*, "curved," and *cheilos*, "lip."

In the high-elevation, misty habitats that *Cyrtochilum tricostatum* inhabits, there are not many oil-producing plants, but the species still appears to be mimicking such flowers to attract oil-collecting bees. The plant does not actually produce any oil, so it is another case of pollination by deceit.

The flower of the Yellow Spade Orchid has bright yellow, clawed, spoon-shaped sepals and broader, upright, spreading petals. The yellow to orangish lip is spade-shaped, with a thickened callus, and the column is sometimes reddish-brown.

SUBFAMILY	Epidendroideae
TRIBE AND SUBTRIBE	Cymbidieae, Oncidiinae
NATIVE RANGE	Colombia (Cundinamarca department)
HABITAT	Cloud forests, at around 6,600 ft (2,000 m)
TYPE AND PLACEMENT	Epiphytic on moist branches in shade
CONSERVATION STATUS	Not assessed
FLOWERING TIME	March

ELOYELLA CUNDINAMARCAE
THINGUMY ORCHID
(P. ORTIZ) P. ORTIZ, 1979

FLOWER SIZE
¼ in (0.6 cm)

PLANT SIZE
½ × 1 in (1.3 × 2.5 cm),
excluding inflorescence,
which is triangular in
cross section and
½–1 in (1.3–2.5 cm) long

287

The truly diminutive Thingumy Orchid forms a small fan of upright, falcate, overlapping leaves that bury in their middle a round pseudobulb. In the axils, a stem with two to six tiny flowers is produced. Even when present in good numbers and in flower, these plants are often overlooked. The genus was named in honor of Colombian botanist Eloy Valenzuela (1756–1834). One of the little creatures from Moomin Valley, Thingumy (along with his friend Bob) steals things in search of a ruby—an appropriate reference for this tiny little jewel of an orchid.

Actual size

Pollination has not been studied, but the orchid appears to produce oil on the lip callus, so this could be a case of oil-bee pollination. The bees collect the oil and mix it with pollen of non-orchid species and feed this mixture to their larvae.

The flower of the Thingumy Orchid
has white or yellow, linear, spreading
sepals and petals, all equal in size.
The lip is yellow, broad, and
recurved, with a thickened callus in
the middle. The column is arching,
with a pair of lateral wings.

SUBFAMILY	Epidendroideae
TRIBE AND SUBTRIBE	Cymbidieae, Oncidiinae
NATIVE RANGE	Tropical America, from Mexico to the Guianas and Peru and southeastern Brazil
HABITAT	Wet premontane and riverine forests, at up to 3,280 ft (1,000 m)
TYPE AND PLACEMENT	Epiphytic
CONSERVATION STATUS	Not assessed
FLOWERING TIME	Throughout the year

FLOWER SIZE
½ in (1.4 cm)

PLANT SIZE
1–2 × 1–2 in
(2.5–5 × 2.5–5 cm),
excluding axillary,
successively two- or three-
flowered inflorescence
1½–2½ in (3.8–6.5 cm) long

ERYCINA PUMILIO
DWARF FAN ORCHID
(REICHENBACH FILS) N. H. WILLIAMS & M. W. CHASE, 2001

A delightful tiny species, the Dwarf Fan Orchid grows almost exclusively on the smallest twigs—and sometimes leaves—forming a flat fan of leaves and clasping its host with wiry roots. The genus name is based on Mount Eryx in Sicily, although the reason behind the use of this reference for *Erycina* (formerly considered a member of the genus *Psygmorchis*) is unknown. The little orchid is remarkable for its rapid development—the time from germination to flowering can be around six months, and there are photos of plants flowering on the leaves of guava and coffee bushes, which last for less than two seasons.

Oil-collecting bees are the pollinators, attracted by flowers that mimic those of the plant family Malpighiaceae. The bees are deceived, as no oil is produced.

Actual size

The flower of the Dwarf Fan Orchid has small, cupped sepals and spreading, lanceolate petals. The lip is large and elaborately six-lobed. The lip callus bears fingerlike protuberances, and the column has a pair of apical wings. The whole flower is chrome yellow, sometimes with a few reddish spots.

SUBFAMILY	Epidendroideae
TRIBE AND SUBTRIBE	Cymbidieae, Oncidiinae
NATIVE RANGE	Ecuador to Peru
HABITAT	Cloud forests, at 6,900–10,170 ft (2,100–3,100 m)
TYPE AND PLACEMENT	Epiphytic
CONSERVATION STATUS	Not assessed
FLOWERING TIME	Winter to early spring

FLOWER SIZE
1¼ in (3 cm)

PLANT SIZE
3–10 × 1–2 in
(8–25 × 2.5–5 cm),
excluding short
inflorescence about
1 in (2.5 cm) long

FERNANDEZIA SUBBIFLORA
SCARLET MIST ORCHID
RUIZ & PAVON, 1798

289

The Scarlet Mist Orchid, an erect to pendent epiphyte, has elongate stems covered with overlapping leaves in two ranks. If it branches, it produces small plants at the base. The leaf blades are ovate, folded, falcate, and thickly rugose. One to four (sometimes more) lateral inflorescences produce single flowers. These plants are monopodial (growing from the apex rather than producing a new growth from the base each year), a habit type unusual in this subtribe.

The red flowers suggest pollination by short-billed hummingbirds, but observations in the field are lacking. The apex of the column has a hood that presumably channels the bird's beak into the center of the flower, where the pollinaria are attached to it. Included now in *Fernandezia* are white to green flowered species formerly included in *Pachyphyllum* and a brown-and-yellow flowered species formerly in *Raycadenco*.

Actual size

The flower of the Scarlet Mist Orchid is red to orange, with short, triangular sepals and larger, spreading petals. The broad, spreading lip forms a tube with the hooded column. The hood is orange and flaring.

SUBFAMILY	Epidendroideae
TRIBE AND SUBTRIBE	Cymbidieae, Oncidiinae
NATIVE RANGE	São Paulo to Rio de Janeiro (southeastern Brazil)
HABITAT	Atlantic rain forests, up to 3,950 ft (1,200 m)
TYPE AND PLACEMENT	Epiphytic
CONSERVATION STATUS	Not assessed
FLOWERING TIME	June to August (winter)

FLOWER SIZE
¾ in (2 cm)

PLANT SIZE
6–10 × 4–8 in
(15–25 × 10–20 cm),
excluding inflorescence,
which is commonly
pendent, 8–15 in
(20–38 cm) long

290

GOMESA ECHINATA
PORCUPINE ORCHID
(BARBOSA RODRIGUES) M. W. CHASE & N. H. WILLIAMS, 2009

The short stems of the Porcupine Orchid carry subcylindrical, compressed elongate pseudobulbs that are partly covered at the base by two-ranked, overlapping leafless sheaths. At the tip, the pseudobulbs carry two (rarely one) oblong-lanceolate folded leaves. A long pendent, racemose or weakly branched inflorescence is densely covered with elliptic bracts and up to 50 flowers. The lip encircles the column, a feature that is rare in this subtribe, which usually has open flowers.

The flowers have the same color spectrum as the oil-producing Malpighiaceae plant family but produce no oil, although they deceive *Centris* and related bees into believing that oil is available. The orchid's common name refers to the species name, *echinata*, meaning "prickly" in Latin.

The flower of the Porcupine Orchid has narrow, pale yellow lateral sepals and a broad, hooded upper sepal. The similarly yellow petals are large and spreading. The lip has sidelobes that are upright and bright yellow, and the midlobe is dark reddish-brown with a pair of large teeth.

Actual size

SUBFAMILY	Epidendroideae
TRIBE AND SUBTRIBE	Cymbidieae, Oncidiinae
NATIVE RANGE	Southeastern Brazil (Atlantic Forest)
HABITAT	Moderately wet forests, at 330–3,950 ft (100–1,200 m)
TYPE AND PLACEMENT	Epiphytic
CONSERVATION STATUS	Not threatened
FLOWERING TIME	September to October (spring)

GOMESA FORBESII
SHINY FOREST SPRITE
(HOOKER) M. W. CHASE & N. H. WILLIAMS, 2009

FLOWER SIZE
2 in (5 cm)

PLANT SIZE
8–12 × 6–10 in
(20–30 × 15–25 cm),
excluding erect to
arching inflorescence
10–16 in (25–41 cm) long

291

The Shiny Forest Sprite produces flattened ovoid pseudobulbs with one or two narrowly lanceolate leaves on top. The inflorescence can bear up to 20 showy flowers. Originally a member of the large genus *Oncidium*, this and its related species were recently moved to *Gomesa*. The species name is in honor of the English botanist and plant collector John Forbes (1799–1823), while the common name refers to the glossy texture of the flowers, which are easily spotted in the wet Atlantic Forest of Brazil.

The species is pollinated by oil-collecting bees that mistake the flowers for those of the unrelated family Malpighiaceae. The lip callus appears to produce some oil, but it is not enough to reward the bees, which would normally collect the oil from Malpighiaceae blooms and mix it with pollen to feed their young.

Actual size

The flower of the Shiny Forest Sprite has ovate, bright yellow sepals and petals that are heavily spotted with chestnut brown. The petals are much broader than the sepals. The lip has two small sidelobes and a widely spreading midlobe, again yellow with abundant chestnut-brown markings.

SUBFAMILY	Epidendroideae
TRIBE AND SUBTRIBE	Cymbidieae, Oncidiinae
NATIVE RANGE	Southeastern and eastern Brazil, from Rio Grande do Sul to Minas Gerais
HABITAT	Atlantic rain forests, at 165–3,950 ft (50–1,200 m)
TYPE AND PLACEMENT	Epiphytic
CONSERVATION STATUS	Not assessed but locally abundant
FLOWERING TIME	April to July (fall to winter)

FLOWER SIZE
2–3 in (5–8 cm)

PLANT SIZE
7–12 × 5–8 in
(18–30 × 13–20 cm),
excluding erect to arching
basal inflorescence, which is
longer than the leaves,
14–26 in (36–66 cm) long

GOMESA IMPERATORIS-MAXIMILIANI
BROWN DANCING LADY
(REICHENBACH FILS) M. W. CHASE & N. H. WILLIAMS, 2009

292

The Brown Dancing Lady produces large, flattened, ovoid pseudobulbs that are tightly clustered and in part enveloped in dry bracts. They carry up to three lanceolate, leathery leaves and an upright to arching, branched inflorescence (panicle) with up to 40 musty-scented flowers that have a glossy texture. As is the case for many "dancing ladies" (a frequent name for species in this subtribe), the species is pollinated by oil-collecting bees, but here there is no oil present, so this is a case of deceit pollination.

The species was previously known as *Oncidium crispum*, but DNA studies have shown that it belongs instead in the genus *Gomesa*. Because there was already a species named *G. crispa*, this orchid was instead named for Ferdinand Maximilian Joseph (1832–67), Archduke of Schönbrunn, Austria, who undertook a botanical expedition to Brazil before becoming Emperor Maximilian of Mexico.

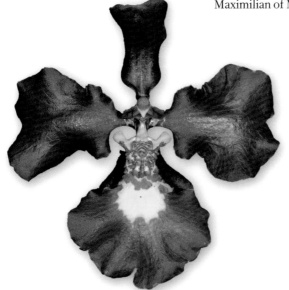

The flower of the Brown Dancing Lady has brown, wavy-edged sepals and petals with a narrow, yellow margin—the sepals narrower and recurved, the petals broad and spreading. The lip is broad at the apex, narrowing to a yellow, warty-callused base.

Actual size

SUBFAMILY	Epidendroideae
TRIBE AND SUBTRIBE	Cymbidieae, Oncidiinae
NATIVE RANGE	Eastern Brazil (Espírito Santo to Rio Grande do Sul) and Misiones (northeastern Argentina)
HABITAT	Cool Atlantic Forest habitats, at 165–4,600 ft (50–1,400 m)
TYPE AND PLACEMENT	Epiphytic or terrestrial
CONSERVATION STATUS	Not assessed
FLOWERING TIME	Fall and winter

FLOWER SIZE
¼ in (2 cm)

PLANT SIZE
8–12 × 6–10 in,
excluding inflorescence

GOMESA RECURVA
GREEN FOXTAIL ORCHID
R. BROWN, 1815

293

The narrowly ovoid, laterally compressed pseudobulbs of the Green Foxtail Orchid carry two upright, linear-oblanceolate, sharp-tipped, folded leaves. From a leaf axil, the plant produces a dense, arching inflorescence with numerous nodding, fragrant flowers that are pollinated by bees searching for oil, which is present only in small volumes. This could be a reward, but given how little is present it may just function as an attractant—another instance of pollination by deceit.

Some authors place only four to six species in the genus *Gomesa*, whereas others include a wide range of species that all share pollination by oil-collecting bees, even though their floral traits are highly divergent. In spite of differing morphologies, such species all share fused lateral sepals, an almost exclusive characteristic for these orchids.

Actual size

The flower of the Green Foxtail Orchid
has spreading, ligulate, green petals and sepals with wavy margins. The much shorter green lip has a flat limb and ridged, white callus with red-orange rim. The column has a white anther cap.

SUBFAMILY	Epidendroideae
TRIBE AND SUBTRIBE	Cymbidieae, Oncidiinae
NATIVE RANGE	Southeastern Brazil to Paraguay and northern Argentina
HABITAT	Moist slopes and hill crests, at around 4,920 ft (1,500 m)
TYPE AND PLACEMENT	Epiphytic
CONSERVATION STATUS	Not assessed
FLOWERING TIME	May

FLOWER SIZE
1 in (2.5 cm)

PLANT SIZE
10–18 × 3–5 in
(25–46 × 8–13 cm),
excluding arching,
typically branched
inflorescence
30–80 in (76–203 cm) long

GRANDIPHYLLUM DIVARICATUM
BRISTLED DANCING LADY
(LINDLEY) DOCHA NETO, 2006

The yellowish-green, globose-flattened pseudobulbs of the Bristled Dancing Lady are topped with a single oblong, leathery, upright leaf. From the base of a pseudobulb, a large, branched inflorescence is produced, each branch carrying 5–15 often scented flowers. The genus name derives from the Latin *grandis*, "large," and the Greek *phyllon*, "leaf." Its species are often referred to as "mule-ear oncidiums," referring to the genus *Oncidium* in which they were previously placed. The species name refers to the divaricating (branching) inflorescence, and the common name points to the bristle-covered lip callus and the resemblance of *Oncidium* flowers to dancing ladies.

Pollination is likely by oil-collecting bees. This is based on the resemblance of the flowers to those of *Oncidium*, which are documented to be pollinated by such bees.

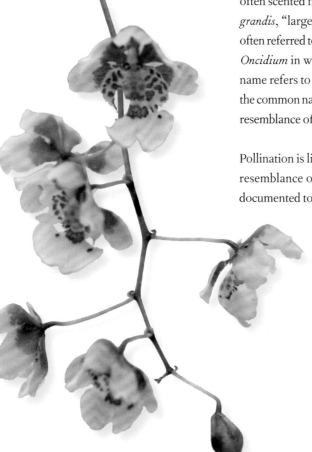

The flower of the Bristled Dancing Lady has spreading yellow sepals and petals with reddish-brown bases. The yellow lip has three lobes and a bristly, cushion-like callus, often all spotted with reddish-brown. The column has a pair of brown-spotted wings.

Actual size

SUBFAMILY	Epidendroideae
TRIBE AND SUBTRIBE	Cymbidieae, Oncidiinae
NATIVE RANGE	Central to southwestern Mexico (Jalisco, Guerrero, Mexico, Morelos, Michoacán, and Oaxaca states)
HABITAT	Steep hillsides and along creeks in oak-pine forests or in dense humid deciduous forests, at 4,920–7,200 ft (1,500–2,200 m)
TYPE AND PLACEMENT	Epiphytic
CONSERVATION STATUS	Not assessed
FLOWERING TIME	January to March (winter to spring)

FLOWER SIZE
½ in (1.3 cm)

PLANT SIZE
2–3 × 1 in (5–8 × 2.5 cm),
including inflorescences
½–1½ in long (1.3–3.8 cm)

HINTONELLA MEXICANA
HINTON'S ORCHID
AMES, 1938

295

The miniature Hinton's Orchid has small globose pseudobulbs, each with three or four basal leaf-bearing sheaths and carrying a single, sulcate terminal leaf. From the base of the pseudobulb, the plant makes one or two arching racemes with up to six small, sweetly fragrant, delicate but long-lasting flowers. The genus was named in honor of the metallurgist turned botanist George Boole Hinton (1882–1943), who collected botanical specimens from 1931 through 1941, including many orchids, in some of the most inaccessible parts of Guerrero, Michoacán, and Mexico states.

Pollination has not been studied, but the presence of oil and glandular hairs (elaiophores) in the lip cavity indicates that the species is pollinated, like the related genus *Ornithocephalus*, by bees of the Anthophoridae family. The insects mix the oil with pollen of other plants to feed their larvae.

Actual size

The flower of Hinton's Orchid has spreading white petals and sepals that form a bell. The white lip has some yellow spots or stripes and is broadly triangular, with rounded sides, a rounded tip, a central stalked ridge, and a hair-filled cavity.

SUBFAMILY	Epidendroideae
TRIBE AND SUBTRIBE	Cymbidieae, Oncidiinae
NATIVE RANGE	Northwestern South America, from Venezuela to Peru
HABITAT	Cloud forests, at 5,250–7,875 ft (1,600–2,400 m)
TYPE AND PLACEMENT	Epiphytic on exposed mossy shrubs
CONSERVATION STATUS	Not assessed and due to its small size easily overlooked, so may be more common than presumed
FLOWERING TIME	Throughout the year

FLOWER SIZE
½ in (1.3 cm)

PLANT SIZE
¾–1½ × ¾–1½ in
(2–3.8 × 2–3.8 cm),
excluding erect
inflorescence
1⁹⁄₁₆–2⅜ in (4–6 cm) tall

296

HOFMEISTERELLA EUMICROSCOPICA
SQUID ORCHID
(REICHENBACH FILS) REICHENBACH FILS, 1852

Actual size

On exposed mossy branches, the tiny Squid Orchid forms a tuft of linear, fleshy, unifacial leaves, which have the same structure on both sides as their normally unfolded surfaces are fused. In the axil of one of the leaves, a highly flattened raceme with successively opening flowers is formed. The inflorescence is much bigger than the plant, and the flowers, too, are comparatively large. As the inflorescence is green, leaflike, and flattened, it may carry out photosynthesis. The common name refers to the resemblance of the flowers to the body of a squid with its arms outstretched.

Flowers of the genus *Hofmeisterella* resemble those of *Telipogon*, to which it is related. As in the latter genus, the lip of the flowers probably resembles a female tachinid fly, but this has not been studied. If the same syndrome as in *Telipogon* is operating, then male flies would attempt to mate with the lip, during which process pollen is transferred.

The flower of the Squid Orchid has yellow-green, linear sepals and petals that are held upright. The lip is yellow with red suffusion and shaped like an arrowhead. The column is winged and forward projecting with a pointed apex.

SUBFAMILY	Epidendroideae
TRIBE AND SUBTRIBE	Cymbidieae, Oncidiinae
NATIVE RANGE	Mexico, Central America, Caribbean islands, and much of northern South America, even found occasionally in southern Florida
HABITAT	Low to mid-elevation scrub, often on cultivated guava or citrus; considered a weed in coffee and cacao plantations
TYPE AND PLACEMENT	Epiphytic on shrubs or small trees
CONSERVATION STATUS	Not threatened
FLOWERING TIME	Winter to spring

FLOWER SIZE
¾–1¼ in (2–3 cm)

PLANT SIZE
4–7 × 3–4 in
(10–18 × 8–10 cm),
excluding erect-arching
inflorescence
20–40 in (51–102 cm) long

IONOPSIS UTRICULARIOIDES
BLADDERWORT ORCHID
(SWARTZ) LINDLEY, 1826

297

A miniature epiphyte that is usually found on small twigs without mosses or lichens, the Bladderwort Orchid occurs in many habitats over an incredible geographic range. With its copious aerial roots and weedy proclivities, the species is obviously adaptable, often forming large colonies and tinting well-colonized trees with a lavender-pink hue when in full bloom. The genus name *Ionopsis* refers to resemblance to a violet, from the Greek *ion* ("violet"), whereas the species epithet implies a similarity to the blooms of the genus *Utricularia*, aquatic carnivores, or bladderworts—hence the common name.

The flowers appear on a panicle, often in great quantity on older plants, emerging basally from reduced pseudobulbs. They can vary in color, ranging from white to pale pink, lavender, or lilac, usually with purplish nectar guides on the labellum. Like many twig epiphytes they may be naturally short-lived.

The flower of the Bladderwort Orchid is small, with the sepals fused into a tube. The petals are reduced and not particularly noticeable. The showy, two-lobed lip is the most prominent feature, often with a bluish or lavender-pink hue.

Actual size

SUBFAMILY	Epidendroideae
TRIBE AND SUBTRIBE	Cymbidieae, Oncidiinae
NATIVE RANGE	Central Mexico, from Veracruz to Oaxaca states
HABITAT	Seasonally wet forests, now mostly found in coffee and old citrus plantations, at 3,300–5,250 ft (1,000–1,600 m)
TYPE AND PLACEMENT	Epiphytic on twigs
CONSERVATION STATUS	Not assessed, but unlikely to be of conservation concern
FLOWERING TIME	May to June (spring)

FLOWER SIZE
¾ in (2 cm)

PLANT SIZE
4–6 × 3–4 in
(10–15 × 8–10 cm),
excluding sometimes branched
arching-pendent inflorescence
8–24 in (20–61 cm) long

298

LEOCHILUS CARINATUS
MEXICAN FINGER ORCHID
(KNOWLES & WESTCOTT) LINDLEY, 1842

A small epiphyte growing on small branches (a twig epiphyte), the Mexican Finger Orchid has rounded, compressed pseudobulbs, each carrying a pair of elliptic-lanceolate, conduplicate leaves. The wiry, arching to pendent inflorescences carry many fragrant flowers, followed by plantlets (*keikis*) that root, forming small colonies of interlinked plants in small trees or shrubs. The genus name comes from the Greek words *leios*, "smooth," and *cheilos*, "lip," referring to the texture of the lip in most of the genus species, but not *Leochilus carinatus*, which has a tuberculate callus.

Pollination of other *Leochilus* species has been shown to be by small bees and polybiine (papermaking) wasps that are attracted by the nectar in the cavity at the base of the lip. This species has the same cavity, filled with nectar, so it is probably also pollinated by halictid (or sweat) bees.

Actual size

The flower of the Mexican Finger Orchid has spreading, dull yellow-brown lateral sepals and a hooded upper sepal. The dark brown striped petals point forward, flanking the column. The light yellow lip is pear-shaped, with a crest of fingerlike papillae and reddish-brown blotches.

SUBFAMILY	Epidendroideae
TRIBE AND SUBTRIBE	Cymbidieae, Oncidiinae
NATIVE RANGE	Central Mexico (Veracruz and Oaxaca states)
HABITAT	Known from a small area on twigs in full sun, at 6,600–7,875 ft (2,000–2,400 m)
TYPE AND PLACEMENT	Epiphytic on twigs
CONSERVATION STATUS	Not assessed, but with a narrow distribution, so likely to be under threat of extinction in the wild
FLOWERING TIME	May to June (late spring to summer)

LEOCHILUS LEIBOLDII
TREEFROG ORCHID
REICHENBACH FILS, 1845

FLOWER SIZE
½ in (1.3 cm)

PLANT SIZE
3–5 × 2–3 in
(8–13 × 5–8 cm),
excluding pendent inflorescence
5–7 in (13–18 cm) long, which
is generally longer than the
leaves

299

The tiny twig epiphyte Treefrog Orchid grows on small branches of shrubs and small trees, producing a cluster of elliptic, slightly flattened pseudobulbs that are subtended by a few overlapping leafy sheaths and a single linear leaf on the top. A pendent or arching inflorescence produces up to 12 flowers. The floral morphology of this species is bizarre, one of the most unusual in a family known for unusually shaped flowers. It has sometimes been placed in its own genus, *Papperizia*, but is embedded in *Leochilus*.

Given that *L. leiboldii* has green to greenish-white flowers and weakly scented flowers at night, it is possible that the plant is moth pollinated. It has a deep, hair-filled nectar cavity formed by the base of the lip and a pair of column arms.

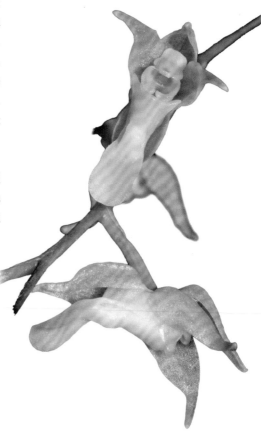

The flower of the Treefrog Orchid has green dorsal sepals and petals that form a hood over the column. The two lateral sepals are fused into a keel behind the cream to white lip, which has two lateral wings and a spoon-shaped apex.

Actual size

SUBFAMILY	Epidendroideae
TRIBE AND SUBTRIBE	Cymbidieae, Oncidiinae
NATIVE RANGE	Peruvian Amazon
HABITAT	Lowlands in wet tropical forests
TYPE AND PLACEMENT	Epiphytic
CONSERVATION STATUS	Not threatened
FLOWERING TIME	Late fall or winter

FLOWER SIZE
¾ in (2 cm)

PLANT SIZE
10–18 × ¾–1¼ in
(25–46 × 2–3 cm),
including short, often
terminal, inflorescences
1–2 in (2.5–5 cm) long

300

LOCKHARTIA BENNETTII
BENNETT'S BRAIDED ORCHID
DODSON, 1989

Actual size

Lockhartia flowers are usually yellow and produce oil that is collected by *Centris* bees searching for oils to mix with pollen and feed to their larvae. Bennett's Braided Orchid, however, has a white flower resembling a small *Cattleya* orchid, with a lip wrapping around the column, although it too produces oil on its lip.

The classic vegetative features of this relatively recent discovery are characteristic of *Lockhartia*, with overlapping foliage resembling a braided belt. These plants are able to continue producing flowers at or near the tip of each stem over several years. The "braiding" of these stems represents just the overlapping bases of leaves that are typical for this subtribe, which are produced without the formation of pseudobulbs. The species name refers to David Bennett, an orchid enthusiast who described many new species of Peruvian orchids.

The flower of the Bennett's Braided Orchid has broad, overlapping, white or cream segments and a lip that surrounds the column. It also has an internal oil-secreting cavity and a deep red to red-brown interior.

SUBFAMILY	Epidendroideae
TRIBE AND SUBTRIBE	Cymbidieae, Oncidiinae
NATIVE RANGE	Eastern and southern Brazil
HABITAT	Tropical wet forests, at 1,640–3,300 ft (500–1,000 m)
TYPE AND PLACEMENT	Epiphytic
CONSERVATION STATUS	Not threatened
FLOWERING TIME	Spring and summer

FLOWER SIZE
¾ in (2 cm)

PLANT SIZE
8–14 × ½–1 in
(20–36 × 1.3–2.5 cm),
excluding short terminal
inflorescences
1–2 in (2.5–5 cm) long

LOCKHARTIA LUNIFERA
HALF-MOON BRAIDED ORCHID
(LINDLEY) REICHENBACH FILS, 1852

301

The Half-moon Braided Orchid is well known due to the novelty of its closely overlapping short leaves. Flowers appear over many years on between one and three-flowered short racemes with large pale green concave floral bracts near the apex of the often-pendent growths. The species derives its name *lunifera* from the half-moon-shaped lateral lobes of its lip, which arch up over the column. The genus is named for David Lockhart (1786–1845), the first superintendent of the botanic gardens in Trinidad, who collected the material upon which the genus was based.

Like other species of *Lockhartia*, the Half-moon Braided Orchid produces oil on its complicated lip callus, which attracts oil-collecting *Centris* bees. The flowers are also thought to mimic those of the unrelated family Malpighiaceae, which has similarly bright yellow flowers that are pollinated by oil-collecting bees.

Actual size

The flower of the Half-moon Braided Orchid is bright yellow with chestnut-brown to reddish markings in the center of the lip and the upwardly curving lateral lobes of the lip. The flower orientation is random, with some upside down as well as with the lip lowermost.

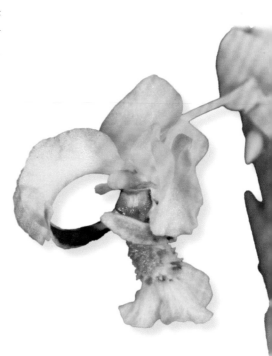

SUBFAMILY	Epidendroideae
TRIBE AND SUBTRIBE	Cymbidieae, Oncidiinae
NATIVE RANGE	Mesoamerica and northwestern South America
HABITAT	Wet forests, from sea level to 985 ft (300 m)
TYPE AND PLACEMENT	Epiphytic on trees
CONSERVATION STATUS	Not assessed
FLOWERING TIME	February to March (spring)

FLOWER SIZE
1 in (2.5 cm)

PLANT SIZE
7½–14 × 1½–2½ in
(19–35 × 4–6.5 cm), excluding
inflorescence, which is about as
long as the leaves and pendent

302

MACRADENIA BRASSAVOLAE
FIREWORKS ORCHID
REICHENBACH FILS, 1852

The Fireworks Orchid has elongate pear-shaped pseudobulbs. Each pseudobulb carries a single oblong, leathery leaf, and from its base a pendent raceme up to 12 in (30 cm) long with many sweetly fragrant flowers is produced, together creating an effect like a fireworks display. The members of this genus are twig orchids that grow on the smallest branches of trees and shrubs.

Like that of the closely related genera *Notylia* and *Macroclinium*, pollination of *Macradenia* species is carried out by male euglossine bees that seek floral fragrance compounds. Given the small size of the flowers relative to the euglossine bees that visit them, the pollinaria are most likely attached to the feet of the bees as they move around the inflorescence, seeking a good place to try to scrape the fragrance compounds from the floral tissue.

The flower of the Fireworks Orchid has linear, yellow-edged, reddish-brown (chestnut) sepals and petals, and a much shorter, narrow, trilobed lip, with a linear terminal lobe and broad sidelobes surrounding the column.

Actual size

SUBFAMILY	Epidendroideae
TRIBE AND SUBTRIBE	Cymbidieae, Oncidiinae
NATIVE RANGE	Costa Rica and Panama
HABITAT	Wet forests, from sea level to 1,300 ft (400 m)
TYPE AND PLACEMENT	Epiphytic
CONSERVATION STATUS	Not assessed
FLOWERING TIME	December to January (winter)

FLOWER SIZE
⅜ in (1 cm)

PLANT SIZE
1–1½ × 1–1½ in
(2.5–3.8 × 2.5–3.8 cm),
excluding tightly clustered
pendent inflorescence
1½–2½ in (3.8–6.5 cm) long

MACROCLINIUM ALLENIORUM
PINK FAN ORCHID
DRESSLER & PUPULIN, 1996

303

The tiny Pink Fan Orchid produces a flat fan of overlapping leaves that continues to grow from its apex rather than forming new side growths every year. Plants produce a raceme that sometimes branches and carries several relatively large, spidery flowers in tight clusters. The genus name reflects the long, pointed tip of the column, from the Greek *makros*, "long," and *kline*, "couch," which refers to the flat structure on which the pollinarium sits. The species name honors the American botanist Paul H. Allen (1911–63), who studied the orchids of Panama.

Macroclinium species are pollinated by male euglossine bees—an improbable relationship, given the relatively huge size of the insects compared to the small flowers. The pollinaria are attached to the feet of the bees as they climb over the inflorescences looking for fragrance compounds to collect.

The flower of the Pink Fan Orchid has long, thin, pointed, greenish-pink petals and sepals, the petals being pink spotted. The pink lip is sometimes also spotted and shortly stalked and has a tongue-shaped limb. The thin column has an enlarged anther cap.

Actual size

SUBFAMILY	Epidendroideae
TRIBE AND SUBTRIBE	Cymbidieae, Oncidiinae
NATIVE RANGE	Southeastern Brazil
HABITAT	Wet forests, at 1,640–2,625 ft (500–800 m)
TYPE AND PLACEMENT	Epiphytic
CONSERVATION STATUS	Not threatened
FLOWERING TIME	Summer

FLOWER SIZE
4 in (10 cm)

PLANT SIZE
8–15 × 6–8 in
(20–38 × 15–20 cm),
excluding single-flowered
inflorescence
12–20 in (30–51 cm) long

304

MILTONIA SPECTABILIS
SPECTACULAR BIG LIP
LINDLEY, 1837

Spectacular Big Lip is a wonderfully variable species in its coloration. The plants form loose clusters of elongate flattened pseudobulbs with thin folded leaves, and flower stems covered in large bracts emerge from the bases of the pseudobulbs. The genus was named in honor of Charles William Wentworth Fitzwilliam, Viscount Milton (1786–1857), a patron of horticulture and orchid enthusiast. The species name, *spectabilis*, is Latin for "notable" or "remarkable."

The dark purple-colored "form" of *Miltonia spectabilis*, formerly var. *moreliana*, is now considered to be a distinct species, *M. moreliana*. Both species have the flower morphology that is typically found in yellow flowers with brown spots—as encountered in most species of genus *Oncidium*—which is associated with pollination by oil-collecting bees. The flowers do not produce oil, so these plants are pollinated by deceit.

Actual size

The flower of Spectacular Big Lip has pink to lavender, lanceolate sepals and petals, a large, broad, flat lip with purple veins that bleed out onto the surface near the base, and a yellow callus just in front of the column.

SUBFAMILY	Epidendroideae
TRIBE AND SUBTRIBE	Cymbidieae, Oncidiinae
NATIVE RANGE	Cundinamarca and Norte de Santander departments of Colombia
HABITAT	Wet forests, at 3,950–5,250 ft (1,200–1,600 m)
TYPE AND PLACEMENT	Epiphytic
CONSERVATION STATUS	Not threatened
FLOWERING TIME	At any time, but more often in late summer to fall

MILTONIOPSIS PHALAENOPSIS
BUTTERFLY-LIPPED ORCHID
(LINDEN & REICHENBACH FILS) GARAY & DUNSTERVILLE, 1976

FLOWER SIZE
2½ in (6.5 cm)

PLANT SIZE
8–14 × 4–7 in
(20–36 × 10–18 cm),
including erect to arching
inflorescence 7–12 in
(18–30 cm) long, which is
about as long as the leaves

305

The Butterfly-lipped Orchid, which carries between three and five flowers per stem, has one of the most curiously patterned lips in the orchid family. Bees have been reported to pollinate other species in this genus, but the exact reasons for such a striking lip pattern, presumably an elaborate set of nectar guides, are unclear. Also, the lip is totally flat when most bee-pollinated species have a lip that to a degree surrounds the column. Pollination by bees at night has also been reported.

The genus name refers to the genus *Miltonia* (-*opsis* means "similar to") in which this species was previously classified. The species name is a reference to the genus *Phalaenopsis*—the moth orchids—named for their similarity to, but not pollination by, moths.

The flower of the Butterfly-lipped Orchid is almost perfectly flat and has white (rarely pale pink) sepals and petals. The extraordinary lip is large and shaped like a butterfly with a complex set of purple-red and yellow markings on a white background.

Actual size

SUBFAMILY	Epidendroideae
TRIBE AND SUBTRIBE	Cymbidieae, Oncidiinae
NATIVE RANGE	Venezuela, Colombia, Ecuador, and Panama
HABITAT	Wet forests, at 1,300–3,300 ft (400–1,000 m)
TYPE AND PLACEMENT	Epiphytic
CONSERVATION STATUS	Not threatened
FLOWERING TIME	Summer to fall

FLOWER SIZE
4 in (10 cm)

PLANT SIZE
12–20 × 8–10 in
(30–51 × 20–25 cm),
excluding arching-erect
inflorescence
10–20 in (25–51 cm) long

MILTONIOPSIS ROEZLII
PANSY ORCHID
(BULL) GODEFROY-LEBEUF, 1889

Hailing from lower elevations than most other members in the genus, the Pansy Orchid is also one of the most spectacular species in South America. One or more inflorescences can bear seven to ten flowers at a time, producing an amazing sight. This species is also well known for its gray green leaves and highly flattened pseudobulbs. The genus name has the ending *-opsis*, meaning "similar to," in this case to the genus *Miltonia*, with which the species of *Miltoniopsis* share their unusual flat flowers.

The petals have characteristic contrasting brilliant purple eyespots, giving the flowers the appearance of a face, similar to the face-like flowers of pansies—hence the common name. Pollination is by large bees, and some observations have indicated that this might even take place at night.

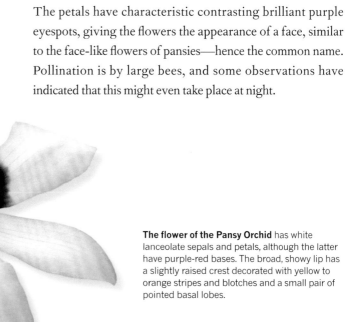

The flower of the Pansy Orchid has white lanceolate sepals and petals, although the latter have purple-red bases. The broad, showy lip has a slightly raised crest decorated with yellow to orange stripes and blotches and a small pair of pointed basal lobes.

Actual size

SUBFAMILY	Epidendroideae
TRIBE AND SUBTRIBE	Cymbidieae, Oncidiinae
NATIVE RANGE	Central America (from southern Mexico to Honduras) and northern Brazil
HABITAT	Humid and seasonally dry forests, swamps, and coffee plantations, below 5,900 ft (1,800 m)
TYPE AND PLACEMENT	Epiphytic on trees and shrubs
CONSERVATION STATUS	Not assessed
FLOWERING TIME	April to May (spring)

NOTYLIA BARKERI

GREEN KNOB ORCHID

LINDLEY, 1838

FLOWER SIZE
¼ in (0.8 cm)

PLANT SIZE
5–9 × 4–8 in
(13–23 × 10–20 cm),
excluding pendent
inflorescences
6–10 in (15–25 cm) long

307

The curious miniature Green Knob Orchid has a cluster of ellipsoid, flattened pseudobulbs, each basally enveloped in sheathing leaves and topped with a single tongue-shaped leaf. An axillary, rarely branched inflorescence carries many densely packed, fragrant flowers arranged in tight spirals. There are two forms, with either free or fused lateral sepals, sometimes recognized as distinct species. The genus name comes from the Greek *notos*, "back," and *tylo*, "knob," referring to the swollen apex of the column that points backward. The common name also refers to this prominent aspect of the flowers of this species.

Male euglossine bees searching for floral fragrances are the pollinators, with several species of *Notylia* often pollinated by the same bee species (*Euglossa viridissima*). However, because pollinia are deposited on different parts of the bee's body, the species remain separate.

The flower of the Green Knob Orchid has cupped, green to orange-green sepals, the upper one arching over the flower. The spreading, darkly spotted petals are falcate. The lip is usually white or at least paler and shorter than the sepals and arrowhead-shaped.

Actual size

SUBFAMILY	Epidendroideae
TRIBE AND SUBTRIBE	Cymbidieae, Oncidiinae
NATIVE RANGE	Northwestern Colombia (Western Cordillera, Chocó, and Valle del Cauca regions)
HABITAT	Forests, at 4,300 ft (1,300 m)
TYPE AND PLACEMENT	Epiphytic
CONSERVATION STATUS	Not assessed
FLOWERING TIME	November

FLOWER SIZE
¼ in (0.7 cm)

PLANT SIZE
3–4 × 1 in
(8–10 × 2.5 cm),
excluding arching-pendent
inflorescence
4–6 in (10–15 cm) long

NOTYLIOPSIS BEATRICIS
HAPPY ORCHID
P. ORTIZ, 1996

The Happy Orchid is a tiny epiphyte with small, sheath-enveloped pseudobulbs topped with a single leaf and a pendent, few-flowered raceme growing from the axil of the leaf. The genus name comes the purported floral similarity of its member species to the genus *Notylia* (the Greek *–opsis* means "looking like"). However, based on DNA studies, the two genera are not closely related, so the similarity is perhaps due to adaptation to similar pollinators. The species name is derived from the Latin *beatricem*, meaning "one who makes happy," which is also referenced in the common name.

This little orchid was discovered fairly recently, and its floral morphology is unlike any other in the subtribe Oncidiinae. From DNA studies, its closest relative appears to be *Zelenkoa onusta*, which has a similar plant habit but a completely different floral morphology.

The flower of the Happy Orchid has boat-shaped, greenish sepals, the upper one arching over the flower, the lateral two fused and held behind. The similarly colored linear petals are spreading and the bulging lip is slipper-shaped and pinkish-green.

Actual size

SUBFAMILY	Epidendroideae
TRIBE AND SUBTRIBE	Cymbidieae, Oncidiinae
NATIVE RANGE	Ecuador to Peru
HABITAT	Tropical forests, at 3,300–7,545 ft (1,000–2,300 m)
TYPE AND PLACEMENT	Epiphytic at the base of tree trunks or nearly terrestrial
CONSERVATION STATUS	Not assessed
FLOWERING TIME	July to October (summer to fall)

OLIVERIANA BREVILABIA
GREEN LANTERN ORCHID
(CHARLES SCHWEINFURTH) DRESSLER & N. W. WILLIAMS, 1970

FLOWER SIZE
¼ in (2 cm)

PLANT SIZE
8–14 × 5–8 in
(20–36 × 13–20 cm),
excluding basal inflorescence,
which is longer than the leaves
at 16–28 in (41–71 cm) long

309

The Green Lantern Orchid has a pair of sheathing leaflike bracts that envelop an elliptic, flattened pseudobulb bearing two linear, folded leaves. A much-branched (paniculate) inflorescence appears from the basal bract and carries 10–35 flowers. The common name reflects the plant's luminous green flowers, and the genus name is in honor of Daniel Oliver (1830–1916), a Keeper of the Herbarium at the Royal Botanic Gardens, Kew, London.

The genus *Oliveriana* is closely related to *Systeloglossum*, also part of the *Oncidium* alliance (subtribe Oncidiinae). The orchid is probably pollinated by some type of bee, but no pollinator has yet been observed. It has a two-parted stigmatic cavity and pollen masses that are widely spaced on their shared stalk, features usually associated with bird pollination. However, the green color and habitat preferences suggest this is unlikely.

Actual size

The flower of the Green Lantern Orchid has three elongate, green sepals that point directly up or down, and two shorter, forward-arching petals. The green lip is short, three-lobed, and partially fused to a broad, hooded column.

SUBFAMILY	Epidendroideae
TRIBE AND SUBTRIBE	Cymbidieae, Oncidiinae
NATIVE RANGE	Western South America, from southern Colombia to Ecuador
HABITAT	Cloud forests, at 3,950–9,500 ft (1,200–2,900 m)
TYPE AND PLACEMENT	Epiphytic
CONSERVATION STATUS	Not formally assessed
FLOWERING TIME	October to December (spring)

FLOWER SIZE
4 in (10 cm)

PLANT SIZE
10–18 × 8–14 in
(25–46 × 20–36 cm),
excluding arching-erect
inflorescence
12–24 in (30–61 cm) tall

310

ONCIDIUM CIRRHOSUM
CURLY SPIDER ORCHID
(LINDLEY) BEER, 1854

The flower of the Curly Spider Orchid has white, brown-spotted sepals and petals with wavy margins and long, curly tips. The lip has two yellow lateral lobes and a terminal one that is similar to the sepals and petals. There are two antenna-like projections on the lip callus.

The Curly Spider Orchid forms clusters of oblong-ovoid, flattened pseudobulbs that are basally enveloped by two or three pairs of overlapping, leaflike sheaths. The pseudobulbs each carry a single, linear-oblong to elliptic-oblong leaf. From the base of a mature pseudobulb, an upright raceme, sometimes with small branches, emerges. This can bear up to 20 white, spidery flowers. Its common and species names are derived from the Latin *cirratus*, meaning "curled."

The flowers of this species are likely to be pollinated by bumblebees or carpenter bees, which would be required to push through the opening between the lip and column, although this has not been verified by observations in nature and no nectar is produced. *Oncidium cirrhosum* was formerly placed in the genus *Odontoglossum*, which is now united with *Oncidium* because of a close genetic relationship between the species.

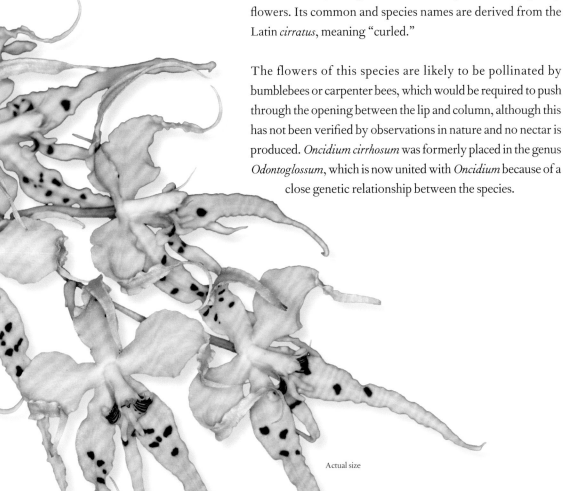

Actual size

SUBFAMILY	Epidendroideae
TRIBE AND SUBTRIBE	Cymbidieae, Oncidiinae
NATIVE RANGE	Western South America, from Antioquia department in Colombia to Peru
HABITAT	Edges of forests, at 2,300–9,850 ft (700–3,000 m)
TYPE AND PLACEMENT	Epiphytic
CONSERVATION STATUS	Not formally assessed
FLOWERING TIME	December to March (summer to fall)

ONCIDIUM HARRYANUM
HARRY'S MOUNTAIN ORCHID
(REICHENBACH FILS) M. W. CHASE & N. H. WILLIAMS, 2008

FLOWER SIZE
4 in (10 cm)

PLANT SIZE
10–18 × 8–12 in
(25–46 × 20–30 cm),
excluding arching
inflorescence
12–18 in (30–46 cm) tall

311

Harry's Mountain Orchid produces tight clusters of ovoid-elliptic, ribbed, and laterally flattened pseudobulbs, which are subtended by several leaf-bearing sheaths and topped by two oblong-elliptic to narrowly oblong leaves. From the base of a pseudobulb, an almost upright stem carries up to a dozen fragrant, long-lasting, waxy flowers. The species and common name refer to Harry James Veitch (1840–1924), a prominent English horticulturalist. Veitch was also one of the early promoters of the Great International Horticultural Exhibition, which subsequently became the Chelsea Flower Show.

Oncidium harryanum most likely attracts bees for pollination, but this has not been observed in the wild. There is no obvious reward for a pollinator, so the process must involve deceit for attraction of the insects. The species is also among those formerly placed in the now obsolete genus *Odontoglossum*.

The flower of Harry's Mountain Orchid has yellowish-green to tan sepals and petals that have a purple to reddish overlay. The lip has short sidelobes, several callus projections and hairs, and is broadly lobed with purple and white markings.

Actual size

SUBFAMILY	Epidendroideae
TRIBE AND SUBTRIBE	Cymbidieae, Oncidiinae
NATIVE RANGE	Venezuela, Colombia, Ecuador, and Peru
HABITAT	Moderately wet forests, at 2,460–5,900 ft (750–1,800 m)
TYPE AND PLACEMENT	Epiphytic
CONSERVATION STATUS	Not threatened
FLOWERING TIME	August to October (late summer to fall)

FLOWER SIZE
2½ in (6.5 cm)

PLANT SIZE
10–18 × 8–12 in
(25–46 × 20–30 cm),
excluding lateral, sometimes
branched inflorescence
30–72 in (76–183 cm) tall

312

ONCIDIUM HASTILABIUM
STRIPED STAR ORCHID
(LINDLEY) BEER, 1854

The Striped Star Orchid produces large, laterally compressed, furrowed pseudobulbs that have between two and four leaf-bearing bracts below and usually two on top. The often-branched inflorescence arises from one of the bracts at the base of the pseudobulb and can reach great lengths with 10–30 large, fragrant starry flowers. This is one of the biggest species in a subtribe of 1,600 members and forms large plants that make a magnificent sight when in flower. The species name refers to the shape of the lip, which is arrowhead shaped, or hastate.

Pollination has not been observed. The column has the swollen base typical of flowers that are pollinated by oil-collecting bees (they hold onto this while collecting the oil), so they could also be the pollinators of this species.

The flower of the Striped Star Orchid has narrowly lanceolate, white to cream sepals and petals that are outstretched and marked by reddish-purple stripes. The white lip has a dark purple base and crest and is three-lobed, with the small, narrow sidelobes projecting forward.

Actual size

SUBFAMILY	Epidendroideae
TRIBE AND SUBTRIBE	Cymbidieae, Oncidinae
NATIVE RANGE	Western South America, from Colombia to Peru
HABITAT	Semidry forests with a rainy season in fall, at 985–6,600 ft (300–2,000 m)
TYPE AND PLACEMENT	Epiphytic
CONSERVATION STATUS	Not assessed, but widespread and therefore unlikely to be of conservation concern
FLOWERING TIME	July–August (summer)

ONCIDIUM HYPHAEMATICUM

RED-EYE ORCHID

REICHENBACH FILS, 1869

FLOWER SIZE
1¼ in (3.2 cm)

PLANT SIZE
10–18 × 8–14 in
(25–46 × 20–36 cm),
excluding inflorescence
24–75 in (61–191 cm) long,
which is much longer
than the leaves

313

The charming Red-eye Orchid has ovate, flattened, ridged pseudobulbs that are subtended by one or two leafy bracts and topped with one oblong-elliptic leaf—sometimes two leaves. The long, branched inflorescence at the base of a mature pseudobulb bears many sweetly scented flowers. The genus name comes from the Greek *onkos,* "small tumor," referring to the complex lip callus of its member plants. The common and species name refer to the same thing: a hyphema caused by bleeding into the eye, making it red-spotted—like the petals and sepals of *Oncidium hyphaematicum.*

The species is pollinated by oil-collecting bees, which attempt to collect oil from the lip callus. There is no oil produced, so this is a case of deceit pollination. Normally, the bees mix the oil with pollen from another non-orchid species, which they feed to their larvae.

The flower of the Red-eye Orchid has spreading, reddish-brown-blotched yellow sepals and petals. The lip is broad and bright yellow with a ridged callus and a notched tip. The column is also bright yellow with a pair of wings near the apex.

Actual size

SUBFAMILY	Epidendroideae
TRIBE AND SUBTRIBE	Cymbidieae, Oncidiinae
NATIVE RANGE	Bolivia, near La Paz
HABITAT	Wet forests, at around 4,600 ft (1,400 m)
TYPE AND PLACEMENT	Epiphytic
CONSERVATION STATUS	Not assessed
FLOWERING TIME	September (spring)

FLOWER SIZE
½ in (1.3 cm)

PLANT SIZE
3–5 × 2–3 in
(8–13 × 5–8 cm),
excluding arching
inflorescence 5–8 in
(13–20 cm) long

314

ONCIDIUM LUTZII
LITTLE SCORPION ORCHID
(KÖNIGER) M. W. CHASE & N. H. WILLIAMS, 2008

The Little Scorpion Orchid has ovoid, highly flattened pseudobulbs, basally enveloped by several overlapping leaf-bearing sheaths, topped with a single, linear-lanceolate, weakly petiolate leaf. It produces many flowers, each stem flowering repeatedly, with all flowers arranged facing in the same direction. The genus name is derived from the Greek *onkos*, "small tumor," referring to the swollen lip callus of the member species. The common name comes from the arched column that makes the flower resemble a scorpion.

The species was previously placed in *Sigmatostalix*, but DNA study has shown that genus to be part of *Oncidium*. Oil-producing glandular hairs (elaiophores) on the complex lip callus provide an oil reward for the pollinators, which are meliponine stingless bees. The insects mix the oil with pollen from other species as food for their brood.

Actual size

The flower of the Little Scorpion Orchid has highly reflexed, green petals and sepals spotted reddish-brown. The column arches upward, with a striped stalk and an enlarged apex. The lip is bright yellow, fleshy, lobed, and folded to produce a cavity into which glands secrete oil.

SUBFAMILY	Epidendroideae
TRIBE AND SUBTRIBE	Cymbidieae, Oncidiinae
NATIVE RANGE	Ecuador and Peru
HABITAT	Cloud forests, at 6,600–9,850 ft (2,000–3,000 m)
TYPE AND PLACEMENT	Epiphytic, but sometimes terrestrial on steep slopes
CONSERVATION STATUS	Not assessed
FLOWERING TIME	July to August (winter)

ONCIDIUM MULTISTELLARE
HOODED STAR ORCHID
(REICHENBACH FILS) M. W. CHASE & N. H. WILLIAMS, 2008

FLOWER SIZE
2 in (5 cm)

PLANT SIZE
10–18 × 8–15 in
(25–46 × 20–38 cm),
excluding erect-arching
lateral inflorescence
8–30 in (20–76 cm) long

315

The Hooded Star Orchid produces tight clusters of flattened, sharply edged pseudobulbs that each has two or three leaf-bearing bracts below and usually two leaves on top. From the uppermost leaf-bearing bracts, a branched inflorescence emerges, carrying 10–50 starry flowers. The species name is Latin for "covered with many stars," and the common name refers to this and the toothed hood over the end of the column.

The plant has a false nectary at the base of the column, a feature it shares with related species pollinated by bumblebees that probe such a cavity. As they explore, the bees encounter the toothlike lip projections and bump into the sticky disc on the pollinia, which becomes attached to their head and is then removed on a subsequent flower visit. The flowers have a strong, somewhat unpleasant fragrance.

The flower of the Hooded Star Orchid has greenish-yellow, outstretched lanceolate sepals and petals with reddish-brown stripes. The white lip is spade-shaped, with a pointed tip and large teeth on the brown-striped callus. The column is hooded and white with a yellow basal cavity.

Actual size

SUBFAMILY	Epidendroideae
TRIBE AND SUBTRIBE	Cymbidieae, Oncidiinae
NATIVE RANGE	Colombia
HABITAT	Cloud forests, at 6,600–7,900 ft (2,000–2,400 m)
TYPE AND PLACEMENT	Epiphytics
CONSERVATION STATUS	Not assessed
FLOWERING TIME	May to August (spring to summer)

FLOWER SIZE
2½ in (6.5 cm)

PLANT SIZE
10–18 × 12–18 in
(25–46 × 30–46 cm),
excluding arching lateral
inflorescence
14–24 in (36–61 cm) long

316

ONCIDIUM NOBILE
NOBLE SNOW ORCHID
(REICHENBACH FILS) M. W. CHASE & N. H. WILLIAMS, 2008

In its cool foggy forest habitats the Noble Snow Orchid remains well hidden until it opens its large, white, spectacular flowers, which have long been held in high regard by orchid fanciers. In the past, the species was placed in the genus *Odontoglossum*, which is based on the Greek words *odon*, "tooth," and *glossa*, "tongue," alluding to the toothlike lip projections.

The plant produces a faint fragrance that at times smells sweet and at other times somewhat sharp and disagreeable. The partial fusion of the lip to the column creates a false nectary, which means that to reach it a bee (most likely a bumblebee) must overcome the impediments created by the complex toothlike lip callus. In struggling to get past, the bee contacts the sticky disc that projects over the lip callus and pulls out the pollinia.

The flower of the Noble Snow Orchid has broadly lanceolate, projecting, white sepals and petals. The lip is also white and weakly trilobed, the broad apical lobe itself bilobed with a complex, red-striped, yellow callus. The column is winged and red spotted.

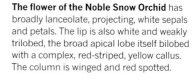

Actual size

SUBFAMILY	Epidendroideae
TRIBE AND SUBTRIBE	Cymbidieae, Oncidiinae
NATIVE RANGE	Southern Mexico, Guatemala, Belize, Honduras, Costa Rica, and Panama
HABITAT	Rain forests, gardens, and abandoned plantations, at 650–5,900 ft (200–1,800 m)
TYPE AND PLACEMENT	Epiphytic on shrubs and old citrus trees
CONSERVATION STATUS	Not formally assessed
FLOWERING TIME	July to August (summer)

FLOWER SIZE
⅛ in (0.5 cm)

PLANT SIZE
3–5 × 3–5 in
(8–13 × 8–13 cm),
excluding inflorescence
4–7 in (10–18 cm) long, which
is slightly longer than the
leaves

ORNITHOCEPHALUS INFLEXUS
CURVED BIRDHEAD ORCHID
LINDLEY, 1840

317

Actual size

The tiny Curved Birdhead Orchid grows, attached by its hairy wiry roots, on small branches and twigs, and forms a fan of flattened curving leaves. From the axils of the leaves, racemes are produced, beset with many, almost translucent flowers. The genus name is derived from the Greek *ornis*, "bird," and *kephale*, "head," referring to the shape of the column in this and other *Ornithocephalus* species, which resembles the head of a long-beaked bird. In fact, the entire flower looks like a tiny egret taking off. The common and species names both refer to the flexed or curving shape of the leaves.

The green dot in the middle of the lip is an oil gland, attracting ground-nesting anthophorid bees. They gather this oil and mix it with pollen (collected from other plants) as food for their larvae.

The flower of the Curved Birdhead Orchid has short, white sepals and much larger, clawed, broad-limbed, white petals. The lip is also white and three-lobed, the rounded sidelobes flanking a green callus, the basal lobe pointing downward. The column is slender and arches over the lip.

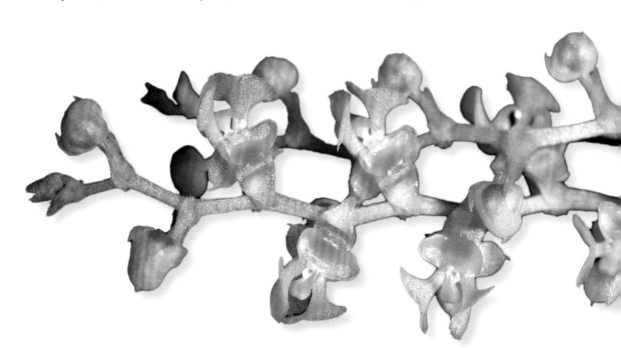

SUBFAMILY	Epidendroideae
TRIBE AND SUBTRIBE	Cymbidieae, Oncidiinae
NATIVE RANGE	Northern and western South America (Colombia to Peru)
HABITAT	Rain forests and cloud forests, at 2,950–9,200 ft (900–2,800 m)
TYPE AND PLACEMENT	Terrestrial on steep clay banks in leaf litter
CONSERVATION STATUS	Not assessed, but occurring only locally
FLOWERING TIME	February to March (late winter to early spring)

FLOWER SIZE
1¼ in (3 cm)

PLANT SIZE
8–14 × 8–12 in
(20–36 × 20–30 cm),
excluding inflorescence,
which is erect,
12–20 in (30–51 cm) tall

318

OTOGLOSSUM BREVIFOLIUM
RUFFLED DANCING LADY
(LINDLEY) GARAY & DUNSTERVILLE, 1976

The showy Ruffled Dancing Lady has smooth, ovoid or pear-shaped, flattened pseudobulbs enveloped by the bases of leafy sheaths, each bearing a single, narrowly elliptic, rigid, leathery leaf. Between each pseudobulb is an elongate rhizome, which enables the plant to run along the surface of the soil and cover a large area in a mass of stems and leaves. The bases of the lateral leafy bracts are folded and enclose the base of an inflorescence with tubular floral bracts and 6–20 flowers.

Pollination of the species is by oil-collecting bees (*Centris* and related genera) that are fooled into visiting the showy flowers by the shiny appearance of the lip callus, although no oil is present. The genus name refers to the earlike sidelobes of the lip, from the Greek words *otos*, for "ear," and *glossa*, for "tongue" (meaning the lip).

The flower of the Ruffled Dancing Lady has yellow-edged, wavy, spreading, chestnut-brown sepals and petals with a thin yellow margin. The yellow lip has a laterally compressed disc with a clawlike base, and there are reddish-brown markings on the base of the midlobe.

Actual size

SUBFAMILY	Epidendroideae
TRIBE AND SUBTRIBE	Cymbidieae, Oncidiinae
NATIVE RANGE	Tropical America, from Costa Rica to Bolivia
HABITAT	Cloud forests, at 3,950–6,600 ft (1,200–2,000 m)
TYPE AND PLACEMENT	Epiphytic
CONSERVATION STATUS	Not assessed, but occurring in many populations throughout tropical America and thus not likely to be of conservation concern
FLOWERING TIME	Throughout the year

FLOWER SIZE
1½ in (3.8 cm)

PLANT SIZE
3–6 × 3–5 in
(8–15 × 8–13 cm),
excluding stems
3–30 in (8–76 cm) long

OTOGLOSSUM GLOBULIFERUM
VINING DISC ORCHID
(KUNTH) N. H. WILLIAMS & M. W. CHASE, 2001

319

The vining, entangled stems of the little Vining Disc Orchid cover the host branches and are rooted by the widely spaced plantlets, each forming a small, flattened globose pseudobulb topped with two folded leaves. The inflorescence emerges from the stem at the base of a pseudobulb, holding a single flower— sometimes two blooms. The common and species names refer to the flat, round pseudobulbs.

The generic name comes from the Greek *oto*, "ear," and Latin *glossum*, "tongue," referring to the auriculate (eared) lip of these species. This group of species was previously part of *Oncidium*, but DNA studies have shown that species of this section are closer to *Otoglossum* than to *Oncidium*. Pollination is by oil-collecting bees that (vainly) attempt to collect what looks like oil from the lip callus— this is a case of deceit pollination.

The flower of the Vining Disc Orchid has bright yellow sepals and petals with basal, reddish-brown bars. The lip has small basal sidelobes, and the midlobe is broad and slightly notched at the tip. The tuberculate callus has red dots.

Actual size

SUBFAMILY	Epidendroideae
TRIBE AND SUBTRIBE	Cymbidieae, Oncidiinae
NATIVE RANGE	Southeastern and eastern Brazil to Misiones province (Argentina), particularly abundant in the Mantiqueira Mountains
HABITAT	Atlantic rain forests, up to 4,920 ft (1,500 m)
TYPE AND PLACEMENT	Epiphytic on small mossy tree trunks
CONSERVATION STATUS	Not assessed
FLOWERING TIME	August (winter)

FLOWER SIZE
⅛ in (0.5 cm)

PLANT SIZE
1–3 × 1–2 in
(2.5–7.5 × 2.5–5 cm),
excluding erect inflorescence
2–4 in (5–10 cm) long,
which is taller than the leaves

320

PHYMATIDIUM DELICATULUM
FAIRY AIR PLANT
LINDLEY, 1833

The tiny Fairy Air Plant orchid forms a clump of falcate, awl-shaped, tapering, twisting, and asymmetric leaves that are triangular in cross section. In winter, finely hairy, zigzag inflorescences with many flowers appear with widely spaced flowers. The plant looks like it could be a small bromeliad of the genus *Tillandsia*, commonly referred to as air plants, hence the common name. The genus name comes from the Greek *phyma*, meaning "tumor," and the diminutive *–idium*, referring to the swollen base of the column. The species name reflects the delicate nature of these denizens of shady sites in wet forests.

The species of genus *Phymatidium* secrete oil from the hairs on the lip callus. The pollinators are bees that collect this oil to mix with pollen as food for their larvae.

The flower of the Fairy Air Plant is tiny, with white, narrow, spreading sepals and petals, the latter slightly longer. The lip folds below the short column to form a shallow cup, and the limb is arrowhead shaped. The swollen base of the column is green.

Actual size

SUBFAMILY	Epidendroideae
TRIBE AND SUBTRIBE	Cymbidieae, Oncidiinae
NATIVE RANGE	Central America, from Chiapas (Mexico) to Guatemala, and from Costa Rica to Colombia
HABITAT	Premontane forests in deep shade, at 1,080–4,600 ft (330–1,400 m)
TYPE AND PLACEMENT	Epiphytic
CONSERVATION STATUS	Not assessed
FLOWERING TIME	Summer

FLOWER SIZE
1⅜ in (3.5 cm)

PLANT SIZE
2–4 × 2–3 in
(5–10 × 5–7.5 cm),
including arching, single-
flowered inflorescence
1–2 in (2.5–5 cm) long

PLECTROPHORA ALATA
CORNUCOPIA ORCHID
(ROLFE) GARAY, 1967

321

The small Cornucopia Orchid has round pseudobulbs, obscured by the leaf bases that wrap around them, with the leaves arranged like a fan. The bulb apex bears a single fleshy elliptic-oblong, folded leaf, and an inflorescence with a single, long-spurred flower grows from its base. The lip is shaped like a horn of plenty or cornucopia, hence the common name. The genus name derives from the Greek words *plektron*, "spur," and *phoros*, "bearing," in reference to the long spur produced by the lip base. The species name, *alata*, from the Latin for "winged," refers to the lateral sepals, which project from the flowers like wings.

Male euglossine bees are reported to be the pollinators of *Plectrophora alata*, but the presence of a spur probably means that the visitors are seeking a nectar reward rather than collecting floral fragrance compounds. Nectar has never been observed in the spur.

The flower of the Cornucopia Orchid has recurved white sepals that are fused at the base to form a long, curved spur. The white petals and lip form a basket, with its margins ruffled and red-orange stripes within.

Actual size

SUBFAMILY	Epidendroideae
TRIBE AND SUBTRIBE	Cymbidieae, Oncidiinae
NATIVE RANGE	Brazil (near Rio de Janeiro)
HABITAT	Seasonally arid areas, often on cactus
TYPE AND PLACEMENT	Epiphyte
CONSERVATION STATUS	Not assessed
FLOWERING TIME	Spring to fall

FLOWER SIZE
1¼–1½ in (3–4 cm)

PLANT SIZE
1–2 × 2–3 in
(2.5–5 × 5–8 cm),
excluding short inflorescence
2–3 in (5–8 cm) tall

PSYCHOPSIELLA LIMMINGHEI
LITTLE BUTTERFLY ORCHID
(E. MORREN EX LINDLEY) LÜCKEL & BRAEM, 1982

322

Actual size

The miniature creeping Little Butterfly Orchid, the only species in its genus, is, as the name suggests, a close relative of the plants of the genus *Psychopsis*, though highly reduced in size. Small flattened, overlapping heart-shaped pseudobulbs with red-veined and mottled leaves clasp their substrate.

The intricate flowers are large, considering the tiny plants from which they emerge. They are somewhat different from typical *Psychopsis* plants in not bearing petals and sepals resembling antennae and appearing more like the oil-rewarding mimics of the Oncidiinae subtribe. The plants also differ from *Psychopsis* species in that the inflorescences do not produce flowers successively. Indeed, all *Psychopsis* species were once treated as a section (*Glanduligera*) of the genus *Oncidium*, until taxonomists realized their many differences.

The flower of the Little Butterfly Orchid is pale yellow with reddish-brown patterning on all segments and lip, particularly near the lip margin, where a zone of spots occurs. The column bears feathery arms on each side that have a minute gland at each tip.

SUBFAMILY	Epidendroideae
TRIBE AND SUBTRIBE	Cymbidieae, Oncidiinae
NATIVE RANGE	Trinidad and Tobago, Panama, French Guiana, Surinam, Venezuela, Colombia, and northern Brazil
HABITAT	Lower montane wet forests, at 1,640–2,625 ft (500–800 m)
TYPE AND PLACEMENT	Epiphytic
CONSERVATION STATUS	Not threatened
FLOWERING TIME	Throughout the year

PSYCHOPSIS PAPILIO
NORTHEASTERN BUTTERFLY ORCHID
(LINDLEY) H. G. JONES, 1975

FLOWER SIZE
6 in (15 cm)

PLANT SIZE
10–16 × 3–5 in
(25–41 × 8–13 cm),
excluding single-flowered
inflorescence, which is erect and
much longer than the leaves,
24–50 in (61–127 cm) long

323

This orchid is famous for its large flowers, which resemble a butterfly. Its antenna-like dorsal sepal and lateral petals, and lateral sepals that resemble wings, have provided the Northeastern Butterfly Orchid with an obvious common name. In spite of this appearance, pollination is by oil-collecting bees, attracted by its production of small amounts of oil on a complicated lip callus. Other highly fanciful theories have included pollination by male butterflies that think the flower is a female of their species and mad male butterflies that think it is a butterfly invading their territory.

All six *Psychopsis* species share the same flower shape and coloration. Flowers emerge successively on long scapes, replacing spent flowers shortly after they fade, often for several years.

The flower of the Northeastern Butterfly Orchid has a yellow background overlaid with red blotching. The dorsal sepal and lateral petals are narrow, similar, and antenna-like, and the lateral sepals are much broader. The lip is large and ringed with ocher and has a broad, yellow center with a complex callus.

Actual size

SUBFAMILY	Epidendroideae
TRIBE AND SUBTRIBE	Cymbidieae, Oncidiinae
NATIVE RANGE	Mexico, Guatemala, Honduras, El Salvador, Costa Rica, and Panama
HABITAT	Humid forests and rocky areas, at 6,600–10,500 ft (2,000–3,200 m)
TYPE AND PLACEMENT	Terrestrial, occasionally epiphytic
CONSERVATION STATUS	Not threatened
FLOWERING TIME	Winter and spring

FLOWER SIZE
1½ in (3.8 cm)

PLANT SIZE
20–35 × 12–20 in
(51–89 × 30–51 cm),
excluding erect
inflorescence
35–68 in (89–173 cm) tall

324

RHYNCHOSTELE BICTONIENSIS
MOTHER OF HUNDREDS
(BATEMAN) SOTO ARENAS & SALAZAR, 1993

The Mother of Hundreds orchid is one of two species in its genus that grow mostly on the ground. The other, *Rhynchostele uroskinneri,* is larger and much more handsome, with more brightly colored flowers. Unfortunately, that species does not make hybrids easily, whereas *R. bictoniensis* has been widely used in artificial hybrids—hence the common name. In nature, the plant forms large clusters of pseudobulbs, from which emerge a sturdy stem bearing 20–50 strange-smelling flowers (like "overheated electronics" to some). The genus name comes from the Greek words *rhynchos*, "beak," and *stele*, "column," referring to the beaked apex of the column.

The pollinators of the Mother of Hundreds are unknown. Unlike many species of the subtribe Oncidiinae, it is not pollinated by oil-collecting bees deceived into visiting flowers mimicking members of the Malpighiaceae family, as such plants do not grow at such high elevations.

The flower of Mother of Hundreds is star-shaped, with sepals and petals in shades of brown to tan, usually with brown spotting. The spade-shaped lip is pink to deep purple with a boat-shaped, basal callus. The arching column has a pair of apical wings.

Actual size

SUBFAMILY	Epidendroideae
TRIBE AND SUBTRIBE	Cymbidieae, Oncidiinae
NATIVE RANGE	Central and southwestern Mexico (states of Durango, Nayarit, Jalisco, Michoacán, Guerrero, Mexico, Morelos, Oaxaca)
HABITAT	Mixed pine and oak forests, cliffs, and rocky slopes, at 4,600–10,500 ft (1,400–3,200 m)
TYPE AND PLACEMENT	Epiphytic
CONSERVATION STATUS	Not formally assessed, but may be under threat from overharvesting
FLOWERING TIME	January to April (winter to spring)

FLOWER SIZE
1⅜–2¾ in (3.6–7.1 cm)

PLANT SIZE
4–8 × 3–6 in
(10–20 × 8–15 cm),
excluding inflorescence,
which is arching pendent and
6–10 in (15–25 cm) long

RHYNCHOSTELE CERVANTESII
SQUIRREL ORCHID
(LEXARZA) SOTO ARENAS & SALAZAR, 1993

325

The beautiful and curiously marked dwarf Squirrel Orchid has clustered glaucous, ovoid-angular pseudobulbs that are often brown spotted. Each pseudobulb is topped with a single oblong, folded leaf. An arching or pendent inflorescence with linear, brown papery bracts and up to ten fragrant flowers is produced after a semidry period in early winter.

The common name is a translation of the ancient Zapotec name *guièe-dẑîl-ndẑìẑ*, which means, "squirrel orchid flower." The unusual concentric markings on the sepals and petals have resulted in another common name: "bull's eye orchid." Pollination of the Squirrel Orchid has not been investigated, but given its floral morphology, specifically the open shape, colors, and markings, a bee would be expected. White flowers with dark markings are the most common form, but there are also pale to bright pink individuals.

The flower of the Squirrel Orchid has spreading white sepals and broader petals with reddish-brown markings. The white lip has some basal red markings and is stalked and spade-shaped, with a swollen, yellow callus. The column is broadly winged.

Actual size

SUBFAMILY	Epidendroideae
TRIBE AND SUBTRIBE	Cymbidieae, Oncidiinae
NATIVE RANGE	Guerrero state (Mexico)
HABITAT	Cliffs facing east, at 3,300–3,950 ft (1,000–1,200 m)
TYPE AND PLACEMENT	Terrestrial or lithophytic
CONSERVATION STATUS	Not assessed
FLOWERING TIME	September to November (fall)

FLOWER SIZE
2 in (5 cm)

PLANT SIZE
15–25 × 12–20 in
(38–64 × 30–51 cm),
excluding erect
axillary inflorescence
25–75 in (64–191 cm) tall

RHYNCHOSTELE LONDESBOROUGHIANA
YELLOW AXE ORCHID
(REICHENBACH FILS) SOTO ARENAS & SALAZAR, 1993

326

Actual size

The flower of the Yellow Axe Orchid has yellow
sepals and petals that have dense, red-brown
banding, almost bull's-eye-like. The lip is bright
yellow with a few brown bars, two short
sidelobes, and a knob-like callus. The yellow
column curves over the lip callus.

The Yellow Axe Orchid has a narrow distribution in the state
of Guerrero, western Mexico, where it occurs on crystalline
rocks in full sun, shedding its leaves during the dry season.
The plant makes tight clusters of egg-shaped pseudobulbs
with two to three lanceolate leaves on top and produces a
tall inflorescence with 15–30 beautiful flowers. The genus
name comes from the Greek words for *rhyncos*, "beak," and
stele, "column," while the common name alludes to the axe-
blade shape of the lip.

Among the species of *Rhynchostele*, *R. londesboroughiana* is an
oddball due to its bright yellow flowers, which at an earlier stage
placed it in its own genus, *Mesoglossum*. DNA analyses link the
plant to *Rhynchostele*, but it does exhibit the floral characteristics
of pollination by oil-collecting bees, traits that it evolved
independently among the species of *Rhynchostele*.

SUBFAMILY	Epidendroideae
TRIBE AND SUBTRIBE	Cymbidieae, Oncidiinae
NATIVE RANGE	Guayanas, Ecuador, southeastern Brazil
HABITAT	Cloud forests, at 1,640–5,900 ft (500–1,800 m)
TYPE AND PLACEMENT	Epiphytic
CONSERVATION STATUS	Not assessed
FLOWERING TIME	March to May (spring)

RODRIGUEZIA VENUSTA
FRAGRANT ANGEL ORCHID
(LINDLEY) REICHENBACH FILS, 1852

FLOWER SIZE
1½ in (3.7 cm)

PLANT SIZE
6–10 × 4–6 in
(15–25 × 10–15 cm),
excluding inflorescence,
which is arching to pendent,
7–12 in (18–30 cm) long

327

The elegant Fragrant Angel Orchid has narrowly ovoid pseudobulbs subtended by leaflike sheaths, each carrying a single linear-lanceolate leathery leaf. Each pseudobulb produces between one and three arching or pendent racemes carrying up to 12 sweet, citrus-scented flowers. Their gracefulness is reflected in the scientific name, which comes from the Latin word *venustus*, meaning "graceful" or "charming." These flowers generally live for only three to four days.

Male and female euglossine bees pollinate the species, attracted by the nectar secreted by special cells on a horn at the base of the lip. Extra-floral nectaries occur on the edges of young leaves and the bracts subtending the flower buds. Ants of the genus *Crematogaster* are reported to visit these nectaries, which presumably discourages passing herbivores.

The flower of the Fragrant Angel Orchid is white, with spreading lateral sepals and a slightly hooded upper sepal. Petals are smaller and flank the column. The lip is three-lobed, with a swollen, yellow callus, and the lateral lobes overlap the emarginate midlobe.

Actual size

SUBFAMILY	Epidendroideae
TRIBE AND SUBTRIBE	Cymbidieae, Oncidiinae
NATIVE RANGE	Guatemala and Belize (Central America) south to Peru
HABITAT	Seasonally dry forests, from sea level to 985 ft (300 m)
TYPE AND PLACEMENT	Epiphytic
CONSERVATION STATUS	Not threatened
FLOWERING TIME	Winter to spring

FLOWER SIZE
1¼ in (3 cm)

PLANT SIZE
10–15 × 12–24 in
(25–38 × 30–61 cm),
excluding the erect-arching
inflorescences,
12–30 in (25–76 cm) long

328

ROSSIOGLOSSUM AMPLIATUM
TURTLE-SHELL ORCHID
(LINDLEY) M. W. CHASE & N. H. WILLIAMS, 2008

Actual size

A handsome, vigorous plant of lowland, deciduous forests, the Turtle-shell Orchid is widely admired for its large, flat pseudobulbs (fancifully imagined to look like a turtle's shell) and branching sprays of delightful yellow flowers. These blooms caused it to be treated as a member of genus *Oncidium* when it was described (by Lindley, in 1833). In subtribe Oncidiinae, species with yellow flowers and this general shape are pollinated by oil-collecting bees and were all formerly placed in *Oncidium*. However, DNA study has shown that this flower type is solely an adaptation to pollination, which has evolved repeatedly in the subtribe. The Turtle-shell Orchid is one of these independent cases.

The genus was named for John Ross, a nineteenth-century collector of orchids in Mexico. The species name is derived from the Latin word for widened, again a reference to the broad pseudobulbs of *Rossioglossum ampliatum*.

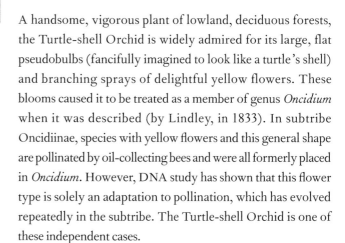

The flower of the Turtle-shell Orchid has small, yellow sepals spotted with brown, whereas the petals are larger, flat, and brighter with spotting at their bases. The midlobe of the lip lobes is much broader than the sidelobes and flat, with a central, lumpy callus with reddish-brown markings.

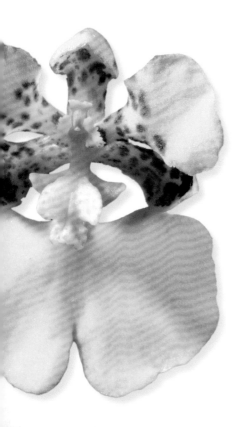

SUBFAMILY	Epidendroideae
TRIBE AND SUBTRIBE	Cymbidieae, Oncidiinae
NATIVE RANGE	Mexico (Chiapas state), Guatemala, and Belize
HABITAT	Deciduous, cool wet forests, at 4,920–8,900 ft (1,500–2,700 m)
TYPE AND PLACEMENT	Epiphytic
CONSERVATION STATUS	Not assessed
FLOWERING TIME	November to January

ROSSIOGLOSSUM GRANDE
TIGER'S MOUTH
(LINDLEY) GARAY & G. C. KENNEDY, 1976

FLOWER SIZE
7–9 in (18–23 cm)

PLANT SIZE
8–12 × 8–10 in
(20–30 × 20–25 cm),
excluding erect-arching
inflorescence, which is
longer than the leaves,
10–16 in (25–41 cm) long

329

The striking, large-flowered Tiger's Mouth, named for its coloration and the pair of prominent teeth on the callus in the center of the lip (*boca del tigre* in Spanish), is possibly the best-known member of the genus *Rossioglossum*. Most species occur at higher elevations in seasonally dry deciduous forests, where their large size and bright colors with a glossy, lacquered finish make an impressive display.

Like many other species in the Oncidiinae subtribe, these orchids engage in oil-reward deception, mimicking other, more common flowers that actually reward *Centris* bees with oil for their pollination services. There seems to be a small amount of oil produced on the complex shiny lip callus, but it would not appear that there is enough oil production to actually serve as reward for the bees.

The flower of the Tiger's Mouth has bright yellow sepals and petals that are overlaid with reddish-brown barring that is almost solid at the base. The lip is paler yellow, with a prominent reddish-brown-barred lip callus with several teeth. The column is yellow and winged.

Actual size

SUBFAMILY	Epidendroideae
TRIBE AND SUBTRIBE	Cymbidieae, Oncidiinae
NATIVE RANGE	Bolivia and southern Peru
HABITAT	Cloud forests, at around 5,900 ft (1,800 m)
TYPE AND PLACEMENT	Epiphytic on mossy branches
CONSERVATION STATUS	Not assessed
FLOWERING TIME	June and November to December

FLOWER SIZE
⅜ in (1 cm)

PLANT SIZE
1–2 × 1 in
(2.5–5 × 2.5 cm),
excluding pendent
inflorescence
1–2 in (2.5–5 cm) long, which
is about as long as the leaves

330

SEEGERIELLA PINIFOLIA
PINE-NEEDLE ORCHID
SENGHAS, 1997

This tiny orchid has a fan of stiff leaves that resemble pine needles embedded in moss—hence the common and species names. When young, the plants are fan-shaped, but, as they mature, their flattened leaves are replaced by ones that are nearly round. The inflorescence is an umbel, with 4–5 simultaneously opening flowers. The genus is named for Hans-Gerhardt Seeger, an orchid grower at the botanical garden of the University of Heidelberg, where this species was originally described from a plant in cultivation.

Little is known about the biology of the fascinating *Seegeriella pinifolia* in the wild, but it could be speculated that, like the related species of *Macroclinium* and *Noytlia*, it is pollinated by fragrance-collecting male euglossine bees. A second species is known from Ecuador.

Actual size

The flower of the Pine-needle Orchid has spreading, pale green to white sepals and petals. The base of the anchor-shaped white lip is fused with the column to its halfway point and then curves away at a right angle. The head of the column is enlarged.

SUBFAMILY	Epidendroideae
TRIBE AND SUBTRIBE	Cymbidieae, Oncidiinae
NATIVE RANGE	Northern South America
HABITAT	Wet forests, at 650–2,300 ft (200–700 m)
TYPE AND PLACEMENT	Epiphytic
CONSERVATION STATUS	Not assessed
FLOWERING TIME	May to July

SOLENIDIUM LUNATUM
LUNAR ORCHID
(LINDLEY) SCHLECHTER, 1914

FLOWER SIZE
¾ in (2 cm)

PLANT SIZE
6–15 × 4–6 in
(15–38 × 10–15 cm),
excluding erect to arching
unbranched inflorescence
8–18 in (20–46 cm) long,
which is longer than the leaves

331

The clump-forming Lunar Orchid has flattened, ellipsoid pseudobulbs that are topped with one or two fleshy leaves. From the base of a mature pseudobulb, a raceme with 8–24 flowers is formed. The genus name comes from the Greek *solen*, "channel," and the diminutive *–idion*, referring to the narrow, somewhat tubular base of the lip. The common and species names both refer to the lip lamina, which is shaped like a crescent moon.

Much is unknown about the biology of this species in the wild, including pollination. There is neither an obvious reward offered by the flowers nor a nectar spur, both of which usually mean that some sort of bee attracted by deceit is the most likely pollinator. The flowers are not adapted for moth or butterfly pollination, which require a long nectar spur.

Actual size

The flower of the Lunar Orchid has similar, spreading, red-brown-spotted sepals and petals. The base of the white lip forms a grooved, elongate, hairy callus, and there are generally red-brown spots on the broadened apex. The anther cap has an upturned front edge.

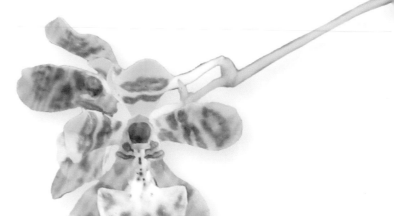

SUBFAMILY	Epidendroideae
TRIBE AND SUBTRIBE	Cymbidieae, Oncidiinae
NATIVE RANGE	Ecuador
HABITAT	Tropical forests, at 1,640–4,920 ft (500–1,500 m)
TYPE AND PLACEMENT	Epiphytic
CONSERVATION STATUS	Not formally assessed
FLOWERING TIME	June to August

FLOWER SIZE
1¼ in (3.2 cm)

PLANT SIZE
6–10 × 4–6 in
(15–25 × 10–15 cm),
excluding erect-arching
inflorescence 12–20 in
(30–51 cm) long, which
greatly exceeds the leaves

332

SYSTELOGLOSSUM ECUADORENSE
BIRD-KNEE ORCHID
(GARAY) DRESSLER & N. H. WILLIAMS, 1970

Actual size

The peculiar Bird-knee Orchid forms tight clusters of highly flattened, ovate-oblong pseudobulbs, partly enveloped by leafy sheaths and each topped by a single, oblong leaf. It makes a flattened, loosely branched panicle, with each branch holding a cluster of between one and four flowers that open several at a time over a long period. The genus name comes from the Greek *systellein*, "fuse together," and *glossa*, "tongue," referring to the fusion of the lip (tongue) to the extended base of the column. The common name refers to the long column foot, which has a reversed bird-like joint (knee) with the pair of lateral sepals.

Pollination of these strangely shaped flowers has not been studied. With their nectar spur and dull color, however, it seems likely that a night-flying moth is the pollinator.

The flower of the Bird-knee Orchid has a prominent column foot to which two sepals fuse to form a bilobed "tongue," while the third sepal and two petals cover the column. The lip sits closely below the hooded column. All parts are dull green to reddish-green.

SUBFAMILY	Epidendroideae
TRIBE AND SUBTRIBE	Cymbidieae, Oncidiinae
NATIVE RANGE	Northwestern South America, Ecuador, Colombia, and northwestern Venezuela
HABITAT	Cloud forests, at 7,545–9,850 ft (2,300–3,000 m)
TYPE AND PLACEMENT	Epiphytic
CONSERVATION STATUS	Not assessed
FLOWERING TIME	March to September (spring and fall)

TELIPOGON HAUSMANNIANUS
MARIPOSA ORCHID
REICHENBACH FILS, 1861

FLOWER SIZE
2 in (5 cm)

PLANT SIZE
2–3 × 2–4 in
(5–8 × 5–10 cm),
excluding erect-arching
inflorescences
2–4 in (5–10 cm) tall

333

Finding one of these small beauties emerging from the mists of their native cloud forests is one of the greatest thrills for any orchid lover. Sadly, the Mariposa Orchid is difficult to maintain in cultivation. It has a short stem, covered by a few elliptic leaves, and large cupped petals with a netted pattern, reminiscent of the mariposa lily (genus *Calochortus*) or a butterfly (*mariposa* in Spanish)—hence its common name. The genus name comes from the Greek *telos*, "end," and *pogon*, "beard," referring to the hairs around the base of the column, which play an important role in pollination.

Telipogon hausmannianus is pollinated by male tachinid flies, which mistake the flowers for females of their species. Pollinia are then placed on the feet of the aroused visitors, whose subsequent amorous visits effect pollen transfer.

The flower of the Mariposa Orchid has three narrow sepals almost hidden behind the broad petals and lip. The color is yellow, with reddish-purple netted venation. More darkly colored basal spots on the petals or lip are covered in long dark hairs.

Actual size

SUBFAMILY	Epidendroideae
TRIBE AND SUBTRIBE	Cymbidieae, Oncidiinae
NATIVE RANGE	Northwestern South America, Ecuador, Colombia, and northwestern Venezuela
HABITAT	Cloud forests, at 4,300–7,545 ft (1,300–2,300 m)
TYPE AND PLACEMENT	Epiphytic
CONSERVATION STATUS	Not assessed
FLOWERING TIME	Throughout the year

FLOWER SIZE
¼ in (0.6 cm)

PLANT SIZE
1–2 × 1–2 in
(2.5–5 × 2.5–5 cm),
excluding arching-erect
inflorescence
3–4 in (7.5–10 cm) long

334

TELIPOGON WILLIAMSII
BEARDED LADY
P. ORTIZ, 2008

Actual size

This little epiphyte grows on small moss-covered branches with a few leaves and no obvious pseudobulb. The inflorescence carries up to five diminutive, flylike flowers produced individually over a period of several months. The lip of the species is so like a female tachinid fly that it is visited by males that attempt copulation and pollinate the flower during their passion. Pollinia are attached to their legs, and the pollen masses contact the stigma during the next amorously deceitful visit.

The Bearded Lady was previously considered to be in the genus *Stellilabium*, but in 2005 its members were shown to be merely smaller species of *Telipogon*, and the two genera were then combined. They share the same bearded or warty patch on the base of the lip, which is referenced in the genus name, from the Greek *telos*, "end," and *pogon*, "beard."

The flower of the Bearded Lady has two tiny lateral sepals, a larger, recurving middle sepal, two spreading petals, and a larger lip with a warty callus. The color is yellowish-green with reddish-brown spots and a large, warty area on the lip in front of the hooked column.

SUBFAMILY	Epidendroideae
TRIBE AND SUBTRIBE	Cymbidieae, Oncidiinae
NATIVE RANGE	Dominican Republic
HABITAT	Dry cactus scrub and dry subtropical forests
TYPE AND PLACEMENT	Epiphytic
CONSERVATION STATUS	Not assessed, but rare with restricted range; several sites have been made nature reserves, and reintroduction programs are in action
FLOWERING TIME	December to July (winter to summer)

TOLUMNIA HENEKENII
HISPANIOLAN BEE ORCHID
(M. R. SCHOMBURGK EX LINDLEY) NIR, 1994

FLOWER SIZE
¾ in (2 cm)

PLANT SIZE
2–4 × 2–4 in
(5–10 × 5–10 cm),
excluding erect, rarely
branched inflorescence
8–15 in (20–38 cm) long

335

The dwarf Hispaniolan Bee Orchid has a short stem with a fan of a few flat, tooth-edged leaves and no pseudobulb. An axillary zigzag raceme successively produces blooms, forming a new bud when the old flower has withered. The orchid is sometimes placed in its own genus, *Hispaniella*, named for the island of Hispaniola, where it occurs. The genus name *Tolumnia* derives from Tolumnius, one of the Rutulians who were adversaries of Aeneas in Virgil's *Aeneid*.

The strangely constructed lip resembles a female bee, attracting male centris bees (*Centris insularis*) for pollination by attempted copulation. There are two types of male bee behavior: a "hit-and-fly" approach and a "mounting-and-caressing" maneuver. Only the latter results in the desired effect, the other probably a result of these bees' territorial nature.

Actual size

The flower of the Hispaniolan Bee Orchid is almost all lip, which is dark reddish-brown—almost black—with a yellow margin and covered in hairs. The central callus is pale orange. There are small, pale yellow, reflexed sepals and spreading, yellow petals.

SUBFAMILY	Epidendroideae
TRIBE AND SUBTRIBE	Cymbidieae, Oncidiinae
NATIVE RANGE	Antillean islands Cuba and Hispaniola
HABITAT	Cloud forests and rain forests, at 2,950–4,920 ft (900–1,500 m)
TYPE AND PLACEMENT	Epiphytic
CONSERVATION STATUS	Not assessed
FLOWERING TIME	August to September (summer)

FLOWER SIZE
1 in (2.5 cm)

PLANT SIZE
3–6 × 3–6 in
(8–15 × 8–15 cm),
excluding laxly erect
inflorescence 10–24 in
(25–61 cm) long, which is
much longer than the leaves

TOLUMNIA TUERCKHEIMII
ANTILLEAN DANCING LADY
(COGNIAUX) BRAEM, 1986

336

Actual size

The Antillean Dancing Lady produces a downward-facing fan of flattened red-spotted or red-suffused leaves that typically hangs on the underside of small branches. The genus, named for Tolumnius, a Rutulian soothsayer in Virgil's *Aeneid*, was originally segregated from the large genus *Oncidium* because of its flattened leaves and lack of pseudobulbs—a move later supported by analyses of DNA. The species name is in honor of Hans von Türckheim (1853–1920), a German lawyer and plant collector. The common name refers to a fanciful resemblance of the large lip to a dancing figure.

Like most species in genus *Oncidium*, *Tolumnia tuerckheimii* is deceitfully pollinated by oil-collecting bees. They mistake the flowers for those of the distantly related plant family Malpighiaceae, which reward bees with oil that is mixed with pollen to feed their larvae.

The flower of the Antillean Dancing Lady has recurved, reddish-brown-spotted, yellow sepals and similarly colored, spreading, wavy-margined petals, and a three-lobed lip with a fingered callus and a broad middle lobe. The column has a pair of terminal wings.

SUBFAMILY	Epidendroideae
TRIBE AND SUBTRIBE	Cymbidieae, Oncidiinae
NATIVE RANGE	Northern and western South America
HABITAT	Wet forests, at 3,300–5,250 ft (1,000–1,600 m)
TYPE AND PLACEMENT	Epiphytic
CONSERVATION STATUS	Not assessed
FLOWERING TIME	July to November (summer to fall)

FLOWER SIZE
1½ in (4 cm)

PLANT SIZE
3–8 × 2–3 in
(8–20 × 5–8 cm),
excluding pendent
inflorescence
2–3 in (5–8 cm) long

337

TRICHOCENTRUM PULCHRUM
BEAUTIFUL SPOTTED ORCHID
POEPPIG & ENDLICHER, 1836

The small Beautiful Spotted Orchid forms a tuft of fleshy, stiff, linear-oblong, red-spotted leaves. It has a pendent inflorescence, which usually holds one flower that opens among the leaves. Two flowers may appear in rare cases. There is no pseudobulb, and for this reason the species of this group are often referred to as mule-ears (*orejas de burro* in Spanish). The genus name is derived from the Greek words *trichos*, for "hair," and *kentron*, "spur," in reference to the often long and slender spurs of the member species.

Trichocentrum pulchrum belongs to a species complex in which the similar species can be identified by their spur length, lip keel, and column wings. Pollination is by euglossine bees, and it is likely that the large, but empty, spur of this species attracts both male and female insects searching for nectar.

The flower of the Beautiful Spotted Orchid has similar, cream to white petals and sepals, the upper sepal slightly hooded, and all with pinkish-red spots. The yellow lip is also red-spotted, angular and with two parallel ridges in the middle. The spur is long and upturned.

Actual size

SUBFAMILY	Epidendroideae
TRIBE AND SUBTRIBE	Cymbidieae, Oncidiinae
NATIVE RANGE	Northern Bolivia and Peru
HABITAT	Tropical rain forests, shady positions in the tree canopy, at 1,640 ft (500 m)
TYPE AND PLACEMENT	Epiphytic
CONSERVATION STATUS	Not assessed
FLOWERING TIME	October

FLOWER SIZE
2½ in (6.5 cm)

PLANT SIZE
24–40 × ½ in
(61–102 × 1.3 cm),
per pendent individual growth,
excluding arching-erect
inflorescence
8–14 in (20–36 cm) long

338

TRICHOCENTRUM STACYI
RAT-TAILED OCELOT
(GARAY) M. W. CHASE & N. H. WILLIAMS, 2001

The Rat-tailed Ocelot produces up to 20 scentless, showy flowers, borne on a zigzag stem that emerges from a cluster of bract-covered, ovoid-cylindrical pseudobulbs, each topped with a long, pendent, terete (round with a groove) leaf. The genus name derives from the Greek *trichos*, "hair," and *kentron* "spur," referring to the long nectar spur observed in some species in the genus—although not in the spur-less Rat-tailed Ocelot. The common name refers to the shape of the leaves and the spotted, ocelot-like appearance of the flowers. The species name honors John Stacy, the American orchid specialist who discovered the plant in Bolivia.

Mimicking species in the plant family Malpighiaceae, *Trichocentrum stacyi* is pollinated by *Centris* bees that attempt to collect what they believe is oil on the lip callus of the flowers. No oil, however, is present.

The flower of the Rat-tailed Ocelot is yellow and heavily spotted with reddish-brown. The sepals and petals are wavy along the edge. The lip is three-lobed, with smaller sidelobes, a heavily fringed and lumpy callus, and a stalked midlobe. The column has a pair of apical wings.

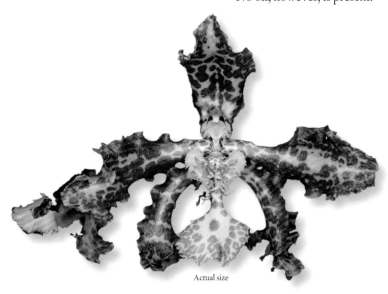

Actual size

SUBFAMILY	Epidendroideae
TRIBE AND SUBTRIBE	Cymbidieae, Oncidiinae
NATIVE RANGE	Colombia, Ecuador, Peru, and Bolivia
HABITAT	Dryer exposed mossy slopes in cloud forests, at 5,900–9,200 ft (1,800–2,800 m)
TYPE AND PLACEMENT	Terrestrial on soil and rocks and epiphytic on lower branches of shrubs
CONSERVATION STATUS	Not threatened
FLOWERING TIME	Late fall to spring

FLOWER SIZE
¾ in (2 cm)

PLANT SIZE
4–6 × 3–5 in
(10–15 × 8–13 cm),
excluding erect inflorescence
9–15 in (23–38 cm) tall, which
is much taller than the rest
of the plant

TRICHOCEROS ANTENNIFER
ANDEAN FLY ORCHID
(HUMBOLDT & BONPLAND) KUNTH, 1816

339

The Andean Fly Orchid has several distantly spaced ovoid pseudobulbs with succulent leaves. These plants produce a tall stem with the flowers clustered near the top, which permits them to stand clear of the shrubby vegetation in which they grow. The genus name is from the Greek words *trichos*, "hair," and *keras*, "horn," referring to the two erect hairy arms of the lip that stand alongside the column. This is also the basis for the species name, which alludes to these same two horns looking like the antennae of a fly.

Many orchid flowers named after an insect are often not in fact pollinated by that insect (moth orchids, genus *Phalaenopsis*, for instance, are pollinated by bees in spite of their mothlike appearance). These fly orchids are pollinated by male tachinid flies that mistake the flowers for females.

Actual size

The flower of the Andean Fly Orchid has pale olive sepals and petals setting off a purple-brown-spotted, hairy lip, with antennae-like sidelobes. The column is short and covered in hairs, with the sticky disc projecting up and over the stigmatic cavity.

SUBFAMILY	Epidendroideae
TRIBE AND SUBTRIBE	Cymbidieae, Oncidiinae
NATIVE RANGE	Ecuador to Peru
HABITAT	Cool wet forests, at 4,920–9,850 ft (1,500–3,000 m)
TYPE AND PLACEMENT	Epiphytic
CONSERVATION STATUS	Not formally assessed, but it is widespread and locally common so probably not of conservation concern
FLOWERING TIME	August to October (winter and spring)

FLOWER SIZE
3 in (8 cm)

PLANT SIZE
4–10 × 2–4 in
(10–25 × 5–10 cm), excluding
arching-pendent single-
flowered inflorescence
4–10 in (10–25 cm) long

340

TRICHOPILIA SANGUINOLENTA
BLOOD-STAINED CAP
(LINDLEY) REICHENBACH FILS, 1867

The flower of the Blood-stained Cap has spreading, greenish-yellow, red-spotted sepals and petals, the upper one curved over the column. The lip is white with a ruffled margin and marked with red lines and spots in its basal portion. The anther has a feathered margin.

Pseudobulbs of this species are enveloped in dry leafless sheaths and clustered tightly together, each topped by a single leaf with a wavy margin. From a mature pseudobulb, one or (rarely) two strongly scented and long-lasting flowers are produced. The genus name is derived from the Greek *trichos*, "hair," and *pilios*, "cap," a reference to the apex of the column, which has a fringed cap over the anther. The common and species name both refer to the (blood) red spotting on the flower (from the Latin *sanguinolentus*, "bloody").

The Blood-stained Cap is pollinated by male and female euglossine bees (as is the case for several other species of *Trichopilia*), which visit the flowers looking for nectar. No reward is offered, even though the fragrance and flower markings suggest the opposite.

Actual size

SUBFAMILY	Epidendroideae
TRIBE AND SUBTRIBE	Cymbidieae, Oncidiinae
NATIVE RANGE	Costa Rica to northern Colombia
HABITAT	Montane, seasonally dry forest margins
TYPE AND PLACEMENT	Epiphytic
CONSERVATION STATUS	Vulnerable to poaching
FLOWERING TIME	February to April

FLOWER SIZE
4–4¾ in (10–12 cm)

PLANT SIZE
10–18 × 4–6 in
(25–46 × 10–15 cm),
excluding the pendent
inflorescence, which is
4–8 in (10–20 cm) long

341

TRICHOPILIA SUAVIS
PINK-SPOTTED HOOD ORCHID
LINDLEY & PAXTON, 1850

Despite bearing large, intensely fragrant, and vividly colored blossoms, the Pink-spotted Hood Orchid is, surprisingly, only occasionally cultivated. The medium-sized plants grow low on thick forest branches, where the pendent spikes attract male euglossine bees, although it is not clear if they are collecting floral fragrance compounds, as is typical for these insects.

The genus is closely related to *Psychopsis* and *Psychopsiella*, a group of species with oil-producing bee-attracting blooms that could not differ more from those of *Trichopilia*—evidence that pollination drives evolution of floral features, even in closely related orchids. The common name of the genus refers to the hood on the apex of the column that covers the anther cap, a reference that occurs in the genus name also, from the Greek word *pilos*, "felt," a material used to make hoods.

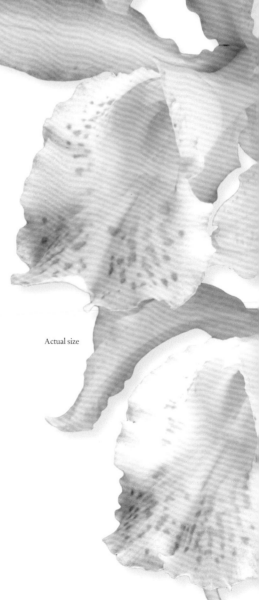

Actual size

The flower of the Pink-spotted Hood Orchid has white or cream tepals (often marked with pale pink) and a tubular lip, typical of many orchids pollinated by euglossine bees. The labellum is crisped around its margin and has vivid pink to red-pink markings on its interior surface.

SUBFAMILY	Epidendroideae
TRIBE AND SUBTRIBE	Cymbidieae, Oncidiinae
NATIVE RANGE	Tropical America from Costa Rica and Trinidad to Bolivia and southern Brazil
HABITAT	Trees at the edge of lowland forests in full sun, up to 3,300 ft (1,000 m), often moving onto cultivated citrus and guava
TYPE AND PLACEMENT	Epiphytic
CONSERVATION STATUS	Not assessed, but widely distributed and thus unlikely to be of conservation concern
FLOWERING TIME	Throughout the year, but more likely spring to summer

FLOWER SIZE
⅛ in (0.3 cm)

PLANT SIZE
2–3 × 2–3 in
(5–8 × 5–8 cm), excluding
erect to arching inflorescence
3–6 in (8–15 cm) long, which is
usually longer than the leaves

342

TRIZEUXIS FALCATA
TRIPLET ORCHID
LINDLEY, 1821

Actual size

The tiny Triplet Orchid, one of the most widespread species in the American tropics, has a flattened pseudobulb topped with a fan of flat, fleshy leaves. The cluster of tiny flowers is so tight that distinguishing individual flowers is difficult. The common name refers to the three sepals, which are partially fused and make up most of what is observable in the partially opening blooms. The same feature occurs in the genus name, derived from the Greek *tri–*, "three," and *zeuxine*, "yoked." The species name refers to the shape of the leaves, which are falcate (curved and tapering to a point).

Plants of the Triplet Orchid produce many capsules, and the species is reported to be self-pollinating. Small, stingless trigonid bees are also reported to be pollinators, but there appears to be no nectar reward.

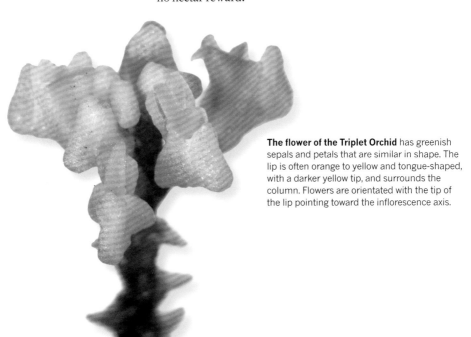

The flower of the Triplet Orchid has greenish sepals and petals that are similar in shape. The lip is often orange to yellow and tongue-shaped, with a darker yellow tip, and surrounds the column. Flowers are orientated with the tip of the lip pointing toward the inflorescence axis.

SUBFAMILY	Epidendroideae
TRIBE AND SUBTRIBE	Cymbidieae, Oncidiinae
NATIVE RANGE	Western South America, from Colombia to Bolivia
HABITAT	Grass and brush on rocky and lateritic clay soils, often on steep slopes, at 6,600–9,200 ft (2,000–2,800 m)
TYPE AND PLACEMENT	Terrestrial, sometimes in rocky terrain
CONSERVATION STATUS	Not assessed
FLOWERING TIME	Spring to early summer

FLOWER SIZE
2 in (5 cm)

PLANT SIZE
25–50 × 20–30 in
(64–127 × 51–76 cm),
excluding erect inflorescence
30–60 in (76–152 cm) tall,
which is taller than the leaves

343

VITEKORCHIS EXCAVATA
ATAHUALPA'S GOLDEN ORCHID
(LINDLEY) ROMOWICZ & SZLACHETKO, 2006

Atahualpa's Golden Orchid is a large plant that forms enormous clusters of four-angled pseudobulbs, which are enveloped in leaf-bearing sheaths and topped by one or two elongate leaves. A many-branched inflorescence forms from the base of a mature pseudobulb and bears more than 100 scentless flowers.

Formerly a species of *Oncidium*, this plant belongs to a group of species in the subtribe Oncidiinae that grow in a more isolated position without close relatives. The genus is named for Austrian botanist Ernst Vitek (1953–), currently director of the botany department at the Natural History Museum in Vienna, which holds the dried orchid collection of the renowned German orchid taxonomist, Heinrich Gustav Reichenbach (1823–89). The common name refers to Atahualpa, the last king of the Incas, who died at Cajamarca in Peru, where the orchid grows in abundance.

The flower of Atahualpa's Golden Orchid has yellow, reddish-brown-blotched sepals and petals, the two lateral sepals recurved. The lip has small sidelobes and a large midlobe with a central notch and warty callus, and the column has a pair of apical wings.

Actual size

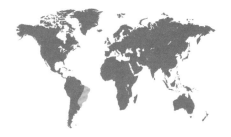

SUBFAMILY	Epidendroideae
TRIBE AND SUBTRIBE	Cymbidieae, Oncidiinae
NATIVE RANGE	Costa Rica, Ecuador, northeastern (state of Pernambuco) and eastern Brazil, south to Paraguay, and Misiones province (Argentina)
HABITAT	Atlantic rain forests
TYPE AND PLACEMENT	Epiphytic
CONSERVATION STATUS	Not assessed
FLOWERING TIME	October to December (spring)

FLOWER SIZE
1 in (2.5 cm)

PLANT SIZE
6–10 × 3–5 in
(15–25 × 8–13 cm),
excluding pendent
inflorescence
4–8 in (10–20 cm) long

344

WARMINGIA EUGENII
WARMING'S ORCHID
REICHENBACH FILS, 1881

The miniature Warming's Orchid has clustered, conical pseudobulbs, each of which holds a single, leathery, narrowly elliptic, folded leaf. The pendent raceme can have up to 35 flowers, each bloom subtended by small bracts. The flowers have a peculiar smell, reminiscent of melting metal or ozone, which possibly attracts euglossine bees for pollination. The bees collect the floral fragrance and modify it to serve as sexual attractants.

The genus and species are named for Danish botanist Eugenius Warming (1841–1924), founding father of plant ecology and an orchid enthusiast. *Warmingia* has a strange distribution: Central America (one species), eastern South America (one species), and eastern Brazil to Paraguay and Argentina (one species). It seems likely that the genus occurs in between, but due to the small size of the plants, it would be easily overlooked.

The flower of Warming's Orchid has white-translucent, spreading petals and sepals. The petals and lip have notched margins, and the white lip is triangular with a yellow central blotch. The column has a pair of apical wings.

Actual size

SUBFAMILY	Epidendroideae
TRIBE AND SUBTRIBE	Cymbidieae, Oncidiinae
NATIVE RANGE	Southwestern Ecuador and northwestern Peru
HABITAT	Seasonally arid areas, often on cactus
TYPE AND PLACEMENT	Epiphytic
CONSERVATION STATUS	Not threatened
FLOWERING TIME	February to April, but can bloom at most times

ZELENKOA ONUSTA
HARRY'S ORCHID
(LINDLEY) M. W. CHASE & N. H. WILLIAMS, 2001

FLOWER SIZE
1¼–2 in (3–5 cm)

PLANT SIZE
4–8 × 4–6 in
(10–20 × 10–15 cm),
excluding erect to
arching inflorescence
8–15 in (20–38 cm) long

345

Well adapted to arid, almost desertlike conditions, with tough coriaceous leaves and bulbs, Harry's Orchid is one of the few species that is naturally epiphytic on cacti. Once included in the genus *Oncidium*, to which it is distantly related, it is now placed alone in its own genus, which honors Harry Zelenko (1928–), a much-admired botanical artist, author, and enthusiast of the subtribe Oncidiinae.

The graceful, branching stems bear many long-lasting, bright yellow flowers on an arching stem. Although superficially resembling the oil-producing bee-attracting flowers of standard species of *Oncidium*, these blooms are reported to attract carpenter bees, *Xylocopa*, which presumably are seeking nectar, although none is present. There are reports of *Zelenkoa onusta* from Panama, Colombia, and Venezuela, but these are either erroneous or based on plants escaped from cultivation.

The flower of Harry's Orchid is uniformly bright yellow with oblanceolate sepals and larger, rounder petals. The labellum sometimes has orange spots on the central lip callus and is deeply trilobed, with a cleft midlobe.

Actual size

SUBFAMILY	Epidendroideae
TRIBE AND SUBTRIBE	Cymbidieae, Oncidiinae
NATIVE RANGE	Southeastern Brazil, in the states of Espírito Santo and Rio de Janeiro
HABITAT	Atlantic rain forests, at sea level to 3,300 ft (1,000 m)
TYPE AND PLACEMENT	Epiphytic
CONSERVATION STATUS	Not assessed
FLOWERING TIME	July to August (summer)

FLOWER SIZE
¾ in (2 cm)

PLANT SIZE
3–5 × 3–5 in
(8–13 × 8–13 cm),
excluding arching-pendent
inflorescence
4–8 in (10–20 cm) long

346

ZYGOSTATES GRANDIFLORA
LONG-BEAKED ORCHID
(LINDLEY) MANSFELD, 1938

The miniature Long-beaked Orchid produces a fan of basally overlapping leaves from which emerges a many-flowered inflorescence with peculiarly upside-down flowers. The genus name is based on the Greek words *zygo*, "yoked," and *states*, "standing," signifying a scale or balance, which in some species (though not this one) is a reference to the appearance of a pair of appendages at the base of the column.

Zygostates grandiflora is unique among all orchids for its relatively long pollinarium (pollen masses plus associated structures), which feature is reflected in the common name. The pollinarium extends across the cavity on the lip base and places the sticky disc near the apex of the lip, where it attaches the pollinarium to the bee. Oil-collecting bees visit the flowers, which have oil-producing glandular hairs in the lip cavity.

The flower of the Long-beaked Orchid is mostly white and has clawed sepals, with the lateral two curved down and the upper one forming a hood over the column. The petals are smaller and folded back. The lip is folded and upturned with a central, green, oil-producing cavity.

Actual size

SUBFAMILY	Epidendroideae
TRIBE AND SUBTRIBE	Cymbidieae, Coeliopsidinae
NATIVE RANGE	Northwestern South America
HABITAT	Very wet cloud forests and lower montane forests, at 3,300–6,070 ft (1,000–1,850 m)
TYPE AND PLACEMENT	Epiphytic
CONSERVATION STATUS	Not assessed
FLOWERING TIME	October to January (fall to winter)

LYCOMORMIUM SQUALIDUM

GOBLIN WOLF ORCHID

(POEPPIG & ENDLICHER) REICHENBACH FILS, 1852

FLOWER SIZE
2¼ in (5.6 cm)

PLANT SIZE
35–55 × 18–30 in
(89–140 × 46–76 cm),
excluding pendent stem
9–15 in (23–38 cm) long

347

The large, clump-forming Goblin Wolf Orchid has pear-shaped, flattened pseudobulbs that are partially enclosed in dry sheaths, topped by two or three lanceolate, plicate leaves with a short stem. A pendent, one-sided raceme arises at the base of a mature pseudobulb bearing tubular bracts and 8–12 fleshy flowers. The general shape of the flower is like that of the Holy Ghost Orchid, *Peristeria elata*, which is the national flower of Panama, although the latter does not share the same dark spotting.

The plant's curious genus and common names are derived from the Greek words *lykos*, meaning "wolf," and *mormos*, "goblin," though the exact reason for this is unclear. The species name refers to the dark coloring on the flower. Pollination is by fragrance-collecting male euglossine bees.

The flower of the Goblin Wolf Orchid has white sepals with dark reddish-purple spotting. They form a cup around the flower, tightly enclosing the white and heavily spotted saccate lip. Petals are similar to the sepals but spreading, and the column is white.

Actual size

SUBFAMILY	Epidendroideae
TRIBE AND SUBTRIBE	Cymbidieae, Coeliopsidinae
NATIVE RANGE	Central America and northwestern South America, Costa Rica to Ecuador
HABITAT	Humid, deciduous mountain forests, at the edge of shaded grassland, or on rocky outcrops, at about 3,300–3,600 ft (100–1,100 m)
TYPE AND PLACEMENT	Lithophytic or terrestrial, or epiphytic at the base of mossy tree trunks
CONSERVATION STATUS	Not formally assessed and still locally abundant, but threatened by overcollecting
FLOWERING TIME	July to August (summer)

FLOWER SIZE
2 in (5 cm)

PLANT SIZE
25–40 × 20–30 in
(64–102 × 51–76 cm),
excluding erect, basal
inflorescence, which is
much taller than the leaves,
35–55 in (89–140 cm) long

348

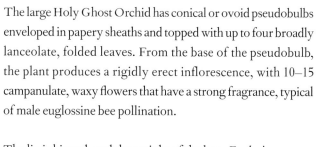

PERISTERIA ELATA
HOLY GHOST ORCHID
HOOKER, 1831

The large Holy Ghost Orchid has conical or ovoid pseudobulbs enveloped in papery sheaths and topped with up to four broadly lanceolate, folded leaves. From the base of the pseudobulb, the plant produces a rigidly erect inflorescence, with 10–15 campanulate, waxy flowers that have a strong fragrance, typical of male euglossine bee pollination.

The lip is hinged, and the weight of the bee, *Euplusia concava*, when landing on the lip causes the hinge to tip and the bee to be thrown against the column, where it picks up the pollinarium while struggling to get free. The orchid, which is the national flower of Panama, has a shape said to resemble a white dove sitting on a nest, hence its other common name, the Dove Orchid. The generic name is close to the Greek word for "dove"—*peristéri*.

The flower of the Holy Ghost Orchid has broad, white sepals and petals that form an enclosed cup around the saccate, white lip and beaked column. The lip has two lateral lobes that are spotted purple.

Actual size

SUBFAMILY	Epidendroideae
TRIBE AND SUBTRIBE	Cymbidieae, Stanhopeinae
NATIVE RANGE	Northwestern South America, from Suriname to Peru
HABITAT	Wet forests, at 2,625–6,600 ft (800–2,000 m)
TYPE AND PLACEMENT	Canopy epiphyte
CONSERVATION STATUS	Not assessed
FLOWERING TIME	March to June (late winter and spring)

FLOWER SIZE
2 in (5 cm)

PLANT SIZE
24–32 × 16–24 in
(60–80 × 40–60 cm),
excluding inflorescence

ACINETA SUPERBA
PARAKEET ORCHID
(KUNTH) REICHENBACH FILS, 1863

349

An epiphyte growing high in the forest canopy, the Parakeet Orchid makes tight clusters of ovoid to cylindrical olive green, glossy, vertically furrowed pseudobulbs. These carry up to four folded, broadly lanceolate leaves, surrounded by several basal sheaths that turn brown with age. From their base grows the pendent inflorescence— which can be up to 28 in (70 cm) long with 10–30 flowers.

The waxy flowers—often cream with dark red or maroon spots—have a delightful, strong scent that attracts brightly colored male euglossine bees of the genera *Eufriesia* or *Eulaema*. The bees collect and absorb the floral fragrance compounds, which probably serve as precursors of sex pheromones. The lip is immobile, hence the scientific name, derived from the Greek word *akinetos*, meaning "moving."

The flower of the Parakeet Orchid is spotted and varies in color from cream to pink or salmon. The three similar sepals are hooded, and the petals parallel to the column and the short lip, which is trilobed with the sidelobes folded up.

Actual size

SUBFAMILY	Epidendroideae
TRIBE AND SUBTRIBE	Cymbidieae, Stanhopeinae
NATIVE RANGE	Northern South America
HABITAT	Rain forests, at 165–1,640 ft (50–500 m)
TYPE AND PLACEMENT	Terrestrial in shade
CONSERVATION STATUS	Not evaluated
FLOWERING TIME	Spring, but February and September reported south of the equator, and March in cultivation

FLOWER SIZE
2 in (5 cm)

PLANT SIZE
10–16 × 3–6 in
(25–40 × 8–15 cm),
excluding inflorescence, which
is shorter than the leaf

BRAEMIA VITTATA
BRAEM'S ORCHID
(LINDLEY) JENNY, 1985

350

Braem's Orchid, the only species in its genus, grows at low elevations in the Orinoco and Amazon River basins, where it occurs on trunks and lower upright branches. Ovoid, ridged pseudobulbs produce a single folded (plicate), lanceolate leaf, from the base of which grows a raceme of up to 25 flowers with a sweet scent of vanilla and chocolate. The roots are unusual among orchids in producing upright branches that collect debris (so-called "trashbasket" roots). On the basis of its flower morphology, the species was separated from the genus *Polycycnis*, which was later supported by DNA analyses.

The orchid appears to be pollinated by male euglossine bees, as its distinctive pollinarium has reportedly been found on a bee attracted to baits impregnated with artificial fragrance compounds. The genus is named for the Belgian orchidologist, Guido Braem (1944–).

The flower of Braem's Orchid has three narrow, brown sepals and petals with yellow edges. The yellow lip is arrowhead-shaped with points on each side and brown venation—its orientation is variable. The column is slender and arching.

Actual size

SUBFAMILY	Epidendroideae
TRIBE AND SUBTRIBE	Cymbidieae, Stanhopeinae
NATIVE RANGE	Southeastern Brazil
HABITAT	Lowland wet tropical forests
TYPE AND PLACEMENT	Epiphytic
CONSERVATION STATUS	Not threatened
FLOWERING TIME	Spring to summer

CIRRHAEA DEPENDENS
TENDRIL ORCHID
(G. LODDIGES) LOUDON, 1830

FLOWER SIZE
1¼–1½ in (3–4 cm)

PLANT SIZE
15–25 × 5–9 in
(38–64 × 13–23 cm),
excluding pendent
inflorescence
12–18 in (30–46 cm) long

351

The Tendril Orchid is at the peak of delicious fragrance production in the cooler hours of the morning, which coincides with the peak of activity for its insect pollinators. Like most other members of this subtribe, the species produces chemically complex fragrances that are scraped from the lip and made into sex pheromones by male euglossine bees.

Cirrhaea is named for its tendril-shaped (curved) column, which attaches pollen to the legs of the bees. By placing pollen on different parts of the bee, co-occurring species can utilize the same pollinator without mixing up their pollen. In this species, the pollen masses are initially too large to fit into the narrow stigmatic cavity. After about an hour, the pollinia shrink enough to fit easily, and the bee is likely to have reached the flowers of another plant by then, preventing self-pollination.

The flower of the Tendril Orchid is waxy, long lasting, and variable in color, usually olive to plum or orange green with irregular spots or bars on the petals, sepals, and the oddly shaped three-lobed lip.

Actual size

SUBFAMILY	Epidendroideae
TRIBE AND SUBTRIBE	Cymbidieae, Stanhopeinae
NATIVE RANGE	Trinidad and Tobago, French Guiana, Surinam, Guyana, Venezuela, Peru, and Brazil
HABITAT	Low to mid elevation tropical wet forests and cultivated guava and citrus groves
TYPE AND PLACEMENT	Epiphytic
CONSERVATION STATUS	Not threatened
FLOWERING TIME	Summer

FLOWER SIZE
5 in (13 cm)

PLANT SIZE
10–18 × 8–12 in
(25–46 × 20–31 cm),
excluding pendent
inflorescence
8–14 in (20–36 cm) long

352

CORYANTHES SPECIOSA
BEAUTIFUL BUCKET ORCHID
HOOKER, 1831

The Beautiful Bucket Orchid displays one of the most remarkable and mind-boggling pollination mechanisms. Attracted to the large, intricate, pendent blooms by an indescribably complex aroma, fragrance-collecting male euglossine bees accidentally fall into the bucket-shaped lip, which is filled with liquid from faucet-like glands at the base of the column. The only way out of the bucket is via a small opening near the rear of the flower, where the bees pick up (or deposit) pollinia as they squeeze their way through the narrow opening.

The genus name is derived from the Greek words *korys*, meaning "helmet," and *anthos*, "flower," a reference to the shape of the lip. The species name, *speciosa*, is Latin for "showy." The plants grow in aerial ant colonies, where formic acid produced by the hosts creates acidic conditions, to which the plants are adapted.

The flower of the Beautiful Bucket Orchid is large, complex, and variable in color but generally with various tawny shades of yellow to brown, often spotted with reddish-brown. The sepals and petals are winglike and reflexed, leaving the pouch-like lip as the prominent floral feature.

Actual size

SUBFAMILY	Epidendroideae
TRIBE AND SUBTRIBE	Cymbidieae, Catasetinae
NATIVE RANGE	Southeastern Panama, Colombia, and Venezuela
HABITAT	Lowland wet tropical forests, at around 1,300–2,600 ft (400–800 m)
TYPE AND PLACEMENT	Epiphytic
CONSERVATION STATUS	Not threatened
FLOWERING TIME	Fall

FLOWER SIZE
Up to 6¾ in (17 cm)

PLANT SIZE
15–25 × 10–18 in
(38–64 × 25–46 cm),
excluding inflorescence,
which is arching-pendent,
8–15 in (20–38 cm) long

CYCNOCHES CHLOROCHILON
GREEN-LIPPED SWAN ORCHID
KLOTZSCH, 1838

353

The Green-lipped Swan Orchid, so-called for the long, graceful curve of its column, is one of the larger and more robust species in a genus of mostly warm-growing, low-elevation deciduous epiphytes with showy, fragrant, and sexually dimorphic flowers. The genus name, *Cycnoches*, is from the Greek words *kyknos*, "swan," and *auchen* "neck."

Elongating rapidly and gradually fattening their leafy cigar-shaped pseudobulbs during the rainy season, plants become completely leafless during the dry season. Flower spikes usually appear near the apex of mature pseudobulbs just before the leaves are shed. With similar male and female flowers, this species may be the least sexually dimorphic of the genus *Cycnoches*. Other species in related genera, such as *Catasetum*, can bear many smaller male blooms with projectile pollinia. These dimorphic genera are all pollinated by fragrance-collecting male euglossine bees.

The flower of the Green-lipped Swan Orchid is non-resupinate, waxy in appearance, and star-shaped. It is usually pale green to greenish-orange with a large white lip and darker green lip callus.

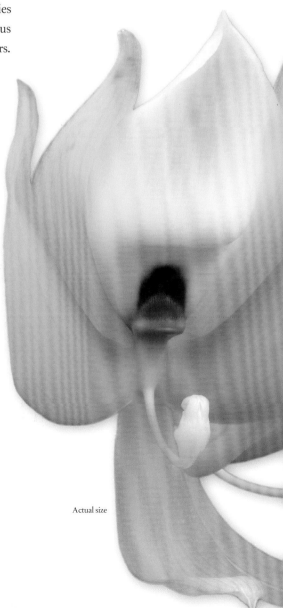

Actual size

SUBFAMILY	Epidendroideae
TRIBE AND SUBTRIBE	Cymbidieae, Stanhopeinae
NATIVE RANGE	El Chocó region of Colombia
HABITAT	Mid-elevation cloud forests
TYPE AND PLACEMENT	Epiphyte
CONSERVATION STATUS	Not threatened
FLOWERING TIME	Spring

FLOWER SIZE
Up to 6¼ in (16 cm)

PLANT SIZE
8–12 × 3–5 in
(20–30 × 8–13 cm), excluding
pendent inflorescence
4–12 in (10–30 cm) long

EMBREEA RODIGASIANA
AL'S ORCHID
(CLAESSENS EX COGNIAUX) DODSON, 1980

354

An outrageous spectacle when in bloom, with enormous flowers, Al's Orchid is one of only two species in a genus close to the vegetatively larger species of *Stanhopea*. It differs by having a single flower per inflorescence and a lip with four distinctive horns versus many flowers and a winged (hornless) lip

Embreea species display the classic, fragrance-collecting, male euglossine bee syndrome. In these flowers, the bees begin to scrape the tissue at the base of the lip, where they lose their grip, fall off, and, guided by the column horns, contact the head of the column and remove the pollen masses. Due to the colossal floral dimensions, one of the larger bee species, probably in the genus *Eulaema*, could be the likely pollinator. The common name refers to the United States orchidologist Alvin Embree, after whom the genus was named.

The flower of Al's Orchid has cream or pale green petals and sepals with variable reddish-purple to chestnut spotting and blotching on the much broader sepals. The lip, peppered with small, reddish-purple spots, is distinctively shaped, with four horns resembling a hatchet.

Actual size

SUBFAMILY	Epidendroideae
TRIBE AND SUBTRIBE	Cymbidieae, Stanhopeinae
NATIVE RANGE	Mountainous regions of southern Mexico and adjacent Guatemala
HABITAT	Mid-elevation cloud forests
TYPE AND PLACEMENT	Epiphytic, rarely lithophytic or terrestrial
CONSERVATION STATUS	Not threatened
FLOWERING TIME	Spring to summer

GONGORA GALEATA
HELMET ORCHID
(LINDLEY) REICHENBACH FILS, 1854

FLOWER SIZE
Up to 1¼–2 in (3–5 cm)

PLANT SIZE
10–15 × 8–12 in
(25–38 × 20–30 cm), excluding
pendent inflorescence
6–10 in (15–25 cm) long

355

The Helmet Orchid has shorter inflorescences than most species in the genus *Gongora* but can produce many of them in a single flush from the bases of glossy, furrowed pseudobulbs.

The fragrance produced by these plants is collected by male orchid bees, which in turn use it to produce a pheromone that attracts females. The composition of the fragrance varies from plant to plant and differs depending on the time of day and floral age, which indicates the attraction of several euglossine bee species. The coloration of the peculiar flowers can range from brown and greenish-yellow to pinkish. The common name refers to the concave shape of the lip, which looks like a helmet. The genus name is in honor of a former Spanish viceroy of New Grenada (mostly modern Colombia), Antonio Caballero y Góngora (1723–96).

The flowers of the Helmet Orchid are yellow to brown and non-resupinate, with the lateral sepals recurved, the dorsal sepal curved underneath like a protective helmet, and the petals narrow and adnate to the column. The fleshy, angled lip is the most prominent feature.

Actual size

SUBFAMILY	Epidendroideae
TRIBE AND SUBTRIBE	Cymbidieae, Stanhopeinae
NATIVE RANGE	Panama
HABITAT	Humid lowland forests
TYPE AND PLACEMENT	Epiphytic or terrestrial
CONSERVATION STATUS	Not assessed
FLOWERING TIME	October to December (fall)

FLOWER SIZE
1¾ in (4.5 cm)

PLANT SIZE
10–18 × 4–7 in
(25–46 × 10–18 cm),
excluding inflorescence,
which is about as long
as the leaf

HORICHIA DRESSLERI
DRESSLER AND HORICH'S ORCHID
JENNY, 1981

Dressler and Horich's Orchid—the only species in the genus *Horichia*—produces a cluster of small, ovoid pseudobulbs, each with a single, broad, plicate, stemmed leaf. From the base of each pseudobulb there are one to three upright inflorescences of up to 20 flowers. The species has a narrow range in lowland forests in Panama and is likely to occur in adjacent parts of Costa Rica. The genus is named after Clarence Horich (1930–94), a German orchid specialist in Costa Rica, and the species for Robert Dressler, an American scientist who has worked extensively on Orchidaceae.

The orchid's floral fragrance is largely due to a compound called paradimethoxybenzene, which probably attracts male euglossine bees, although pollination of *H. dressleri* has not been studied. The bees are known to collect fragrance compounds from the surfaces of flowers in the subtribe Stanhopeinae.

Actual size

The flower of Dressler and Horich's Orchid has narrow, folded, brown sepals and two narrower, clawed, yellow-tipped, brown petals. The yellow lip has three narrow lobes, with the lateral ones recurved. The greenish column has a hook-like appendage at its apex.

SUBFAMILY	Epidendroideae
TRIBE AND SUBTRIBE	Cymbidieae, Stanhopeinae
NATIVE RANGE	Southeastern Brazil, from Espírito Santo to Paraná states
HABITAT	Wet forests, at 2,300–3,300 ft (700–1,000 m)
TYPE AND PLACEMENT	Terrestrial, in leaf litter and other detritus
CONSERVATION STATUS	Not assessed
FLOWERING TIME	January to April (late summer to fall)

HOULLETIA BROCKLEHURSTIANA
OCELOT ORCHID
LINDLEY, 1841

FLOWER SIZE
3½ in (9 cm)

PLANT SIZE
25–40 × 8–12 in
(64–102 × 20–30 cm),
including erect
inflorescence
15–28 in (38–71 cm) tall

357

The Ocelot Orchid bears tight clusters of strongly ribbed pear-shaped pseudobulbs that each carries a single pleated leaf with a long stalk (petiole). The plant produces an upright raceme with up to ten persistent, fragrant flowers. The scent attracts male fragrance-collecting euglossine bees, and the species *Euglossa chalybeata* has been observed pollinating the plants in nature. After collecting fragrance compounds from several plants, the bees move them to another site on their bodies, where they are converted in a sex pheromone used to attract a female.

The genus is named in honor of French orchid collector and grower Jean-Baptiste Houllet (1815–90), who found this species in Brazil and later became director of the Jardin des Plantes in Paris. Thomas Brocklehurst, the inspiration for the species name, was a distinguished amateur orchid grower in Manchester, England, in the late nineteenth century.

Actual size

The flower of the Ocelot Orchid faces downward and has cream, red-spotted parts, the upper sepal cupped, the petals clawed at the base. The lip is heavily spotted and has two fang-like lateral lobes. The column projects forward.

SUBFAMILY	Epidendroideae
TRIBE AND SUBTRIBE	Cymbidieae, Stanhopeinae
NATIVE RANGE	Colombia and Peru
HABITAT	Wet forests, at 4,920–5,900 ft (1,500–1,800 m)
TYPE AND PLACEMENT	Epiphytic
CONSERVATION STATUS	Not assessed
FLOWERING TIME	Spring

FLOWER SIZE
3½ in (9 cm)

PLANT SIZE
12–20 × 2–4 in
(30–51 × 5–10 cm), including
erect inflorescences 8–14 in
(20–36 cm) long, which
are shorter than the leaves

HOULLETIA IOWIANA
BLACK WALNUT ORCHID
REICHENBACH FILS, 1874

358

The large Black Walnut Orchid has pear-shaped pseudobulbs, each carrying a single, plicate, lanceolate leaf. The erect inflorescence is enveloped by tubular bracts and has two fleshy, non-resupinate flowers at the tip. The blooms have a strong scent of fresh black walnut (*Juglans nigra*)—hence the plant's common name—and are pollinated by male euglossine bees that collect floral fragrance compounds from the lip. In this species, *Eufriesea mariana* bees have been observed as pollinators, with the pollinia attaching to their legs.

The genus was named for French orchidologist Jean-Baptiste Houllet (1815–90), director of the Jardin des Plantes in Paris. There are two groups of species in the genus *Houlletia*, one with resupinate flowers and one with non-resupinate flowers. The latter, including this species, have in the past been segregated into the genus *Jennyella*.

The flower of the Black Walnut Orchid is yellow to cream and not resupinate, with three equally sized, cupped sepals, two spreading, shortly clawed petals and an upfacing, three-lobed lip that has a horseshoe shape with an arching midlobe and flaring sidelobes.

Actual size

SUBFAMILY	Epidendroideae
TRIBE AND SUBTRIBE	Cymbidieae, Stanhopeinae
NATIVE RANGE	Trinidad and Venezuela (the Guiana Shield) to Amapá state, Brazil
HABITAT	Dense, tall rain forests, at 1,970–2,300 ft (600–700 m)
TYPE AND PLACEMENT	Epiphytic
CONSERVATION STATUS	Not assessed
FLOWERING TIME	Late summer to fall

KEGELIELLA HOUTTEANA

SOARING EAGLE

(REICHENBACH FILS) L. O. WILLIAMS, 1942

FLOWER SIZE
1½ in (3.8 cm)

PLANT SIZE
8–14 × 5–10 in
(20–36 × 13–25 cm),
excluding pendent
inflorescence
8–14 in (20–36 cm) long

359

The Soaring Eagle has tight clusters of cone-shaped, glossy, wrinkled pseudobulbs, each topped with one or two pleated leaves that are maroon below and stalked. A pendent inflorescence covered in little brown hairs can bear up to a dozen flowers, which are pollinated by male, fragrance-collecting euglossine bees. The common name refers to the shape of the column, the apex of which looks vaguely like the prominently beaked head of a bird of prey.

The genus is named in honor of German gardener and naturalist Hermann Kegel (1819–56), who discovered *Kegeliella houtteana* in Suriname, where he contracted a tropical disease from which he never recovered. The species name is for Louis Benoît van Houtte (1810–76), who had a nursery in Ghent, where the Soaring Eagle flowered for the first time in horticulture.

The flower of the Soaring Eagle has cream, cupped, red-striped or red-spotted sepals and smaller, similar, forward-projecting petals. The lip is yellow and three-lobed, with two upright, broad sidelobes and a forward-pointing midlobe. The green column arches forward with two thin, apical wings.

Actual size

SUBFAMILY	Epidendroideae
TRIBE AND SUBTRIBE	Cymbidieae, Stanhopeinae
NATIVE RANGE	Mesoamerica, from Chiapas state (Mexico) to Panama
HABITAT	Rain forests, at 3,950–5,600 ft (1,200–1,700 m)
TYPE AND PLACEMENT	Terrestrial on steep embankments and fallen trees
CONSERVATION STATUS	Not assessed
FLOWERING TIME	Spring

FLOWER SIZE
2 in (5 cm)

PLANT SIZE
14–30 × 14–25 in
(36–76 × 36–64 cm),
excluding pendent
inflorescence
12–24 in (30–61 cm) long

360

LACAENA SPECTABILIS
BEAUTIFUL HELEN
(KLOTZSCH) REICHENBACH FILS, 1854

The Beautiful Helen is one of only two species in the genus *Lacaena*, which was most likely named for *Lakaina*, one of the names of Helen of Troy in Greek mythology. Her legendary beauty also inspired the common name and the species name, which is the Latin for "outstanding." The plant has egg-shaped, ridged pseudobulbs that carry two or three pleated, elliptic, stalked apical leaves. The inflorescence bears 10–25 highly fragrant flowers.

Pollination is similar to what occurs in other flowers that face downward. A male euglossine bee enters the flowers upside down and scrapes the midlobe of the lip to collect floral fragrance compounds. When the bee releases its grip, it cannot fly immediately because of the limited space, so it falls, hits the apex of the column, and picks up pollen masses with the back of its head.

Actual size

The flower of Beautiful Helen is cup-shaped and pale pink with darker pink spotting. Ovate petals surround the hooded column and a much more darkly spotted trilobed lip. The raised sidelobes surround the column, and the triangular apical lobe projects forward.

SUBFAMILY	Epidendroideae
TRIBE AND SUBTRIBE	Cymbidieae, Stanhopeinae
NATIVE RANGE	Northwestern South America, south to Peru
HABITAT	Wet forests at 3,300–3,950 ft (1,000–1,200 m)
TYPE AND PLACEMENT	Epiphytic, rarely terrestrial on steep slopes
CONSERVATION STATUS	Least concern
FLOWERING TIME	July to August (summer)

LUEDDEMANNIA PESCATOREI

GOLDEN-BROWN FOX-TAIL ORCHID

(LINDLEY) LINDEN & REICHENBACH FILS, 1854

FLOWER SIZE
1 in (2.5 cm)

PLANT SIZE
18–36 × 12–25
(46–91 × 30–64 cm),
excluding pendent
inflorescence
20–40 in (51–102 cm) long

361

The large Golden-brown Fox-tail Orchid produces a cluster of egg-shaped, grooved pseudobulbs that carry two to four elliptic, plicate, clasping leaves. It has a long, pendulous inflorescence that grows directly underneath the plant and is densely set with numerous (up to 50) sweetly scented flowers, each subtended by an elliptic bract. The bracts, flower stalks, and outer surfaces of the sepals are covered with black-brown globular bodies, for which the function is unknown.

The species is widespread from Venezuela to Peru, but collections are scarce. It can be found at lower elevations around Machu Picchu, where numerous tourists will have encountered it. The pollinator of the orchid is not known, but the shape and fragrance indicate some type of bee as a visitor.

The flower of the Golden-brown Fox-tail Orchid has brownish-orange sepals and similar but smaller, forward-curved, bright yellow petals, together forming a tube with the bright yellow column and lip. The lip has an arrowhead-shaped midlobe.

Actual size

SUBFAMILY	Epidendroideae
TRIBE AND SUBTRIBE	Cymbidieae, Stanhopeinae
NATIVE RANGE	Northern South America
HABITAT	Wet forests in shade, mostly on moss-covered understory trees, from sea level to 4,920 ft (1,500 m)
TYPE AND PLACEMENT	Epiphytic
CONSERVATION STATUS	Not assessed
FLOWERING TIME	April to May

FLOWER SIZE
4 in (10 cm)

PLANT SIZE
8–12 × 6–10 in
(20–30 × 15–25 cm),
excluding pendent
inflorescence 3–8 in
(8–20 cm) long

362

PAPHINIA CRISTATA
WHITE-BEARDED
STAR ORCHID
(LINDLEY) LINDLEY, 1843

Relatively small plants with comparatively large and stunning flowers, *Paphinia* species thrive in sultry, humid conditions. Their star-shaped flowers appear on sharply pendent stems and can occur in great profusion. Most of the 15 described species have only been discovered since the 1980s.

Although closely related to the powerfully fragrant species of *Stanhopea*, which achieve pollination by attracting male euglossine bees that harvest fragrance compounds, *Paphinia* species are not particularly fragrant. Nonetheless, they do manage to attract euglossine bees, presumably because there are compounds present that human noses (and gas chromatographs) cannot detect. The lip of the White-bearded Star Orchid is adorned with a filamentous crest or comb from which it gets its species and common names, from the Latin word *cristatus*, meaning "crested." *Paphia* was a local Cypriot name for Aphrodite.

The flower of the White-bearded Star Orchid has white parts so extensively overlaid with maroon markings that they appear almost solidly colored. The lip has an impressive apical, white beard, and the column is yellow with an inflated apex.

Actual size

SUBFAMILY	Epidendroideae
TRIBE AND SUBTRIBE	Cymbidieae, Stanhopeinae
NATIVE RANGE	Central America and northwestern South America
HABITAT	Rain forests and cloud forests, at up to 4,920 ft (1,500 m)
TYPE AND PLACEMENT	Epiphytic
CONSERVATION STATUS	Not assessed
FLOWERING TIME	February to April (spring)

POLYCYCNIS BARBATA
BEARDED SWAN-ORCHID
(LINDLEY) REICHENBACH FILS, 1855

FLOWER SIZE
2 in (5.1 cm)

PLANT SIZE
10–18 × 5–8 in
(25–46 × 13–20 cm),
excluding inflorescence,
which is arching-pendent,
8–14 in (20–36 cm) long

363

The Bearded Swan-orchid has ovate pseudobulbs, each of which forms a single plicate, elliptic-lanceolate leaf. A pendent, racemose inflorescence displays many fragrant, short-lived flowers, with purple markings on a white or yellow background. The genus name refers to the curved column, shaped like the neck of a swan (from the Greek *polys*, "many" and *kyknos*, "swan"). The species name, *barbata*, means "bearded" in Latin.

Euglossine bees (*Eulaema speciosa*) are the specific pollinators of this species. When the bee lands on the lip, it moves to the base and starts scratching to collect the fragrance. The weight of the bee pulls the flower down, so that when the bee retreats, the pollinia, at the tip of the swan-neck-shaped column, are hooked beneath the scutellum of the bee. In the next flower visited, the pollinia are removed from this position.

The flower of the Bearded Swan-orchid has three similar, spreading sepals and two clawed, spreading petals. The lip is three-lobed with the sidelobes upturned, and the central lobe is diamond-shaped and covered with white hairs.

Actual size

SUBFAMILY	Epidendroideae
TRIBE AND SUBTRIBE	Cymbidieae, Stanhopeinae
NATIVE RANGE	Costa Rica, Colombia, and probably Panama
HABITAT	Cloud forests, at 5,900–7,200 ft (1,800–2,200 m)
TYPE AND PLACEMENT	Epiphytic (rarely terrestrial)
CONSERVATION STATUS	Not formally assessed, but apparently rare in Costa Rica
FLOWERING TIME	Late spring to summer

FLOWER SIZE
½ in (1.25 cm)

PLANT SIZE
7–12 × 4–6 in
(18–30 × 10–15 cm),
excluding pendent
inflorescence 8–14 in
(20–36 cm) long

364

SCHLIMIA JASMINODORA
HELMET ORCHID
PLANCHON & LINDEN, 1852

The Helmet Orchid forms a cluster of pseudobulbs along a short stem, each topped with a broadly ovate, plicate leaf. The inflorescence is pendent and emerges from the base of a mature pseudobulb. It carries several jasmine-scented flowers that are visited by male *Euplusia* (euglossine) bees. These insects scratch the lip to obtain fragrance compounds, during which time the pollinarium is attached to the scutellum of the bee.

The common name refers to the shape of the lip, whereas the genus name honors the Belgian botanist Louis Joseph Schlim (1819–63), who collected plants in tropical America and was cousin of the orchidologist and genus describer Jean Jules Linden (1817–98). The genus has often been spelled "*Schlimmia*," by Linden among others, but it was later corrected and is thus to be written with a single "m."

The flower of the Helmet Orchid has a large, cup-shaped lip uppermost. The sepal points directly downward, and from the bottom of the lip the narrow petals emerge. All parts are white to cream with a few pink spots.

Actual size

SUBFAMILY	Epidendroideae
TRIBE AND SUBTRIBE	Cymbidieae, Stanhopeinae
NATIVE RANGE	Costa Rica and Panama
HABITAT	Lower elevation wet forests
TYPE AND PLACEMENT	Epiphytic, rarely lithophytic or terrestrial
CONSERVATION STATUS	Not assessed
FLOWERING TIME	July to October (summer to fall)

SIEVEKINGIA FIMBRIATA
RAGGED-EAR ORCHID
REICHENBACH FILS, 1886

FLOWER SIZE
Up to 1 in (2.5 cm)

PLANT SIZE
8–12 × 4–6 in
(20–30 × 10–15 cm),
excluding pendent
inflorescence
6–10 in (15–25 cm) long

365

The Ragged-ear Orchid makes dense clusters of ovoid pseudobulbs, each bearing a single stalked leaf. The inflorescence bears up to 15 weakly scented but colorful flowers. The common name refers to the petals of the species, which are shaped like the ear of a dog but with deeply fringed edges. The species name is from the Latin *fimbriatus*, "fringed," again a reference to the petals. The genus name honors Dr. Friedrich Sieveking, who was mayor of the German city of Hamburg and instructor of the renowned orchidologist H. G. Reichenbach, who described this genus and species.

Pollination is by male euglossine bees that collect floral fragrances from the downward-facing flowers. Pollinia are attached to their legs as they scrape the lip. The plants are frequently associated with ant nests, which gives them protection from insect pests.

The flower of the Ragged-ear Orchid is usually brilliant golden yellow with elliptic sepals and similar petals, fringed on their margins. The lip is broad and cupped, with prominent keels, and trilobed, with reddish-brown markings.

Actual size

SUBFAMILY	Epidendroideae
TRIBE AND SUBTRIBE	Cymbidieae, Stanhopeinae
NATIVE RANGE	Western Colombia to northwestern Ecuador
HABITAT	Rain forests, at sea level to 325 ft (100 m)
TYPE AND PLACEMENT	Epiphytic
CONSERVATION STATUS	Not assessed, but rare in the wild
FLOWERING TIME	December to January (spring)

FLOWER SIZE
1⅝ in (4 cm)

PLANT SIZE
10–18 × 8–16 in
(25–46 × 20–41 cm),
including the erect
inflorescence

366

SOTEROSANTHUS SHEPHEARDII
YELLOW-TAIL ORCHID
(ROLFE) JENNY, 1986

Plants of the Yellow-tail Orchid have clusters of ribbed, sheath-enveloped pseudobulbs each topped by two or three red-backed, pleated leaves. The erect inflorescence is as long as the leaves and carries a dense inflorescence with non-resupinate, fragrant flowers spread around all sides of the stem, giving it a plumelike or taillike appearance, hence the common name. The derivation of the genus name is unclear, beyond the "flower" meaning of the Greek word *anthos*.

Soterosanthus shepheardii has long been an enigmatic species, rare in the wild and cultivation. Pollination has not been studied, but the distinctive pollinia have been found on a male euglossine bee, *Euglossa crassipunctata*. The species, therefore, fits the male-euglossine pattern observed in nearly all members of the subtribe Stanhopeinae.

The flower of the Yellow-tail Orchid has the lip uppermost, with broad, yellow, blunt petals, pointed sepals and a cupped, broad, usually red-spotted lip, hooded over the column, which is green-winged and has a yellow anther cap with a long pointed tip.

Actual size

SUBFAMILY	Epidendroideae
TRIBE AND SUBTRIBE	Cymbidieae, Stanhopeinae
NATIVE RANGE	Mexico
HABITAT	Seasonally wet forests, at 1,970–4,920 ft (600–1,500 m)
TYPE AND PLACEMENT	Epiphytic
CONSERVATION STATUS	Not threatened
FLOWERING TIME	July to October (summer to fall)

STANHOPEA TIGRINA
SPOTTED BULL
BATEMAN EX LINDLEY, 1838

FLOWER SIZE
8 in (20 cm)

PLANT SIZE
10–18 × 4–5 in
(25–46 × 10–13 cm),
excluding pendent
inflorescence
5–8 in (13–20 cm) long

367

The Spotted Bull produces complex flowers, looking like a descending bird of prey with its talons outstretched. They are also renowned for their short-lived nature and strong fragrance. The genus *Stanhopea* is named for Philip Henry Stanhope (1781–1855), President of the Medico-Botanical Society of London and Fellow of the Royal Society. The common name comes from the name in Spanish, *torito*, meaning "little bull." The species name, *tigrina*, refers to the tiger-like flower markings.

The sweet, complex fragrance attracts male euglossine bees as pollinators, collecting the fragrance compounds from the lip and using them to produce a sex pheromone to attract females. As the bees scrape the waxy surface of the lip, they fall through the flowers, meeting the pollinia as they plummet. A subsequent visit to another flower repeats the fall, and pollination takes place.

The flower of the Spotted Bull is generally creamy yellow with dark purplish-brown markings that can vary from small spots to large blotches. The sepals and petals are reflexed, and the lip is complex and bears appendages that direct the pollinator toward the column tip.

Actual size

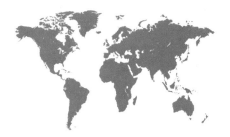

SUBFAMILY	Epidendroideae
TRIBE AND SUBTRIBE	Cymbidieae, Zygopetalinae
NATIVE RANGE	Southern Ecuador
HABITAT	Wet forests, at 2,626–5,900 ft (800–1,800 m)
TYPE AND PLACEMENT	Epiphytic
CONSERVATION STATUS	Not assessed
FLOWERING TIME	July to August (winter)

FLOWER SIZE
1¼ in (3 cm)

PLANT SIZE
6–10 × 4–8 in
(15–25 × 10–20 cm),
including inflorescence

AETHEORHYNCHA ANDREETTAE
STRANGE SNOUT ORCHID
(JENNY) DRESSLER, 2005

368

Actual size

The flower of the Strange Snout Orchid has
creamy white, linear, spreading sepals forming
a funnel-shaped chamber at the base. Petals
are white, flanking the column. The trilobed lip
is white with yellow and red blotches and spots
in the throat and midlobe. The lateral lobes
enfold the column.

The Strange Snout Orchid produces a fan of around eight
overlapping leaves. The leaf blades are oblanceolate. Flowers
appear singly from the axils of lower leaves, the flower stalk
enveloped at the base by the leaves. More than one inflorescence
can occur at a time, but they are always one-flowered. The genus
name derives from the Greek words *aethes*, "strange," and
rhyncos, "snout," which also gives this species its common name.

This is the only species in the genus—it was originally placed in
Chondrorhyncha but was segregated from it because of the finer
aspects of the floral parts. The species name is in honor of Father
Angel Andreetta, of the Salesian Catholic ministry. He was
a pioneer in native orchid cultivation in Ecuador, sparking
enthusiasm among many Ecuadoreans, who previously
had ignored their native orchids.

SUBFAMILY	Epidendroideae
TRIBE AND SUBTRIBE	Cymbidieae, Zygopetalinae
NATIVE RANGE	Colombia, Venezuela, Peru, and Brazil
HABITAT	Lower elevation wet forests
TYPE AND PLACEMENT	Epiphytic, rarely lithophytic
CONSERVATION STATUS	Not threatened
FLOWERING TIME	Year round

FLOWER SIZE
Up to 2½ in (6.5 cm)

PLANT SIZE
10–18 × 5–8 in
(25–46 × 13–20 cm),
excluding erect-arching
inflorescence
10–15 in (25–38 cm) long

AGANISIA CYANEA
BLUE WATER ORCHID
(LINDLEY) REICHENBACH FILS, 1869

369

Its large lavender to blue flowers help to make the Blue Water Orchid one of the most coveted of orchid species. Growing low on the trunks of trees in steamy lowland jungles along rivers, including the Amazon, *Aganisia* species have been reported to survive submergence in floods for weeks during the rainy season, with only the flowers appearing above the water surface—hence the common name of this species.

The plants produce a long stem between their fusiform pseudobulbs and often climb up their hosts almost as if they are trying to avoid being submerged. The genus name comes from the Greek word *aganos*, meaning "desirable," in reference to its attractive appearance. The pollinator of *A. cyanea* is unknown, but, given the shape of the flowers, male euglossine bees are the potential candidates.

The flower of the Blue Water Orchid is usually pale blue but can be darker, often tinted with lavender, pink, or mauve. The lip is pandurate, trilobed, and variable in color from white to blue, purple, bronze, or even reddish with darker purple spots.

Actual size

SUBFAMILY	Epidendroideae
TRIBE AND SUBTRIBE	Cymbidieae, Zygopetalinae
NATIVE RANGE	Surinam, Colombia, and Peru
HABITAT	Wet forests, at 985–2,300 ft (300–700 m)
TYPE AND PLACEMENT	Epiphytic
CONSERVATION STATUS	Not assessed, but probably not threatened
FLOWERING TIME	September to November (spring to early summer)

FLOWER SIZE
2 in (5 cm)

PLANT SIZE
8–12 x 6–10 in
(20–30 x 15–25 cm),
excluding arching to
pendent inflorescence
8–12 in (20–30 cm) long

BATEMANNIA ARMILLATA
GHOST ORCHID
REICHENBACH FILS, 1875

The Ghost Orchid produces ovoid, slightly four-angled pseudobulbs with a series of dry bracts below and a pair of broadly lanceolate leaves on top. The inflorescence emerges from the base of a pseudobulb and carries between two and eight ghostly flowers that appear to be suspended in air. The species name is Latin for "wearing bracelets," a reference to the thickened callus in the middle of the lip. The genus is named in honor of Englishman James Bateman (1811–97), who published a series of spectacular illustrated books on orchids.

The species has not been studied in nature, but its floral morphology is similar to other bee-pollinated species that have a basal cavity on the lip, suggesting the presence of nectar. While the insects, attracted to the flowers by their color and sweet fragrance, look for the absent nectar, pollination takes place.

The flower of the Ghost Orchid has bright pure green, lanceolate, outstretched sepals and petals. The lip is white and trilobed—the two lateral lobes surrounding the column, and the midlobe larger with a cream callus forming a pronounced ridge.

Actual size

SUBFAMILY	Epidendroideae
TRIBE AND SUBTRIBE	Cymbidieae, Zygopetalinae
NATIVE RANGE	Border region of Ecuador and Peru (Cordillera del Cóndor)
HABITAT	Cloud forests and humid dark rain forests, at 330–4,920 ft (100–1,500 m)
TYPE AND PLACEMENT	Epiphytic in shade on trunks of trees
CONSERVATION STATUS	Near threatened
FLOWERING TIME	November to June

BENZINGIA CAUDATA

ACKERMAN'S ORCHID

(ACKERMAN) DRESSLER, 2010

FLOWER SIZE
1 in (2.5 cm)

PLANT SIZE
6–10 × 2–4 in
(15–25 × 5–10 cm),
excluding inflorescence,
which is shorter than
the leaves

371

Ackerman's Orchid is one of many species found in the highly biodiverse Cordillera del Cóndor. It is a pendent epiphyte producing a small fan of sheathing linear leaves, which have nipple-like (protruding) epidermal cells that make them sparkle. The nine species in this genus were previously assigned to other genera because of their diverse flower shapes, but their sparkling cells and growth form support placing them together. A pendent inflorescence from the axils of the lower leaves bears one flower.

Benzingia caudata was originally described by the Puerto Rican orchid expert, James Ackerman (1950–). Later, the US botanist Calaway H. Dodson (1928–) established the genus *Ackermania* for it, but as that name had already been used for a fungal genus, it was assigned to *Benzingia*, named for the US orchid ecologist, David Benzing (1937–). The orchid is pollinated by fragrance-collecting male euglossine bees.

Actual size

The flower of Ackerman's Orchid is white with similar, spreading petals and sepals, the upper sepal hooded over the flower. The cup-shaped lip is yellow with red markings and has a pointed tip. The column arches over the lip.

SUBFAMILY	Epidendroideae
TRIBE AND SUBTRIBE	Cymbidieae, Zygopetalinae
NATIVE RANGE	Ecuadorian and Peruvian Andes
HABITAT	Lower elevation wet forests
TYPE AND PLACEMENT	Epiphyte
CONSERVATION STATUS	Not threatened
FLOWERING TIME	Spring to summer

FLOWER SIZE
2–3 in (5–7.5 cm)

PLANT SIZE
10–18 × 9–20 in
(25–46 × 23–51 cm),
including multiple,
short single-flowered
inflorescences
3–5 in (8–13 cm) long

372

CHAUBARDIA HETEROCLITA
RED-STRIPED FAN ORCHID
(POEPPIG & ENDLICHER) DODSON & D. E. BENNETT, 1989

This species is characterized by its fan shape with broad leaves that conceal a vestigial pseudobulb at their bases. This shape is also the basis for the common name of the Red-striped Fan Orchid. The plant grows in wet forests with year-round rain, which makes water storage in pseudobulbs obsolete.

The solitary flowers, which arise from the leaf axils, are large and showy. They have narrow, tawny segments with reddish stripes and a pale violet blue lip with a callus displaying a rigid fimbriate margin, similar to *Huntleya*, a closely related genus into which this species has been assigned in the past. Although pollination has not been observed, male euglossine bees are certainly implicated due to the sweet, complex fragrance and floral morphology. The German botanist and orchidologist H.G. Reichenbach (1823–89) named the genus in honor of Louis Athanase Chaubard (1781–1854), a French botanist.

The flower of the Red-striped Fan Orchid has yellowish, lanceolate, acuminate tepals, usually overlaid with rusty red stripes. The labellum is rounded and recurved, usually pale violet blue with a fimbriate crest that surrounds the column.

Actual size

SUBFAMILY	Epidendroideae
TRIBE AND SUBTRIBE	Cymbidieae, Zygopetalinae
NATIVE RANGE	Southern Ecuador to Peru
HABITAT	Forests, at 1,640–3,950 ft (500–1,200 m)
TYPE AND PLACEMENT	Epiphyte
CONSERVATION STATUS	Unassessed
FLOWERING TIME	July to September

FLOWER SIZE
1½ in (3.8 cm)

PLANT SIZE
3–5½ × 2⅜–5 in
(8–14 × 6–13 cm),
excluding inflorescence

CHAUBARDIELLA HIRTZII
TIGER-STRIPED ORCHID
DODSON, 1989

373

The Tiger-striped Orchid has a short stem held on a tree by a cluster of wiry white roots, enveloped in spirally arranged, lanceolate, basally clasping leaves without a pseudobulb. In the axil of a leaf, a single, large upside-down (non-resupinate) flower emerges, with an unpleasant scent. This species grows in wet, moderately hot tropical areas, and for this reason it has no need to store water, and so has no pseudobulbs and relatively thin leaves.

Male euglossine bees visit the flowers to collect fragrance compounds, and the hook-shaped pollinaria are deposited on their legs. No other reward is offered to pollinators. The species is named in honor of Alexander Hirtz (1945–), an Ecuadorean orchidophile and expert on Andean orchids.

The flower of the Tiger-striped Orchid is non-resupinate and has similar, spreading, brown-spotted or striped sepals and petals. The lip curves over the flower as a hood and has a similar tiger-striped pattern inside. The column is short and bears two short arms.

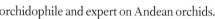

Actual size

SUBFAMILY	Epidendroideae
TRIBE AND SUBTRIBE	Cymbidieae, Zygopetalinae
NATIVE RANGE	Venezuela to northern Brazil (the Guiana Shield)
HABITAT	Rain forests, at 650–1,970 ft (200–600 m)
TYPE AND PLACEMENT	Epiphytic on damp, mossy tree trunks or sometimes on mossy rocks
CONSERVATION STATUS	Not assessed
FLOWERING TIME	November to March (winter to spring)

FLOWER SIZE
¼ in (0.8 cm)

PLANT SIZE
2–4 × 3–5 in
(5–10 × 8–13 cm),
excluding erect
inflorescence
4–5 in (10–13 cm) tall

374

CHEIRADENIA CUSPIDATA
HAND ORCHID
LINDLEY, 1853

Actual size

The little-known Hand Orchid is the sole species in its genus. It is a small plant with a short rhizome and clustered pseudobulbs, each carrying two to five velvety, elliptical leaves at the apex. From an axil of a leaf, a raceme arises carrying two to four flowers. The genus name comes from the Greek words *cheiro*, "hand," and *aden*, "gland," referring to the fan of fingerlike calli (the gland) on its lip, which resembles the digits on a hand. The species name is from the Latin *cuspidatus*, meaning "tapering to a sharp point"—another reference to the calli.

Pollination of these tiny flowers has not been studied. However, given their similarity in form to those of species that are pollinated by fragrance-collecting male euglossine bees, the same could be expected here as well.

The flower of the Hand Orchid has white with red-spotted or red-barred spreading sepals and petals, all more or less the same shape and size with a pointed tip. The lip is deeply cupped, held below the yellow, winged column with fingerlike lobes on its margin.

SUBFAMILY	Epidendroideae
TRIBE AND SUBTRIBE	Cymbidieae, Zygopetalinae
NATIVE RANGE	Colombia to northwestern Venezuela
HABITAT	Cloud forests, at 4,300–6,600 ft (1,300–2,000 m)
TYPE AND PLACEMENT	Epiphytic on moss-covered branches
CONSERVATION STATUS	Not formally assessed but possibly in a category of threat
FLOWERING TIME	June (summer)

CHONDRORHYNCHA ROSEA
PINK SNOUT ORCHID
LINDLEY, 1846

FLOWER SIZE
3⅜ in (8.5 cm)

PLANT SIZE
6–12 × 6–13 in
(15–30 × 15–33 cm),
excluding inflorescence,
which is usually shorter
than the leaves

375

The Pink Snout Orchid has short stems without pseudobulbs and grows in perpetually wet regions. The leaves are clustered and oblanceolate, with the bases enfolding the next leaf. From the base of one or two of these leaves, a slender inflorescence appears with three or four bracts and a single flower. The common name refers to the beak-like apex of the column.

Male euglossine bees of the genus *Eulaema* visit the flowers. The bee enters and crawls up to the base of the lip, looking for sites of floral fragrance production, and the pollinium is placed on the end of the scutellum of the insect as it backs out. In the past, many other orchids were considered members of this genus, but it is now limited to only six or seven species.

The flower of the Pink Snout Orchid
has two linear-cupped sepals that stand
upright, and a middle sepal that points
forward. The petals are broad and, with the
lip, form a tube around the stout column.

Actual size

SUBFAMILY	Epidendroideae
TRIBE AND SUBTRIBE	Cymbidieae, Zygopetalinae
NATIVE RANGE	Colombia to northern Ecuador
HABITAT	Cloud forests, at 3,950–4,920 ft (1,200–1,500 m)
TYPE AND PLACEMENT	Epiphytic
CONSERVATION STATUS	Not assessed
FLOWERING TIME	Variable, generally throughout the year

FLOWER SIZE
2½ in (6.5 cm)

PLANT SIZE
10–18 × 10–14 in
(25–46 × 25–36 cm),
including arching
single-flowered inflorescences
3–5 in (8–13 cm) long

376

CHONDROSCAPHE AMABILIS
ENCHANTING FAN ORCHID
(SCHLECHTER) SENGHAS & G. GERLACH, 1993

The short stem, with many aerial roots, of the Enchanting Fan Orchid has a fan of oblanceolate leaves and no pseudobulb—an indication that the plant grows in habitats with no pronounced drought. The genus name comes from the Greek words *chondros* and *skyphos*, for "cartilage" and "bowl," a reference to some aspect of the lip callus that is unclear. The species name—Latin for "lovable"—refers to the beauty of the flowers.

The pollinator of *Chrondroscaphe amabili*s is not known, but related species are pollinated by euglossine bees. In species with a similar lip shape, the bees land on the middle of the lip and probe its base, placing their tongues into the backward-swept, rolled-up lateral sepals. This is thought to mimic a nectar spur, although there is no nectar present.

The flower of the Enchanting Fan Orchid has pale yellow, spreading sepals and back-curling petals with a frilly edge. The large, red-spotted lip forms an open "mouth" with the column and also has a beardlike edge.

Actual size

SUBFAMILY	Epidendroideae
TRIBE AND SUBTRIBE	Cymbidieae, Zygopetalinae
NATIVE RANGE	Tropical America, from Mexico to Peru, and Trinidad and Jamaica
HABITAT	Humid forests in shade, at 100–3,300 ft (30–1,000 m)
TYPE AND PLACEMENT	Epiphytic
CONSERVATION STATUS	Not assessed
FLOWERING TIME	Throughout the year

CRYPTARRHENA LUNATA
HIDDEN MOON ORCHID
R. BROWN, 1816

FLOWER SIZE
⅛ in (0.3 cm)

PLANT SIZE
2–6 × 3–5 in
(5–15 × 8–12 cm),
excluding inflorescence
4–6 in (10–15 cm) long

377

The Hidden Moon Orchid is the only *Cryptarrhena* species that lacks a pseudobulb. Instead, it has a short stem topped by folded leaves that clasp the stem at their bases. From the center, a simple, short-bracted, arching to pendent inflorescence emerges, with 12 to 30 flowers arranged in a closely spaced cylinder. The yellowish lip contrasts with the green flower, reminiscent of a crescent moon—hence the common and species names. The hood on the apex of the column, which hides the anther from view, provides the genus name (Greek *krypto*, "hidden," and *harren*, "stamen").

No observations on pollination of *Cryptarrhena* species are known. Given the combination of floral traits and color, and its lack of any obvious reward, it is difficult to imagine what sort of insect might be operating as pollinator.

The flower of the Hidden Moon Orchid has green, spreading petals and sepals, and an anchor-shaped, yellow to orange lip. The column protrudes and has a hooded apex. There is no nectar spur.

Actual size

SUBFAMILY	Epidendroideae
TRIBE AND SUBTRIBE	Cymbidieae, Zygopetalinae
NATIVE RANGE	Ecuador and northern Peru
HABITAT	Lowland rain forests, at up to 4,300 ft (1,300 m)
TYPE AND PLACEMENT	Epiphytic
CONSERVATION STATUS	Not assessed
FLOWERING TIME	Spring and summer

FLOWER SIZE
¼ in (0.8 cm)

PLANT SIZE
8–20 × 3–6 in
(20–51 × 8–15 cm),
including single-flowered
inflorescences
½–1 in (1.3–2.5 cm) long

378

DICHAEA CALYCULATA
WHITE SPIDERWORT ORCHID
POEPPIG & ENDLICHER, 1836

Actual size

The upright stem of the White Spiderwort Orchid carries two ranks of linear leaves with a sheathing base. As they age, longer stems may become arching to pendent. Inflorescences are produced simultaneously from the bases of several leaves. The genus name is Greek—*di*–, "two," and *keio*, "to divide"—and refers to the two ranks of leaves typical of these plants. The common name refers to the similarity of this species to a member of the Commelinaceae family (spiderworts). Below each flower is a small bract that partially surrounds the bud as it is developing, which feature gives the species its scientific name, from the Latin *calyculus*, meaning "a bract."

As in all *Dichaea* species, the flowers are pollinated by male euglossine bees. They collect floral fragrance compounds, which are used to make a pheromone that attracts female bees.

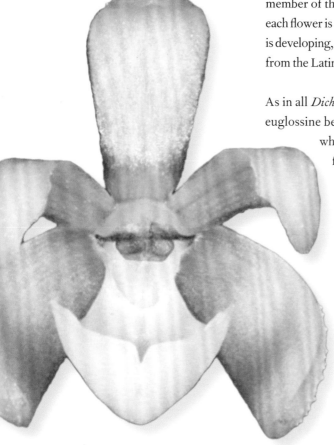

The flower of the White Spiderwort Orchid has similar, forwardly projecting sepals and petals. The white lip is cupped, with three lobes. The column forms a short hood that is often purple tinged, with the two pollinia held below.

SUBFAMILY	Epidendroideae
TRIBE AND SUBTRIBE	Cymbidieae, Zygopetalinae
NATIVE RANGE	Colombia to northern Ecuador
HABITAT	Rain forests, at 2,790–4,100 ft (850–1,250 m)
TYPE AND PLACEMENT	Epiphytic
CONSERVATION STATUS	Not assessed
FLOWERING TIME	Throughout the year

FLOWER SIZE
⅜ in (1 cm)

PLANT SIZE
10–25 × 3–5 in
(25–64 × 8–13 cm),
including single-flowered
inflorescences
½–1 in (1.3–2.5 cm) long

DICHAEA RUBROVIOLACEA
RED WANDERING-JEW ORCHID

DODSON, 1989

379

This cane-like orchid is completely enveloped by overlapping leafy sheaths, the blades yellowish-green, stiff, keeled, and in two ranks. At any time of the year, single-flowered stems can appear from the leaf axils, each subtended by an ovate floral bract. Originally growing upright, the plants, as they age and the stem becomes longer, can become arching to pendent. The common name points to the similarity of the species to another plant, the Wandering Jew—a member of the Commelinaceae or spiderwort family.

The flower of *Dichaea rubroviolacea* is pollinated by a species of euglossine bees. The males collect floral fragrance compounds, which are used to produce a pheromone that attracts females. The flower does not produce nectar.

Actual size

The flower of the Red Wandering-Jew Orchid has spreading, reddish-purple sepals and petals of similar size and shape, the bases of which are marked with darker red spots and stripes against a paler red-purple background. The lip is similar in color, projecting forward and spade-shaped.

SUBFAMILY	Epidendroideae
TRIBE AND SUBTRIBE	Cymbidieae, Zygopetalinae
NATIVE RANGE	Southern Colombia and northern Brazil
HABITAT	Lowland Amazon rain forests, at 330–1,640 ft (100–500 m)
TYPE AND PLACEMENT	Epiphytic
CONSERVATION STATUS	Not assessed
FLOWERING TIME	Summer

FLOWER SIZE
1¾ in (4.5 cm)

PLANT SIZE
12–22 × 10–18 in
(30–56 × 25–46 cm),
excluding arching
inflorescence, which is
shorter than the leaves,
2–5 in (5–13 cm) long

380

GALEOTTIA NEGRENSIS
RIO NEGRO TIGER ORCHID
SCHLECHTER, 1925

The pseudobulbs of the Rio Negro Tiger Orchid have two basal sheaths and are topped with one to three upright, lanceolate leaves. From the base of a mature bulb a short inflorescence carries between two and five flowers, appearing at the same time as the new leaves. The species is named for the Rio Negro, one of the largest tributaries of the Amazon River in Brazil (where it was first found).

Although no studies of pollination have been published for this species, its floral chemistry is typical of species pollinated by male euglossine bees, which scrape floral compounds from floral tissues to use them in sexual pheromones. Pollinaria of another species, *Galeottia grandiflora*, were found on male euglossine bees that were lured to baits of artificial fragrance compounds.

Actual size

The flower of the Rio Negro Tiger Orchid has broad, spreading, red-purple spotted, yellow sepals and petals, the upper sepal somewhat cupped and the lateral sepals twisted. The white lip is red-spotted, cupped, and deeply dissected with long teeth.

SUBFAMILY	Epidendroideae
TRIBE AND SUBTRIBE	Cymbidieae, Zygopetalinae
NATIVE RANGE	Venezuela, Guyana, Brazil, Peru, Bolivia, and Trinidad and Tobago
HABITAT	Wet forests, at 2,000–4,300 ft (600–1,300 m)
TYPE AND PLACEMENT	Epiphytic
CONSERVATION STATUS	Not threatened
FLOWERING TIME	Spring to summer

HUNTLEYA MELEAGRIS
GUINEA FOWL ORCHID
LINDLEY, 1837

FLOWER SIZE
4¾–6 in (12–15 cm)

PLANT SIZE
12–20 × 12–25 in
(30–51 × 30–64 cm),
including short,
single-flowered inflorescences
5–8 in (13–20 cm) long

381

The stately, often stunning Guinea Fowl Orchid has large, showy, and aromatic flowers with distinctive characteristics. Leaves are arranged in a graceful fan shape and make a perfect frame for the spectacular star-shaped flowers that emerge singly on short stems from the leaf axils.

The flowers are large and flat as well as colorful, usually with concentric spotted patterns and a glossy sheen, like the feathers of a guinea fowl, hence the common name. Male euglossine bees foraging for fragrance compounds pollinate the species. Its labellum bears a crest of rigid papillae that subtend the column and probably steer the pollinator to its desired position on the flower, where pollinia will adhere effectively. In nature, the lips of the flowers can be heavily damaged by the foraging bees in their efforts to collect the fragrances.

The flower of the Guinea Fowl Orchid is star-shaped, with broad petals and sepals that taper to an acuminate tip. It is generally mottled purplish-brown but set off by a white and yellow center and a whitish, brown-tipped lip, with a crest of curving papillae.

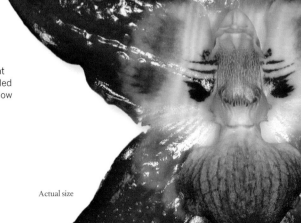

Actual size

SUBFAMILY	Epidendroideae
TRIBE AND SUBTRIBE	Cymbidieae, Zygopetalinae
NATIVE RANGE	Colombia (department of Amazonas)
HABITAT	Rain forests, at 330–650 ft (100–200 m)
TYPE AND PLACEMENT	Epiphytic
CONSERVATION STATUS	Not assessed
FLOWERING TIME	Late winter to spring

FLOWER SIZE
¾ in (2 cm)

PLANT SIZE
4–8 × 3–6 in
(10–20 × 8–15 cm),
excluding inflorescence,
which is shorter than
the leaves

IXYOPHORA CARINATA
FAUN ORCHID

(ORTIZ VALDIVIESO) DRESSLER, 2005

The Faun Orchid occurs in wet forests and does not have pseudobulbs to store water for dry periods. A short stem is covered by overlapping, sheathing leaves, in the axils of which are formed one or two single-flowered inflorescences. The genus name comes from the Greek for having a narrow waist (*ixys*, "waist," and *phorein*, "bearing") in reference to the narrow middle part of the pollinarium (the pollen-bearing structure in these orchids).

The flower, with its curled, backward-projecting sepals, resembles the horned head of a faun, a creature from Greek mythology. Like its close relatives in the genus *Chaubardiella*, for which pollination has been studied, this unstudied species probably forms false nectar horns (the backward-projecting sepals) that are probed by nectar-seeking bees.

Actual size

The flower of the Faun Orchid has two reflexed, spur-like, pale yellow sepals and one similar sepal pointed forward. The pale yellow petals are falcate and spreading, and the whitish lip has a wine-red central blotch with a wide opening partly surrounding the broad column.

SUBFAMILY	Epidendroideae
TRIBE AND SUBTRIBE	Cymbidieae, Zygopetalinae
NATIVE RANGE	Colombia and northern Venezuela
HABITAT	Dense wet forests, at 3,300–5,900 ft (1,000–1,800 m)
TYPE AND PLACEMENT	Epiphytic on tree trunks
CONSERVATION STATUS	Not assessed
FLOWERING TIME	May to September (summer to fall), but flowers may also occasionally appear at other times

KEFERSTEINIA GRAMINEA

FRINGED GRASS ORCHID

(LINDLEY) REICHENBACH FILS, 1852

FLOWER SIZE
1½ in (4 cm)

PLANT SIZE
6–10 × 6–10 in
(15–25 × 15–25 cm),
including arching-pendent,
single-flowered inflorescence
2–4 in (5–10 cm) long

383

The lovely Fringed Grass Orchid produces tight clusters of short stems carrying a fan of four or five linear-lanceolate leaves with several basal leafless bracts. From the axils of these bracts, up to 20 short inflorescences with one flower each appear, with no detectable scent. The genus was named in honor of Herr Keferstein of Kröllwitz, Germany, who was well known during the peak of orchidologist Heinrich Gustav Reichenbach's activity (1850–89). The common name alludes to the grasslike leaves and heavily fringed lip.

Euglossine bees are the pollinators of this species, as they are of other species in this and related genera. Here, pollen becomes attached to the base of the antennae when the bee is forced to twist its body around due to the prominent tooth on the underside of the column.

Actual size

The flower of the Fringed Grass Orchid has purple-spotted, pale green lanceolate sepals and petals of similar size. The lip is broad and curves around the column with complex folds and a finely fringed, wavy margin. The callus has two raised folds.

SUBFAMILY	Epidendroideae
TRIBE AND SUBTRIBE	Cymbidieae, Zygopetalinae
NATIVE RANGE	Northern South America, south to Peru and northern Brazil, the Guianas, Trinidad, and Puerto Rico
HABITAT	Humid, montane, deciduous, and evergreen forests
TYPE AND PLACEMENT	Terrestrial (rarely epiphytic) in low to moderate shade
CONSERVATION STATUS	Not assessed, but widespread
FLOWERING TIME	Winter to spring

FLOWER SIZE
1 in (2.5 cm)

PLANT SIZE
6–20 × 5–10 in
(15–51 × 13–25),
excluding inflorescence,
which is usually about
as long as the leaves

KOELLENSTEINIA GRAMINEA
GRASSLEAF ORCHID
(LINDLEY) REICHENBACH FILS, 1856

384

Actual size

The Grassleaf Orchid is a clump-forming species, usually occurring as a terrestrial, although it can also grow epiphytically in wetter habitats. It has no pseudobulb. Each new growth forms a fan of grasslike, folded leaves that wrap around the short stem. A lateral, unbranched inflorescence bears up to 15 small, cupped flowers. Until they observe the brightly colored flowers, many people think the species is a grass—hence the species name. The genus name honors an Austrian military man, Captain Kellner von Koellenstein, who was a collector of tropical plants in the nineteenth century.

Pollination has not been reported in *Koellensteinia graminea*. However, the similarity of its flowers to others in the subtribe that are pollinated by male euglossine bees suggests that they pollinate this species, too.

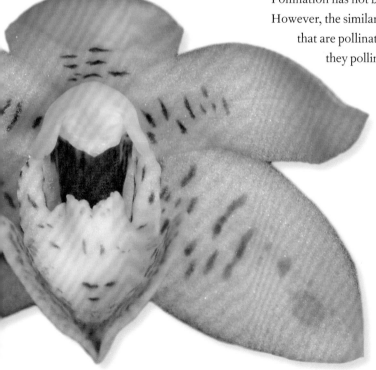

The flower of the Grassleaf Orchid has similar, flat, incurved, creamy white sepals and petals with almost concentric red barring. The similarly colored lip is trilobed; the lateral lobes form a tube with the lateral column-lobes and the median lobe is arrow-shaped with a warty, yellow callus.

SUBFAMILY	Epidendroideae
TRIBE AND SUBTRIBE	Cymbidieae, Zygopetalinae
NATIVE RANGE	Eastern Brazil, Serra do Mar from Rio de Janeiro to Espírito Santo
HABITAT	Atlantic rain forests in shady and humid places, at up to 2,300 ft (700 m)
TYPE AND PLACEMENT	Epiphytic or lithophytic
CONSERVATION STATUS	Not formally assessed, but probably threatened by deforestation
FLOWERING TIME	October to January (spring to early summer)

FLOWER SIZE
2 in (5 cm)

PLANT SIZE
9–15 × 6–12 in
(23–38 × 15–30 cm),
excluding inflorescence,
which is nearly always
shorter than the leaves

PABSTIA JUGOSA

HAWK ORCHID

(LINDLEY) GARAY, 1873

385

The Hawk Orchid produces a cluster of ovoid, slightly compressed, nearly smooth pseudobulbs, each topped with two or three elongate-lanceolate leaves. On new growths, an upright inflorescence grows, with up to four highly fragrant flowers. The genus is closely related to *Zygopetalum*, from which it differs in minor characteristics.

The purple-spotted petals contrast strongly with the white sepals, and together with the column, which forms the "head," these resemble a bird of prey in flight, hence the common name. The genus name honors the renowned Brazilian orchidologist, Guido Pabst (1914–80), founder of the Herbarium Bradeanum in Rio de Janeiro. Pollination of this species has not been studied, but it is likely that the pollinators are male euglossine bees seeking floral fragrance compounds.

The flower of the Hawk Orchid has fleshy, spreading, white sepals and petals, the latter with reddish-purple spots and bars. The white, blue-marked lip is fan-shaped with the edges curved down. The anther cap is large and conspicuous.

Actual size

SUBFAMILY	Epidendroideae
TRIBE AND SUBTRIBE	Cymbidieae, Zygopetalinae
NATIVE RANGE	Southeastern Brazil from Rio de Janeiro to Santa Catarina state
HABITAT	Shady sites in seasonally dry forests
TYPE AND PLACEMENT	Terrestrial
CONSERVATION STATUS	Not assessed
FLOWERING TIME	November to December (late spring to early summer)

FLOWER SIZE
¼ in (2 cm)

PLANT SIZE
8–16 × 6–10 in
(20–41 × 15–25 cm),
excluding erect inflorescence
12–25 in (30–64 cm) tall

386

PARADISANTHUS MICRANTHUS
SMALL PARADISE ORCHID
(BARBOSA RODRIGUEZ) SCHLECHTER, 1918

The Small Paradise Orchid is a terrestrial orchid with small ovoid pseudobulbs, each bearing between one and three elongate lanceolate leaves. It produces an unbranched stem (rarely a small side branch) carrying between 8 and 20 flowers. The genus name comes from the Greek words *paradeisos*, "paradise" or "garden," and *anthos*, "flower." The blooms, although not large, are attractive, hence the common name. The species name is the Greek for "small flower."

Pollination has not been studied in any of the five species of *Paradisanthus*, but they are similar to other orchids pollinated by fragrance-collecting male euglossine bees and offer no other reward. The Atlantic Forest area in which the Small Paradise Orchid occurs is a renowned biodiversity hotspot, which has largely been converted to farming.

The flower of the Small Paradise Orchid has similar, spreading, green to yellow, red-spotted sepals and petals, the sepals being the larger. The white lip is spade-shaped and has a cavity just below the column and a blue-marked callus that forms a channel with the column.

Actual size

SUBFAMILY	Epidendroideae
TRIBE AND SUBTRIBE	Cymbidieae, Zygopetalinae
NATIVE RANGE	Colombia
HABITAT	Extremely foggy, montane forests
TYPE AND PLACEMENT	Epiphytic
CONSERVATION STATUS	Not assessed
FLOWERING TIME	March to July (spring to summer)

FLOWER SIZE
3 in (7.5 cm)

PLANT SIZE
12–20 in (30–51 cm),
including single-flowered
inflorescence,
4–8 in (10–20 cm) long,
which is shorter than
the leaves

PESCATORIA COELESTIS
BLUE SKIES
(REICHENBACH FILS) DRESSLER, 2005

387

Blue Skies produces six to ten oblong-lanceolate leaves arranged in a fan shape. It has short inflorescences with sheathing bracts, each with a solitary, fleshy, highly fragrant, and long-lived flower. These blooms are pollinated by male euglossine bees that collect floral fragrance compounds from which they make sex pheromones to attract females. Female euglossine bees are not attracted to the original compounds collected from the lips of these flowers.

In the past, the species was commonly placed in *Bollea*, but DNA studies show that it is properly considered in *Pescatoria*. The common and species names both refer to its spectacular color, but white and variously reddish-purple variants are also known. The genus is named for Jean-Pierre Pescatore (1793–1855), a Luxembourgian-French businessman, philanthropist, and orchid collector.

The flower of Blue Skies has spreading, blue to reddish-purple sepals and petals, all similar and with light and dark bands of color. The column is broad and forms a hood over the ridged, undulating lip with a bright yellow, ridged center.

Actual size

SUBFAMILY	Epidendroideae
TRIBE AND SUBTRIBE	Cymbidieae, Zygopetalinae
NATIVE RANGE	Minas Gerais and Espírito Santo states of Brazil
HABITAT	Mid-elevation cool, wet tropical forests
TYPE AND PLACEMENT	Epiphytic, sometimes lithophytic on mossy rocks
CONSERVATION STATUS	Not threatened
FLOWERING TIME	Spring to summer

FLOWER SIZE
1¾ in (4.5 cm)

PLANT SIZE
3–6 × 3–5 in
(8–15 × 8–13 cm),
excluding pendent
inflorescence, which
is shorter than the leaves,
2–4 in (5–10 cm) long

PROMENAEA STAPELIOIDES
BLACK-SPOTTED ORCHID
(LINK & OTTO) LINDLEY, 1843

The species name of the distinctive Black-spotted Orchid, which grows in cool, shady forested regions in the Brazilian Atlantic Forest, refers to its resemblance to African carrion flowers of the genus *Stapelia*—although the plant does not smell like a dead animal or attract flies as pollinators. Instead, this diminutive orchid, with glossy pseudobulbs and usually a pair of gray-green elongate apical leaves, emits a sweet, complex fragrance from its comparatively large flowers.

Male euglossine bees pollinate the orchid, collecting fragrance compounds from its lip and using them to make pheromones to attract females. A number of attractive hybrids have been produced using this species and others in the genus, the name of which is based on that of a Greek priestess Promeneia, who helped to guard the oracle at Dodona.

The flower of the Black-spotted Orchid is proportionately large and showy. Sepals and petals are yellow green or pale green, usually heavily overlaid with purplish-brown markings, sometimes coalescing into bars or concentric patterns. The lip color approaches black, especially in the center.

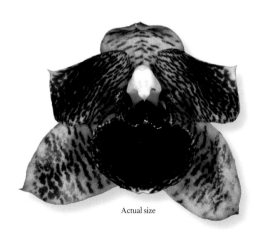

Actual size

SUBFAMILY	Epidendroideae
TRIBE AND SUBTRIBE	Cymbidieae, Zygopetalinae
NATIVE RANGE	Northern South America, from Trinidad and Guyana to Ecuador and Peru
HABITAT	Wet forests at 985–5,900 ft (300–1,800 m)
TYPE AND PLACEMENT	Epiphytic on small shrubs and abandoned citrus and cocoa trees
CONSERVATION STATUS	Not formally assessed, but locally common
FLOWERING TIME	Throughout the year

STENIA PALLIDA
FALSE SLIPPER ORCHID
LINDLEY, 1837

FLOWER SIZE
2 in (5 cm)

PLANT SIZE
6–10 × 6–8 in
(15–25 × 15–20 cm)
including the arching/pendent
single-flowered
inflorescence, which is
2–4 in (5–10 cm) long

389

The small False Slipper Orchid produces a fan of tightly packed leaves, from which an axillary or basal inflorescence with a single pendulous flower emerges. The orchid grows in permanently wet forests, so does not produce any water storage structures, such as pseudobulbs. The genus name refers to the long, narrow, pollen masses—unusual in this subtribe of orchids. The species name reflects the pale color of the flowers.

The slipper-like lip of *Stenia pallida* has a similar function in pollination as in the true slipper orchids (Cypripedioideae). Here, the insects—fragrance-collecting male euglossine bees—enter the opening at the apex of the lip, then pass inside, looking for the source of the floral fragrance. Once they are inside, the sides of the lip have a toothed structure that prevents them from escaping except through the opening just below the pollinia.

The flower of the False Slipper Orchid has spreading, white to cream petals and sepals. The yellow lip is bilobed and the two lobes form a slipper-like sack with papillae inside and small red spots outside.

Actual size

SUBFAMILY	Epidendroideae
TRIBE AND SUBTRIBE	Cymbidieae, Zygopetalinae
NATIVE RANGE	Costa Rica, Cuba, Honduras, Panama, Colombia, Venezuela, and Ecuador
HABITAT	Mid-elevation wet tropical forests
TYPE AND PLACEMENT	Epiphytic
CONSERVATION STATUS	Not threatened
FLOWERING TIME	Spring to summer

FLOWER SIZE
2⅜ in (6 cm)

PLANT SIZE
8–15 × 6–10 in
(20–38 × 15–25 cm), including
single-flowered, erect
inflorescences, which
are shorter than the leaves,
4–6 in (10–15 cm) long

WARCZEWICZELLA DISCOLOR
CEDAR-CHIP ORCHID
(LINDLEY) REICHENBACH FILS, 1852

A remarkably variable species, with some beautiful forms and unusual colors, the Cedar-chip Orchid has, like many other orchids of subtribe Zygopetalinae, fan-shaped growths and no pseudobulbs, indicative of it coming from continuously wet habitats. Erect to slightly arching inflorescences arise from leaf axils and carry exquisite solitary flowers with a tubular lip that, in some forms, is an intensely dark, cobalt violet-blue, whereas in others it approaches rosy pink.

The flower shape and unusual fragrance are compatible with pollination by male euglossine bees that are seeking floral fragrance compounds, from which they make pheromones to attract females. The floral fragrance has been variously described, but cedarwood or cedar chips are often invoked, hence the common name. The genus was named after Józef Warszewicz (1812–66), an eminent Polish botanist.

Actual size

The flower of the Cedar-chip Orchid has sepals that are mostly sharply recurved and usually pale green to cream. The petals are similar or whiter, but they can be flushed with pale violet. The tubular lip is highly variable but often dark cobalt blue.

SUBFAMILY	Epidendroideae
TRIBE AND SUBTRIBE	Cymbidieae, Zygopetalinae
NATIVE RANGE	Colombia, Venezuela, and Panama
HABITAT	Rain forests, at 165–1,640 ft (50–500 m)
TYPE AND PLACEMENT	Epiphytic
CONSERVATION STATUS	Not threatened
FLOWERING TIME	December to July, but can flower at any time

FLOWER SIZE
2½ in (6.5 cm)

PLANT SIZE
8–12 × 6–10 in
(20–30 × 15–25 cm),
including erect-arching,
single-flowered inflorescence
3–5 in (7.5–13 cm) long

WARCZEWICZELLA MARGINATA
PURPLE-LIPPED FAN ORCHID
REICHENBACH FILS, 1852

391

The Purple-lipped Fan Orchid produces flowers singly on short stems between the leaves, with no pseudobulbs to store water because it grows in perpetually wet forests. The genus was named after Józef Warszewicz Ritter von Rawicz (1812–66), a Polish botanist, biologist, and plant and animal collector who was particularly interested in orchids, which he supplied to the great German orchidologist H. G. Reichenbach for description.

The fragrant flowers, with pale violet markings, have a broad, saclike lip and prominent nectar guides. They are typical nectar-deceit blooms, with the lateral sepals incurved and projecting backward to form a false nectar spur to which the tongue of a euglossine bee gains access through a hole on each side of the lip. The species attracts both male and female bees searching for nectar.

The flower of the Purple-lipped Fan Orchid has white, lanceolate petals and dorsal sepals, with the lateral sepals narrower and projecting behind. The lip has lateral margins curled around the column, and there are several central, violet keels with a white-ridged basal callus and a darker purple margin.

Actual size

SUBFAMILY	Epidendroideae
TRIBE AND SUBTRIBE	Cymbidieae, Zygopetalinae
NATIVE RANGE	Tropical South America, from Venezuela to northern Argentina and southeastern Brazil
HABITAT	Shaded understory of dense wet forests, at 1,970–3,300 ft (600–1,000 m)
TYPE AND PLACEMENT	Terrestrial (rarely epiphytic) in leaf matter
CONSERVATION STATUS	Not assessed
FLOWERING TIME	July to September (summer to early fall)

FLOWER SIZE
1½ in (3.8 cm)

PLANT SIZE
25–40 × 12–20 in
(64–102 × 30–51 cm),
excluding erect inflorescence
30–50 in (76–127 cm) tall

392

WARREA WARREANA
BEAUTY OF THE FOREST
(LODDIGES EX LINDLEY) SCHWEINFURTH, 1955

The Beauty of the Forest forms tight clusters of ovoid pseudobulbs that are hidden by several large leaf-bearing bracts with one to three terminal leaves, all of which are narrowly lanceolate with multiple folds. The inflorescence is taller than the leaves, holding the flowers well aloft and making these beautiful white to cream-colored blooms easily spotted in the shady reaches of the forests in which they occur, giving rise to the common name. The genus and species names both refer to the English explorer and collector Frederick Warre, who discovered *Warrea warreana* during an expedition to Brazil in around 1820.

Pollination has not been studied, but the strong fragrance and floral features make fragrance-collecting male euglossine bees the likely pollinators. The large lip and heavy pigmentation feature heavily in the attraction of insects to the species.

The flower of the Beauty of the Forest has white to cream sepals and petals that are all broadly lanceolate and forward projecting, sometimes with a pink suffusion. The lip is also white but with a dark purple to rosy pink center and a large, ridged callus.

Actual size

SUBFAMILY	Epidendroideae
TRIBE AND SUBTRIBE	Cymbidieae, Zygopetalinae
NATIVE RANGE	Antioquia department (Colombia)
HABITAT	Well-drained slopes in forest understory, at 6,600–7,875 ft (2,000–2,400 m)
TYPE AND PLACEMENT	Terrestrial in decaying leaf matter in moderate shade
CONSERVATION STATUS	Not assessed
FLOWERING TIME	April to September (spring to fall)

WARREELLA PATULA
PINK WOOD NYMPH
GARAY, 1973

FLOWER SIZE
2 in (5 cm)

PLANT SIZE
15–35 × 18–30 in
(38–89 × 46–76 cm),
excluding erect
inflorescence
25–50 in (64–127 cm) tall

393

The Pink Wood Nymph forms large clusters of obovoid pseudobulbs that bear several leaf-bearing basal bracts and between one and three terminal leaves, all of which have many folds. The extensive root system ranges through the accumulated layers of tree leaves on the forest floor, and its upright inflorescence bears 6–12 open-faced flowers that catch the sun as it breaks through the tree canopy above. The common name refers to its shaded forest habitat, whereas the genus name alludes to its similarity to the related genus *Warrea*. The species name—the Latin word for "wide open"—refers to the flowers not being cupped as they are in members of *Warrea*.

Pollination of *Warreella patula* has not been studied. However, it is likely that it fits into the fragrance-collecting male euglossine bee syndrome.

Actual size

The flower of the Pink Wood Nymph
has pale pink to dark purple-pink,
spreading sepals and petals that
are similarly broadly lanceolate, the
latter often darker. The lip is white and
reflexed with a broad, dark-pink border.
The column is white with a yellow base.

SUBFAMILY	Epidendroideae
TRIBE AND SUBTRIBE	Cymbidieae, Zygopetalinae
NATIVE RANGE	Northern Peru to Bolivia and eastern Brazil
HABITAT	Mountain slopes and ridges with shrubs and rocks, sometimes on roadside banks, at 3,300–8,200 ft (1,000–2,500 m)
TYPE AND PLACEMENT	Terrestrial in wet, mossy sites among rocks
CONSERVATION STATUS	Widespread and locally abundant, but not assessed
FLOWERING TIME	September to October (spring), but potentially throughout the year

FLOWER SIZE
1¾ in (4.5 cm)

PLANT SIZE
15–25 × 12–18 in
(38–64 × 30–46 cm),
excluding erect-arching
inflorescence
20–35 in (51–89 cm) tall

394

ZYGOPETALUM MACULATUM
SPOTTED CAT OF THE MOUNTAIN
(KUNTH) GARAY, 1970

The Spotted Cat of the Mountain can grow into large clumps of tightly clustered, ovoid pseudobulbs with some basal leaf-bearing bracts and two to three apical linear leaves. These plants, which generally grow in places with some shade from nearby trees, can be spectacular, with one or two tall inflorescences per pseudobulb carrying up to 25 flowers. The genus name is based on the Greek words *zygon*, "yoke," and *petalon*, "petal," referring to the way the lip appears to tie together, or yoke, the petals. The species name, which means "spotted" in Latin, refers, as does the common name, to the spots on the sepals and petals.

The species has a pleasingly sweet fragrance dominated by the compounds that are typically associated with fragrance-collecting male euglossine bees. Pollination, though, has yet to be studied in the wild.

The flower of the Spotted Cat of the Mountain has green sepals and petals similar in size and shape, well covered with reddish spots and bars. The lip has a single large lobe, which is white with purple veins, and a large, raised basal callus.

Actual size

SUBFAMILY	Epidendroideae
TRIBE AND SUBTRIBE	Cymbidieae, Zygopetalinae
NATIVE RANGE	Northern South America (Colombia to Surinam, south to Peru and across to eastern Brazil)
HABITAT	Shady sites on the lower portions of trees in wet forests, especially along streams and rivers, at 1,970–3,950 ft (600–1,200 m)
TYPE AND PLACEMENT	Epiphytic or terrestrial, often climbing up onto trunks from the ground
CONSERVATION STATUS	Not assessed
FLOWERING TIME	October to November

FLOWER SIZE
3 in (8 cm)

PLANT SIZE
10–18 × 7–15 in
(25–46 × 18–38 cm),
including arching
inflorescence, which is
about as tall as the leaves

ZYGOSEPALUM LABIOSUM
LARGE-LIPPED STREAM ORCHID
(RICHARD) C. SCHWEINFURTH, 1967

395

Witnessing the seemingly suspended-in-air flowers of the Large-lipped Stream Orchid in the dark forest understory is a memorable experience. The plant produces ovoid pseudobulbs along a creeping stem that climbs over a substrate of rocks, humus mats, and tree trunks. A developing pseudobulb produces an inflorescence that carries between one and three flowers. The genus name derives from the Greek words *zygon*, "yoke," and *sepalon*, "sepals," and is a reference to a relationship to the genus *Zygopetalum*, in which *Zygosepalum* species were once included. The species and common names refer to the prominent lip of this species.

Pollination is by males of the euglossine bee genus *Eulaema*. The insects are attracted by fragrance compounds, which they then collect and store in pouches on their hind legs.

The flower of the Large-lipped Stream Orchid
has pale green to tan, lanceolate-pointed sepals
and petals that either arch up or project
backward. The large lip is white with rosy pink
veins in its center, and the dark pink column has
a prominent hood and a pair of wings.

Actual size

SUBFAMILY	Epidendroideae
TRIBE AND SUBTRIBE	Epidendreae, Agrostophyllinae
NATIVE RANGE	Southeast Asia and Malesia to the Solomon Islands
HABITAT	Lowland tropical rain forests, up to 1,640 ft (500 m)
TYPE AND PLACEMENT	Epiphyte
CONSERVATION STATUS	Not assessed
FLOWERING TIME	Throughout the year

FLOWER SIZE
⅜ in (1.1 cm)

PLANT SIZE
12–16 × 2–3 in
(30–41 × 5–8 cm),
including inflorescence

AGROSTOPHYLLUM STIPULATUM
STRAW ORCHID
(GRIFFITH) SCHLECHTER, 1912

Actual size

The pendent epiphytic Straw Orchid forms a double row of leaves in one plane all along the stem. The bases of the leaves clasp the stem tightly, while the leaf blades, which have emarginate tips, twist outward. Other species in the genus have more grasslike leaves, hence the name—from the Greek words *agrostis*, meaning "grass," and *phyllon*, meaning "leaf."

Old stems drop their leaves, and additional stems grow from their bases. Tight clusters of flowers appear from the ends of the stem. Their structure suggests moth pollination, although some species in New Guinea are self-pollinated, which may be the case for *Agrostophyllum stipulatum*—at least when pollinators are absent. One species in the genus is used in Borneo as a charm to protect against curses.

The flower of the Straw Orchid has three broad, straw-colored sepals—the upper one slightly hooded, the lateral two flaring—and two narrow petals. The lip has sidelobes and a large median callus that form a small tube with the pink sides of the column.

SUBFAMILY	Epidendroideae
TRIBE AND SUBTRIBE	Epidendreae, Agrostophyllinae
NATIVE RANGE	New Zealand and Chatham Islands
HABITAT	Lowland to montane forests
TYPE AND PLACEMENT	Epiphytic on thick branches or tree trunks, or lithophytically on rocky banks
CONSERVATION STATUS	Common, not threatened
FLOWERING TIME	February to May (fall)

EARINA AUTUMNALIS
EASTER ORCHID
(G. FORSTER) HOOKER FILS, 1853

FLOWER SIZE
½ in (1.3 cm)

PLANT SIZE
10–20 × 2–4 in
(25–51 × 5–10 cm),
including short terminal
inflorescence 2–4 in
(5–10 cm) long

397

Actual size

The Easter Orchid flowers during the New Zealand fall, at Easter time, hence the scientific and common names. Its Maori name *Raupeka* can mean "droop," no doubt in reference to the drooping stems that are clothed in basally sheathing, folded, falcate leaves. Stems are terminated by an upturned inflorescence that is sheathed with bracts, each protecting a pure white, strongly and sweetly perfumed flower.

What pollinates this orchid is still debated. Crane flies and thrips have been seen on the flower, but the shape and white color, with a central spot of yellow, suggest that some sort of bee is more likely to be the effective pollinator. Most orchids pollinated by crane flies and thrips are green, not white, as is the case with the Easter Orchid.

The flower of the Easter Orchid is relatively simple and small. It has white, similar, spreading sepals and petals and a cup-shaped, yellow-blotched, white lip. Although not individually showy, the flowers when clustered atop the leafy stems put on an attractive show.

SUBFAMILY	Epidendroideae
TRIBE AND SUBTRIBE	Epidendreae, Bletiinae
NATIVE RANGE	Northern Caribbean, from southern Florida and the Bahamas to Cuba, Hispaniola, and Puerto Rico
HABITAT	Thickets, open pineland savannas, hammocks, and seasonally dry deciduous forests, at sea level to 2,460 ft (750 m)
TYPE AND PLACEMENT	Terrestrial in thick litter or humus over limestone
CONSERVATION STATUS	Not formally assessed, but of conservation concern in Florida and likely also elsewhere in the Caribbean
FLOWERING TIME	September to November (late summer to early winter)

FLOWER SIZE
¼ in (2 cm)

PLANT SIZE
10–18 × 1–2 in
(25–46 × 2.5–5 cm),
including erect inflorescence

BASIPHYLLAEA CORALLICOLA
CARTER'S ORCHID
(SMALL) AMES, 1924

Actual size

The terrestrial Carter's Orchid grows from an underground globose tuber. A single leaf (rarely two leaves), which is slightly fleshy and linear to narrowly elliptic, emerges at the base, followed by a raceme with up to ten nodding flowers. These plants are so inconspicuous that they are undoubtedly more common than the few records indicate. Flowers often do not open and instead self-pollinate.

The genus and species names are based on its basal leaf and occurrence on coralline limestone, respectively. In Florida, it was named for the Pennsylvanian botanist J. J. Carter, who, as he traveled through the pinelands of southern Miami-Dade County in 1903, was the first to see this orchid. Since then, the plant has been seen on few occasions. It has the odd habit of lying dormant through its normal flowering season, sometimes for several years, before reappearing.

The flower of Carter's Orchid has white to cream, lanceolate, flat sepals and linear petals that project forward. The lip is white, often tinged pink to purple, and three-lobed, with the side lobes forming an open tube with the column wings. The callus is ridged.

SUBFAMILY	Epidendroideae
TRIBE AND SUBTRIBE	Epidendreae, Bletiinae
NATIVE RANGE	Florida, through Central America to northern South America and the Caribbean islands
HABITAT	Open grasslands, pine forest edges, and clearings
TYPE AND PLACEMENT	Terrestrial
CONSERVATION STATUS	Not threatened in most of its range but considered threatened in Florida
FLOWERING TIME	Late spring, usually May

BLETIA PURPUREA
PINE PINK
(LAMARCK) A. DE CANDOLLE, 1840

FLOWER SIZE
2–2⅜ in (5–6 cm)

PLANT SIZE
15–25 × 10–20 in
(38–64 × 25–51 cm),
excluding inflorescence,
which is erect to arching and
20–38 in (51–97 cm) tall

399

A common and highly adaptable orchid, the Pine Pink occurs over an extraordinarily large range and has brilliantly colored flowers on long, typically branched stems. Plants found in Florida are often cleistogamous, meaning their flowers barely open and form seeds without pollination. This is thought to have evolved in the northernmost portion of the orchid's range, where its pollinator may be rare or even absent. Carpenter and euglossine bees (males and females seeking food, not floral fragrances) are pollinators outside Florida.

The common name refers to its occurrence in pine forests and its color. The species has pseudobulbs, but they are sometimes completely buried in the soil, so may not have been observed. The Aztecs in Mexico made a glue from the psuedobulbs, and it is still used to repair wooden objects in modern Mexico.

The flower of the Pine Pink is variable in color but generally appears in shades of pink or purple to almost pure white with spreading sepals and petals covering the column. The lip is trilobed with raised keels that range from white to orange.

Actual size

SUBFAMILY	Epidendroideae
TRIBE AND SUBTRIBE	Epidendreae, Bletiinae
NATIVE RANGE	Panama, Colombia, and Venezuela
HABITAT	Seasonally wet forests, at 2,300–5,600 ft (700–1,700 m)
TYPE AND PLACEMENT	Epiphytic, occasionally terrestrial
CONSERVATION STATUS	Not threatened
FLOWERING TIME	April to May (spring)

FLOWER SIZE
3 in (8 cm)

PLANT SIZE
12–16 × 5–9 in
(30–41 × 13–23 cm),
excluding arching-pendent
inflorescence
10–14 in (25–36 cm) long

CHYSIS AUREA
GOLDEN CHESTNUT ORCHID
LINDLEY, 1837

400

The Golden Chestnut Orchid bears long, often pendulous pseudobulbs with between four and eight, often deciduous plicate leaves. The inflorescence emerges in the basal half of the pseudobulb and bears 4 to 15 large, sweetly fragrant flowers. The genus name comes from the Greek word for "fusion" or "melting," an allusion to the way the pollinia in some of these orchids seem to dissolve into a single mass. The common and species names refer to the color of the flowers, from the Latin *aurea* for "gold."

Pollination of *Chysis aurea* has not been studied, but the shape, color, and sweet fragrance of the flowers are suited for pollination by some form of bee, possibily euglossines. In particular, the lip has nectar guides (darker sets of lines and spots leading an insect deeper into the flower), although no nectar is present.

The flower of the Golden Chestnut Orchid is yellow to chestnut brown, often darker toward the tips. Sepals and petals are similar and broadly lanceolate. The lip is three-lobed, with the sidelobes wrapping around the column and the centrally keeled midlobe striped or spotted.

Actual size

SUBFAMILY	Epidendroideae
TRIBE AND SUBTRIBE	Epidendreae, Bletiinae
NATIVE RANGE	Central and southeastern United States, from the southern Midwest and Arizona to Maryland and Florida
HABITAT	Variable, from swamps to desert canyons, usually in dry soil in forests or on sandstone up to 1,970 ft (600 m), although some higher occurrences have been documented
TYPE AND PLACEMENT	Terrestrial, parasitic, often in *Juniperus* litter over limestone
CONSERVATION STATUS	Not threatened globally, but rare or endangered in much of its range, often threatened by mining activities
FLOWERING TIME	May to August (spring to summer)

HEXALECTRIS SPICATA

CRESTED CORAL ROOT

(WALTER) BARNHART, 1904

FLOWER SIZE
2 in (5 cm)

PLANT SIZE
8–31 in (20–80 cm)
high (no leaves)

401

The leafless Crested Coral Root lacks chlorophyll and parasitizes fungi to obtain its nutrients, which come through a mutualistic (beneficial to both) relationship between soil fungi and nearby surrounding photosynthetic plants. An underground knobby rhizome with no true roots produces a pinkish-brown or yellowish unbranched stem with large bracts bearing up to 25 strongly colored flowers, which smell faintly of baby powder. The lip is trilobed and has wavy margins.

The species is close to *Basiphyllaea* and may in fact be an achlorophyllous member of that genus, which is photosynthetic. The common name comes from its resemblance to coralroot orchids of the genus *Corallorhiza*, but it is not related; their similarity is due to convergence. Pollination has not been reported, but the coloration and structure of the flowers are compatible with some form of bee pollination.

The flower of the Crested Coral Root has three spreading, yellow-striped, reddish-purple sepals and two spreading, smaller petals. The lip is pink with seven crests on the midlobe and two upturned lateral lobes.

Actual size

SUBFAMILY	Epidendroideae
TRIBE AND SUBTRIBE	Epidendreae, Calypsoinae
NATIVE RANGE	North America, Quebec to Georgia and west to Oklahoma
HABITAT	Deciduous temperate woods
TYPE AND PLACEMENT	Terrestrial
CONSERVATION STATUS	Not threatened, but generally rare
FLOWERING TIME	Late spring, usually May

FLOWER SIZE
1¼–1½ in (3–4 cm)

PLANT SIZE
4–6 in (10–15 cm),
leaves withered or absent when
in bloom, inflorescence
8–12 in (20–30 cm) tall

APLECTRUM HYEMALE
PUTTY ROOT
(MUHLENBERG EX WILLDENOW) NUTTALL, 1818

The beautiful silver-streaked, purple-backed, ovate, pleated leaf of the Putty Root emerges in the fall long after the plant has bloomed. It photosynthesizes through the winter, building stores of energy to support a spring blooming. The small, subtly hued flowers appear in late spring on a modest upright raceme of between 4 and 12 blooms. The leaf generally withers as the plant blooms.

The common name comes from the paired underground corms, which yield mucilage when crushed and were used by early Americans and indigenous peoples as an adhesive. Plants often grow beneath beech and sugar maple trees, and pollination is by a small halictid bee (*Lasioglossum*), but with no reward, so this is a case of deceit pollination. Eating of the leaves by deer and competition by invasive plants are the major threats to this species.

The flower of the Putty Root appears not to open fully. Sepals and petals are olive and tipped a dull plum purple. The lateral sepals spread widely, while the dorsal sepal and petals cover the column. The lip is white with a plum tip and a frilly margin.

Actual size

SUBFAMILY	Epidendroideae
TRIBE AND SUBTRIBE	Epidendreae, Calypsoinae
NATIVE RANGE	Circumboreal—northern Europe, Asia, and North America
HABITAT	Boreal, mostly evergreen forests
TYPE AND PLACEMENT	Terrestrial
CONSERVATION STATUS	Not threatened
FLOWERING TIME	Early spring, as snow melts, later farther north

CALYPSO BULBOSA

FAIRY SLIPPER

(LINNAEUS) OAKES, 1842

FLOWER SIZE
1⅜–1½ in (3.5–4 cm)

PLANT SIZE
Usually only one basal
leaf 2–3 in (5–8 cm) long,
held flat on the soil
surface, with an erect
inflorescence
4–6 in (10–15 cm) tall

403

A beautiful spring-flowering species from the far northern latitudes of our globe, *Calypso bulbosa* was named in honor of the seductive sea nymph, Calypso, from Homer's *Odyssey*. It has several distinctive regional varieties over its vast range. Preferring shady habitats in old-growth evergreen forests, the diminutive plants bear a single, dark green, ovoid, pleated leaf with a short inflorescence and one vanilla-scented bloom. When Linnaeus originally described the species, it was treated in the genus *Cypripedium* due to its pouch-like lip, which is superficially similar to those of the lady's slipper orchids.

As the species name implies, there is a subterranean corm, which was collected as food by Native Americans in northwestern North America. The flowers of some forms have yellow tufts at the opening of their lip, which are thought to be mimicking pollen. Bumblebees are their usual pollinators.

The flower of the Fairy Slipper varies by region, but sepals and petals are generally pink, sometimes white. In variety *americana* the pouch-like lip is streaked internally with vivid maroon and has yellow hairs. Other forms have heavily red-spotted lips with a white tuft of hairs.

Actual size

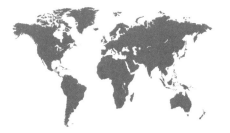

SUBFAMILY	Epidendroideae
TRIBE AND SUBTRIBE	Epidendreae, Calypsoinae
NATIVE RANGE	Mexico, Honduras, and Guatemala
HABITAT	Rain forests in partial shade
TYPE AND PLACEMENT	Mostly terrestrial, sometimes epiphytic and lithophytic
CONSERVATION STATUS	Not threatened
FLOWERING TIME	Summer

FLOWER SIZE
2 in (5 cm)

PLANT SIZE
10–18 × 6–10 in
(25–46 × 15–25 cm),
with an erect-arching
inflorescence shorter
than the leaves,
6–10 in (15–25 cm) long

COELIA BELLA
BEAUTIFUL EGG ORCHID
(LEMAIRE) REICHENBACH FILS, 1861

The tough, sturdy, and adaptable Beautiful Egg Orchid is common and abundant in many locations. Capable of growing high up in trees as an epiphyte, it is even more plentiful as a terrestrial or lithophyte over much of its range. The plants have shiny pseudobulbs that are ovoid, or egg-shaped—hence the common name—with between three and five lanceolate apical leaves. Inflorescences emerge from the base of the plants and bear colorful flowers that emit a sweet fragrance.

Coelia bella usually has 6–12 blooms per spike, and these are often nestled in among the pseudobulbs or hidden beneath plentiful foliage. The pleasing scent implies a bee pollination syndrome, which makes sense in terms of floral morphology, although no studies have so far documented what sort of bees (which receive no reward) might be involved.

Actual size

The flower of the Beautiful Egg Orchid is crystalline in texture, usually white, with sepals larger than the petals and tipped with brilliant rose. The lip, usually sulfur yellow, is gullet-shaped with a recurving midlobe.

SUBFAMILY	Epidendroideae
TRIBE AND SUBTRIBE	Epidendreae, Calypsoinae
NATIVE RANGE	Canada to Guatemala
HABITAT	Shady woodlands
TYPE AND PLACEMENT	Terrestrial
CONSERVATION STATUS	Not threatened
FLOWERING TIME	Late summer to fall

FLOWER SIZE
Up to 1¼ in (3 cm)

PLANT SIZE
Leafless, non-green plants
with an erect inflorescence
8–30 in (20–76 cm) tall

CORALLORHIZA MACULATA
SPOTTED CORALROOT
RAFINESQUE, 1817

405

Completely lacking chlorophyll, this extraordinary plant was once thought to live directly off decaying vegetation in the soil (as a saprophyte) but is now understood to derive its water and nutrition from an association with a fungus that itself is exchanging minerals for carbohydrates with trees and other nearby plants. Brilliant maroon-red to yellowish stems hold up to 50 flowers. The underground root mass is shaped like coral, hence the common and genus names.

The flowers are visited by several kinds of insects, including halictid bees and some flies, although no reward is offered. If pollination by insects fails, then the pollen masses twist on their stalks and contact the stigma (female receptive area), resulting in heavy seed production. Dried rhizomes were used medicinally by Native Americans and now appear in modern herbal remedies that treat skin diseases, night sweats, and many infectious diseases.

The flower of the Spotted Coralroot is usually maroon-red or sometimes yellowish or whitish suffused with red. The labellum is white with maroon spots.

Actual size

SUBFAMILY	Epidendroideae
TRIBE AND SUBTRIBE	Epidendreae, Calypsoinae
NATIVE RANGE	Ecuador to Peru
HABITAT	Cloud forests and open woodland, at 4,920–8,530 ft (1,500–2,600 m)
TYPE AND PLACEMENT	Terrestrial
CONSERVATION STATUS	Not assessed
FLOWERING TIME	Fall

FLOWER SIZE
1¼ in (3.2 cm)

PLANT SIZE
18–26 × 10–16 in
(46–66 × 25–41 cm),
excluding erect inflorescence
20–30 in (51–76 cm) tall,
which is longer than the leaves

406

GOVENIA TINGENS
SPOTTED PIXIE ORCHID
POEPPIG & ENDLICHER, 1836

The stem of the Spotted Pixie Orchid, its base swollen into a corm-like pseudobulb, is enveloped by several tubular sheaths and topped by two, more or less opposite, elliptic to oval, plicate leaves held horizontally. An erect pyramidal raceme, with many, closely spaced flowers, grows from the base of the stem inside the sheaths. The genus name is in honor of J. R. Gowen, a nineteenth-century English botanist and plant collector in Assam. The epithet *tingens* means "to dye," referring to the spotting or stripes of color. Members of *Govenia* are called *duendecito* in Spanish, meaning "pixie," hence the common name.

The flower structure would be suited to pollination by some sort of large bee, but this has not been studied. In Mexico, a glue used to repair wooden objects is made from an extract of the stems.

The flower of the Spotted Pixie Orchid has two falcate, spreading, whitish lateral sepals and a hooded upper sepal. The petals are partly enclosed by the hooded sepal. The lip is cup-shaped and held under the column. All parts are spotted or striped reddish-pink.

Actual size

SUBFAMILY	Epidendroideae
TRIBE AND SUBTRIBE	Epidendreae, Calypsoinae
NATIVE RANGE	Temperate East Asia, from the western Himalayas through China to the Russian Far East, Japan, and Korea
HABITAT	Forests and forest margins, thickets and grassy slopes, and shaded places in valleys and ravines, at 3,300–9,850 ft (1,000–3,000 m)
TYPE AND PLACEMENT	Terrestrial
CONSERVATION STATUS	Not assessed
FLOWERING TIME	June to July (summer)

OREORCHIS PATENS
MOUNTAIN ORCHID
(LINDLEY) LINDLEY, 1858

FLOWER SIZE
¾ in (2 cm)

PLANT SIZE
10–16 × 12–20 in
(25–41 × 30–51 cm),
including erect
inflorescence

407

The conical corms of the Mountain Orchid (in effect, underground pseudobulbs) carry up to two apical, linear leaves that gradually taper at the base into an indistinct stalk. A tall raceme, with several sheaths in its lower half, carries up to 30 densely packed flowers. The plant's Chinese name, *shan lan*, translates as "orchid of the mountain," which is also the common name in English and reflects the genus name, derived from the Greek words *oros*, "mountain," and *orchis*, "orchid." The flowers greatly resemble those of the genus *Corallorhiza*, although, by contrast, its species have no leaves and depend completely on fungi for food.

Two marmalade hoverflies, *Episyrphus balteatus* and *Sphaerophoria menthastri*, have been observed as the pollinators. In Chinese traditional medicine, the leaves of *Oreorchis patens*, combined with other herbs and a shed snakeskin, are used to cure rubella (German measles).

The flower of the Mountain Orchid has orange to green flowers with spreading lateral sepals. The upper sepals and petals form a hood over the arching column and white, three-lobed lip, which often has some purplish spots. The sidelobes are linear and flank the column.

Actual size

SUBFAMILY	Epidendroideae
TRIBE AND SUBTRIBE	Epidendreae, Calypsoinae
NATIVE RANGE	Central and eastern United States (Illinois to Florida)
HABITAT	Deciduous, mixed, or coniferous woodlands, often in sandy, acid oak-pine forests and along stream banks, at sea level to 2,625 ft (800 m)
TYPE AND PLACEMENT	Terrestrial in humus-rich soil
CONSERVATION STATUS	Not listed as nationally endangered, but is endangered in some states (for example, New York)
FLOWERING TIME	June to September (summer to early fall)

FLOWER SIZE
½ in (1.3 cm)

PLANT SIZE
4–26 in (10–66 cm);
there are no leaves
at the time of flowering

TIPULARIA DISCOLOR
CRANE FLY ORCHID
(PURSH) NUTTALL, 1818

In September, the underground tubers of the Crane Fly Orchid produce a solitary, often-spotted broad leaf that is purple below. The leaf persists through the winter and disappears in spring as the trees start to leaf out. In the summer, an upright, red-stemmed inflorescence produces up to 40 flowers.

The flowers are pollinated by the Armyworm Moth, *Pseudaletia unipuncta*, which probes the long nectar spur for the nectar produced by the blooms and picks up the pollen masses on its eyes. The column is slightly twisted to the left or right, such that the pollinia are deposited on only one eye. The common and scientific names refer to the general resemblance of the flower to a crane fly (genus *Tipula*). The tubers were harvested at the time of flowering and consumed by Native Americans as a food.

The flower of the Crane Fly Orchid has spreading, greenish-tan to greenish-red sepals and similar but smaller petals. The lip is greenish-white and three-lobed—the sidelobes are shorter and embrace the column, and the middle lobe is elongate with a basal opening leading to the long nectar spur.

Actual size

SUBFAMILY	Epidendroideae
TRIBE AND SUBTRIBE	Epidendreae, Calypsoinae
NATIVE RANGE	Patchy distribution in mountains in temperate Asia: Assam, China (northern Fujian and Jiangxi provinces), eastern Taiwan, and Japan
HABITAT	Bamboo and coniferous forests, at 5,900–6,600 ft (1,800–2,000 m)
TYPE AND PLACEMENT	Terrestrial
CONSERVATION STATUS	Not formally assessed
FLOWERING TIME	June to July (summer)

FLOWER SIZE
1½ in (4 cm)

PLANT SIZE
5–8 in (13–20 cm),
just a leafless cluster of
flowers and a pinkish-brown
stem, sometimes covered by
leaf litter with only the
flowers revealed

YOANIA JAPONICA
DEMON QUELLER
MAXIMOWICZ, 1873

409

The spectacular leafless Demon Queller has an underground fleshy, branched, tuberculate rhizome, which produces an upright inflorescence with 3 to 12 long-stalked flowers. The branching and scaly rhizome lives in exclusive association with a fungus, from which it gets all its minerals and sugars. The genus name is in honor of Udagawa Yōan (1798–1846), a Japanese physician and botanical artist.

In Japan, the common name is *Shōki-ran*, the orchid of Shōki, a character from Taoist mythology, who vanquishes malevolent demons and other evil beings and protects the emperor. This orchid appears from nowhere in deep, dark forests and, with its beautiful flowers, is said to chase dark and evil forces away. Pollination has not been studied, but the flower's bright color and the complex lip shape with a nectar spur suggest that a bee would be involved.

The flower of the Demon Queller has spreading, red-veined, rounded pink sepals and forward-pointing petals that cover the column and lip. The white to cream lip is saccate, with a blunt, forward-projecting spur and a recurving, yellow upper margin that has some dark brown spots.

Actual size

SUBFAMILY	Epidendroideae
TRIBE AND SUBTRIBE	Epidendreae, Laeliinae
NATIVE RANGE	Serra do Sincorá, Chapada Diamantina region, Bahia state, Brazil
HABITAT	Montane forests, at 2,950–4,600 ft (900–1,400 m)
TYPE AND PLACEMENT	Epiphyte in sunny places
CONSERVATION STATUS	Not assessed, but rare and highly restricted
FLOWERING TIME	December to February (summer/wet season)

FLOWER SIZE
1⅛ in (3 cm)

PLANT SIZE
3½–6 × ¾–1⅛ in
(9–15 × 2–3 cm),
excluding inflorescence

410

ADAMANTINIA MILTONIOIDES
DIAMANTINA ORCHID
VAN DEN BERG & C. N. GONÇALVES, 2004

Actual size

The Diamantina Orchid is an epiphyte with clustered pseudobulbs bearing usually one (rarely two), leathery, olive green leaves. From the joint of the leaf and pseudobulb, it produces a long pendent inflorescence with many large, brightly colored flowers. This species and genus, only described in 2004, have been found just twice at a classic site, known for its many orchids and of great botanical interest. Initially, only a photograph was taken, but the second time a botanical specimen and DNA sample were prepared.

The placement of this orchid was uncertain until DNA analyses confirmed its relationship to *Isabelia* and *Leptotes* in subtribe Laeliinae. The original describers speculated that, based on its color and morphology, some type of bee—perhaps a bumblebee (*Bombus*)—could be the pollinator.

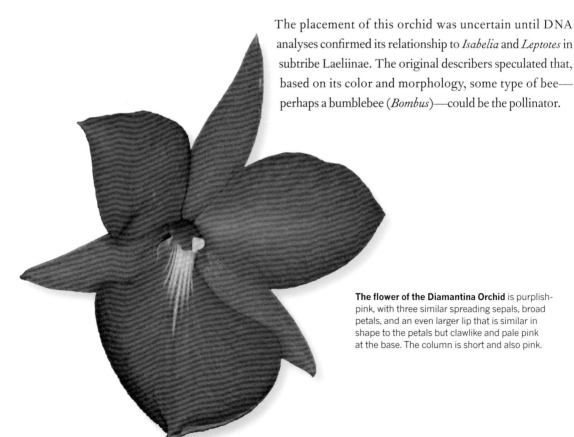

The flower of the Diamantina Orchid is purplish-pink, with three similar spreading sepals, broad petals, and an even larger lip that is similar in shape to the petals but clawlike and pale pink at the base. The column is short and also pink.

SUBFAMILY	Epidendroideae
TRIBE AND SUBTRIBE	Epidendreae, Laeliinae
NATIVE RANGE	Mexico
HABITAT	Pastures, open woodlands, and on lava flows, usually attached to large oaks or rocks, at 4,920–8,900 ft (1,500–2,700 m)
TYPE AND PLACEMENT	Epiphytic or lithophytic, on trunks of large oaks or on rocks beneath them
CONSERVATION STATUS	Not assessed
FLOWERING TIME	April to July (spring to early summer)

ALAMANIA PUNICEA
CRIMSON WOOD SPRITE
LEXARZA, 1825

FLOWER SIZE
½ in (1.3 cm)

PLANT SIZE
1½–3 × 1½–2 in
(4–7.5 × 4–5 cm),
excluding inflorescence

411

The Crimson Wood Sprite forms a cluster of elongate-ovoid pseudobulbs, attached to the substrate by fleshy roots. The two or three leaves are broadly attached to the ends of the pseudobulb. The inflorescence—a cluster of between two and fifteen brilliant red flowers—is on a separate stem or at the top of the pseudobulbs. *Alamania punicea* is the only species of its genus. While a member of the *Encyclia* group, it is not closely related to any of these genera.

Because of the shape and color of the flowers, especially their dark pollinia, hummingbirds are the probable pollen carriers, but pollination has not been studied for this species. It has a nectar spur formed by the lip and partially fused to the column base, but nectar has not been reported.

The flower of the Crimson Wood Sprite
has three red sepals and two similar, slightly narrower petals. The petal-like lip is also red, apart from the base, which is yellow and shortly basally lobed. The cream column bears two dark red pollinia atop the orange stigma.

Actual size

SUBFAMILY	Epidendroideae
TRIBE AND SUBTRIBE	Epidendreae, Laeliinae
NATIVE RANGE	Mexico, Guatemala, Belize, El Salvador, Honduras, Nicaragua, Costa Rica, Colombia, Venezuela, and Jamaica
HABITAT	Seasonally dry foothill forests, at 2,600–4,900 ft (800–1,500 m)
TYPE AND PLACEMENT	Epiphytic
CONSERVATION STATUS	Not threatened
FLOWERING TIME	Late winter to early spring

FLOWER SIZE
¼ in (0.8 cm)

PLANT SIZE
20–30 × 4–5 in
(51–76 × 10–13 cm),
excluding terminal erect
inflorescence, which
extends beyond the leaves,
6–12 in (15–30 cm) long

412

ARPOPHYLLUM GIGANTEUM
CANDLESTICK ORCHID
HARTWEG EX LINDLEY, 1840

Commonly encountered in the foothills of Central America, the Candlestick Orchid is well known to locals, who call it *pico de curvo* (raven's beak) or *masorquilla* (little corn orchid) and often use its leaves for medicinal purposes, including the treatment of dysentery. The plant has long, narrow, scimitar-like leaves and beautiful, jewellike, spirally arranged flowers on showy, candle-like inflorescences that arise from a sheath at the pseudobulb apex. The genus name, from the Greek words *arpe*, "sickle," and *phyllon*, "leaf," refers to the shape of the leaves.

Despite the orchid's familiarity and popularity, pollination has not been documented. However, due to the vibrant colors, copious nectar, and darkly colored pollen masses, the probable pollinators are birds, which are thought to be less likely to remove dark pollinia from their beaks.

The flower of the Candlestick Orchid is non-resupinate and forms part of a tight cylindrical spiral of blooms. The tepals are various shades of pink or pale purple with a darker, rounded labellum of often-brilliant magenta or purple and a dark purple column.

Actual size

SUBFAMILY	Epidendroideae
TRIBE AND SUBTRIBE	Epidendreae, Laeliinae
NATIVE RANGE	Southwestern Mexico (Guerrero and Oaxaca states)
HABITAT	Evergreen cloud forests on the highest peaks of the Sierra Madre del Sur, at 7,875–10,170 ft (2,400–3,100 m)
TYPE AND PLACEMENT	Epiphytic
CONSERVATION STATUS	Highly restricted distribution—probably of concern, due to logging
FLOWERING TIME	December to March (winter to early spring)

FLOWER SIZE
2 in (5 cm)

PLANT SIZE
Each growth
6–10 × 3–6 in
(15–25 × 8–15 cm),
excluding inflorescence

413

ARTORIMA ERUBESCENS
BLUSHING FROST ORCHID
(LINDLEY) DRESSLER & G. E. POLLARD, 1971

The Blushing Frost Orchid has a thick rhizome with roots all along, which produces widely spaced pseudobulbs carrying between two and four oblong-lanceolate leaves. The inflorescence is 3 ft (1 m) long, branches at its end, and bears six to nearly a hundred fragrant, pink to purplish-pink flowers, which form spectacular branching masses high in the trees. The flowers appear when night temperatures drop slightly below freezing, giving it its common name.

The orchid is mostly collected from the wild by local people and used in home altars at Christmas, resulting in it having vernacular names in Spanish—*uña de gavilán* ("hawk's claw")—and other native languages. It was thought to be a member of *Encyclia*, but DNA analyses confirm that the species is much closer to the dwarf, hummingbird-pollinated *Alamania*, also a Mexican endemic. Pollination is probably by bees.

The flower of the Blushing Frost Orchid has three similar, spreading sepals, two broad petals, and a trilobed lip with a swollen, yellow callus and auricles surrounding the column. No nectar or spur is present.

Actual size

SUBFAMILY	Epidendroideae
TRIBE AND SUBTRIBE	Epidendreae, Laeliinae
NATIVE RANGE	Southern Mexico to Nicaragua
HABITAT	Seasonally dry oak forests
TYPE AND PLACEMENT	Epiphytic, sometimes lithophytic
CONSERVATION STATUS	May be endangered due to overcollecting
FLOWERING TIME	Spring

FLOWER SIZE
2–3 in (5–8 cm)

PLANT SIZE
6–10 × 3–6 in
(15–25 × 8–15 cm),
excluding terminal
inflorescence
8–12 in (20–30 cm) tall

414

BARKERIA SPECTABILIS
SHOWY OAK ORCHID
BATEMAN EX LINDLEY, 1842

The species name of the Showy Oak Orchid is Latin for
"showy," and this is one of the showiest members of a genus
of mostly small orchids. Although each flower is short-lived,
there can be up to 8–12 flowers on a terminal raceme that is often
twice as long as the plant is tall. The genus name is in honor of
George Barker (1776–1845), an English horticulturalist who first
imported these plants from Mexico.

Found exclusively in seasonally dry areas on oaks (hence the
common name), *Barkeria* species are deciduous in the winter
months, often resembling a bundle of twigs. Although closely
allied to the large genus *Epidendrum*, with which artificial
hybrids have been created, *Barkeria* has been found in DNA
studies to be distinct. Pollination is by carpenter bees, which
must push their way forcibly into the small space between the
lip and column.

The flower of the Showy Oak Orchid is usually
pale lilac or purple, with lanceolate sepals and
petals arranged like a fan. The spade-shaped lip
often bears reddish markings and is partially
fused to the column.

Actual size

SUBFAMILY	Epidendroideae
TRIBE AND SUBTRIBE	Epidendreae, Laeliinae
NATIVE RANGE	Mexico, Guatemala, Belize, El Salvador, Honduras, Nicaragua, Costa Rica, Panama, French Guiana, Surinam, Guyana, Venezuela, and Colombia
HABITAT	Low to mid-elevation coastal wet forests
TYPE AND PLACEMENT	Epiphytic
CONSERVATION STATUS	Not threatened
FLOWERING TIME	Summer to fall

BRASSAVOLA CUCULLATA

DADDY LONGLEGS

(LINNAEUS) R. BROWN, 1813

FLOWER SIZE
6¼–8 in (16–20 cm)

PLANT SIZE
10–18 × 3–5 in
(25–46 × 8–13 cm),
including short
terminal inflorescence,
which emerges partway
along the pendent leaves

415

The spidery blossoms with narrow, ribbonlike, white to yellow-brown petals and sepals make *Brassavola cucullata* an outstandingly beautiful species and are responsible for its common name. The species name, *cucullata*, means "hoodlike" in Latin and refers to the way the column is partially covered, or hooded, by the lip. The night-fragrant flowers attract moths for pollination. Often such flowers offer a nectar reward in a long spur that projects from the base of the lip, but in *Brassavola* species the nectary is hidden instead in a hollow tube buried in the elongate ovary.

While most species of *Brassavola* have superb scents, this orchid offers a fragrance that some people have described as disgusting, with others reporting light citrusy notes. Flowers appear singly, occasionally doubly, on inflorescences and sometimes have a reddish cast when first opening, usually fading to pure white as they mature.

The flower of the Daddy Longlegs is usually white with a yellow tinge on the ends of all parts, and sometimes a reddish-orange tinge when it first opens. It has long, narrow, pointed sepals and petals, with a fringed lip bearing a central lobe nearly as long and ribbonlike as the other segments.

Actual size

SUBFAMILY	Epidendroideae
TRIBE AND SUBTRIBE	Epidendreae, Laeliinae
NATIVE RANGE	Widespread in lowland forests, from Mexico to Brazil and the Caribbean
HABITAT	Seasonally dry forests, often near the coast, sea level to 1,640 ft (500 m)
TYPE AND PLACEMENT	Epiphytic
CONSERVATION STATUS	Not threatened
FLOWERING TIME	Throughout the year

FLOWER SIZE
4 in (10 cm)

PLANT SIZE
5–9 × 1–1½ in
(13–23 × 2.5–3.8 cm),
excluding erect to arching
6–10 in (15–25 cm) long
inflorescence

416

BRASSAVOLA NODOSA
LADY OF THE NIGHT
(LINNAEUS) LINDLEY, 1831

The Lady of the Night produces clusters of thick, almost round pencil-like leaves, sometimes covering the trunks of trees with what looks like a massive shaggy carpet. Between four and twelve flowers are produced in tight clusters that emit a wonderful sweet fragrance at night, which gives the plant its common name, although some have tried to tie its scent to the cheap perfume worn by women of ill repute. The genus name is after the Italian physician Antonio Musa Brasavola (1500–55).

Although not bearing a long spur like other moth-pollinated flowers such as species of the genus *Angraecum*, including the Madagascar Comet Orchid, the members of *Brassavola* have instead a nectar tube that runs underneath the ventral surface of the ovary. This serves the same purpose as a spur.

The flower of the Lady of the Night has green to whitish-cream, narrowly lanceolate, spidery sepals and petals. Its large, prominent tubular lip flares into a heart shape, sometimes with a few reddish spots deep in the throat.

Actual size

SUBFAMILY	Epidendroideae
TRIBE AND SUBTRIBE	Epidendreae, Laeliinae
NATIVE RANGE	Jamaica
HABITAT	Low to mid-elevation coastal wet forests
TYPE AND PLACEMENT	Epiphytic
CONSERVATION STATUS	Not threatened
FLOWERING TIME	Spring to summer

BROUGHTONIA SANGUINEA
BLOOD-RED
JAMAICAN ORCHID
(SWARTZ) R. BROWN, 1813

FLOWER SIZE
1½–2⅜ in (4–6 cm)

PLANT SIZE
5–8 × 2–3 in
(13–20 × 5–8 cm),
excluding erect-arching
terminal inflorescence
10–24 in (25–61 cm) long

417

Although the color of the small, brilliant-flowered Blood-red Jamaican Orchid leads some to claim hummingbird pollination, the plant instead displays a bee pollination syndrome. The floral morphology of the species, such as the lack of a nectar spur, is unsuitable for bird pollination but is similar to many bee-pollinated species. These tough, succulent plants are often found in full sun near the beach, and thus require strong light and tropical heat to thrive. The genus was named in honor of Arthur Broughton (*circa* 1758–96), an English botanist who worked in Jamaica.

The successively flowering spikes appear from the apex of plump, round pseudobulbs, which usually bear two tough leathery leaves. The flowers, although small, are long lasting and intensely colored, which has resulted in them being artificially hybridized with other members of this subtribe to impart to their hybrids brilliant color on small-sized plants.

The flower of the Blood-red Jamaican Orchid is not always red: pink, yellow, purple, and other colors, including bicolored forms, are known. Flowers are flat with pointed sepals, rounded petals, and a broad, flat lip, usually with a yellow center.

Actual size

SUBFAMILY	Epidendroideae
TRIBE AND SUBTRIBE	Epidendreae, Laeliinae
NATIVE RANGE	State of Bahia (Brazil)
HABITAT	Hot lowland scrub near the beach
TYPE AND PLACEMENT	Epiphytic
CONSERVATION STATUS	Threatened by collecting and habitat degradation
FLOWERING TIME	Summer to fall

FLOWER SIZE
2⅜–4 in (6–10 cm)

PLANT SIZE
4–8 × 3–6 in
(10–20 × 8–15 cm),
excluding erect
inflorescence
2–3 in (5–8 cm) long

CATTLEYA ACLANDIAE
LADY ACKLAND'S ORCHID
LINDLEY, 1840

418

A dwarf member of its genus, Lady Ackland's Orchid clings to rough-barked trees at sea level. It bears two leaves at the apex of each slender, cylindrical pseudobulb, and the flowers, usually borne one or two at a time, are large in proportion to the rest of the plant and highly (sweetly) fragrant. The plant was named for Lady Lydia Ackland (1786–1856), an English orchid grower who first successfully flowered the species in Europe.

Growing near the seashore among scrubby vegetation, plants are vigorous, adaptable, and long-lived. These features, as well as the novelty of the spotted pattern, make the orchid a popular parent of many hybrids, creating progeny with compact size, vigor, and unusual coloration. Pollination is by euglossine and other large bees, which visit the flowers in search of nectar, although in this case none is available.

The flower of Lady Ackland's Orchid is large, with yellow-green, olive, or pale brownish segments with purplish-brown blotches. The three-lobed lip has the sidelobe wrapped around the column and is variable in color but usually rose purple, or lilac pink with darker purple markings.

Actual size

SUBFAMILY	Epidendroideae
TRIBE AND SUBTRIBE	Epidendreae, Laeliinae
NATIVE RANGE	Southern and southeastern Brazil to Misiones province (northern Argentina)
HABITAT	Atlantic rain forests, at 2,130–5,480 ft (650–1,670 m)
TYPE AND PLACEMENT	Epiphytic on moss-covered trees or mossy rocks
CONSERVATION STATUS	Not assessed
FLOWERING TIME	April to June (fall and early winter)

CATTLEYA COCCINEA
SCARLET CATTLEYA
LINDLEY, 1836

FLOWER SIZE
1½–3 in (3–8 cm)

PLANT SIZE
4–6 × 3–5 in
(10–15 × 8–13 cm),
excluding inflorescence
4–10 in (10–25 cm),
just a little longer
than the leaves

419

Formerly placed in the genus *Sophronites*, the Scarlet Cattleya is a beautiful miniature orchid with a relatively large flower. It forms a cluster of closely packed pseudobulbs, each bearing a single, elliptic, leathery leaf. The leaf midvein is red, and from the base of a leaf an inflorescence grows with a single, long-lasting, scarlet flower. This species has been extensively used in hybridization with other species of *Cattleya* to create bright red, large-flowered cultivars.

It has been suggested that hummingbirds may pollinate the Scarlet Cattleya, but the shape of the lip and column does not appear to fit this hypothesis. Because bees can see red, they seem to be the more likely pollinators for this species, which has no nectar spur or obvious place for a hummingbird to probe.

The flower of the Scarlet Cattleya is brilliant red to orange (rarely yellow) with broad, spreading sepals and even larger petals. The lip is the same color but has yellow markings inside and is smaller, forward projecting, and surrounds the column.

Actual size

SUBFAMILY	Epidendroideae
TRIBE AND SUBTRIBE	Epidendreae, Laeliinae
NATIVE RANGE	Venezuela
HABITAT	Dense montane forests, high in the canopy
TYPE AND PLACEMENT	Epiphytic
CONSERVATION STATUS	Threatened by collecting and habitat degradation
FLOWERING TIME	Spring, usually March to May

FLOWER SIZE
Up to 8 in (20 cm)

PLANT SIZE
14–20 × 6–10 in
(36–51 × 15–25 cm), including
short, erect inflorescence,
which is slightly shorter
than the leaves and
8–12 in (20–30 cm) tall

CATTLEYA MOSSIAE
EASTER ORCHID
C. PARKER EX HOOKER, 1838

The flower of the Easter Orchid is large and
varies in color and form. Typical forms are
lavender, with narrow lanceolate sepals, broad,
forward-projecting petals, and a flaring lip
variably marked with darker purple and
brilliant yellow spots in the throat.

Native to Venezuela, the Easter Orchid has the distinction of
also being the country's national flower—no mean feat in a
country with many endemic beautiful species, including seven
other showy species of *Cattleya*. On a single inflorescence up to
five large, beautifully colored, and deliciously fragrant flowers
grow, which are produced at Easter (hence the common name).

Cattleya mossiae was the second of the showy single-leafed
species of the genus *Cattleya* to be discovered (*C. labiata* was
the first), but it was readily imported and became the earliest
species available to enthusiasts in the 1830s. Information about
its locality was, perhaps deliberately, misrepresented, leading
to the plant being lost for 70 years before it was rediscovered.
Pollination of this and most other species of *Cattleya* is by nectar-
seeking euglossine bees.

Actual size

SUBFAMILY	Epidendroideae
TRIBE AND SUBTRIBE	Epidendreae, Laeliinae
NATIVE RANGE	Colombia, Venezuela, Guyana, French Guiana, Surinam, Brazil, and Trinidad and Tobago
HABITAT	Coastal, seasonally dry forests or riparian woodlands
TYPE AND PLACEMENT	Epiphytic, sometimes lithophytic on cliffs
CONSERVATION STATUS	Not threatened
FLOWERING TIME	Usually January to February, but can bloom as late as May in the Northern Hemisphere

CAULARTHRON BICORNUTUM

VIRGIN MARY

(HOOKER) RAFINESQUE, 1837

FLOWER SIZE
Up to 4 in (10 cm)

PLANT SIZE
10–18 × 8–12 in
(25–46 × 20–30 cm),
excluding erect to arching
terminal inflorescence
10–16 in (25–41 cm) tall

421

The distinctive white-flowered Virgin Mary is often found near the seashore or around river systems, in open, sunny locations. Up to 20 fragrant, waxy flowers are borne on a terminal raceme (rarely weakly branched). The flower's lip has two hollow, hornlike projections from which its species name, *bicornutum* (meaning "two-horned"), is derived. The genus name refers to the obviously jointed stem, from the Greek *kaulos*, "stem," and *arthron*, "joint."

Species of *Caularthron* have a mutualistic relationship with biting ants, providing a hollow center in their large fusiform pseudobulbs for the ants to colonize. In return, the ants patrol the plants, protecting them from predation by insects and even mammals and birds. Any slight disturbance brings the ants out immediately in great numbers to chase away any intruder, no matter how large. Pollination is reported to be by carpenter bees.

The flowers of the Virgin Mary form mostly white clusters at the tip of a long stem and have ovate-acuminate petals and sepals. The trilobed lip is peppered with cinnamon spots and bears two yellow to orange, hornlike projections.

Actual size

SUBFAMILY	Epidendroideae
TRIBE AND SUBTRIBE	Epidendreae, Laeliinae
NATIVE RANGE	Serra do Cipo region in the Brazilian state of Minas Gerais
HABITAT	Open cerrado habitat, at about 4,600 ft (1,400 m)
TYPE AND PLACEMENT	Epiphytic on *Vellozia* bushes
CONSERVATION STATUS	Not assessed, but likely to be threatened due to small area of occupancy and low dispersability
FLOWERING TIME	July to October (winter to spring)

FLOWER SIZE
1⁄16 in (0.2 cm)

PLANT SIZE
1–2 × 1–2 in
(2.5–5 × 2.5–5 cm),
excluding inflorescence
typically 1 in (2.5 cm) tall

CONSTANTIA CIPOENSIS
CIPO ORCHID
PORTO & BRADE, 1935

Actual size

The flower of the Cipo Orchid has similar, spreading sepals and petals—all parts are white or creamy white. The lip is recurved at the upper edge and projects forward, with a yellow blotch at its base. The column has a pair of short wings.

The Cipo Orchid is one of six species in the genus *Constantia*, named for the wife of Brazilian botanist João Barbosa Rodrigues (1842–1909), the authority of his day on Brazilian orchids. It has round, slightly flattened pseudobulbs, carrying one or two apical ovate leaves. An inflorescence bearing a single flower emerges from the top of the pseudobulb. Though small, the flower is large for the size of the plant and emits a scent in the twilight hours.

The fragrance attracts the large Bamboo-nesting Carpenter Bee, *Xylocopa artifex*, which nests among the branches of *Vellozia piresiana* and *V. compacta* (Velloziaceae) on which the orchid grows. Because the bees are territorial and pollen dispersal between plants is limited, seed production is low. *Constantia* species all have narrow distributions, often being known from a single mountain or area in southeastern Brazil.

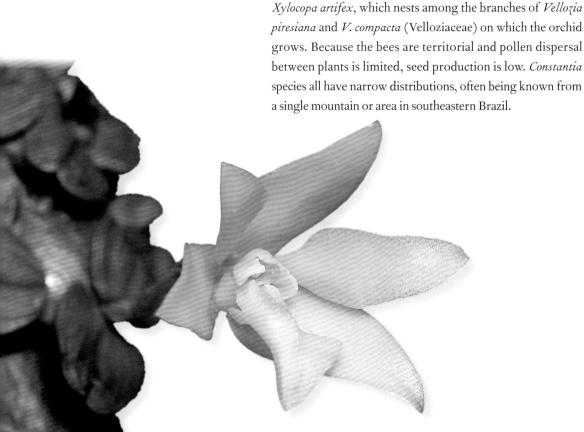

SUBFAMILY	Epidendroideae
TRIBE AND SUBTRIBE	Epidendreae, Laeliinae
NATIVE RANGE	Northern South America, Jamaica, Trinidad
HABITAT	Lowland rain forests, up to 2,300 ft (700 m)
TYPE AND PLACEMENT	Epiphytic
CONSERVATION STATUS	Not assessed
FLOWERING TIME	October to May (fall to spring)

DIMERANDRA STENOPETALA
PINK PANSY ORCHID
(HOOKER) SCHLECHTER, 1922

FLOWER SIZE
Up to 1 in (2.5 cm)

PLANT SIZE
8–15 × 4–6 in
(20–38 × 10–15 cm),
including the short
terminal inflorescence

423

The Pink Pansy Orchid produces an elongate, thickened stem with nodes from which lanceolate leaves with clasping bases grow. An inflorescence covered in tight bracts develops from the stem apex and carries numerous flowers that look somewhat like pink pansies, hence their common name. The genus name refers to the arms on each side of the apex of the column (Greek *di*, for "pair," and *andros*, for "man," referring to the anther). The species name refers to the relatively narrow petals of this species, although they are wider than those of many other species in the genus.

The open nature of these flowers, the absence of a spur, and their color suggest that some type of bee is the pollinator. There are, however, no reported observations of pollination for this species.

Actual size

The flower of the Pink Pansy Orchid is fairly flat, with spreading, bright pink sepals and petals. The wide lip is pink with a white blotch at its base underneath the column, which has two wings at its apex.

SUBFAMILY	Epidendroideae
TRIBE AND SUBTRIBE	Epidendreae, Laeliinae
NATIVE RANGE	Mexico to Nicaragua, also in Jamaica and Cuba
HABITAT	Mixed lowland forests, often on oaks
TYPE AND PLACEMENT	Epiphytic or lithophytic
CONSERVATION STATUS	Not threatened
FLOWERING TIME	Spring

FLOWER SIZE
1¼ in (3 cm)

PLANT SIZE
2–3 × 1 in
(5–8 × 2.5 cm),
excluding short,
single-flowered terminal
inflorescence
1–2 in (2.5–5 cm) long

424

DINEMA POLYBULBON
STRING OF BEADS
(SWARTZ) LINDLEY, 1831

The String of Beads is a diminutive rambling plant, with pea-sized, glossy pseudobulbs strung along a creeping stem (hence its common name). The plants grow quickly, often forming a dense, exuberant mat in a short time. One to three leaves crown the apex of each miniature stem, and in spring most of the newest growths produce short apical inflorescences, emerging from a tiny sheath or spathe, each bearing a single flower that emits a strong, delicious honey-like fragrance. The genus name refers to the two thin erect column wings, from the Greek, *di*, "double," and *nema*, "thread."

Highly adaptable, the species occurs in habitats that vary from steamy lowland jungles to cooler, higher-elevation cloud forest conditions. Pollination has not been studied, but the sweetly fragrant flowers most likely attract small bees. No reward appears to be offered.

The flower of the String of Beads has narrow sepals and petals of a golden brown color with lighter margins and tips. The broader lip is pure white, and the column and tiny sidelobes of the lip are flushed with purple.

Actual size

SUBFAMILY	Epidendroideae
TRIBE AND SUBTRIBE	Epidendreae, Laeliinae
NATIVE RANGE	Cuba, Haiti, Dominican Republic, and Puerto Rico
HABITAT	Seasonally dry, deciduous forests, at 330–2,625 ft (100–800 m)
TYPE AND PLACEMENT	Epiphyte
CONSERVATION STATUS	Not threatened
FLOWERING TIME	Spring

DOMINGOA HAEMATOCHILA
BLOODY-LIPPED ORCHID
(REICHENBACH FILS) CARABIA, 1943

FLOWER SIZE
1¼ in (3 cm)

PLANT SIZE
4–8 × 1–1½ in
(10–20 × 2.5–3.8 cm),
excluding arching or pendent
terminal inflorescence
4–8 in (10–20 cm) long

425

A small group of charming epiphytes native to the Caribbean and central Mexico, *Domingoa* derives its name from Santo Domingo, on the island of Hispaniola, where it was first discovered. *Domingoa haematochila*, with its leathery, lanceolate leaves, is well adapted for the hot and dry desertlike conditions to which it is often exposed. Plants on Mona Island (Puerto Rico) often experience temperatures well above 100°F (38°C).

The darkly hued flowers, proportionately large for the size of the plant, appear in succession over a long period and have an intense fragrance. Pollinators are bees seeking nectar, which is absent. The bees force their way into the small space between the lip and column, where their thorax breaks open an apical chamber, releasing glue to which the pollen masses (pollinia) then adhere.

Actual size

The flower of the Bloody-lipped Orchid has yellow to olive petals and sepals, often rimmed and finely striped with blood-red. Although the sepals are spread widely, the petals cover the column. The flattened lip is boldly blood-red.

SUBFAMILY	Epidendroideae
TRIBE AND SUBTRIBE	Epidendreae, Laeliinae
NATIVE RANGE	Central and southwestern Mexico
HABITAT	Dry oak and pine-oak forests, at 3,300–6,600 ft (1,000–2,000 m)
TYPE AND PLACEMENT	Epiphytic
CONSERVATION STATUS	Not assessed
FLOWERING TIME	April to June (spring to early summer)

FLOWER SIZE
3 in (8 cm)

PLANT SIZE
12–20 × 10–15 in
(30–51 × 25–38),
excluding terminal
erect-arching inflorescence
25–35 in (64–89 cm) long

ENCYCLIA ADENOCAULA
LITTLE PURPLE DREIDEL
(LEXARZA) SCHLECHTER, 1918

The Little Purple Dreidel (in Spanish, *trompillo morado*, "purple nightshade") is a striking and fragrant species that produces from the apex of pseudobulbs large branched inflorescences of bright, beautiful, clustered flowers that appear to dance in the breeze. The genus name comes from the Greek for "encircling," *enkyklein*, a reference to the lip that wraps around the column.

The species name refers to the prominent but small warty growths on the inflorescence, from the Greek *aden*, "gland," and *caulo*, "stem." The common name is probably a reference to the cone-like pseudobulbs, which resemble the shape of a spinning top (or dreidel) used in a Jewish gambling game.

Pollination of the colorful and fragrant flowers in genus *Encyclia* is assumed to be by bees, although no reward is offered. Despite the abundance and showiness of the blooms, however, no pollination studies have been published.

Actual size

The flower of the Little Purple Dreidel has narrowly lanceolate, rich pink to purple petals and sepals. These produce a star shape around the similarly colored column and trilobed lip, which has darker nectar guides and a white base, with the basal lobes surrounding the column.

SUBFAMILY	Epidendroideae
TRIBE AND SUBTRIBE	Epidendreae, Laeliinae
NATIVE RANGE	From Mexico throughout Central America to Panama
HABITAT	Tropical, semi-deciduous and oak forests, at 330–3,300 ft (100–1,000 m)
TYPE AND PLACEMENT	Epiphytic
CONSERVATION STATUS	Not assessed
FLOWERING TIME	June to August (summer)

FLOWER SIZE
2 in (5 cm)

PLANT SIZE
15–25 × 12–22 in
(38–64 × 30–56 cm),
excluding terminal erect-
arching inflorescence
12–80 in (30–203 cm) long

427

ENCYCLIA ALATA
BUTTERFLY ORCHID
(BATEMAN) SCHLECHTER, 1914

The Butterfly Orchid produces tight clusters of ovoid, cone-shaped pseudobulbs carrying between one and three apical, linear-lanceolate leaves that can be darkly pigmented. Over its large range, the plant varies considerably in shape and flower color. The common name comes from the honey-like fragrance, which attracts butterflies, although not as pollinators. The species name is Latin for "winged," probably referring to the lateral lobes of the lip, which make the column appear to be winged.

Pollination has not been studied in this or any related species, but the shape, color, and sweet fragrance suggest some sort of bee. In cultivation in the Philippines, the plant is reported to be visited by bees. General floral features, including nectar guides, function to attract and orient pollinators despite no nectar being present.

The flower of the Butterfly Orchid has spoon-shaped sepals and petals that are greenish to chestnut-brown, paler basally. The cream to yellow lip has three lobes— the middle lobe is the largest, with reddish-purple veins, and the sidelobes flare widely and surround the yellow column.

Actual size

SUBFAMILY	Epidendroideae
TRIBE AND SUBTRIBE	Epidendreae, Laeliinae
NATIVE RANGE	Mexico, Guatemala, Belize, El Salvador, Honduras, Nicaragua, Costa Rica, Panama, French Guiana, Surinam, Guyana, Venezuela, and Colombia
HABITAT	Lower elevation, seasonally dry forests, at 330–1,640 ft (100–500 m)
TYPE AND PLACEMENT	Epiphytic, occasionally lithophytic
CONSERVATION STATUS	Threatened by overcollection
FLOWERING TIME	Late spring to summer

FLOWER SIZE
3½ in (9 cm)

PLANT SIZE
16–30 × 10–18 in
(41–76 × 25–46 cm),
excluding erect,
terminal inflorescence,
20–32 in (51–81 cm) tall

428

ENCYCLIA CORDIGERA
FLOWER OF THE INCARNATION
(KUNTH) DRESSLER, 1964

The sturdy, vigorous Flower of the Incarnation is extremely variable in flower color, with many lovely forms found in nature. Flower spikes emerge from an apex of glossy, large ovoid pseudobulbs, each bearing a pair of leathery leaves. The erect spikes hold usually between six and twelve large, highly fragrant flowers well above the foliage, making a superb show. The genus name is based on the Greek word *enkyklein*, "encircle," referring to the lip, which is partially fused to and wraps around the column. The common name is simply a reference to the "incarnate" beauty of the flowers.

Encyclia cordigera is a mostly lowland species from seasonally dry areas, where it can form huge clusters of massive pseudobulbs. Large bees, especially carpenter bees, pollinate the plants, seeking nectar, which is never present.

The Flower of the Incarnation is variable in color but generally has brownish to olive petals and sepals, though purplish in some forms. The broad, flaring lip is often of a rich pink, rose, or purple, and sometimes white with a purple blotch at the center.

Actual size

SUBFAMILY	Epidendroideae
TRIBE AND SUBTRIBE	Epidendreae, Laeliinae
NATIVE RANGE	Florida and the Bahamas
HABITAT	Hammocks along rivers, cypress swamps, oak forests, mangroves, from sea level to about 80 ft (25 m)
TYPE AND PLACEMENT	Epiphytic, on live oaks or other evergreen trees
CONSERVATION STATUS	Locally common, but wild collecting is prohibited
FLOWERING TIME	June to September (summer)

ENCYCLIA TAMPENSIS

FLORIDA BUTTERFLY ORCHID

(LINDLEY) SMALL, 1913

FLOWER SIZE
1½ in (4 cm)

PLANT SIZE
10–18 × 8–16 in
(25–46 × 20–41 cm),
excluding inflorescence,
which is erect and
20–30 in (51–76 cm) long

429

The Florida Butterfly Orchid forms a tight cluster of ovoid pseudobulbs, each with a single narrow leaf. In some cases, there is a second leaf or a leaf-bearing bract at the base of the pseudobulb. Mature plants produce a much-branched panicle with 6 to 25 sweetly scented flowers. The genus name comes from the Greek for "encircle" (*enkyklein*), in reference to the lip that wraps around the column in these species.

Although the orchid can be locally abundant in many parts of Florida, it is a protected species, and collecting is prohibited. It is also not necessary to take these orchids from the wild as many commercial growers sell plants grown from seed, produced by crossing particularly attractive wild-occurring forms. Despite its common name, the species is reportedly pollinated by small bees attracted to its sweetly fragrant flowers, rather than butterflies.

The flower of the Florida Butterfly Orchid
has similar, spreading, greenish-brown petals
and sepals. The lip is white with a central
purple blotch and trilobed, with the sidelobes
enveloping the column and the basal lobe
projecting forward.

Actual size

SUBFAMILY	Epidendroideae
TRIBE AND SUBTRIBE	Epidendreae, Laeliinae
NATIVE RANGE	Tropical America, from Mexico and the Caribbean to Peru and Brazil
HABITAT	Wet or semi-deciduous seasonally dry forests, at sea level to 3,300 ft (1,000 m)
TYPE AND PLACEMENT	Epiphytic and sometimes lithophytic on bare rocks
CONSERVATION STATUS	Not assessed, but the species is widespread and locally common so not of conservation concern
FLOWERING TIME	April to December (winter to early spring)

FLOWER SIZE
5 in (13 cm)

PLANT SIZE
20–32 × 6–8 in
(51–81 × 15–20 cm),
excluding erect-arching
inflorescence
8–12 in (20–30 cm) long, which
generally exceeds the leaf

430

EPIDENDRUM CILIARE
CATTLE EGRET ORCHID
LINNAEUS, 1759

The Cattle Egret Orchid has large ovoid pseudobulbs covered with overlapping papery sheaths, each topped by one or two elliptic, blunt, leathery leaves. The inflorescence appears at the top of a newly maturing pseudobulb, carrying 4–12 highly perfumed flowers in two rows. The pseudobulbs are similar to those found in members of the genus *Cattleya*, but this orchid belongs to the more than 1,500 species in the megagenus *Epidendrum*, established by Linnaeus in 1759 for all epiphytic orchids, most of which were later transferred to other genera.

The genus name is simply derived from the Greek words *epi–*, "on," and *dendron*, "tree." The plant is called *garcita* ("cattle egret") in Spanish, which reflects its common English name. Pollination is carried out by night-flying hawk moths. They probe the nectary that is buried deeply in the flower stem.

The flower of the Cattle Egret Orchid
has greenish, narrow petals and sepals,
the upper sepal rolled inward and the petals
arching forward. The column and lip are
white and form a short, fused tube with
the lip apex split into three feathery lobes.

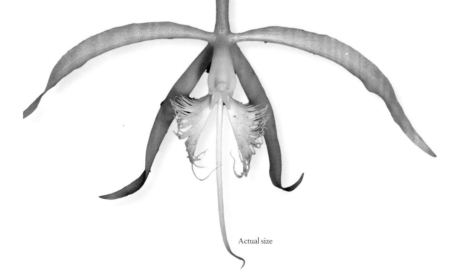

Actual size

SUBFAMILY	Epidendroideae
TRIBE AND SUBTRIBE	Epidendreae, Laeliinae
NATIVE RANGE	Northern South America (Colombia to northern Brazil) and Trinidad
HABITAT	Open slopes, rock faces, and disturbed sites such as roadsides
TYPE AND PLACEMENT	Terrestrial or lithophytic
CONSERVATION STATUS	Not assessed, but widespread and locally common
FLOWERING TIME	Throughout the year

FLOWER SIZE
1 in (2.5 cm)

PLANT SIZE
24–50 × 6–10 in
(61–127 × 15–25 cm),
including terminal erect
inflorescence

EPIDENDRUM IBAGUENSE
CRUCIFIX ORCHID
KUNTH, 1816

431

The highly variable Crucifix Orchid has pencil-thick, reedlike stems supported by white fleshy roots. The stems lack any swelling as typically seen with the pseudobulbs that are present in some other species of *Epidendrum*, such as *E. ciliare*. Terminal, simple, or occasionally branched racemes emerge at any time of the year and have a cluster of flowers at the tip that open over long periods. Flowering stems often produce keikis (plantlets), which can form another plant when a stem falls over.

The common name was given by the missionaries in Ibagué, Colombia, who first encountered this species, as its upright lip resembles a cross. The Ibagué area is also responsible for the species name, although the orchid is much more widely distributed. Pollination is by butterflies that mistake these nectarless plants for similarly colored milkweeds (*Asclepias*) and *Lantana* species.

The flower of the Crucifix Orchid can be orange-red to pinkish-purple (rarely white) and bears its lip upright, with lanceolate, spreading sepals and petals. The column is fused with the lip to form a tube. The lip is three-lobed, with all lobes about the same size and the margins irregularly fringed.

Actual size

SUBFAMILY	Epidendroideae
TRIBE AND SUBTRIBE	Epidendreae, Laeliinae
NATIVE RANGE	Southwestern Mexico (Guerrero, Jalisco, Oaxaca states), along the Pacific coast
HABITAT	Pine-oak forests, at 4,920–5,600 ft (1,500–1,700 m)
TYPE AND PLACEMENT	Epiphytic
CONSERVATION STATUS	Not assessed
FLOWERING TIME	June to August (summer)

FLOWER SIZE
1¼ in (3 cm)

PLANT SIZE
8–15 × 5–8 in
(20–38 × 13–20 cm),
excluding the nodding to
pendent inflorescence
5–8 in (13–20 cm) long

432

EPIDENDRUM MARMORATUM
MARBLED ORCHID
A. RICHARD & GAELOTTI, 1845

Actual size

The Marbled Orchid has stout, cylindrical pseudobulbs and reddish stems bearing several leathery pairs of terminal leaves. It produces a terminal, downward arching, densely flowered raceme with many beautiful flowers. The genus dates back to Carl Linnaeus (1707–78), who used it initially for all the epiphytic orchids he encountered. The name refers to the fact that these plants grow on trees (Greek *epi*, "on," plus *dendron*, "tree"). The genus *Epidendrum* is one of the largest in the orchid family with more than 1,500 species.

Pollination has not been reported, but most species of the genus are pollinated by butterflies or moths. There is a central nectar spur formed by the fused lip and column, into which the tongue of a pollinator would be inserted.

The flower of the Marbled Orchid has white and red striped, spreading petals and sepals. The white lip also has red markings near its margins, with seven to nine prominent ridges. It forms a short tube with the column.

SUBFAMILY	Epidendroideae
TRIBE AND SUBTRIBE	Epidendreae, Laeliinae
NATIVE RANGE	Ecuador
HABITAT	Cloud forests, at 4,920–8,200 ft (1,500–2,500 m)
TYPE AND PLACEMENT	Epiphytic
CONSERVATION STATUS	Not threatened
FLOWERING TIME	Mostly summer but can bloom anytime

EPIDENDRUM MEDUSAE
MEDUSA-HEAD ORCHID
(REICHENBACH FILS) PFITZER, 1889

FLOWER SIZE
2¼–4 in (7–10 cm)

PLANT SIZE
10–16 × 4–6 in
(25–41 × 10–15 cm),
excluding short, pendent
terminal inflorescence, which
is slightly longer than the
stems, 2–3 in (5–8 cm) long

433

The pendent Medusa-head Orchid resides in the cooler reaches of Ecuadorian cloud forests, its fleshy chain-like foliage hanging gracefully in virtually constant moist conditions from mossy, shady branches. The bizarre flowers, usually borne one to three at a time on short, terminal inflorescences, are variable but often have a ruby-garnet to maroon coloration with a remarkable, raggedly fringed lip, which, likened to the snake-haired head of the monster Medusa from Greek mythology, gives the plant its scientific and common name.

In this species, the lip is extensively fused to the column, as is typical for nearly all species of *Epidendrum*, and the apex of the column has a narrow slit suited to the tongue of a butterfly or moth, which are frequently found to be the pollinators of *Epidendrum*. It seems likely that these flowers are adapted for moths due to their dull color.

The flower of the Medusa-head Orchid can be variable but generally bears yellow to olive segments overlaid with a flush of maroon. The broad, oval lip is sometimes greenish but often deep dark red or maroon and ringed with a distinctive fringe.

Actual size

SUBFAMILY	Epidendroideae
TRIBE AND SUBTRIBE	Epidendreae, Laeliinae
NATIVE RANGE	Mexico through Central America to Colombia and northern Venezuela
HABITAT	Wet pine and cloud forests, at 1,300–5,900 ft (400–1,800 m)
TYPE AND PLACEMENT	Epiphytic on tree trunks, where it forms mats
CONSERVATION STATUS	Widespread, but local populations may be under threat due to habitat destruction
FLOWERING TIME	Usually twice a year, irregularly so, but mostly in fall to spring

FLOWER SIZE
⅝ in (1.5 cm)

PLANT SIZE
2–4 × ¼–1 in
(5–10 × 2–2.5 cm),
including single-flowered
inflorescence

EPIDENDRUM PORPAX
BEETLE ORCHID
REICHENBACH FILS, 1855

434

Actual size

The little Beetle Orchid can form a large mat of flattened stems that carry up to eight ovate-lanceolate leaves, arranged in two rows opposite each other. Each stem produces a single, terminal, fleshy, fragrant flower from an ovate-lanceolate, spathe-like bract. The glossy lip resembles the shield-wings of a beetle, hence the common name. The species name comes from the Greek word for the handle of a shield, *porpax*, which alludes to the shape of the paired leaves.

Pollination of *Epidendrum porpax* has not been studied in nature, but it is assumed that with its dull color and sweetly fragrant flowers this species is most likely pollinated by some form of night-flying moth not attracted by color. The plant has a nectary formed by the fused lip and column that appears to be suited to a long-tongued pollinator, although no reward is offered.

The flower of the Beetle Orchid has a heart-shaped, maroon lip with three green warts at the base, which is fused to the column, forming a narrow tube. The narrow, lanceolate, green, spreading sepals and upturned petals frame the lip and column.

SUBFAMILY	Epidendroideae
TRIBE AND SUBTRIBE	Epidendreae, Laeliinae
NATIVE RANGE	Central America (Costa Rica) to northwestern South America (Ecuador)
HABITAT	Rain and cloud forests, at 1,640–6,900 ft (500–2,100 m)
TYPE AND PLACEMENT	Epiphytic and terrestrial
CONSERVATION STATUS	Not assessed
FLOWERING TIME	November to December (winter)

EPIDENDRUM WALLISII
GREATER PURPLE-WART ORCHID
REICHENBACH FILS, 1875

FLOWER SIZE
1½ in (3.8 cm)

PLANT SIZE
18–30 × 3–8 in
(46–76 × 8–20 cm),
including inflorescence
4–8 in (10–20 cm) long

435

The Greater Purple-wart Orchid forms clumps of reed-like stems, with leaves arranged on two sides and no pseudobulbs. It is the largest-flowered of a group of species that all have stems covered with purple warts, the function of which is unknown. Flowering starts at the top of the plant in the first year of a stem and then moves downward over the next several seasons. The species was named in honor of the prodigious German plant collector Gustav Wallis (1830–78), who introduced more than 1,000 species to European horticulture, among them many orchids.

As for many species of *Epidendrum*, pollination of *E. wallisii* is most likely by butterflies. The lip is fused to the column to produce a channel leading into the stem below the flower parts, into which the insect inserts its long tongue to reach the nectar.

The flower of the Greater Purple-wart Orchid has outstretched, yellow to green, oblanceolate sepals and petals. The trilobed lip is cream to white with purplish-red markings and a bright yellow to orange blotch immediately in front of the entrance to the nectar spur.

Actual size

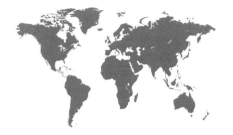

SUBFAMILY	Epidendroideae
TRIBE AND SUBTRIBE	Epidendreae, Laeliinae
NATIVE RANGE	Mesoamerica, from southern Mexico to Costa Rica
HABITAT	Humid forests, at 650–7,545 ft (200–2,300 m)
TYPE AND PLACEMENT	Epiphytic on tree trunks or epilithic on granite cliffs
CONSERVATION STATUS	Not assessed
FLOWERING TIME	January to April (winter to spring, after the dry season)

FLOWER SIZE
3½ in (9 cm)

PLANT SIZE
15–26 × 8–15 in
(38–66 × 20–38 cm),
excluding the terminal
inflorescence 5–8 in
(13–20 cm) long

GUARIANTHE SKINNERI
SAN SEBASTIAN ORCHID
(BATEMAN) DRESSLER & W. E. HIGGINS, 2003

The sometimes branched, creeping rhizome of the San Sebastian Orchid produces cylindrical pseudobulbs with a larger diameter near the leaves than at the base; each bears two centrally folded leaves at its apex. In the middle of the leaves, there is a large, thin floral sheath that encloses the apical bud, which in turn produces an inflorescence bearing up to 15 fragrant, brilliant purple-fuchsia flowers.

This orchid is the national flower of Costa Rica. Its genus name comes from the Costa Rican word for an epiphytic orchid (*guaria*), and the Greek *anthe*, "flower." Its local Costa Rican name is *la guaria morada*, the "darkly colored orchid," while in Guatemala, it is known as the *flor de San Sebastián*. Pollination by bees, probably carpenter (*Xylocopa*) and solitary bees (*Thygater*), is likely, although specific reports are lacking.

The flower of the San Sebastian Orchid has three lanceolate, bright purple sepals and two much broader petals. The lip is darker purple with a white central area, tube-shaped, and folded around the column.

Actual size

SUBFAMILY	Epidendroideae
TRIBE AND SUBTRIBE	Epidendreae, Laelinae
NATIVE RANGE	Southern Mexico (Jalisco state) to Guatemala
HABITAT	Oak forests, at 4,920–6,400 ft (1,500–1,950 m)
TYPE AND PLACEMENT	Epiphytic
CONSERVATION STATUS	Not assessed, but rare and protected in Mexico
FLOWERING TIME	Throughout the year

HAGSATERA ROSILLOI
MEXICAN BEE ORCHID
R. GONZÁLEZ, 1974

FLOWER SIZE
1⅜ in (3.5 cm)

PLANT SIZE
6–10 × 4–6 in
(15–25 × 10–15 cm),
excluding erect
terminal inflorescence
4–6 in (10–15 cm) long

437

The Mexican Bee Orchid forms upright, almost climbing stems with cylindrical pseudobulbs spaced along its length, each topped with a fleshy, folded leaf. From the tip, an inflorescence with up to ten nodding, fragrant flowers emerges. The genus is named for the highly esteemed Mexican orchidologist Eric Hágsater (1945–), founder and director of the herbarium of the Asociación Mexicana de Orquideología. The species name is in honor of another Mexican orchidologist, Salvador Rosillo de Velasco (1905–87).

The resemblance of the lip to an insect had led to speculation that the flower might be pollinated by a male bee mistaking it for a female and attempting to mate with it. This hypothesis, however, has still to be confirmed by field studies of pollination. The common name derives from this insect resemblance.

Actual size

The flower of the Mexican Bee Orchid has green, forward-spreading sepals and smaller, forward-pointing petals. The lip is three-lobed, with the midlobe largest and the base a dark red color. The center of the lip is velvety, and there is no nectar cavity.

SUBFAMILY	Epidendroideae
TRIBE AND SUBTRIBE	Epidendreae, Laeliinae
NATIVE RANGE	Mexico (Jalisco, Michoacán, Guerrero, and México states)
HABITAT	Cool pine and mixed-pine deciduous forests, at 5,250–7,900 ft (1,600–2,400 m)
TYPE AND PLACEMENT	Epiphytic
CONSERVATION STATUS	Not threatened
FLOWERING TIME	September to November (fall)

FLOWER SIZE
⅜ in (1 cm)

PLANT SIZE
1–2 × ½–¾ in
(2.5–5 × 1.3–2 cm),
excluding arching-erect
single-flowered inflorescence
½–1 in (1.3–2.5 cm) long

438

HOMALOPETALUM PACHYPHYLLUM
MEXICAN GREEN WOOD NYMPH
(L. O. WILLIAMS) DRESSLER, 1964

Actual size

The flower of the Mexican Green Wood Nymph
is translucent white to green, with the dorsal
sepal and petals forming a hood covering the
column and nectary on the base of the lip.
The lateral sepals are spreading, and the
forward-projecting lip is sometimes spotted
purplish red.

The Mexican Green Wood Nymph has proportionally large,
translucent flowers on short stems emerging from the base of
closely spaced ovoid pseudobulbs arranged in a chain, each
bearing a single fleshy leaf. Inhabiting seasonally dry forests of
the Pacific slopes of Mexico, these plants are so small that they
can be easily missed. The genus name comes from the Greek
words *omalos*, "even," and *petalon*, "petal," alluding to the lack
of clear distinction of the flower parts. The species name refers
to the thick leaves, from the Greek words *pachys*, "fat," and
phyllo, "leaf."

Pollination of the species is unstudied, but the shape of the flower
and its sweet fragrance suggest pollination by a bee. A reward
is provided in a nectar-containing cavity on the base of the lip.

SUBFAMILY	Epidendroideae
TRIBE AND SUBTRIBE	Epidendreae, Laeliinae
NATIVE RANGE	Eastern Brazil (Atlantic Forest) to Misiones province (Argentina)
HABITAT	Mid-elevation tropical wet forests
TYPE AND PLACEMENT	Epiphytic and lithophytic
CONSERVATION STATUS	Not threatened
FLOWERING TIME	Winter to spring

FLOWER SIZE
⅜–⅝ in (1–1.5 cm)

PLANT SIZE
1–2 × 1 in
(2.5–5 × 2.5 cm),
including short,
pendent single-flowered
inflorescence

ISABELIA VIRGINALIS
FAIRY BASKET
BARBOSA RODRIGUES, 1877

439

Actual size

One of three remarkable species in the genus, the Fairy Basket is miniature, with pretty flowers. The wiry rhizome bears many small ovoid bulbs in succession, each with a single, narrow, needlelike leaf at its apex. A netlike, interwoven mat of fibers (giving this small plant its common name) covers most of the plant, probably serving some protective function. While mostly epiphytic, plants are also found growing in debris-filled cracks of sandstone outcrops.

The genus was named after Isabel, Princess Imperial of Brazil (1846–1921) and daughter of Emperor Pedro II. Pollination is unstudied, but the color and flower morphology is suited to pollination by a small bee. Flower orientation is irregular in this species (the lip can be up or down), which suggests that whatever insect pollinates the flowers, it is able to maneuver itself into the correct position.

The flower of the Fairy Basket is small, with white sepals and petals sometimes flushed with pale pink. The lip is also white, and the lip callus and column are lavender purple, making the tiny blossoms appear pinkish from a distance.

SUBFAMILY	Epidendroideae
TRIBE AND SUBTRIBE	Epidendreae, Laeliinae
NATIVE RANGE	Hidalgo (state), Mexico
HABITAT	Tree trunks on forest edges, at 4,920–6,600 ft (1,500–2,000 m)
TYPE AND PLACEMENT	Epiphytic
CONSERVATION STATUS	Possibly extinct in the wild but remains in cultivation
FLOWERING TIME	Late October to December (fall)

FLOWER SIZE
3 in (8 cm)

PLANT SIZE
25–40 × 12–20 in
(64–102 × 30–51 cm),
excluding inflorescence,
which is usually twice as
long as the leaves

440

LAELIA GOULDIANA
HALLOWEEN ORCHID
REICHENBACH FILS, 1888

The Halloween Orchid has a creeping horizontal stem (rhizome) on which elongate, somewhat flattened pseudobulbs are clustered and topped with two or three lanceolate leaves. From the apex of each growth, an upright inflorescence carries up to ten, large, mildly fragrant blooms. The plants are epiphytes, usually on oaks high enough in the mountains to experience winter frosts.

In Mexico the orchid is called *flor de muerto* ("flower of death") or *calaverita* ("little skull") because it flowers around Halloween—hence the common name. Like many other orchids, it was commonly collected from the wild to decorate graves for the traditional Day of the Dead celebrations and has now become rare, possibly extinct, in the wild. Pollination has not been reported, but related species from Mexico are recorded as pollinated by bumblebees (*Bombus*).

The flower of the Halloween Orchid has spreading, bright purple to rosy-purple sepals and petals, the petals somewhat wider. The lip is trilobed, with the sidelobes wrapping around the column, the median lobe darker, with a yellow callus and red spots or stripes, and curved outward.

Actual size

SUBFAMILY	Epidendroideae
TRIBE AND SUBTRIBE	Epidendreae, Laeliinae
NATIVE RANGE	Eastern and southern Mexico
HABITAT	Dry, open oak forests, at 4,600–7,875 ft (1,400–2,400 m)
TYPE AND PLACEMENT	Epiphytic on mossy branches
CONSERVATION STATUS	Threatened due to overharvesting for horticultural and religious purposes
FLOWERING TIME	May to August (spring to summer)

FLOWER SIZE
8 in (20 cm)

PLANT SIZE
10–20 × 5–8 in
(25–51 × 13–20 cm)
tall, including terminal
inflorescence
5–8 in (13–20 cm) long

LAELIA SPECIOSA
MAYFLOWER ORCHID
(KUNTH) SCHLECHTER, 1914

The showy Mayflower Orchid has short, round pseudobulbs topped with one or two leaves that are fleshy and tinged purple. By the end of the dry season in April, the old pseudobulb has withered and an inflorescence is produced on a newly developing pseudobulb. It bears one to four bright pink, strongly scented, large flowers, making the plant sought-after among orchid-growing enthusiasts. The plant grows slowly, sometimes maturing in 16 to 19 years, high in the mountains, where the weather is cool with occasional frosts in the winter months.

It is known locally as *flor de todos santos* ("all-saints flower"). Mexican villagers make candy from a starchy paste ground from the pseudobulbs, mixed with sugar, lemon juice, and egg white and poured into wooden molds, forming little decorative animals, fruit, and skulls for the Day of the Dead.

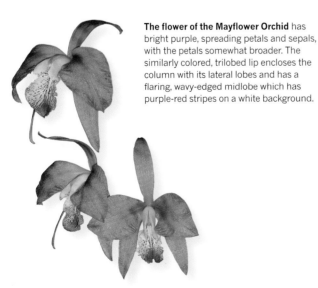

The flower of the Mayflower Orchid has bright purple, spreading petals and sepals, with the petals somewhat broader. The similarly colored, trilobed lip encloses the column with its lateral lobes and has a flaring, wavy-edged midlobe which has purple-red stripes on a white background.

Actual size

SUBFAMILY	Epidendroideae
TRIBE AND SUBTRIBE	Epidendreae, Laeliinae
NATIVE RANGE	Eastern to southern Brazil and Paraguay
HABITAT	Subtropical rain forests and coastal seasonally dry forests, at 1,640–2,950 ft (500–900 m)
TYPE AND PLACEMENT	Epiphytic
CONSERVATION STATUS	Not formally assessed
FLOWERING TIME	August to October (late winter and spring)

FLOWER SIZE
1½ in (3.8 cm)

PLANT SIZE
3–6 × ¼ in
(8–15 × 0.8 cm),
including arching-pendent
inflorescence 2–3 in (5–8 cm)
long, which is shorter
than the leaves

442

LEPTOTES BICOLOR
BRAZILIAN DWARF VANILLA
LINDLEY, 1833

The floriferous miniature Brazilian Dwarf Vanilla produces pencil-like, often red-spotted leaves covered at their bases by thin sheaths with a grooved surface. Between one and three relatively large, fragrant flowers that last for 10 to 12 days are formed on a short inflorescence arising from the base of a leaf. The genus name, which comes from the Greek word *leptos*, meaning "delicate," alludes to the seemingly fragile beauty of these plants in flower.

The pollinator of *Leptotes bicolor* has never been observed in the field, but based on the color and morphology of the flowers, bees may be the likely agent, although they gain no reward. The plant is only distantly related to the genus *Vanilla*, but extracts from its seed capsules, which contain vanillin, are used in Brazil to flavor milk, tea, candy, ice cream, and sorbet.

The flower of the Brazilian Dwarf Vanilla has strap-shaped, forward-curving, white to cream petals and sepals. The white lip is spade-shaped with a pair of rounded basal lobes, downturned margins, and a center marked with purple pink. The column is greenish-purple.

Actual size

SUBFAMILY	Epidendroideae
TRIBE AND SUBTRIBE	Epidendreae, Laeliinae
NATIVE RANGE	Southern Mexico, Guatemala, Honduras, and El Salvador
HABITAT	Seasonally dry tropical forests, at 1,300–4,920 ft (400–1,500 m)
TYPE AND PLACEMENT	Epiphytic and lithophytic
CONSERVATION STATUS	Not threatened
FLOWERING TIME	Mostly spring and summer

MEIRACYLLIUM TRINASUTUM
PURPLE DWARF
REICHENBACH FILS, 1854

FLOWER SIZE
1 in (2.5 cm)

PLANT SIZE
2–3 × 1–2 in
(5–8 × 2.5–5 cm),
including short
inflorescences
1 in (2.5 cm) long

443

The miniature, creeping Purple Dwarf is one of the two tiny species in the genus *Meiracyllium*. It has tough, coriaceous oblong-elliptic leaves with acuminate tips and without pseudobulbs. Presumably, the thick leaves take over the water storage function of pseudobulbs. The plant eventually forms a mat of overlapping leaves from which emerge short stems of gemlike little flowers that emit a strong cinnamon aroma. These flowers are short-lived, usually open for about a week. The genus name comes from the Greek word *meirakyllion*, meaning "stripling" (youth), a reference to the small size of the species.

The genus is allied to *Cattleya* and its relatives, a relationship that seems unlikely given the small size of the species and lack of pseudobulbs. The floral fragrance is similar to that found in other species pollinated by male euglossine bees, but pollination has not been studied in this species.

The flower of the Purple Dwarf has pale lavender sepals—the median one projecting over the column—darker purple petals, and a labellum decorated with amethyst purple spots. The lip surrounds the column, which is pointed. White and red forms are also known.

Actual size

SUBFAMILY	Epidendroideae
TRIBE AND SUBTRIBE	Epidendreae, Laeliinae
NATIVE RANGE	Mexico, Guatemala, Belize, Honduras, Costa Rica, and Venezuela
HABITAT	Seasonally dry forests, in full sun, at 650–1,970 ft (200–600 m)
YPE AND PLACEMENT	Epiphytic
CONSERVATION STATUS	Not threatened
FLOWERING TIME	July to October (summer to fall)

FLOWER SIZE
3 in (8 cm)

PLANT SIZE
15–25 × 6–10 in
(38–64 × 15–25 cm),
excluding terminal
erect inflorescence
80–200 in (203–508 cm) tall

444

MYRMECOPHILA TIBICINIS
ANT-LOVING PIPER
(BATEMAN EX LINDLEY) ROLFE, 1917

The giant Ant-loving Piper is a dry-forest plant that has evolved a defense against predation by animals. Its elongate pseudobulbs are hollow, with a basal opening created by an ant colony that serves as bodyguards, protecting the plant from herbivores (and naive orchid collectors). If their home is touched, the ants rush outside to defend it, a strategy reflected in the genus name, derived from the Greek words *myrmeco*, "ant," and *phila*, "loving." The species name, *tibicinis*, is Latin for a "flute-player" or "piper," hence, also, the common name. The ants pack older, abandoned pseudobulbs with debris, which, as it decays, provides some nutrients for the plant.

Bees, attracted by the bright colors and sweet fragrance, are the pollinators, although no reward is offered. A closely related species, *Myrmecophila christinae*, is pollinated by a food-seeking euglossine bee, *Eulaema polychroma*.

The flower of the Ant-loving Piper is variable in color and form but generally bears lanceolate, purplish sepals and petals with an undulate margin and darker tips. The lip is three-lobed, with the two lateral lobes bearing darker nectar guides. The midlobe is smaller and darker with a yellow central spot.

Actual size

SUBFAMILY	Epidendroideae
TRIBE AND SUBTRIBE	Epidendreae, Laeliinae
NATIVE RANGE	Southern Mexico
HABITAT	Mid-elevation seasonally dry forests
TYPE AND PLACEMENT	Epiphytic
CONSERVATION STATUS	Threatened by overcollecting
FLOWERING TIME	Winter to spring

FLOWER SIZE
2¾–3½ in (7–9 cm)

PLANT SIZE
8–15 × 5–9 in
(20–38 × 13–23 cm),
excluding pendent
inflorescence
6–8 in (15–20 cm) long

445

PROSTHECHEA CITRINA
TULIP ORCHID
(LEXARZA) HIGGINS, 1998

Like so many other large-flowered, highly ornamental New World species, the Tulip Orchid was imported in the thousands to European greenhouses (then called "stove houses"), where most perished. The common name in Mexico is *lemoncito* (little lemon), and bunches of cut flower stems are sold along the roadsides by children.

The pendent plants have large oblong-ovoid pseudobulbs and usually a pair of fleshy leaves that have an unusual glaucous sheen. The citrus-scented (hence the species name *citrina*) flowers hang from the apex of the bulbs on short stems and are pollinated by nectar-seeking euglossine bees. The plants are considered locally valuable and have been overcollected in their native range as ornamental plants, as a painkilling medicine to treat wounds, and as a source of mucilage (glue) for repairing wooden objects.

The flower of the Tulip Orchid is typically brilliant yellow (sometimes greenish-yellow), with a large, tubular lip and a white margin. There are often white and greenish nectar guides in the throat of the lip.

Actual size

SUBFAMILY	Epidendroideae
TRIBE AND SUBTRIBE	Epidendreae, Laeliinae
NATIVE RANGE	Caribbean region (including Florida) and Mexico to northern South America
HABITAT	Tropical evergreen and deciduous oak forests, from sea level to 6,200 ft (1,900 m)
TYPE AND PLACEMENT	Epiphytic
CONSERVATION STATUS	Apparently secure due to its frequency and widespread distribution, so not of conservation concern except in Florida where it is considered endangered
FLOWERING TIME	Throughout the year

FLOWER SIZE
3 in (8 cm)

PLANT SIZE
10–20 × 7–10 in
(25–51 × 18–25 cm),
excluding erect terminal
inflorescence, which
grows 8–15 in (20–38 cm)
longer than the leaves

446

PROSTHECHEA COCHLEATA
COCKLESHELL ORCHID
(LINNAEUS) W. E. HIGGINS, 1998

The Cockleshell Orchid has smooth, ovoid to elliptical, slightly flattened pseudobulbs that are enveloped basally by overlapping, dry sheaths and topped with two or three elliptic-lanceolate leaves. From the top of a mature pseudobulb, an upright inflorescence is produced that holds up to 20 upside-down—lip uppermost—scentless flowers that open over a long period of up to six months. The genus name comes from the Greek word *prostheke*, "appendage," referring to the short, pointed growth on the back of the anther.

Prosthechea cochleata is the national flower of Belize, where it is known as the Black Orchid (*orquídea negra*). The plant is fairly commonly cultivated for its unusual, long-lasting flowers, which are pollinated in nature by wasps, although they receive no reward. In Central America, mucilage is extracted from the pseudobulbs and used as glue for repairing wooden objects.

Actual size

The flower of the Cockleshell Orchid has twisted and downward-pointing, greenish-yellow sepals and petals, with an upright, hoodlike, yellow lip that has bold stripes of reddish-purple fusing solidly at the margins. The column is fat and has some purple spots basally.

SUBFAMILY	Epidendroideae
TRIBE AND SUBTRIBE	Epidendreae, Laeliinae
NATIVE RANGE	Southern North America (eastern and central Mexico)
HABITAT	Dry oak forests, 3,300–3,950 ft (1,000–1,200 m)
TYPE AND PLACEMENT	Epiphytic
CONSERVATION STATUS	Not assessed
FLOWERING TIME	May to July (late spring to summer)

PROSTHECHEA MARIAE
MARY'S WEDDING ORCHID
(AMES) W. E. HIGGINS, 1997

FLOWER SIZE
3 in (8 cm)

PLANT SIZE
5–8 × 5–10 in
(13–20 × 13–25 cm),
excluding erect-arching
terminal inflorescence
2–10 in (5–25 cm) long

447

This beautiful species was only first discovered in 1937, growing in Mexico near the border with Texas. This was an area in which orchids were not expected and had been little investigated by botanists. The plant was named for Mary, the wife of the discoverer of the species, Eric Östlund (1875–1938), and its large white lip suggests that it is a good flower for weddings. The genus name derives from the Greek word for "appendage," *prostheke*, referring to the short projection on the back of the column. The plant forms tight clusters of cone-shaped pseudobulbs that have two to three gray-green apical leaves. The inflorescence emerges from the top of the pseudobulb and holds two to five flowers.

Pollination of these large blooms has not been studied, but a bee would be expected, given the sweet fragrance produced during daylight hours. No nectar is present.

The flower of the Mary's Wedding Orchid has bright, pure green to yellow, outstretched sepals and petals. The white lip is weakly trilobed, the sidelobes enfolding the column and the bilobed, strikingly green-veined midlobe having a yellow patch in its throat.

Actual size

SUBFAMILY	Epidendroideae
TRIBE AND SUBTRIBE	Epidendreae, Laeliinae
NATIVE RANGE	Eastern Brazil, southern Bahia to Espírito Santo
HABITAT	Rocky places with permanent water seepage, at sea level to 3,300 ft (1,000 m)
TYPE AND PLACEMENT	Lithophytic on rocks or epiphytic on the base of *Vellozia* bushes
CONSERVATION STATUS	Not assessed but likely to be of concern due to its limited range
FLOWERING TIME	April to September (winter to spring)

FLOWER SIZE
1½ in (3.8 cm)

PLANT SIZE
12–20 × 10–14 in
(30–51 × 25–36 cm),
excluding erect terminal
inflorescence, which grows
20–40 in (51–102 cm)
longer than the leaves

PSEUDOLAELIA VELLOZICOLA
VELLOZIA ORCHID
(HOEHNE) PORTO & BRADE, 1935

The Vellozia Orchid grows on rocks and the bases of bushes of *Nanuza plicata*, a plant characteristic of rocky outcrops and coastal vegetation in eastern Brazil. The plant forms mat-like communities of interlocking spindle-shaped pseudobulbs connected by elongate horizontal rhizomes. Each pseudobulb carries between three and seven lanceolate leaves, and the rhizomes are covered with bracts that soon become papery and frayed, like feathery fringes. A tall inflorescence holds aloft up to 15 brightly colored flowers.

The genus name is derived from the Greek word *pseudes*, "false," and the genus *Laelia*, referring to the similarity between the genera, although they are only distantly related. In fact, *Pseudolaelia* is close to the dwarf genus *Isabelia*, which also grows in eastern Brazil. Pollination of this species is by bumblebees, which probe the cavity formed at the base of the lip, even though no nectar is present.

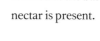

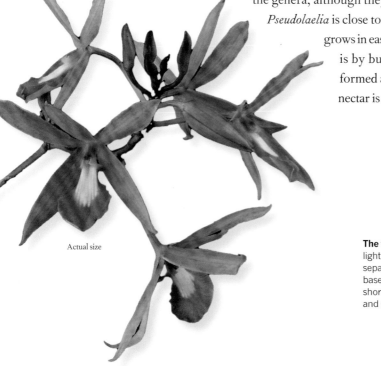

Actual size

The flower of the Vellozia Orchid has similar, light purple or lavender, spreading, oblanceolate sepals and petals. The column is fused to the base of the purple, three-lobed lip, its lateral lobes short and spreading and its terminal lobe broad and spade-shaped with a white-cream center.

SUBFAMILY	Epidendroideae
TRIBE AND SUBTRIBE	Epidendreae, Laeliinae
NATIVE RANGE	Southeastern Mexico to northern Honduras and Guatemala
HABITAT	Sunny places among bushes and thorny acacia in dry forests on limestone, at sea level to 1,640 ft (500 m)
TYPE AND PLACEMENT	Epiphytic
CONSERVATION STATUS	Not formally assessed
FLOWERING TIME	May to August (summer)

FLOWER SIZE
7 in (18 cm)

PLANT SIZE
10–18 × 3–4 in
(25–46 × 8–10 cm),
including single-flowered
inflorescence, which is
mostly shorter than
the leaves

449

RHYNCHOLAELIA DIGBYANA
QUEEN OF THE NIGHT
(LINDLEY) SCHLECHTER, 1918

The flattened, closely spaced, elongate pseudobulbs of the Queen of the Night have a single elliptic, upright, fleshy leaf that is covered in a gray, dustlike substance. From the top of the pseudobulb, an upright inflorescence is formed, bearing a single flower, with a long stalk into which a long nectar cavity is embedded. The genus name comes from the Greek word *rhynchos*, "beak"—referring to the long stalk of the flowers—and its floral similarity to the genus *Laelia* (a related group in the same subtribe).

Pollination is by night-flying moths that are initially attracted by the pervasive lemon scent of the flowers. The nocturnal nature of these large blooms is responsible for the plant's common name, while the species name is in honor of a Mr. Digby, an English orchid enthusiast of the period when this species was first described in 1846.

The flower of the Queen of the Night has green, spreading, narrowly lanceolate sepals and petals, which are wider. The greenish-white lip surrounds the column and has a long-fringed margin and a raised callus creating a tube to the nectar cavity.

Actual size

SUBFAMILY	Epidendroideae
TRIBE AND SUBTRIBE	Epidendreae, Laeliinae
NATIVE RANGE	Tropical America from southern North America (Mexico) to southern South America (Bolivia)
HABITAT	Seasonally dry forests, at 1,640–4,920 ft (500–1,500 m)
TYPE AND PLACEMENT	Epiphytic
CONSERVATION STATUS	Not threatened
FLOWERING TIME	January to March (late winter to early spring in Northern Hemisphere, summer in Southern Hemisphere)

FLOWER SIZE
⅜ in (1 cm)

PLANT SIZE
8–12 × ¼–⅜ in
(20–30 × 0.6–0.9 cm),
including terminal
inflorescence
2–4 in (5–10 cm) long

450

SCAPHYGLOTTIS LIVIDA
SPINDLE ORCHID
(LINDLEY) SCHLECHTER, 1918

Actual size

The Spindle Orchid was given its common name in reference to its spindle-shaped pseudobulbs, from the top of which an inflorescence grows, bearing between three and six strangely colored flowers. In some lights, the blooms appear to be green, in others a bluish-gray. The latter color gave rise to the Latin species name, *livida*, meaning "lead-colored." The genus name is derived from the Greek words *skafi*, "trough," and *glotta*, "tongue," alluding to the central groove on the lip. In some individuals, new pseudobulbs emerge from the top of the previous year's growth, resulting in a chain of the spindle-like stems.

Short-tongued insects, such as bees and bee flies, are the likely pollinators of *Scaphyglottis livida*. They are probably attracted by the nectar produced in the groove running down the center of the lip.

The flower of the Spindle Orchid has green to bluish-green sepals and petals, with the dorsal sepal much smaller than the other four. The lip is similar in color and three-lobed, the midlobe broad, with a bilobed apex and a central groove down its basal two-thirds.

SUBFAMILY	Epidendroideae
TRIBE AND SUBTRIBE	Epidendreae, Laeliinae
NATIVE RANGE	Greater Antilles (Jamaica, Cuba, Hispaniola) and the Bahamas
HABITAT	Along steep stream banks at low elevation
TYPE AND PLACEMENT	Terrestrial or epilithic
CONSERVATION STATUS	Not assessed
FLOWERING TIME	February to April (late winter and spring)

TETRAMICRA PARVIFLORA
WALLFLOWER ORCHID
LINDLEY EX GRISEBACH, 1864

FLOWER SIZE
½ in (1.25 cm)

PLANT SIZE
3–8 × 3–5 in
(8–20 × 8–13 cm),
excluding erect
terminal inflorescence
10–18 in (25–46 cm) long

451

From the rhizome of the Wallflower Orchid, which may be slightly below the soil surface, short stems with one or two rigid, fleshy, glaucous, round leaves with a channel emerge. The inflorescence is fine and threadlike with sheathing bracts and produces eight to ten widely spaced and faintly fragrant flowers. The genus name, from the Greek, *tetra*, "four," and *mikros* "small," probably refers to the four pairs of pollinia. The species name alludes to the small size of the flowers, and the common name to a similarity of the plants to the wallflower genus *Erysimum*, a member of the mustard family (Brassicaceae).

Pollination has not been observed, although the flower color and shape suggest that some type of bee is a likely pollinator. The flowers, however, produce no obvious reward, and there is no nectar cavity or spur.

Actual size

The flower of the Wallflower Orchid has lanceolate, slightly cupped, greenish-brown sepals and petals and a showy, three-lobed, pale purple-pink lip with the lobes equal in size and shape and a dark purple spot in the middle.

SUBFAMILY	Epidendroideae
TRIBE AND SUBTRIBE	Epidendreae, Pleurothallidinae
NATIVE RANGE	Mexico to tropical America (south to Argentina and Peru)
HABITAT	Rain forests, at 650–2,625 ft (200–800 m)
TYPE AND PLACEMENT	Epiphytic
CONSERVATION STATUS	Not threatened
FLOWERING TIME	Late winter to early spring

FLOWER SIZE
¼ in (0.6 cm)

PLANT SIZE
3–5 × 1–2 in
(8–13 × 2.5–5 cm),
including inflorescence
1–1½ in (2.5–3.8 cm) long

452

ACIANTHERA PUBESCENS
TOADS ON A LEAF
(LINDLEY) PRIDGEON & M. W. CHASE, 2001

The Toads on a Leaf varies in color, form, flower size, and numbers over its vast geographic area, perhaps one of the broadest of any orchid. Spotted flowers emerge from a sheath where the petiole meets the obovate, darkly spotted leaf and mostly sit upon the leaf. This flowering habit, combined with the dull colors of the blooms, gives the plant its common name. The species name refers to the short hairs that cover the outsides of the flowers.

The number of flowers varies between four and ten, arranged in an alternating pattern on the stem, framed by the distinctive thick leaves. The unpleasant scent of the flowers indicates that the pollinator is probably a fruit fly, fond of rotten or overripe fruit, which clambers over the leaves and blooms looking for the source of the smell.

Actual size

The flower of the Toads on a Leaf Orchid is generally creamy white overlaid with maroon stripes on its dorsal sepal and maroon spots on its fused sepals. The petals are red and filamentous at their tips, while the dark red lip is tonguelike with two upturned basal lobes.

SUBFAMILY	Epidendroideae
TRIBE AND SUBTRIBE	Epidendreae, Pleurothallidinae
NATIVE RANGE	Tropical America, from Mexico to southern Brazil and Peru
HABITAT	Wet mountain and cloud forests, at 1,575–10,170 ft (480–3,100 m)
TYPE AND PLACEMENT	Epiphyte
CONSERVATION STATUS	Not assessed, but relatively common and widespread
FLOWERING TIME	Throughout the year, peak flowering in fall

ANATHALLLIS SCLEROPHYLLA
HARD-LEAVED BONNET ORCHID
(LINDLEY) PRIDGEON & M. W. CHASE, 2001

FLOWER SIZE
½ in (1.2 cm)

PLANT SIZE
4–6 in (10–15 cm) tall,
excluding inflorescence

453

The Hard-leaved Bonnet Orchid produces short branches that each hold a stiff, elliptic or lanceolate leaf and a 12 in (30 cm) long inflorescence with up to 40 flowers. This species used to be considered a member of the genus *Pleurothallis*, but this genus was polyphyletic (derived from more than one ancestral group) and so was split.

The inflorescence is much longer than the leaf, and the sweetly fragrant flowers open simultaneously and attract many insects. Fruit flies (*Drosophila*), small and large wasps (Draconidae, Vespidae), fungus gnats (Sciaridae), and snout beetles have all been observed as visitors on this species. Nectar production has not been demonstrated. Which—if any—of the visiting insects is the effective pollinator is unstudied.

The flower of the Hard-leaved Bonnet Orchid has three minutely hairy, elongate, recurved sepals. The petals are much shorter and enclosed in the sepal tube. The lip is tiny as well and projects only slightly from the flower.

Actual size

SUBFAMILY	Epidendroideae
TRIBE AND SUBTRIBE	Epidendreae, Pleurothallidinae
NATIVE RANGE	Ecuador
HABITAT	Montane forests, at 4,920–6,600 ft (1,500–2,000 m)
TYPE AND PLACEMENT	Epiphyte
CONSERVATION STATUS	Vulnerable
FLOWERING TIME	Throughout the year

FLOWER SIZE
⅛ in (0.5 cm)

PLANT SIZE
2–5 × 1–2 in
(5–10 × 3–5 cm),
excluding inflorescence

ANDINIA PENSILIS
HANGING BAT ORCHID
(SCHLECHTER) LUER, 2000

454

Actual size

Attached to the substrate by fleshy roots, the Hanging Bat Orchid produces a wiry, pendent stem with alternately arranged leaves and delicate, upright inflorescences carrying four to twelve flowers, each subtended by a small bract. The flowers are similar to those of the large genus *Lepanthes*. This probably means that they, too, are pollinated during pseudocopulation by a species of fly. The small *Andinia* orchids have several different floral features, but all are likely to follow the same pollination syndrome. One of the consistent differences between *Andinia* and *Lepanthes* is the presence of dense trichomes (hairs) on the ovary in the former.

The genus name refers to the Andes Mountains, where all the dozen or so species occur. Like nearly all members of this subtribe, the species grow in cool, wet montane forests.

The flower of the Hanging Bat Orchid has three large sepals, each with a cylindrical tip. The petals are minute and narrow, and the lip is folded, with two large sidelobes that encircle the column and resemble the ears of a bat.

SUBFAMILY	Epidendroideae
TRIBE AND SUBTRIBE	Epidendreae, Pleurothallidinae
NATIVE RANGE	Colombia, Venezuela, Ecuador, Peru, and Bolivia
HABITAT	Mid to upper elevation tropical wet forests
TYPE AND PLACEMENT	Epiphytic, sometimes terrestrial or lithophytic
CONSERVATION STATUS	Not threatened
FLOWERING TIME	Can bloom anytime but less likely in summer

BARBOSELLA CUCULLATA
COWLED SPRITE
(LINDLEY) SCHLECHTER, 1918

FLOWER SIZE
About 1 in (2.5 cm)

PLANT SIZE
4–6 × 1 in
(10–15 × 2.5 cm),
excluding erect to arching
single-flowered inflorescence
5–8 in (13–20 cm) tall

455

Actual size

The miniature Cowled Sprite forms vigorous clumps of shiny succulent leaves, each of which can give rise to a proportionally large, single tawny to reddish-brown flower per stem. With a propensity to flower exuberantly in a simultaneous flush, the peculiar, eye-catching blooms, held well above the foliage, put on an impressive show for such a petite orchid.

The flowers differ from those of other genera in the subtribe Pleurothallidinae by their curved column, which forms a hood over a cavity on the base of the lip. As in many orchids, this motile lip maneuvers the pollinators into the correct position to pick up and deposit the pollen, although as yet the specific insect that pollinates these plants is unknown. The common name refers to the hood (cowl) formed by the column, which is also the source of the species name, *cucullata*, derived from the Latin word for "hooded."

The flower of the Cowled Sprite is yellow green to reddish-brown, sometimes approaching pink. It has long, acuminate sepals, outwardly pointing, much shorter, filamentous petals, and a large, downward-sweeping synsepal framing its motile labellum, which has a reddish, open cavity.

SUBFAMILY	Epidendroideae
TRIBE AND SUBTRIBE	Epidendreae, Pleurothallidinae
NATIVE RANGE	Eastern slopes of Andes, central and southern Ecuador
HABITAT	Mid to upper elevation tropical wet forests
TYPE AND PLACEMENT	Terrestrial on slopes and embankments
CONSERVATION STATUS	Not threatened
FLOWERING TIME	Spring

FLOWER SIZE
1½–2 in (4–5 cm)

PLANT SIZE
2–3 × 1 in (5–8 × 2.5 cm),
excluding erect single-
flowered inflorescence
1–2 in (2.5–5 cm) tall

456

BRACHIONIDIUM DODSONII
DODSON'S GREEN MOSS SPRITE
LUER, 1995

A diminutive clump-forming plant with no pseudobulbs and thin leaves, *Brachionidium dodsonii* forms masses in mossy leaf-litter. Although growing in dense clusters, the plants can become almost vinelike and clamber about in the deep mosses, rooting as they progress. Other species in the genus occur at the limit of orchids in the Andes, as high as 12,800 ft (3,900 m). This species is named for the botanist Calaway Dodson (1928–), who has long studied the orchids of the Andes and made a huge contribution to what we know about the orchids of this region.

The thin-textured flowers are small and beautiful but extremely delicate and short-lived—usually no more than a day or two. Pollinators have not been recorded, but like nearly all species of subtribe Pleurothallidinae, the orchid is likely to be visited by small flies.

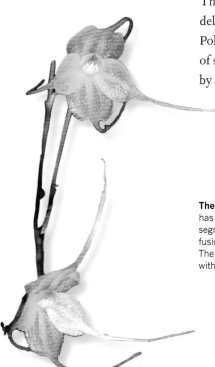

The flower of Dodson's Green Moss Sprite
has four green, similarly shaped, long-tailed
segments, which comprise two sepals (one the
fusion of the two lateral sepals) and two petals.
The small lip is usually short, broad, and smooth
with a pointed tip and a deep green center.

Actual size

SUBFAMILY	Epidendroideae
TRIBE AND SUBTRIBE	Epidendreae, Pleurothallidinae
NATIVE RANGE	Greater Antilles
HABITAT	Cloud forests along ridges and roadsides, at 2,460–8,200 ft (750–2,500 m)
TYPE AND PLACEMENT	Epiphytic, rarely terrestrial
CONSERVATION STATUS	Not assessed but locally common
FLOWERING TIME	December to February (winter to early spring)

DILOMILIS MONTANA
PARROTBEAK ORCHID
(SWARTZ) SUMMERHAYES, 1961

FLOWER SIZE
⅝ in (1.5 cm)

PLANT SIZE
10–18 × 3–4 in
(25–46 × 8–10 cm),
excluding erect terminal
inflorescence, which is
8–14 in (20–36 cm)
taller than the leaves

457

The cane-like stems of the Parrotbeak Orchid are enveloped in a persistent, leafy sheath, with lanceolate, rigid, and leathery blades. The plant forms upright, simple, or branched racemes with stiff floral bracts and up to a dozen violet-scented flowers. The genus name derives from the Greek words *di–*, "two," and *lom–*, "fringe," plus the Latin ending *–ilis*, meaning "pertaining to," together referring to the double-crested lip. The common name refers to the lip shape.

In Puerto Rico, hummingbirds are the reported pollinators, with the flowers said to be visited by the endemic Puerto Rican Emerald hummingbird (*Chlorostilbon maugaeus*). The flowers, however, produce a scent that a bird cannot detect, so it may be that bees visit as well. The flowers also have no tube to direct a beak so are not ideally designed for bird pollination, whereas the shape is like that of many bee-pollinated species.

The flower of the Parrotbeak Orchid has white to cream, spreading sepals and petals, with the upper sepal arched over the flower. The column has a black anther cap and downward-pointing wings that are embraced by the sides of the two-crested lip, which is white, three-lobed, and red-banded.

Actual size

SUBFAMILY	Epidendroideae
TRIBE AND SUBTRIBE	Epidendreae, Pleurothallidinae
NATIVE RANGE	Costa Rica to Ecuador
HABITAT	Montane rain forests, at 2,300–4,600 ft (700–1,400 m)
TYPE AND PLACEMENT	Epiphytic
CONSERVATION STATUS	Not assessed
FLOWERING TIME	June to August (summer)

FLOWER SIZE
⅝ in (1.5 cm)

PLANT SIZE
1–3 × 1 in (2.5–8 × 2.5 cm),
excluding lateral single-
flowered inflorescence
2–4 in (5–10 cm) long

458

DIODONOPSIS ERINACEA
PORCUPINEFISH ORCHID
(REICHENBACH FILS) PRIDGEON & M. W. CHASE, 2001

Actual size

The small Porcupinefish Orchid forms dense clumps of stiff, lanceolate leaves with basal tubular bracts; each leaf bears a single arching-pendent inflorescence with a single flower held just above the leaves. The flowers are remarkable for their complicated morphology, with many spines on their exterior and large flaring sepals with long tail-like extensions terminating in a swollen tip. The spines give the plant its common name and the genus name, from the Greek word for "two-toothed," *diodon* (also the genus name of porcupinefish), while the species name *erinacea* is derived from the Latin for a hedgehog.

Pollination of this species has not been studied, but its morphology is similar to *Masdevallia* (in which it was previously included) and *Pleurothallis*, which are pollinated by various types of flies seeking food or mating sites. Something similar must occur with the Porcupinefish Orchid.

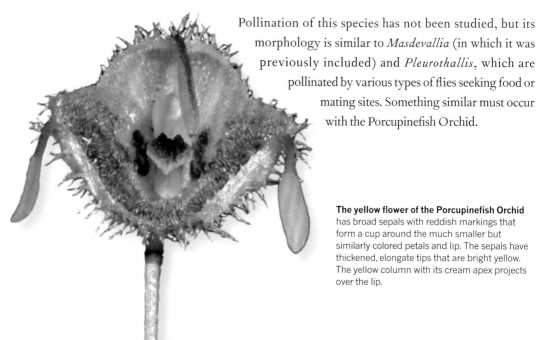

The yellow flower of the Porcupinefish Orchid has broad sepals with reddish markings that form a cup around the much smaller but similarly colored petals and lip. The sepals have thickened, elongate tips that are bright yellow. The yellow column with its cream apex projects over the lip.

SUBFAMILY	Epidendroideae
TRIBE AND SUBTRIBE	Epidendreae, Pleurothallidinae
NATIVE RANGE	Southeastern Ecuador
HABITAT	Cloud forests
TYPE AND PLACEMENT	Epiphytic
CONSERVATION STATUS	Not threatened
FLOWERING TIME	Spring, fall, and winter

FLOWER SIZE
6 in (15 cm)

PLANT SIZE
7–10 × 1–2 in
(18–25 × 2.5–5 cm),
excluding pendent
inflorescence
5–8 in (13–20 cm) long

DRACULA SIMIA
MONKEY-FACE ORCHID
(LUER) LUER, 1978

459

Examined up close, the center of each pendent down-facing flower reveals what appears to be a monkey's face, from which feature both the species name *simia* (from the Latin word for "monkey") and the common name are derived. Species of the genus *Dracula* are largely wet loving, cool cloud forest dwellers with leaves that generally resemble members of *Masdevallia*. This large genus historically included *Dracula* species, and the two are still genetically compatible (and make artificial hybrids).

The long-tailed flowers are produced successively and hang below the foliage. Blooms tend to be warty and hairy with a rounded lip that bears structures reminiscent of the gills on the underside of a mushroom. Although the flowers remind us of a monkey's face, to a pollinating fungus gnat the appearance of this orchid and its fungus-like fragrance suggest that it is a good site to lay its eggs.

The flower of the Monkey-face Orchid consists of three prominent, reddish-brown sepals, each with a long tail and a sparsely haired surface. The petals are small and darkly colored. The lip is movable, white to cream, with structures similar to the gills underneath a mushroom cap.

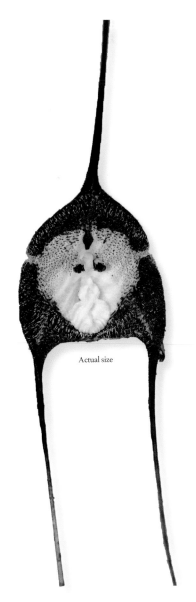

Actual size

SUBFAMILY	Epidendroideae
TRIBE AND SUBTRIBE	Epidendreae, Pleurothallidinae
NATIVE RANGE	Western Ecuador
HABITAT	Cloud forests, at 5,900–7,200 ft (1,800–2,200 m)
TYPE AND PLACEMENT	Epiphytic
CONSERVATION STATUS	Threatened by poaching and collection for horticulture
FLOWERING TIME	Throughout the year

FLOWER SIZE
7 in (18 cm)

PLANT SIZE
8–12 × 1–2 in
(20–30 × 2.5–5 cm)

460

DRACULA VAMPIRA
VAMPIRE DRAGON
(LUER) LUER, 1978

The Vampire Dragon has one of the most evocative names in the orchid family, conjuring up vivid images from horror movies. The genus name is derived from Medieval Latin for "little dragon," *draco, -ula.* Another source of the name is from Vlad III Dracula (1431–76), ruler of Wallachia (modern Romania), whose father was a member of the Order of the Dragon (*Dracul* in Romanian). The association with a vampire derives from Bram Stoker's Gothic novel *Dracula* (1897). The sinister-looking flowers with their bold, nearly black striping are suspended well away from the leaves on arching to pendent stems, contributing to their eerie appearance.

The species of *Dracula* all appear to be pollinated by fungus-eating gnats that are attracted to the lip, which has the shape of a mushroom, complete with "gills" and an appropriate odor. All of this impressive display is artifice.

The flower of the Vampire Dragon consists of three creamy or pale green sepals prominently overlaid with black-brown stripes, each sepal tipped with long tails. The petals are short and pale and flank the column. The lip is ladle-shaped and pinkish-cream with ridges radiating from the center.

Actual size

SUBFAMILY	Epidendroideae
TRIBE AND SUBTRIBE	Epidendreae, Pleurothallidinae
NATIVE RANGE	Guatemala and Costa Rica
HABITAT	Humid evergreen forests, at 2,625–3,600 ft (800–1,100 m)
TYPE AND PLACEMENT	Epiphytic
CONSERVATION STATUS	Not assessed
FLOWERING TIME	December to February (winter)

FLOWER SIZE
⅜ in (1 cm)

PLANT SIZE
3–4 × 1 in
(8–10 × 2.5 cm),
including arching single-
flowered inflorescence

461

DRESSLERELLA PILOSISSIMA

FUZZY PANTOFLE

(SCHLECHTER) LUER, 1978

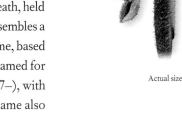

Actual size

The miniature hair-covered Fuzzy Pantofle has short pendent stems, each with a single ovate-lanceolate leaf. It produces a succession of solitary flowers on a stalk with a basal sheath, held just above the leaf. The woolly surface of the sepals resembles a cozy slipper, which is responsible for the common name, based on an old-fashioned term for a slipper. The genus is named for the eminent orchid biologist, Robert L. Dressler (1927–), with the additional Latin diminutive, *-ella*. The species name also refers to its hairiness, from the Latin *pilosus*, meaning "hairy."

The hairs that cloak the plant likely serve as a defense against herbivores. The ends of the dorsal sepal and petals bear scent-producing glands (osmophores) that presumably attract some type of fly, although pollination has not been recorded in nature.

The flower of the Fuzzy Pantofle has cream and red-striped, fused lateral sepals, the upper one long-pointed, gland-tipped, woolly outside, and held over the opening. The petals are similar in color and have thickened tips. The lip is spade-shaped and enclosed within the slipper.

SUBFAMILY	Epidendroideae
TRIBE AND SUBTRIBE	Epidendreae, Pleurothallidinae
NATIVE RANGE	Southeastern Brazil
HABITAT	Tropical wet forests
TYPE AND PLACEMENT	Epiphytic
CONSERVATION STATUS	Not threatened
FLOWERING TIME	Throughout the year, but most often in summer

FLOWER SIZE
1¼–1½ in (3–4 cm)

PLANT SIZE
2–3 × ½ in
(5–8 × 1.3 cm),
including short, erect
single-flowered inflorescences
½ in (1.3 cm) long

462

DRYADELLA EDWALLII
PARTRIDGE IN THE GRASS ORCHID
(COGNIAUX) LUER, 1978

Actual size

The flower of the Partridge in the Grass Orchid
is triangular, consisting of three sharply pointed,
usually brown-spotted sepals with a yellow or
pale green background. The yellow, brown-
spotted petals are situated on the sides of a
broad, curving lip.

The spotted flowers of the Partridge in the Grass Orchid are
often hard to see at first, usually nestled within the tufts of
copious thin leaves. This position of the flowers among the
foliage is responsible for the common name and is also the
reason for the genus name, which derives from the Greek *Dryad*,
a wood nymph, and -e*lla*, a diminutive.

Nearly all species in the genera of the subtribe Pleurothallidinae
are pollinated by some form of fly, which moves around the
flowers looking for food or a place to lay its eggs. The lip of these
flowers is hinged, so when the fly reaches the balance point it is
thrown against the column, where it struggles to free itself and
in doing so dislodges the pollen masses, which become attached
to its back.

SUBFAMILY	Epidendroideae
TRIBE AND SUBTRIBE	Epidendreae, Pleurothallidinae
NATIVE RANGE	Costa Rica
HABITAT	Cloud forests, at about 6,600 ft (2000 m)
TYPE AND PLACEMENT	Epiphytic
CONSERVATION STATUS	Not assessed
FLOWERING TIME	Spring to fall

ECHINOSEPALA STONEI
BURDOCK ORCHID
(LUER) PRIDGEON & M. W. CHASE, 2002

FLOWER SIZE
¾ in (2 cm)

PLANT SIZE
7–12 × 1–1½ in
(18–30 × 2.5–3.8 cm),
including inflorescence,
which is produced
separately on the rhizome,
½–1 in (1.3–2.5 cm) long

463

The little Burdock Orchid forms clusters of upright stems, each with a single upright, thickly leathery elliptical leaf. Basally, these are enveloped in three or four, loosely overlapping, tubular, leaflike sheaths. A short inflorescence arises from the stem between the terminal leaf and the one just behind it and forms a single flower that opens only along two slits on its sides, remaining closed at the tip. Pollination of this bizarre orchid is undocumented. The subtribe Pleurothallidinae has specialized in pollination by various types of flies, and this is also likely to be the case here.

The common name refers to the spiny exterior of these flowers, resembling the bur of Burdock (*Arctium*, in the daisy family, Asteraeae). The genus name is from the Greek words *echinus*, "hedgehog," and *sepalum* "sepal," again referring to the spiny exterior of the sepals.

Actual size

The flower of the Burdock Orchid has densely hairy almost black sepals, two of them fused together. The petals are short and spathulate with ridges, and the lip is stalked, cup-shaped, and trilobed with the sidelobes narrow and the central lobe broad and hairy.

SUBFAMILY	Epidendroideae
TRIBE AND SUBTRIBE	Epidendreae, Pleurothallidinae
NATIVE RANGE	Central America: El Salvador to Costa Rica and perhaps western Panama
HABITAT	Cloud forests, often also on fence post trees along wet pastures, at 4,300–4,920 ft (1,300–1,500 m)
TYPE AND PLACEMENT	Epiphytic
CONSERVATION STATUS	Not assessed
FLOWERING TIME	July to October (late summer and fall), possibly throughout the year

FLOWER SIZE
⅛ in (0.3 cm)

PLANT SIZE
3–4 × 1 in (8–10 × 2.5 cm),
including successively
flowering inflorescence

LEPANTHES COSTARICENSIS
BABYBOOT ORCHID
SCHLECHTER, 1923

464

Actual size

The tiny epiphytic Babyboot Orchid has a short stem enveloped completely by a loose sheath with marginal hairs. From the point where the stem and leaf meet, a two-rowed inflorescence sits on the leaf surface and produces one flower at a time. The genus name comes from the Greek words *lepis*, "scale," and *anthos*, "flower," referring to the small scale-like flowers. The common name is based on the shape of the petals, which resemble a baby's bootee.

Despite their small size, the flowers reveal something incredibly complex on close inspection—reminiscent of Horton, the elephant in the Dr. Seuss book *Horton Hears a Who!*, discovering a world he never realized existed on a speck of dust. Likewise, the tiny fly pollinators of this and other species in the genus are probably completely unknown to most people due to their tiny dimensions.

The flower of the Babyboot Orchid has thin, spreading, yellow to reddish-yellow sepals of equal shape and size, with the central vein stained red. The red to orange petals have two linear lobes that clasp the magenta to orange-red, two-parted lip, under which the column sits.

SUBFAMILY	Epidendroideae
TRIBE AND SUBTRIBE	Epidendreae, Pleurothallidinae
NATIVE RANGE	Ecuador
HABITAT	Cloud forests, at 2,460–3,300 ft (750–1,000 m)
TYPE AND PLACEMENT	Epiphytic
CONSERVATION STATUS	Not threatened
FLOWERING TIME	Late winter and spring

FLOWER SIZE
⅛ in (0.2 cm)

PLANT SIZE
1–2 × ¾–1 in
(2.5–5 × 2–2.5 cm),
including inflorescence
¼–½ in (0.6–1.3 cm) long

465

LEPANTHES VOLADOR
FLYING TRAPEZE ORCHID
LUER & HIRTZ, 1996

With well over 1,000 species, and more being discovered all the time, *Lepanthes* is one of the largest orchid genera. This exquisite species has round leaves with darker veins and small, complex flowers that appear in succession on an erect stem suspended above their leaves. With no pseudobulbs for storing moisture, the plants grow in cloud forests, which are continuously moist. The species name, *volador*, is Spanish for "flying," a reference to the way the flowers are carried above the leaves rather than on their surface, as in other species of *Lepanthes*.

Lepanthes species engage in pseudocopulation with small flies and gnats, which are attracted by a fragrance that mimics the sex pheromone produced by the female insects. *Lepanthes volador*, from the similarity of its flower form, would appear to follow this syndrome, although little is known about its pollination.

Actual size

The flower of the Flying Trapeze Orchid has greenish-yellow, ovate sepals and yellow to orange petals that are bilobed, with the upper lobe having an extended tail and the lower lobe rounded. The red lip is oval and wraps around the orange to purple column.

SUBFAMILY	Epidendroideae
TRIBE AND SUBTRIBE	Epidendreae, Pleurothallidinae
NATIVE RANGE	Colombia (Antioquia department)
HABITAT	Cloud forests, at 4,920–6,070 ft (1,500–1,800 m)
TYPE AND PLACEMENT	Epiphytic
CONSERVATION STATUS	Not threatened
FLOWERING TIME	February to March (late winter and spring)

FLOWER SIZE
⅛ in (0.2 cm)

PLANT SIZE
3–4 × ⅜–⅝ in
(8–10 × 1–1.5 cm),
excluding inflorescence
3–4 in (8–10 cm) long

466

LEPANTHOPSIS PRISTIS
SAWFISH ORCHID
LUER & R. ESCOBAR, 1986

Actual size

The genus name derives from its vegetative similarity to *Lepanthes*. Both sets of species have no pseudobulbs, but instead their leafy growths have delicate stems (ramicauls) characterized by a covering of funnel-shaped bracts, termed lepanthiform sheaths. The flowers of the two genera are different, with the blooms of *Lepanthopsis pristis* arranged in an orderly fashion on two sides of the stem, giving them the appearance of the toothy beak of a sawfish (*pristis* is Greek for "sawfish").

The Sawfish Orchid bears 20–30 flowers at once on a single inflorescence that emerges from the apex of the stem at the base of the leaf. Little is known about pollination, but it has been suggested that aphids are the pollinators. However, aphid behavior (as sedentary sucking insects) suggests that they would not be effective in this role. More likely, some sort of fly is involved.

The flower of the Sawfish Orchid is translucent green or tan with broadly lanceolate, basally fused sepals—collectively slightly cupped—tiny, half-round petals, and a green, triangular lip. The column is short and has two broad sidelobes.

SUBFAMILY	Epidendroideae
TRIBE AND SUBTRIBE	Epidendreae, Pleurothallidinae
NATIVE RANGE	Costa Rica
HABITAT	Humid forests, at 4,920–5,900 ft (1,500–1,800 m)
TYPE AND PLACEMENT	Epiphytic
CONSERVATION STATUS	Not formally assessed
FLOWERING TIME	January to February (late winter to early spring)

FLOWER SIZE
1 in (2.5 cm)

PLANT SIZE
4–7 × ½–1 in
(10–18 × 1.3–2.5 cm),
including basal
inflorescence, which is
shorter than the leaves

MASDEVALLIA CHASEI
MONTEVERDE YELLOW NYMPH
LUER, 1980

467

Actual size

The Monteverde Yellow Nymph forms clusters of stems, each enveloped by basal tubular sheaths and producing a single, upright, leathery, obovate leaf. From the base of each leaf, a slender, erect to arching inflorescence with a single flower is produced; occasionally a second occurs. The species is still known only from the original location just outside the Monteverde Cloud Forest Reserve in Costa Rica. The Reserve is famous for the large number of orchid species recorded there (more than 500 so far).

Masdevallia chasei is named for Professor Mark W. Chase, who—together with Dr. Kerry Walter—discovered this species in 1979. Chase was one of the first plant taxonomists to adopt DNA studies for studying plant evolution.

The flower of the Monteverde Yellow Nymph has three fused, yellow sepals that form a tube, often with a red blotch in the base that has three long, threadlike tips. The short, yellow petals, lip, and column are often tinged red.

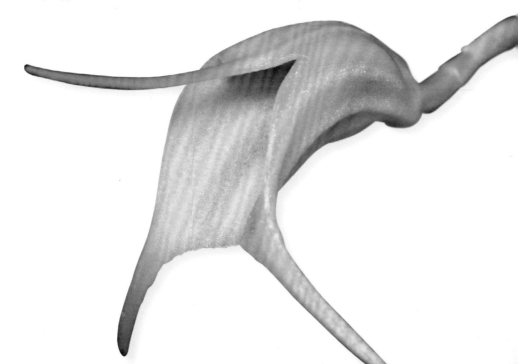

SUBFAMILY	Epidendroideae
TRIBE AND SUBTRIBE	Epidendreae, Pleurothallidinae
NATIVE RANGE	Central and western cordillera of Colombia
HABITAT	Mid-elevation tropical wet forests
TYPE AND PLACEMENT	Epiphytic
CONSERVATION STATUS	Not threatened
FLOWERING TIME	Fall

FLOWER SIZE
2 in (5 cm)

PLANT SIZE
3–5 × 1 in (8–13 × 2.5 cm),
including single-flowered
short erect inflorescences
1–2 in (2.5–5 cm) long

MASDEVALLIA HERRADURAE
HIDDEN SPIDER ORCHID
F. LEHMANN & KRAENZLIN, 1899

The Hidden Spider Orchid has narrow, almost linear, glossy, highly succulent leaves broadening slightly midway through their length. The long-tailed, eye-catching flowers, which emit a strong odor reminiscent of ripe (nearly rotting) coconut, are produced low in the foliage. The common name refers to the fact that the spidery flowers are almost completely hidden by the leaves.

The genus *Masdevallia* is large and distributed throughout the American tropics, particularly at higher elevations, and, with few exceptions, these species are pollinated by flies. In this plant, the flies are looking for the rotting fruit that its fragrance seems to indicate is present. In their search, one of the flies passes the balance point of the hinged lip and is thrown into the column, which results in the pollen masses becoming attached to the insect's back.

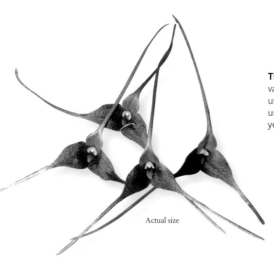

The flower of the Hidden Spider Orchid can vary in color but is most often maroon red. It is usually waxy, with each sepal bearing a long tail, usually fading to yellow at the tip. Short, pale yellow petals and lip surround the column.

Actual size

SUBFAMILY	Epidendroideae
TRIBE AND SUBTRIBE	Epidendreae, Pleurothallidinae
NATIVE RANGE	Peru, only known from Machu Picchu
HABITAT	Cloud forests, open rocky sites, at 6,600–13,100 ft (2,000–4,000 m)
TYPE AND PLACEMENT	Mostly terrestrial, sometimes lithophytic, rarely epiphytic
CONSERVATION STATUS	Threatened by collection for horticulture
FLOWERING TIME	September–December (spring and early summer)

FLOWER SIZE
8 in (20 cm)

PLANT SIZE
6–10 × ¼–1¼ in
(15–25 × 1.9–3.2 cm),
excluding the erect,
single-flowered inflorescence,
10–20 in (25–51 cm) tall

469

MASDEVALLIA VEITCHIANA
VEITCH'S MARVEL
REICHENBACH FILS, 1868

A renowned species from the area surrounding the archaeological site at Machu Picchu, the stunning Veitch's Marvel bears a shockingly vibrant flower produced on sturdy erect stems and held well above the foliage. Often growing in full sun, the leaves are protected from sunburn by surrounding grasses. The flower color sometimes appears uneven or asymmetrical because iridescent purple hairs cover the blooms, creating a dazzling surface sheen. The plant was named for the Veitch Nurseries, based in Devon and London, which in the nineteenth century were the largest family-based nurseries in Europe, introducing many new orchid species into cultivation.

Pollination of *Masdevallia veitchiana* has never been studied. However, at the elevations at which the plant grows, the pollinator is likely to be a bee or hummingbird, although the lack of nectar probably precludes the latter.

The flower of Veitch's Marvel has brilliant orange sepals fused at their bases to produce an elongate, triangular flower outline. The surface is covered with light-catching purple hairs, often producing a shimmering gleam. Petals and lip are darker and much reduced, and form a tube around the column.

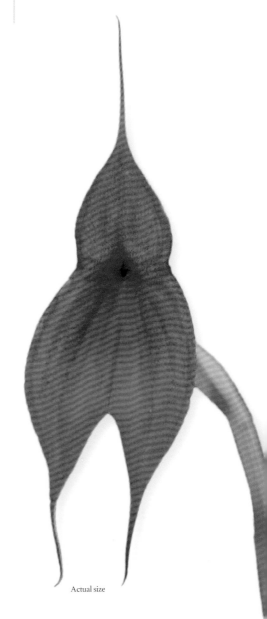

Actual size

SUBFAMILY	Epidendroideae
TRIBE AND SUBTRIBE	Epidendreae, Pleurothallidinae
NATIVE RANGE	Colombia and Venezuela
HABITAT	Wet cool forests, at 6,600–8,200 ft (2,000–2,500 m)
TYPE AND PLACEMENT	Epiphytic
CONSERVATION STATUS	Not threatened
FLOWERING TIME	Winter to spring

FLOWER SIZE
¼–1 in (2–2.5 cm)

PLANT SIZE
4–6 × 1 in
(10–15 × 2.5 cm),
including short
inflorescence
½–1 in (1.3–2.5 cm) long

470

MYOXANTHUS HYSTRIX
SPINY-MOUSE ORCHID
(REICHENBACH FILS) LUER, 1982

Actual size

The Spiny-mouse Orchid grows in upper elevation cloud forests, where it blooms through a large part of the year, producing a new flower once the old one withers. The stems are noteworthy for their strong bristles—the feature to which the species name refers, from the Latin word for porcupine, *hystrix*. The genus name is an obscure reference to the Greek words *myoxos*, "dormouse," and *anthos*, "flower."

The small, oddly colored and shaped flowers have antennal knobs on the petals, which bear scent-producing glands. Like the flowers of nearly all members of the subtribe Pleurothallidinae, these are most likely pollinated by a fly, which lands on the leaf before crawling onto the bloom. The flower has a hinged lip that would at some point tip the fly against the column, causing removal of the pollen as the insect struggles to free itself. No reward is offered.

The flower of the Spiny-mouse Orchid has pale greenish-yellow sepals striped with dark mahogany brown. The petals are triangular and similar in color but spotted with a yellow knob at their tip. The lip is mostly reddish-brown, curled, and hinged.

SUBFAMILY	Epidendroideae
TRIBE AND SUBTRIBE	Epidendreae, Pleurothallidinae
NATIVE RANGE	Highlands of the Dominican Republic (Hispaniola)
HABITAT	Cloud forests, at 3,300–4,600 ft (1,000–1,400 m)
TYPE AND PLACEMENT	Epiphytic on twigs of shrubs
CONSERVATION STATUS	Not threatened
FLOWERING TIME	Spring and summer

FLOWER SIZE
2–2⅜ in (5–6 cm)

PLANT SIZE
4–8 × 2–3 in
(10–20 × 5–8 cm), excluding
terminal, erect-arching
inflorescence, which is
generally longer than the
leaves, 5–12 in (13–30 cm) long

471

NEOCOGNIAUXIA HEXAPTERA
SCARLET CLOUD ORCHID
(COGNIAUX) SCHLECHTER, 1913

With its intense colors and rounded flower, the spectacular Scarlet Cloud Orchid, endemic to the cloud forests of the Dominican Republic, is one of the most sought-after Caribbean species, despite its difficulties in cultivation. The species, a twig epiphyte, often grows low down on lichen-encrusted shrubs. Its pollinia, like those of other bird-pollinated orchids, are darkly colored, making them less noticeable to the bird and therefore less likely to be scraped off before being deposited on another plant. No nectar has been reported, so this is a case of deceit pollination.

The genus is named for Alfred Cogniaux (1841–1916), who is famous for his taxonomic treatments of the orchids of Brazil and the West Indies. The species name is derived from the Greek words *hex*, "six," and *pterus*, "winged," presumably in reference to the six-parted flower.

The flower of the Scarlet Cloud Orchid is brilliant orange red, with broad, rounded sepals and petals. The lip is yellow with sidelobes folded around the column, which is purple red at its apex.

Actual size

SUBFAMILY	Epidendroideae
TRIBE AND SUBTRIBE	Epidendreae, Pleurothallidinae
NATIVE RANGE	Central and South America: Nicaragua, Trinidad, and Colombia to Bolivia and southern Brazil
HABITAT	Wet forests, at 330–8,200 ft (100–2,500 m)
TYPE AND PLACEMENT	Epiphytic
CONSERVATION STATUS	Widespread and common, so unlikely to be of concern
FLOWERING TIME	October to March (in both hemispheres)

FLOWER SIZE
¾ in (2 cm)

PLANT SIZE
6–12 × 1–2 in
(15–30 × 2.5–5 cm),
including inflorescence

OCTOMERIA GRANDIFLORA
LARGE GNAT ORCHID
LINDLEY, 1842

472

Actual size

The Large Gnat Orchid is a clump-forming species with cane-like stems, each carrying a single, narrowly lanceolate, erect leaf. From the point where the leaf and stem meet, a cluster of between one and three flowers appears from a papery, sheathing bract. This is the largest species in the genus, the name of which derives from the Greek words *okto*, "eight," and *meros*, "part," an allusion to the eight pollen masses carried by the plants.

The floral scent and nectar produced by *Octomeria grandiflora* attract only a single species of dark-winged fungus gnat. These flies look for nectar, which is located on the lip base, and as they move toward the base of the lip their thorax contacts the column and removes the pollinia. A subsequent visit deposits the pollen, an event that often results in flies being stuck and, unable to extricate themselves, dying.

The flower of the Large Gnat Orchid has similar, thin, yellow sepals and petals. The yellow lip is lobed, with the lateral lobes curving upward and hinged to the column. There are red markings on the base of the lip and column, and the latter has a white apex.

SUBFAMILY	Epidendroideae
TRIBE AND SUBTRIBE	Epidendreae, Pleurothallidinae
NATIVE RANGE	Southern Brazil
HABITAT	Sea level to 2,950 ft (900 m)
TYPE AND PLACEMENT	Epiphytic on twigs of shrubs
CONSERVATION STATUS	Not threatened
FLOWERING TIME	Late winter to early spring

FLOWER SIZE
⅝ in (1.5 cm)

PLANT SIZE
3–4 × 1 in (8–10 × 2.5 cm),
excluding erect-arching
inflorescence
4–6 in (10–15 cm) long

473

PABSTIELLA MIRABILIS
BABY-IN-A-BLANKET
(SCHLTR.) BRIEGER & SENGHAS, 1976

The Baby-in-a-blanket is an exquisite little orchid that even botanist Rudolf Schlechter recognized when he described it in 1918 as a "remarkable" species, with its pristine white flower and a column emerging from the tubular lip (the column being the "baby" and the lip being the "blanket" of the common name). The species name is Latin for "remarkable," and the genus is named for the renowned Brazilian orchid taxonomist Guido Pabst (1914–80).

Pollination of this orchid has not been studied, but the nectar spur and floral size and color would be compatible with a small bee as the pollinator. Most species of subtribe Pleurothallidinae are pollinated by some form of fly, with food deception figuring prominently, but this species seems to be differently adapted. The other species of *Pabstiella* appear to follow the standard fly-adapted morphology.

Actual size

The flower of the Baby-in-a-blanket is pure crystalline white with a hooded dorsal sepal and recurved synsepal. The petals and lip project forward, surrounding the column. Each petal has a thin, pale pink to purple stripe.

SUBFAMILY	Epidendroideae
TRIBE AND SUBTRIBE	Epidendreae, Pleurothallidinae
NATIVE RANGE	Central America, Mexico (Chiapas state) to Panama
HABITAT	Rain and cloud forests, at 3,300–4,920 ft (1,000–1,500 m)
TYPE AND PLACEMENT	Epiphytic
CONSERVATION STATUS	Not threatened
FLOWERING TIME	April to July (spring and summer)

FLOWER SIZE
⅛ in (0.3 cm)

PLANT SIZE
1¾–2½ × ¼–½ in
(4.5–6.5 × 0.8–1.3 cm),
including terminal
inflorescence
1–1¼ in (2.5–3.2 cm) long

474

PLATYSTELE OVATILABIA
ORANGE WOOD SPRITE
(AMES & C. SCHWEINFURTH) GARAY, 1974

Actual size

The miniature Orange Wood Sprite produces tight clusters of narrowly lanceolate glossy green leaves. From the base of the leaf, covered by two large bracts, an upright to pendent stem bears between two and six star-shaped flowers. The genus name comes from the Greek words *platy*, "wide," and *stili*, "column," referring to the characteristic columns of its member species. The species name, *ovatilabia*, refers to the shape of lip, and the common name to the plant's small size and occurrence in misty, wooded sites.

Although these are small flowers, other species of *Platystele* have even smaller flowers than these. The smallest orchid flowers, measuring just ¹⁄₆₄ in (0.5 mm), are probably in the unrelated genus *Campylocentrum*. The members of the orchid subtribe Pleurothallidinae, including *P. ovatilabia*, are nearly all pollinated by flies, so this species is also likely to be fly-pollinated, although by what sort is entirely speculative.

The flower of the Orange Wood Sprite has translucent, oval-acuminate sepals and petals, which are similar in shape, and a brilliant orange oval lip. The column is short and has two broad side wings with a white to cream anther cap.

SUBFAMILY	Epidendroideae
TRIBE AND SUBTRIBE	Epidendreae, Pleurothallidinae
NATIVE RANGE	Central America: Costa Rica to western Panama
HABITAT	Humid forests, at 4,300–8,200 ft (1,300–2,500 m)
TYPE AND PLACEMENT	Epiphytic
CONSERVATION STATUS	Not assessed
FLOWERING TIME	March to May (spring)

PLEUROTHALLIS PHYLLOCARDIA
HEARTLEAF ORCHID
REICHENBACH FILS, 1866

FLOWER SIZE
¾ in (2 cm)

PLANT SIZE
7–12 × 3–4 in
(18–30 × 8–10 cm),
including inflorescence

475

The Heartleaf Orchid produces tight clusters of slender, upright stems that are enveloped by tubular sheaths and each carrying a single, upright, leathery, heart-shaped leaf. The leaf base encircles a single, successively blooming inflorescence that sits partway up the leaf. The genus name comes from the Greek words for *pleuron*, "rib," and *thallos*, "shoot," referring to the thin stems found in the member species.

Two cavities on the lip of this orchid secrete nectar, and glands on many floral parts produce a fungus-like odor. The pollinator has not been studied, although a closely related and similar Colombian species attracts fungus gnats at night. The broad leaf surface provides a good place for male and female flies to meet, and while there they pick up the pollen masses and pollinate further flowers when they land on other leaves.

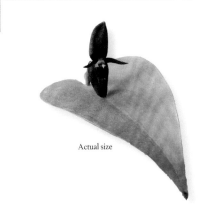

Actual size

The flower of the Heartleaf Orchid has lateral sepals that are fused into a single lower sepal opposite an upright upper sepal. The petals are linear and point sideways. The lip is short, broadly spear-shaped and immediately below the column. Flower color varies from reddish-purple to greenish-cream.

SUBFAMILY	Epidendroideae
TRIBE AND SUBTRIBE	Epidendreae, Pleurothallidinae
NATIVE RANGE	Tropical Americas
HABITAT	Wet montane forests and foothills, at 130–6,600 ft (40–2,000 m)
TYPE AND PLACEMENT	Epiphytic or lithophytic
CONSERVATION STATUS	Not formally assessed but locally abundant and widespread, so likely to be of least concern.
FLOWERING TIME	June to August (summer)

FLOWER SIZE
½–¾ in (1.3–1.9 cm)

PLANT SIZE
6–11 × 3–4 in
(15–28 × 8–10 cm), including
short inflorescence,
½–¾ in (1.3–1.9 cm) long
and borne at the base
of the leaf blade

PLEUROTHALLIS RUSCIFOLIA
GREEN BONNET ORCHID
(VON JACQUIN) ROBERT BROWN, 1813

476

Actual size

The small, clump-forming Green Bonnet Orchid has numerous stems that are basally enveloped in a series of tubular sheaths. Each stem bears a single lanceolate leaf, which is narrower basally and forms a short stem from which a short cluster of flowers is produced. These have a delicate sweet fragrance, even though in some plants the flowers are self-pollinating.

The species is named from the Greek for "ruscus-leafed" as it resembles *Ruscus* (Asparagaceae), a relative of asparagus that also has flowers appearing to emerge from the leaf. Pollination has not been observed, but flies are the most likely pollinators as is the case with most members of the subtribe Pleurothallidinae. The green color and sweet smell of these flowers are also compatible with this hypothesis. *Pleurothallis* was formerly one of the largest genera of orchids, with more than 1,500 species, but recent taxonomic changes have reduced it to around 1,000.

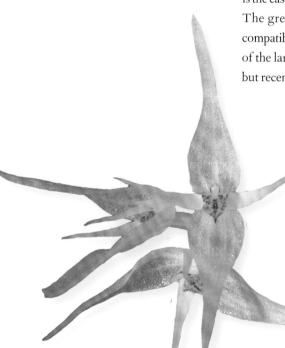

The flower of the Green Bonnet Orchid has two sepals—one often somewhat hooded over the flower, the other similar in size and shape but composed of two lateral sepals. The petals are linear, and the lip is short and arrowhead-shaped.

SUBFAMILY	Epidendroideae
TRIBE AND SUBTRIBE	Epidendreae, Pleurothallidinae
NATIVE RANGE	Colombia, Ecuador, and northwestern Venezuela
HABITAT	Cloud forests, at 5,250–9,850 ft (1,600–3,000 m)
TYPE AND PLACEMENT	Epiphytic or sometimes terrestrial
CONSERVATION STATUS	Not formally assessed
FLOWERING TIME	August to October (fall)

FLOWER SIZE
¾ in (2 cm)

PLANT SIZE
4–6 × 1–1½ in
(10–15 × 2.5–3.8 cm),
excluding inflorescence,
which is erect and
5–10 in (13–25 cm) tall

PORROGLOSSUM MUSCOSUM
CATCHFLY ORCHID
(REICHENBACH FILS) SCHLECHTER, 1920

477

The slender, swollen stem of the Catchfly Orchid is basally enveloped in tubular sheaths and topped with a single, leathery, warty, narrowly elliptic, purple-tinged leaf. A densely pubescent inflorescence appears with a single bract and tubular, overlapping bracteoles, bearing several flowers of which only one is open at any time and held well above the leaf. *Poroglossum* is most closely related to the genus *Masdevallia*, from which it was segregated in 1920 by the German taxonomist, Rudolf Schlechter (1872–1925).

The lip is motile and can close within a second if triggered by a visiting insect, probably a fly, trapping and pressing it against the column, allowing the pollinia to be attached. After about 30 minutes the lip relaxes, and the insect may repeat the process and so transfer pollen to the stigma of a flower on another plant.

Actual size

The flower of the Catchfly Orchid has yellow-green sepals with long-tipped lobes and yellow veins that are fused into a cup. The yellow petals are small and narrow and flank the column. The motile yellow lip has a reddish apex and is stalked and hinged.

SUBFAMILY	Epidendroideae
TRIBE AND SUBTRIBE	Epidendreae, Pleurothallidinae
NATIVE RANGE	Northwestern South America, Venezuela to Bolivia and Peru
HABITAT	Andean forests, at 5,250–11,480 ft (1,600 to 3,500 m)
TYPE AND PLACEMENT	Epiphytic on tree trunks
CONSERVATION STATUS	Not assessed
FLOWERING TIME	February to April and August to October (late winter to early fall, but not in summer)

FLOWER SIZE
2 in (5 cm)

PLANT SIZE
6–10 × 2–3 in
(15–25 × 5–8 cm), excluding
single-flowered inflorescence,
1–3 in (2.5–8 cm) long,
produced at the base of the
leaf just above the stem

RESTREPIA ANTENNIFERA
COCKROACH ORCHID
KUNTH, 1816

The clump-forming Cockroach Orchid has stems that are enveloped at their bases by several purple-spotted sheaths; each stem carries a single erect or spreading ovate-elliptic, leathery leaf with a rounded tip. The inflorescence holds a single nodding flower on a slender stem. The coloration and "antennae" (threadlike petals) resemble those of a cockroach, and the species name *antennifera* refers to this feature (Latin for "antennae bearing").

The orchid's "antennae" are osmophores (scent glands), releasing volatile aminoids and terpenes that attract flies over long distances for pollination. Nearly all members of this large subtribe of orchids are fly-pollinated and typically offer no reward, tricking their pollinators into expecting something that is not present. The genus *Restrepia* is named after José Manuel Restrepo (1781–1863), a Colombian naturalist and historian particularly interested in orchids.

Actual size

The flower of the Cockroach Orchid is striped, with one large sepal (created by the fusion of the lateral sepals). The upper sepal and petals are threadlike with thickened tips. The lip is flat with two pointed sidelobes, and the column arches forward over the base of the lip.

SUBFAMILY	Epidendroideae
TRIBE AND SUBTRIBE	Epidendreae, Pleurothallidinae
NATIVE RANGE	Central and South America, Mexico to Colombia
HABITAT	Oak and broadleaf forests, in and alongside coffee plantations, at 130–4,920 ft (40–1,500 m)
TYPE AND PLACEMENT	Epiphytic
CONSERVATION STATUS	Not assessed but widespread and common
FLOWERING TIME	December to March (winter to spring)

FLOWER SIZE
¾ in (2 cm)

PLANT SIZE
7–12 × 1–1½ in
(18–30 × 2.5–3.8 cm),
including inflorescence

RESTREPIELLA OPHIOCEPHALA
SNAKE'S HEAD
(LINDLEY) GARAY & DUNSTERVILLE, 1966

479

Actual size

The short creeping rhizome of the Snake's Head bears a cluster of stout, upright, cylindrical stems covered in tubular sheaths and topped by a single elliptic-lanceolate, fleshy leaf on a short stalk. Up to four, tightly clustered foul-smelling flowers are formed from the point where the stalk of the leaf meets the stem. Based on differing numbers of pollen masses, the genus was separated from *Restrepia*, to which it is similar but not closely related. The species name is derived from the Greek words *ophis*, "snake," and *kephale*, "head," referring to the resemblance of the flowers to the open mouth of a snake. This is also the source of the common name.

Pollination of the species has not been documented. However, given the foul scent of the flowers, flies looking for a place to lay their eggs are the likely pollinators.

The flower of the Snake's Head has downy sepals, the lower pair fused and held opposite the upper one. In a basal tube formed by the sepals are two short petals and a smooth, short, fleshy lip. Color varies from yellow with pale tan spotting to dark reddish with purple spotting.

SUBFAMILY	Epidendroideae
TRIBE AND SUBTRIBE	Epidendreae, Pleurothallidinae
NATIVE RANGE	Antioquia department (Colombia)
HABITAT	Montane forests, at 3,950–4,920 ft (1,200–1,500 m)
TYPE AND PLACEMENT	Epiphytic
CONSERVATION STATUS	Not assessed
FLOWERING TIME	Most of the year

FLOWER SIZE
¾ in (2 cm)

PLANT SIZE
6–10 × 2–3 in
(15–25 × 5–8 cm),
excluding upright
inflorescence,
8–15 cm (20–38 cm) long,
which exceeds the leaves

SCAPHOSEPALUM GRANDE
YELLOW
MUSTACHE ORCHID
KRAENZLIN, 1922

480

Actual size

The upside-down flower of the Yellow Mustache Orchid has fused, lateral sepals with long tips, the lower sepal keeled and pointing forward. The interior of the synsepal is usually red-tinged with red spots and stripes. The petals and lip are much smaller than the rest of the parts.

The clump-forming Yellow Mustache Orchid has slender stems that are each enveloped by two or three tubular sheaths and topped with an elliptical leaf. From the base of the stems, a lateral raceme with conspicuous bracts and many flower buds is formed, but the flowers open (mostly) one at a time. Flowering can occur over several months to nearly the whole year. When flies visit the flowers, scent areas on the fused sepals guide them into position to contact the pollen masses and stigma.

The genus name refers to the hollow, fused, lateral sepals, from the Greek *skaphos*, meaning "a vessel" or "anything hollowed out," and Latin *sepalum*, "sepal." This structure has a long tail in many species of the genus, and on this species it has a flattened perpendicular surface, resembling a mustache with long tips.

SUBFAMILY	Epidendroideae
TRIBE AND SUBTRIBE	Epidendreae, Pleurothallidinae
NATIVE RANGE	Venezuela, Colombia, and Costa Rica
HABITAT	Cloud forests, at 7,900 ft (2,400 m)
TYPE AND PLACEMENT	Epiphytic
CONSERVATION STATUS	Not assessed
FLOWERING TIME	March and September (observed in cultivation)

FLOWER SIZE
¾ in (2 cm)

PLANT SIZE
4–6 × ¾ in (10–15 × 2 cm),
excluding basal arching
inflorescence
7–12 in (18–30 cm) long

SPECKLINIA DUNSTERVILLEI
DUNSTERVILLE'S FRUITFLY ORCHID
KARREMANS, PUPULIN & GRAVENDEEL, 2015

481

Only described in 2015, *Specklinia dunstervillei* was confused with its relative *S. endotrachys* but differs in several characteristics, especially flower shape. The plant was illustrated in 1963 by Venezuela-based English petroleum technician and orchid enthusiast Galfrid C. K. Dunsterville, and the species was named in his honor. Plants in cultivation are known from Colombia and Costa Rica as well. The genus was named in honor of Veit Rudolph Speckle (d. 1550), a German woodcut maker who produced the illustrations for botanist Leonhart Fuchs' *De Historia Stirpum* (1542), a book of medicinal plants.

The flowers of this and related species are pollinated by nectar-feeding fruit flies (genus *Drosophila*), which are attracted by their ripe-fruit smell. No nectar is present, and as the flies inspect the flower they are tipped into the column, where they pick up the pollen masses as they struggle to free themselves.

Actual size

The flower of Dunsterville's Fruitfly Orchid
has orangish-red, forward-pointing lateral sepals and an arching upper sepal covering the green-winged column. The petals are small, ligulate, and reddish. The lip is also reddish, small, and hinged to the lip base.

SUBFAMILY	Epidendroideae
TRIBE AND SUBTRIBE	Epidendreae, Pleurothallidinae
NATIVE RANGE	Ecuador
HABITAT	Cloud forests, at 4,920–7,900 ft (1,500–2,400 m)
TYPE AND PLACEMENT	Epiphytic, occasionally terrestrial
CONSERVATION STATUS	Not threatened
FLOWERING TIME	September to October (fall)

FLOWER SIZE
¼ in (0.6 cm)

PLANT SIZE
2–3 × ⅝–¾ in
(5–8 × 1.7–2 cm),
excluding erect
inflorescence
3–4 in (8–10 cm) tall

STELIS CILIOLATA
HAIRY MISTLETOE ORCHID
LUER & DALSTRÖM, 2004

482

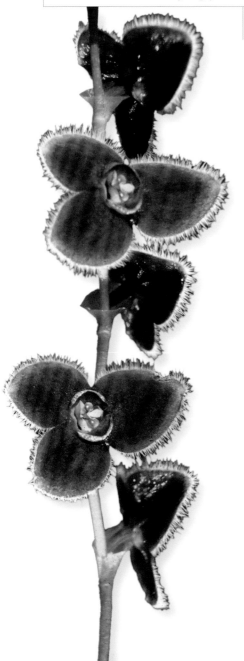

With more than 700 species occurring throughout the American tropics, the miniature Hairy Mistletoe Orchid is one among many with highly similar flowers. As with many such miniatures, close inspection reveals tremendous diversity and variation— in this case the prominent hairs along the margins, referred to in the species name, *ciliolata*, from the Latin for "with a tiny fringe." The genus name is Greek for a type of mistletoe, a parasitic plant, with which these epiphytic plants were often confused when the genus was described in 1799.

The tiny blooms open only during the day. The extremely tiny petals and lip bear small piles of calcium oxalate on their surface, which may refract light in the same way as exposed nectar would, attracting the small flies, such as fungus gnats, that have been reported visiting the flowers.

The flower of the Hairy Mistletoe Orchid is usually dark brownish to reddish-purple with prominent, fused sepals that bear pale hairs along their edges. The petals are tiny and have abruptly flattened surfaces that bear small piles of white crystals.

Actual size

SUBFAMILY	Epidendroideae
TRIBE AND SUBTRIBE	Epidendreae, Pleurothallidinae
NATIVE RANGE	North and Central America, Mexico to El Salvador
HABITAT	Montane tropical forests, at 4,920–8,200 ft (1,500–2,500 m)
TYPE AND PLACEMENT	Epiphytic
CONSERVATION STATUS	Not assessed
FLOWERING TIME	December to July (winter to summer)

FLOWER SIZE
½ in (1.3 cm)

PLANT SIZE
2–3 × ½ in
(5–8 × 1.3 cm), excluding
arching inflorescence
2½–4 in (6.5–10 cm) tall,
which is generally longer
than the leaves

483

STELIS ORNATA
ICICLE ORCHID
(REICHENBACH FILS) PRIDGEON & M. W. CHASE, 2001

The lovely miniature Icicle Orchid has slender, slightly flattened stems that are enveloped in tubular sheaths terminating in a leathery, elliptical leaf. Flowers are formed one at a time on a terminal, lax inflorescence that zigzags and holds flowers on opposite sides. The species is often known in horticulture as *Pleurothallis schiedei* (also often called *P. villosa*), but that species has hairiness inside the flower only and lacks the decorative dangly appendages of *Stelis ornata* (also formerly placed in the genus *Pleurothallis*).

Pollination has not been studied in the field, but the dangling growths probably function as an attractant to pollinators, most likely flies, which most species in the subtribe Pleurothallidinae attract. Other *Stelis* species have calcium oxalate crystals on their petals and lip, which sparkles and attracts in a manner similar to the dangling appendages of *S. ornata*.

Actual size

The flower of the Icicle Orchid has spreading, red-spotted, cream to yellow sepals, fringed with dangling, white, icicle-like appendages. The small petals and lip are spotted like the sepals, with the spots coalescing apically in the former. The arrow-shaped column is winged and dark red.

SUBFAMILY	Epidendroideae
TRIBE AND SUBTRIBE	Epidendreae, Pleurothallidinae
NATIVE RANGE	Central America: Costa Rica to western Panama
HABITAT	Lower rain forests, at 3,950–5,900 ft (1,200–1,800 m)
TYPE AND PLACEMENT	Epiphytic
CONSERVATION STATUS	Not assessed
FLOWERING TIME	December to April (winter to spring)

FLOWER SIZE
¼ in (0.7 cm)

PLANT SIZE
3–5 × ⅜ in (8–13 × 1 cm),
excluding arching-erect
inflorescence
5–8 in (13–20 cm) tall,
which is longer than
the leaves

484

STELIS PILOSA
HAIRY TOILET SEAT ORCHID
PRIDGEON & M. W. CHASE, 2002

The small Hairy Toilet Seat Orchid has upright, fleshy stems enveloped basally by tubular sheaths and topped with a single oblanceolate leaf. From a bract where the leaf meets the stem, it forms an unbranched inflorescence with 6–20 flowers. The hairy sepals are shaped like a toilet seat and hinged top—hence the common name. The genus name is based on the Greek word for a kind of mistletoe, *stelis*—an allusion to the habit of orchids to grow on trees, although not as a parasite.

The species was previously widely known in *Pleurothallis*, but it should be included in *Stelis* based on DNA analyses. Pollination is assumed to be by some sort of fly due to the flower color and shape, but there is no evidence so far of how this process might operate.

The flower of the Hairy Toilet Seat Orchid has fused, pale yellow to green sepals, the laterals fused to form a shallowly sacklike pouch that is ribbed and hairy inside, the margin of the upper sepal lined with longer hairs and narrower. The short column is downcurved and flanked by the minute petals and lip.

Actual size

SUBFAMILY	Epidendroideae
TRIBE AND SUBTRIBE	Epidendreae, Pleurothallidinae
NATIVE RANGE	Tropical America, from southern Mexico and the Caribbean to Bolivia
HABITAT	Wet forests, at 2,625–7,050 ft (800–2,150 m)
TYPE AND PLACEMENT	Epiphytic
CONSERVATION STATUS	Not assessed
FLOWERING TIME	September to May (fall to spring)

TRICHOSALPINX MEMOR
EYELASH ORCHID
(REICHENBACH FILS) LUER, 1983

FLOWER SIZE
⅛ in (0.2 cm)

PLANT SIZE
5–8 × ¼–½ in
(13–20 × 0.8–1.3 cm),
including inflorescence

485

The species of genus *Trichosalpinx* have the same type of unusual sheaths around their stems as the members of *Lepanthes*. The inflorescence of the Eyelash Orchid appears from a bract where the leaf and stem meet and has a dense, two-rowed structure with most flowers opening at the same time. The genus name comes from the Greek words *trichos*, "hair," and *salpinx*, "trumpet," referring to the tubular sheaths with flaring, hairy ends that surround the stems. The species name, from the Latin word for "remembering," is an unclear allusion. The edges of the floral parts all have fine hairs, similar to eyelashes—hence the common name.

Pollination is unknown, but the flower colors and shape suggest that some sort of fly is likely to be attracted. Also, like most species in its subtribe, *T. memor* offers no obvious reward.

The flower of the Eyelash Orchid has lateral sepals fused into a boat shape. The upper sepal arches over the column, which is flanked by the short petals and tongue-shaped lip. Color ranges from brown to purple-brown, while the column is white.

Actual size

SUBFAMILY	Epidendroideae
TRIBE AND SUBTRIBE	Epidendreae, Pleurothallidinae
NATIVE RANGE	Costa Rica to Bolivia, Brazil, and Peru
HABITAT	Wet forests, at 650–6,600 ft (200–2,000 m)
TYPE AND PLACEMENT	Epiphytic
CONSERVATION STATUS	Not threatened
FLOWERING TIME	June to October (summer to fall)

FLOWER SIZE
¾ in (2 cm)

PLANT SIZE
¾–1 × ⅛ in
(2–2.5 × 0.5 cm),
excluding erect inflorescence
1½–2 in (3.8–5 cm) tall

TRISETELLA TRIGLOCHIN
STRIPED BRISTLE ORCHID
(REICHENBACH FILS) LUER, 1980

486

Actual size

A clump-forming, miniature epiphyte, the Striped Bristle Orchid has a threadlike inflorescence that emerges from the base of a leaf and extends far beyond it. The flowers appear to float on these wiry stems. Although individual flowers do not last for long, the species blooms successively. The genus name derives from the Latin *tri-*, "three," and *seta*, "bristle," plus the diminutive suffix *-ella*. Similarly, the species name is from the Greek *glochin*, "a point," referring to the odd shape of the projections on the top and sides of the flowers.

The showiest parts of these small blooms are the sepals, and the lip and petals are much smaller and relatively inconspicuous. Pollination in the wild has not been studied, but the flowers fit well the syndrome of fly pollination.

The flower of the Striped Bristle Orchid
is yellow, overlaid with dark reddish-brown stripes that coalesce on the distal ends of fused sepals. It has prominent yellow tails with a swollen tip, and the lip and petals are much reduced and red.

SUBFAMILY	Epidendroideae
TRIBE AND SUBTRIBE	Epidendreae, Pleurothallidinae
NATIVE RANGE	Western South America, Colombia to Ecuador
HABITAT	Cloud forests, at 3,300–5,900 ft (1,000–1,800 m)
TYPE AND PLACEMENT	Epiphytic or terrestrial
CONSERVATION STATUS	Not assessed
FLOWERING TIME	June to August (summer)

ZOOTROPHION HYPODISCUS
MENAGERIE ORCHID
(REICHENBACH FILS) LUER, 1982

FLOWER SIZE
½ in (1.3 cm)

PLANT SIZE
4–6 × 1–1½ in
(10–15 × 2.5–4 cm),
including inflorescence

487

The name of this peculiar genus is Greek for "a menagerie" and refers to the similarity of the flowers to the heads of exotic animals. The blooms of the Menagerie Orchid have fused sepals that form a ridged chamber, with an opening on each side and hairs and tissue extensions along the veins. The well-developed, erect, highly clustered stems arise from a short rhizome and are enclosed in inflated, compressed sheaths. Each stem bears a single, elliptical, leathery leaf at its tip. An inflorescence, with one or two flowers, appears at the junction of the stem and leaf.

Zootrophion species may have brood-site-deceptive flowers pollinated by fungus gnats that normally lay their eggs on mushrooms. There is no obvious reward offered to a pollinator. Fly eggs have been found inside the flowers of *Z. hypodiscus*.

Actual size

The flower of the Menagerie Orchid has fused, reddish-purple sepals with an opening on both sides and a beaked apex. The lip is hinged to the stout base of the column. The petals are short and solid red, and everything is well hidden inside the sepals.

SUBFAMILY	Epidendroideae
TRIBE AND SUBTRIBE	Epidendreae, Ponerinae
NATIVE RANGE	Southern Mexico to Panama
HABITAT	Wet tropical forests, at 2,950–5,250 ft (900–1,600 m)
TYPE AND PLACEMENT	Epiphytic
CONSERVATION STATUS	Not threatened
FLOWERING TIME	July to September (summer to fall)

FLOWER SIZE
⅛ in (0.5 cm)

PLANT SIZE
15–25 × 2–2½ in
(38–64 × 5–6.5 cm),
including terminal,
lax inflorescence
1–2 in (2.5–5 cm) long

488

ISOCHILUS CHIRIQUENSIS
BRILLIANT GRASS ORCHID

SCHLECHTER, 1922

The Brilliant Grass Orchid bears 10–20 flowers per inflorescence, with the last three to five leaves at the tip of its stems tinged with a similarly bright color. The plants are thin stemmed, bear no pseudobulbs, and are covered with narrow, grassy leaves. Their roots are almost three times thicker than the stems. The genus name comes from the Greek words *iso-*, "same," and *cheilos*, "lip," referring to the similarity of the lip to the other flower parts. When not in flower, the plants are easily mistaken for a grass, hence the common name. The species name refers to its original place of collection, Chiriquí province, Panama.

The flushing of the terminal leaves with bright pink increases the visibility of the flowers to hummingbirds. These pollinators are drawn by the abundant nectar in these flowers.

Actual size

The flower of the Brilliant Grass Orchid is small and cupped, usually brilliant pink to purple with a pair of darker eyespots on its lip. The blooms are arranged in two rows. The sepals, petals, and lip are all similarly shaped and form a tube.

SUBFAMILY	Epidendroideae
TRIBE AND SUBTRIBE	Gastrodieae
NATIVE RANGE	Tropical Asia and Oceania, from Afghanistan to southern China, India, and Taiwan, throughout Southeast Asia and Malesia to northern Australia and east as far as the South Pacific island of Niue
HABITAT	Shady areas under bushes in grasslands, coastal areas, and in bamboo thickets, at sea level to 1,475 ft (450 m)
TYPE AND PLACEMENT	Terrestrial in humus
CONSERVATION STATUS	Not formally assessed but rare throughout its range and often overlooked as it is only above ground for about a month
FLOWERING TIME	August to March in the Southern Hemisphere, April to May in the Northern Hemisphere (spring)

FLOWER SIZE
½ in (1.3 cm)

PLANT SIZE
7–10 in (18–25 cm),
totally leafless and
non-photosynthetic
throughout its life

489

DIDYMOPLEXIS PALLENS
CRYSTAL BELLS
GRIFFITH, 1844

From the Crystal Bells' underground, fleshy, horizontal, tuberous rhizome with many fleshy beaded roots, an erect, light brown inflorescence appears, carrying up to six flowers, generally with only one open at any time. The plant is mycoheterotrophic, meaning it is completely dependent on connections with a fungus to provide it with carbohydrates and minerals. The genus name derives from the Greek word *didymos*, "in pairs," and the Latin word *plexus*, "interwoven," referring to the two column wings that are fused with the base of the lip. The common name is an allusion to the sparkling texture of the white flowers, which appear suddenly in the deep shade.

Insects, possibly bees, are the likely pollinators of the flowers, given their color (especially the yellow lip callus) and shape. However, this has not been observed in the field.

Actual size

The flower of Crystal Bells is bright white and has the upper sepal and petals fused into a single, three-lobed segment, forming a shallow tube with the partially fused lateral sepals. The lip has three lobes, the sidelobes longer and the midlobe with a rough, yellow callus.

SUBFAMILY	Epidendroideae
TRIBE AND SUBTRIBE	Gastrodieae
NATIVE RANGE	Southern and eastern Australia, including Tasmania
HABITAT	Moist situations in sclerophyll forests and woodland, at 100–3,600 ft (30–1,100 m)
TYPE AND PLACEMENT	Terrestrial in heavy leaf litter
CONSERVATION STATUS	Not globally assessed, but least concern in Queensland and near threatened in South Australia
FLOWERING TIME	September to December (spring to summer)

FLOWER SIZE
¾ in (2 cm)

PLANT SIZE
25–50 in tall (64–127 cm),
totally leafless and
non-photosynthetic
throughout its life

490

GASTRODIA SESAMOIDES
CINNAMON BELLS
R. BROWN, 1810

The tall leafless Cinnamon Bells is mycoheterotrophic—totally dependent on a fungus from which it gets all its minerals and carbohydrates. Underground it produces a large carrot-like tuber that in turn gives rise to a brown inflorescence carrying 8 to 30 nodding, cinnamon-scented flowers, hence the common name. The fused sepals have a swollen basal saclike portion, which is responsible for the genus name, from the Greek words *gaster*, "stomach," and *-odes*, "resembling."

Aboriginal peoples ate the large starchy tubers, giving *Gastrodia sesamoides* its occasional name of Potato Orchid. Tubers were often located by observing where the small marsupial bandicoots scratched the ground, pinpointing the underground plants, said to smell sesame-like (giving rise to the species name). They are purported to taste like an insipid, watery beetroot. Dried tubers of other *Gastrodia* species play an important part in Chinese traditional medicine.

The flower of the Cinnamon Bells is bell-shaped, with the basal, swollen portion orangish-tan and the lip uppermost. The sepals are fused and open to expose the petals, which, like the inside of the sepals, are white. The column and lip are short and hidden deep within the sepal tube.

Actual size

SUBFAMILY	Epidendroideae
TRIBE AND SUBTRIBE	Malaxideae, Dendrobiinae
NATIVE RANGE	Nepal, India (Assam), and Vietnam to southern China
HABITAT	Evergreen and deciduous forests on limestone outcrops, at 985–3,300 ft (300–1,000 m)
TYPE AND PLACEMENT	Lithophytic
CONSERVATION STATUS	Not assessed
FLOWERING TIME	December to January (winter)

FLOWER SIZE
1¼ in (3 cm)

PLANT SIZE
3–6 × 1–1½ in
(8–15 × 2.5–3.8 cm),
excluding single-flowered
pendent inflorescence
4–7 in (10–18 cm) long

BULBOPHYLLUM AMBROSIA
SHY HONEY ORCHID
(HANCE) SCHLECHTER, 1919

491

The Shy Honey Orchid has a long stem (rhizome) between each ellipsoidal pseudobulb, which is topped by a single elongate-oblong leaf. From the rhizomes, both old and new, an inflorescence bears a single flower (rarely two), which is often partially hidden under a leaf, and hence the "shy" part of the common name. Ambrosia, from the Greek for "immortal," *ambrotos*, was the food of the gods, and the reference here is to the species' sweet, honey-like, floral fragrance. *Bulbophyllum* derives from the Greek words *bolbos*, "bulb," and *phyllon*, "leaf," referring to the leaf-bearing pseudobulbs, which are not unique to this genus.

The flowers of most species of *Bulbophyllum* smell unpleasant and are pollinated by various types of fly. This species, in contrast, has a sweet smell and attracts a bee, *Apis cerana*, the Asiatic honeybee. However, no nectar is present.

Actual size

The flower of the Shy Honey Orchid has similar, cream with red-purple-striped, oval sepals and acuminate cream petals. The lip is cream with red spots and is attached by a hinge to a long, thin extension of the column. The cream column has a yellow cap.

SUBFAMILY	Epidendroideae
TRIBE AND SUBTRIBE	Malaxideae, Dendrobiinae
NATIVE RANGE	Himalayas, southern China (Yunnan province), and Indochina
HABITAT	Evergreen forests, at 650–6,900 ft (200–2,100 m)
TYPE AND PLACEMENT	Epiphytic
CONSERVATION STATUS	Not assessed, but abundant and widespread
FLOWERING TIME	December to April (winter to spring)

FLOWER SIZE
¼ in (0.8 cm)

PLANT SIZE
5–9 × 1–2 in
(13–23 × 2.5–5 cm),
excluding pendent
inflorescence
6–10 in (15–25 cm) long

492

BULBOPHYLLUM CAREYANUM
ROTTEN BANANA ORCHID
(HOOKER) SPRENGEL, 1826

The Rotten Banana Orchid has a short stem (rhizome) between each globose, weakly ribbed pseudobulb, topped by a single broadly lanceolate leaf. From the base of the pseudobulb, a dense inflorescence with 40 to 60 flowers and many long bracts is produced. The species is named in honor of John Carey (1797–1880), an English botanist. The genus name derives from the Greek for "bulb," *bolbos*, and "leaf," *phyllon*.

Most species of *Bulbophyllum* are pollinated by various types of flies, and, as in the case of the Rotten Banana Orchid, their flowers smell bad. *Bulbophyllum careyanum* is pollinated by a fruit fly (genus *Drosophila*), which is attracted by the prospect of a place to find food and lay its eggs. However, for the unfortunate fruit fly there is nothing here to eat.

The flower of the Rotten Banana Orchid has similar, light brown to reddish-brown sepals and petals that form a cup. The lip is similar in color and attached by a hinge to an extension of the column. The cream column has a pair of long apical arms.

Actual size

SUBFAMILY	Epidendroideae
TRIBE AND SUBTRIBE	Malaxideae, Dendrobiinae
NATIVE RANGE	Central to West Africa
HABITAT	Low elevation forests, at up to 5,900 ft (1,800 m)
TYPE AND PLACEMENT	Epiphytic, rarely lithophytic
CONSERVATION STATUS	Not threatened
FLOWERING TIME	Late winter to early spring

BULBOPHYLLUM FALCATUM
SICKLE-SHAPED FLY ORCHID
(LINDLEY) REICHENBACH FILS, 1861

FLOWER SIZE
⅜ in (1 cm)

PLANT SIZE
4–6 × 1–2 in
(10–15 × 2.5–5 cm),
excluding flattened, erect
to arching inflorescence
7–10 in (18–25 cm) long

493

"Strange" is a word often used to describe the genus *Bulbophyllum*, but it is particularly apt for the bizarre subgenus *Megaclinium*, of which this is the most commonly encountered and widespread species. With their elaborate yet diminutive flowers placed on either side of a flattened sickle-shaped stem reminiscent of a peapod, the members of *Megaclinium* are unique in the configuration of their inflorescences.

Many *Bulbophyllum* species are known for their fetid fragrances, which they use to attract carrion and other flies looking for carcasses or rotting fruit in which to lay eggs. It is unclear which syndrome (carrion or rotting fruit) is operating in this species, but certainly the plant is fly-pollinated. The flattened stem serves as a landing platform for the flies, allowing them to walk about in search of the source of the bad smell.

The flower of the Sickle-shaped Fly Orchid is usually yellow to brown with purple on the lip and yellow-tipped petals. Flowers are arranged in two ranks on both sides of a flattened stem.

Actual size

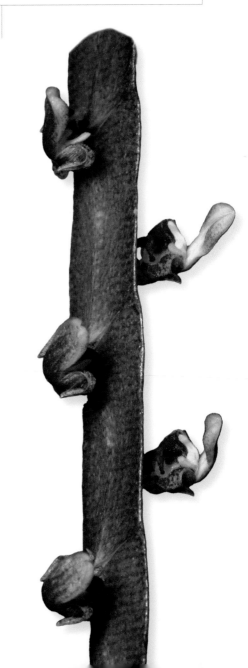

SUBFAMILY	Epidendroideae
TRIBE AND SUBTRIBE	Malaxideae, Dendrobiinae
NATIVE RANGE	Eastern Indonesia (Sulawesi, Maluku Islands), New Guinea, and Solomon Islands
HABITAT	Rain forests, at 650–2,626 ft (200–800 m)
TYPE AND PLACEMENT	Epiphytic on lower limbs or trunks
CONSERVATION STATUS	Not assessed
FLOWERING TIME	April to May (fall)

FLOWER SIZE
6 in (15 cm)

PLANT SIZE
6–9 × 1–2 in
(15–23 × 2.5–5 cm),
excluding erect-arching
single-flowered inflorescence
8–11 in (20–28 cm) tall

BULBOPHYLLUM GRANDIFLORUM
FOUL GIANT
BLUME, 1849

494

The Foul Giant is one of the largest-flowered species in the largest orchid genus (1,900 species)—but it is also one of the most offensive. Its egg-shaped, angular pseudobulbs are borne a short distance apart and topped by a single oblong leaf. An inflorescence with two to three large sheathing bracts bears a single flower.

Pollination has not been studied in detail, but it is clear from the foul smell of the flowers that the plant is attracting a fly looking for a site to deposit its eggs. The large sepals probably serve as landing platforms, allowing the flies to wander about, during which activity they climb onto the hinged lip, pass its balance point, and are thrown into the column. As they extricate themselves, pollinia are deposited on their bodies.

The flower of the Foul Giant is dominated by three large, cream to tan sepals, sometimes with reddish-purple spots, but always with translucent spots and the dorsal sepal bending forward to cover the flower. The petals are highly reduced and green, and the lip is motile, small, and white, all with purple spots.

Actual size

SUBFAMILY	Epidendroideae
TRIBE AND SUBTRIBE	Malaxideae, Dendrobiinae
NATIVE RANGE	India (Arunachal Pradesh and Assam states), Indochina, Indonesia, Malaysia, and the Philippines
HABITAT	Seasonally wet forests, at 650–6,600 ft (200–2,000 m)
TYPE AND PLACEMENT	Epiphytic
CONSERVATION STATUS	Not threatened
FLOWERING TIME	August (summer)

BULBOPHYLLUM LOBBII
SERENITY ORCHID
LINDLEY, 1847

FLOWER SIZE
4 in (10 cm)

PLANT SIZE
8–12 × 3–4 in
(20–30 × 8–10 cm),
excluding erect to pendent,
single-flowered inflorescence
10–15 in (25–38 cm) long

495

The Serenity Orchid thrives in hot, humid, shady conditions, generally on tree trunks and large main branches. It has an extensive system of stems, with 1–2 in (2.5–5 cm) separating large, globular pseudobulbs that each bear a single glossy leaf. Its spectacular flowers often emerge in great profusion. The common name alludes to the flower resembling the meditation, or serenity, pose of Buddha. The species name refers to Thomas Lobb (1817–94), an English botanist and collector of plants in India, Indonesia, and the Philippines.

As found in many *Bulbophyllum* species, the lip of *B. lobbii* is movable and attached to the extended base of the column. This creates a balance point, which, if passed by a fly looking for food, causes the insect pollinator to be thrown into the column, where it picks up the pollinia in its struggle to escape.

The flower of the Serenity Orchid is variable in color, but most forms are yellow, often striped and spotted, or simply spotted, with maroon. The dorsal sepal is upright, and lateral sepals often sweep down. The petals also droop downward, while the spade-shaped lip often bears a yellow crest.

Actual size

SUBFAMILY	Epidendroideae
TRIBE AND SUBTRIBE	Malaxideae, Dendrobiinae
NATIVE RANGE	Thailand, Malaysia, Borneo, Lesser Sunda Islands, and Sumatra
HABITAT	Lowland forests, at sea level to 1,300 ft (400 m)
TYPE AND PLACEMENT	Epiphytic
CONSERVATION STATUS	Not threatened
FLOWERING TIME	Fall and winter

FLOWER SIZE
4¾–6 in (12–15 cm)

PLANT SIZE
5–8 × 1–2 in
(13–20 × 2.5–5 cm),
excluding arching
inflorescence,
6–10 in (15–25 cm) long

496

BULBOPHYLLUM MEDUSAE
MEDUSA'S HEAD FLY ORCHID
(LINDLEY) REICHENBACH FILS, 1861

Named for the Medusa, the most horrific of the three Gorgons in Greek mythology, there is nothing hideous about this species' magnificent inflorescences. More closely resembling festive fireworks than a head of snakes, well-bloomed specimens are a wonder to behold. Hailing from hot and steamy lowland areas, the orchid is one of the few *Bulbophyllum* species that can handle direct sun as well as intense heat. The inflorescences on close inspection can have 15–30 tightly clustered flowers.

With a fetid, fungus-like odor, the plant attracts flies that feed on or lay their eggs in some sort of rotting substrate. There is nothing like this on offer in these flowers, so it is a case of deceit pollination. Flowers live for less than a week and stink for only a short period. In spite of their unpleasant odor, these orchids are nonetheless popular in cultivation.

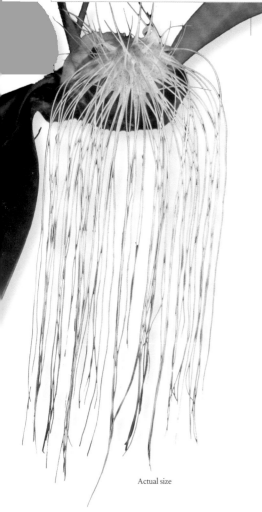

Actual size

The flower of the Medusa's Head Fly Orchid is borne in a dense inflorescence with often dozens of filamentous flowers. They are usually creamy white, occasionally with pale brown spots. Individual flowers have long sepals and petals and a yellow lip with some red markings.

SUBFAMILY	Epidendroideae
TRIBE AND SUBTRIBE	Malaxideae, Dendrobiinae
NATIVE RANGE	Peninsular Thailand and Malaysia to Java and the Sunda Islands
HABITAT	Wet forests on mossy larger limbs and trunks, at 3,300–5,250 ft (1,000–1,600 m)
TYPE AND PLACEMENT	Epiphytic
CONSERVATION STATUS	Not assessed
FLOWERING TIME	February to April (winter to spring)

FLOWER SIZE
1¼ in (3.2 cm), with
forward-projecting sepals
1¼–2 in (3.2–5 cm) long

PLANT SIZE
3–4 × 1–1¼ in
(8–10 × 2.5–3.2 cm),
including inflorescence
½–1 in (1.3–2.5 cm) long

BULBOPHYLLUM MIRUM
MARVELOUS FRINGE ORCHID

J. J. SMITH, 1906

497

This small species produces sharply quadrangular ovoid pseudobulbs with a 1 in (2.5 cm) rhizome separating them. From the rhizome, a short inflorescence emerges that regularly carries two flowers with bizarre petals that have long appendages fluttering in the slightest breeze. The species name comes from the Latin word *mirus*, meaning "wonderful" or "marvelous," referring to these spectacular petals.

Pollination of few of the 1,900 species of the *Bulbophyllum* genus has been studied. In general, however, flies are the likely pollinators, looking either for rotting fruit or animal flesh as sites for egg laying. The exact function of the motile petal appendages is speculative, but it is thought they appeal to the aggregation instinct of most flies (if one fly has found something good, then the other flies want to investigate), causing them to form swarms and so increase pollination frequency.

The flower of the Marvelous Fringe Orchid has a pale, oval, dark-spotted, pink dorsal sepal and longer, lanceolate lateral sepals. The petals are circular and bear about a dozen motile, pale pink to white hairs, and the lip is short and hinged to the extended column base.

Actual size

SUBFAMILY	Epidendroideae
TRIBE AND SUBTRIBE	Malaxideae, Dendrobiinae
NATIVE RANGE	New Guinea
HABITAT	Cool, wet forests, at 2,950–4,600 ft (900–1,400 m)
TYPE AND PLACEMENT	Epiphytic
CONSERVATION STATUS	Not assessed, rare in the wild
FLOWERING TIME	September to October (fall)

FLOWER SIZE
2½ in (6.5 cm)

PLANT SIZE
12–20 × 10–14 in
(30–51 × 25–36 cm),
excluding erect-arching
apical inflorescence
6–10 in (15–25 cm) long

498

DENDROBIUM ALEXANDRAE
ALEXANDRA'S DRAGON ORCHID
SCHLECHTER, 1912

With its spidery flowers and large dramatically colored and shaped lip, looking like little dragons, Alexandra's Dragon Orchid is remarkably striking. Its elongate, spindle-shaped, longitudinally grooved pseudobulbs have papery bracts and between two and five apical leaves. The genus name derives from the Greek words *dendron*, "tree," and *bios*, "life," referring to the plant's tree-based existence, which is also true for most of the 1,500 species of *Dendrobium*. The species name is in honor of Alexandra, wife of the renowned German orchid specialist Rudolf Schlechter (1872–1925).

Four to eight flower buds develop fully but remain closed for several weeks before finally opening. The shape of the flowers and their honey-like scent, although nectarless, indicate that a large bee, perhaps a carpenter bee (genus *Xylocopa*), is the pollinator.

The flower of Alexandra's Dragon Orchid has lanceolate, spreading sepals and petals, cream to yellow with greenish-brown spots. The lip has three lobes, the apical lobe spade-shaped and the laterals curving over the column, all of which are green to cream with reddish-purple stripes or spots.

Actual size

SUBFAMILY	Epidendroideae
TRIBE AND SUBTRIBE	Malaxideae, Dendrobiinae
NATIVE RANGE	Maluku Islands, New Guinea, Australia (Queensland), and the Solomon Islands
HABITAT	High in trees in coastal forests, mangrove swamps, and wet forests, at sea level to 2,625 ft (800 m)
TYPE AND PLACEMENT	Epiphytic
CONSERVATION STATUS	Endangered in Australia and vulnerable overall
FLOWERING TIME	December to March (summer)

FLOWER SIZE
3 in (8 cm)

PLANT SIZE
20–40 × 10–14 in
(51–102 × 25–36 cm),
excluding erect-arching
inflorescence
8–14 in (20–36 cm) long

499

DENDROBIUM ANTENNATUM

ANTELOPE ORCHID

LINDLEY, 1843

The Antelope Orchid produces large, cane-like, longitudinally grooved pseudobulbs with papery bracts and oblong leaves along their length. Inflorescences can appear from the middle of the pseudobulb upward, carrying from 4 to 15 large flowers, with their thin petals erect and twisted, looking either like the horns of an antelope, giving the common name, or antennae, resulting in the species name. The genus name alludes to these orchids living in trees, from the Greek words for tree, *dendron*, and life, *bios*.

The shape of the flowers and their sweet scent indicate that some type of bee, perhaps a colletid bee (genus *Hylaeus*), as recorded for a related Australian species, is the pollinator. No nectar is present in the small cavity on the base of the lip. The function of the twisting, upright narrow petals that gives this species its unusual appearance is unknown.

The flower of the Antelope Orchid has lanceolate white sepals. The petals are narrow, erect, and twisting, starting white and ending greenish-yellow. The lip has three lobes, the apical lobe spade-shaped and the laterals curving upward. All are white with purple stripes, and the column apex is yellow.

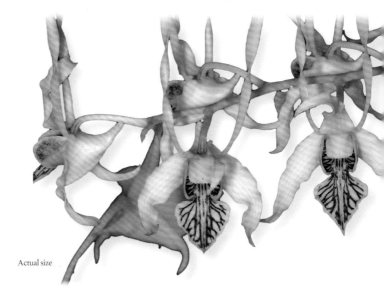

Actual size

SUBFAMILY	Epidendroideae
TRIBE AND SUBTRIBE	Malaxideae, Dendrobiinae
NATIVE RANGE	Myanmar, Laos, Thailand, Vietnam, China, eastern Himalayas, Bangladesh, and Assam state (India)
HABITAT	Sea level to 1,300 ft (400 m)
TYPE AND PLACEMENT	Epiphytic
CONSERVATION STATUS	Threatened due to overcollection for use in herbal medicine
FLOWERING TIME	Late winter to early spring

FLOWER SIZE
1½–2 in (4–5 cm)

PLANT SIZE
8–14 × 6–8 in
(20–36 × 15–20 cm),
including arching-pendent
inflorescence
8–12 in (20–30 cm) long

500

DENDROBIUM CHRYSOTOXUM
FRIED-EGG ORCHID
LINDLEY, 1847

An important medicinal plant in Southeast Asia, the Fried-egg Orchid bears copious (20 or more on a spike), honey-scented blooms. The flowers are collected and dried to produce a delicious medicinal tea said to induce peaceful, dreamless sleep. The leaves are used to treat a range of ailments, especially those associated with diabetes. The plants originate in monsoonal climates with extreme spring and summer seasonal rainfall, and this wet-dry cycle heavily influences their growth and flowering patterns.

The flowers have an exceptionally lacerate margin and appear near the apex of tall, cylindrical, slightly angled, cane-like pseudobulbs, usually in a glorious flush. Unfortunately, the splendid show lasts only between seven and ten days. The flowering coincides with the spring water-splashing festival of the Buddhist Dai people of Yunnan, China, who decorate the roofs of their houses with this orchid.

The flower of the Fried-egg Orchid can be variable in color but is generally brilliant yellow orange with a flat form and waxy sepals and petals. Some plants can have a darker orange to reddish-brown spot in the center of the lip.

Actual size

SUBFAMILY	Epidendroideae
TRIBE AND SUBTRIBE	Malaxideae, Dendrobiinae
NATIVE RANGE	Queensland and New South Wales, Australia
HABITAT	Open rocky slopes near coast and in mountains
TYPE AND PLACEMENT	Lithophytic
CONSERVATION STATUS	Not threatened
FLOWERING TIME	Late winter to early spring

FLOWER SIZE
1–1½ in (2.5–4 cm)

PLANT SIZE
6–15 × 3–5 in long
(15–38 × 8–13 cm),
excluding the erect terminal
long inflorescence, which is
6–10 in (15–25 cm) long

DENDROBIUM KINGIANUM
PINK ROCK ORCHID
BIDWILL EX LINDLEY, 1844

501

Possibly the most common Australian species, the tough and adaptable Pink Rock Orchid can survive in a multitude of environmental conditions. Since the plant grows naturally in harsh, windswept rocky habitats (hence the common name), it flourishes luxuriantly when placed in greenhouse conditions and is commonly cultivated around the world. Each flower spike usually carries 10–20 fragrant, starlike pink blooms, in a tremendous array of forms and colors. In the wild, the plants often create natural hybrids with sympatric species within the same genus, such as *Dendrobium speciosum*.

Given that the Pink Rock Orchid hybridizes with other species in nature, it must share a pollinator with those species, which are visited by a number of bees, such as members of the genus *Trigona* (a small native genus of stingless bees) and introduced honeybees (*Apis*). The flowers of the orchid provide nectar as a reward to pollinators.

The flower of the Pink Rock Orchid is dark pink to white (sometimes purple). The sepals form a triangle of pointed segments with smaller petals and lip in between. The lip is trilobed, often marked with darker pink to purple spots, the sepal bases forming a short spur.

Actual size

SUBFAMILY	Epidendroideae
TRIBE AND SUBTRIBE	Malaxideae, Dendrobiinae
NATIVE RANGE	Chinese Himalayas, eastern Himalayas, India (states of Assam and Sikkim), Nepal, Bhutan, Myanmar, Thailand, Laos, and Vietnam
HABITAT	Broadleaf, evergreen forests and primary montane forests as well as on mossy limestone rocks
TYPE AND PLACEMENT	Lithophytic and terrestrial
CONSERVATION STATUS	Threatened due to overcollection, mostly for its medicinal properties
FLOWERING TIME	Late winter to early spring

FLOWER SIZE
2⅜–3 in (6–8 cm)

PLANT SIZE
10–18 × 4–6 in
(25–46 × 10–15 cm),
including multiple short,
erect to arching inflorescences
2–5 in (5–13 cm) long

502

DENDROBIUM NOBILE
NOBLE ROCK ORCHID
LINDLEY, 1830

A naturally widespread orchid species, the Noble Rock Orchid has a long history of cultivation and medicinal use and is beloved in Asia for its beautiful flowers. The plant is one of the main orchid species exploited in herbal medicine traditions as a cure-all in India, Sri Lanka, and China—so much so that it is endangered in many parts of its range. The species is adapted to monsoonal seasonal rains and winter drought and is deciduous (dropping nearly all of its leaves) before blooming just prior to the advent of spring rains.

With up to four large flowers emerging from mostly terminal nodes on its canes, the plant is among the showiest of orchids that have made their way into the commercial trade. Pollination is by large bees from several genera, and there is a nectar reward in the short spur.

The flower of the Noble Rock Orchid is large, waxy, and extremely variable in color. Typical forms are pale pink suffused with purple on the sepals and petals. The lip is usually white with a dark interior eyespot often ringed in yellow, although pure white and even orange forms are also known.

Actual size

SUBFAMILY	Epidendroideae
TRIBE AND SUBTRIBE	Malaxideae, Dendrobiinae
NATIVE RANGE	Philippines
HABITAT	Mossy, wet, cool oak forests with rhododendrons, at 4,300–8,200 ft (1,300–2,500 m)
TYPE AND PLACEMENT	Epiphytic
CONSERVATION STATUS	Not assessed
FLOWERING TIME	April to May (spring), but in flower almost continuously

DENDROBIUM VICTORIAE-REGINAE
QUEEN VICTORIA BLUE
LOHER, 1897

FLOWER SIZE
1½ in (3.8 cm)

PLANT SIZE
10–16 × 6–10 in
(25–41 × 15–25 cm),
including inflorescence
1–2 in (2.5–5 cm) long,
which grows near the end
of the pseudobulb

503

The Queen Victoria Blue produces cane-like, longitudinally grooved pseudobulbs with papery bracts and lanceolate-ovate leaves along their length. Inflorescences can appear from the middle to the end of the often-pendent pseudobulb, carrying between two and five flowers. The genus name, from the Greek for "tree," *dendron*, and "life," *bios*, refers to the cool mossy forest habitat of the plants. The species name honors Queen Victoria, who was nearing the end of her long reign at the time the species was discovered.

Although pollination has not been studied, the shape of the flowers, their color, and their lack of scent could indicate that they are pollinated by honeyeaters or other birds, as has been found for other species of this *Dendrobium* group. The blue color of the blooms is unusual among orchids and could be mimicking flowers of *Rhododendron* species.

The flower of the Queen Victoria Blue has similar, oblanceolate, blue to lavender-purple sepals and petals with white bases and darker veins. The lip is unlobed and spoon-shaped, and also has a white base with blue stripes. The column is white and winged.

Actual size

SUBFAMILY	Epidendroideae
TRIBE AND SUBTRIBE	Malaxideae, Malaxidinae
NATIVE RANGE	Boreal and temperate Northern Hemisphere: in Europe south to the Alps and Carpathians, in Asia south to Ukraine and throughout Siberia to northern Japan, in North America throughout Alaska to central Canada and south to Minnesota
HABITAT	Sphagnum bogs, at up to 3,600 ft (1,100 m)
TYPE AND PLACEMENT	Terrestrial in acidic swamps
CONSERVATION STATUS	Least concern, although it has not been globally assessed and is declining throughout its range due to habitat loss
FLOWERING TIME	June to September (summer to early fall)

FLOWER SIZE
⅛ in (0.5 cm)

PLANT SIZE
3–6 × 1–2 in
(8–15 × 2.5–5 cm),
including erect
inflorescence

504

HAMMARBYA PALUDOSA
BOG ORCHID
(LINNAEUS) KUNTZE, 1891

The tiny Bog Orchid (the species name, *paludosa*, is Latin for "swamp") has two pseudobulbs, one from the previous season and one from the current. The upper pseudobulb has between one and three ovate fleshy leaves, and an upright inflorescence bears 4–25 flowers. The leaves produce plantlets that fall off and result in large stands, which are unlikely to be noticed due to their small stature.

The genus was named for Hammarby, the town in Sweden where the botanist Carl Linnaeus (1707–78) had his summer home. Charles Darwin studied *Hammarbya paludosa* for his book *On the Various Contrivances by Which British and Foreign Orchids are Fertilized by Insects* (1862), where he noted that the ovary twists 360 degrees, a totally unnecessary change because an orchid lip is developmentally in that position. Fungus gnats pollinate the flowers, attracted by the odor of cut cucumber and a nectar reward.

The flower of the Bog Orchid is totally green and simple in structure, with the lateral sepals pointing up. The petals are shorter and broader and point sideways. The lip is similar in shape to the petals, points upward, and sometimes has darker green veins.

Actual size

SUBFAMILY	Epidendroideae
TRIBE AND SUBTRIBE	Malaxideae, Malaxidinae
NATIVE RANGE	Malaysian Borneo
HABITAT	Wet forests, at 2,625–5,900 ft (800–1,800 m)
TYPE AND PLACEMENT	Terrestrial
CONSERVATION STATUS	Not assessed
FLOWERING TIME	December to March (winter to spring)

FLOWER SIZE
⅜ in (1 cm)

PLANT SIZE
8–12 × 5–8 in
(20–30 × 13–20 cm),
including terminal
inflorescence

CREPIDIUM PUNCTATUM
SPOTTED BOOT ORCHID
(J. J. WOOD) SZLACHETKO, 1995

505

The Spotted Boot Orchid produces upright thin pseudobulbs that are completely enveloped by the leaf bases. The leaf blades are purple-spotted, ovate-elliptic, pleated, and prominently veined. From the apex of the pseudobulb, a single upright, densely flowered inflorescence emerges with flowers that smell of freshly cut cucumber. Species of *Crepidium* are still often treated under the genus *Malaxis*, in which they were formerly included. The genus name is based on the Greek word for a small boot, *krepidion*, alluding to the floral cavity on the lip.

Pollination of none of the nearly 300 species of *Crepidium* has been studied, despite the plants being relatively common and widespread. As is the case in other genera of the subtribe Malaxidinae, the pollinator is assumed to be some sort of fly that is attracted to the nectar produced on the lip.

The flower of the Spotted Boot Orchid is purplish-red and has two upright, rounded sepals and a long, downward-pointing, narrow sepal. Petals are narrow, recurved, or spreading, and the lip is unlobed, has a darker central cavity, and embraces the short, winged column like a cloak.

Actual size

SUBFAMILY	Epidendroideae
TRIBE AND SUBTRIBE	Malaxideae, Malaxidinae
NATIVE RANGE	Tropical South and Southeast Asia, from India and Sri Lanka to southern China, Taiwan, Philippines, and Thailand
HABITAT	Forests and humid places in thickets, at 1,300–5,900 ft (400–1,800 m)
TYPE AND PLACEMENT	Terrestrial
CONSERVATION STATUS	Not formally assessed but widespread
FLOWERING TIME	June to July (summer)

FLOWER SIZE
¼ in (0.8 cm)

PLANT SIZE
8–16 × 10–14 in
(20–41 × 25–36 cm),
including terminal
inflorescence

506

CREPIDIUM PURPUREUM
PURPLE BOOT ORCHID
(LINDLEY) SZLACHETKO, 1995

The terrestrial Purple Boot Orchid generally does not form a pseudobulb. Instead, its stem is often thickened, with a basal cluster of roots and a covering of between three and five basally sheathing, ovate to oblong, long-tipped leaves. From the middle of the rosette, an erect inflorescence emerges, eventually carrying up to 30 or more purplish-red or pale yellowish-green flowers. The genus name comes from the Greek word *krepidion*, "little boot," referring perhaps to a small cavity on the lip. The common name reflects this same origin, and the species name comes from the sometimes-purple color of the flowers.

A liquid secreted on the surface of the center of the lip is most likely nectar, but this has not been confirmed. No pollinator is known, but some sort of gnat is suspected.

The flower of the Purple Boot Orchid has the lip uppermost, with broadly oblong-ovate sepals, the lower one the narrowest. Petals are narrowly linear. The sidelobes curve around the column, with the upper lobe in two parts. The column is short and has two side arms.

Actual size

SUBFAMILY	Epidendroideae
TRIBE AND SUBTRIBE	Malaxideae, Malaxidinae
NATIVE RANGE	Western Central America to northwestern South America (Nicaragua, Costa Rica, Panama, Ecuador, and Colombia)
HABITAT	Wet forests, at 1,970–8,700 ft (600–2,650 m)
TYPE AND PLACEMENT	Terrestrial
CONSERVATION STATUS	Not assessed
FLOWERING TIME	Throughout the year

CROSSOGLOSSA TIPULOIDES
BUG ORCHID
(LINDLEY) DODSON, 1993

FLOWER SIZE
⅜ in (1 cm)

PLANT SIZE
8–12 × 4–6 in
(20–30 × 10–15 cm),
including erect terminal
inflorescence

507

As a juvenile, the Bug Orchid has a fan of several overlapping, elliptical leaves that later grow along two sides of a stem, which their bases envelop; no pseudobulb is present. From the center of the leaves, a keeled or winged stem carries 10 to 40 green flowers that superficially resemble bugs, hence the common name. The genus name derives from the Greek words *krossos*, "fringe," and *glossa*, "lip," referring to the minute serrations of the lip. The species name refers to a resemblance to a crane fly (genus *Tipula*).

Although they look like stinkbugs, it is more likely that the little green flowers, which produce nectar on a groove down the center of their lip, are pollinated by some form of fly or gnat. However, no formal studies have so far been conducted.

Actual size

The flower of the Bug Orchid is green and has narrowly lanceolate, spreading sepals of the same size. The petals are linear and tinged reddish, and the lip is broad, with short basal lobes that grasp the short, straight column.

SUBFAMILY	Epidendroideae
TRIBE AND SUBTRIBE	Malaxideae, Malaxidinae
NATIVE RANGE	North Temperate Zone (Eurasia and North America)
HABITAT	Forests, bogs, marshes, and seeps, from sea level to 2,950 ft (900 m)
TYPE AND PLACEMENT	Terrestrial
CONSERVATION STATUS	Threatened as wetlands are converted to agriculture
FLOWERING TIME	April to July (late spring and early summer)

FLOWER SIZE
⅜ in (1 cm)

PLANT SIZE
8–12 × 3–5 in
(20–30 × 8–13 cm),
including erect
terminal inflorescence
4–8 in (10–20 cm) tall

508

LIPARIS LOESELII
FEN ORCHID
(LINNAEUS) RICHARD, 1817

The Fen Orchid bears a small, oblong pseudobulb and usually only two leaves—and it is not until its relatively tall inflorescences begin to emerge in late spring that it is noticed, if at all. The genus name comes from the Greek word *liparos*, "greasy" or "shiny," referring to the smooth leaves of the species. The common name alludes to its frequent habitat type, and the species name was given by Linnaeus in honor of the German botanist and physician, Johannes Loesel (1607–55), an expert on Prussian flora.

Pollination of *Liparis loeselii* has been studied in the field in several parts of its extensive range, but no pollinating insects have ever been observed. Instead, there are reports that wind and rain can dislodge the pollinia, which brings them into contact with the stigma, effecting pollination.

The flower of the Fen Orchid has narrow, pale olive green sepals and even narrower (almost filamentous) petals. The green lip is much broader and sharply recurved in the middle, with a deep central groove. The column arches forward over the lip and has a yellow cap.

Actual size

SUBFAMILY	Epidendroideae
TRIBE AND SUBTRIBE	Malaxideae, Malaxidinae
NATIVE RANGE	Philippines
HABITAT	Evergreen forests, at 3,300–3,950 ft (1,000–1,200 m)
TYPE AND PLACEMENT	Epiphytic
CONSERVATION STATUS	Not threatened
FLOWERING TIME	Variable, at any time of the year

FLOWER SIZE
⅛ in (0.3 cm)

PLANT SIZE
3–5 x 3–5 in
(8–13 x 8–13 cm),
excluding pendent
inflorescence
6–10 in (15–25 cm) long

OBERONIA SETIGERA
BRISTLY FAIRY KING
AMES, 1912

509

The Bristly Fairy King is a small fan-shaped plant with its leaves flattened in one plane. In most cases, it grows pendently from its host tree, usually on larger branches or trunks. The inflorescence has hundreds of tiny flowers arranged in rings around its axis, each bloom with a dorsal sepal that has an extraordinarily long bristle at its apex, making the inflorescence look like a tiny foxtail. The genus name is derived from Oberon, the fairy king, who was always carefully hidden from view, a reference to the small flowers of the member species.

Pollination has not been studied, and it is anyone's guess what might be pollinating these tiny flowers. There is no obvious fragrance or nectar reward, and the role of the bristles is unknown. There are no other orchid species with similarly long bristles.

The flower of the Bristly Fairy King is rusty red. The dorsal sepal bears a long bristle that becomes nearly white at its tip, and the anther cap is also white. The lip has several pointed lobes.

Actual size

SUBFAMILY	Epidendroideae
TRIBE AND SUBTRIBE	Malaxideae, Malaxidinae
NATIVE RANGE	China (Hainan Island) and Thailand through Malaysia, Indonesia to New Guinea
HABITAT	Evergreen forests, at 3,300–3,950 ft (1,000–1,200 m)
TYPE AND PLACEMENT	Epiphytic
CONSERVATION STATUS	Not threatened
FLOWERING TIME	May to June (late spring to early summer)

FLOWER SIZE
¾ in (1.9 cm)

PLANT SIZE
7–10 x 4–6 in
(18–25 x 10–15 cm),
excluding terminal
inflorescence
6–10 in (15–25 cm) long

510

STICHORKIS LATIFOLIA
SHEEP'S EAR GARLIC
LINDLEY, 1830

The Sheep's Ear Garlic has an oblong pseudobulb, subtended by one broad leaf with another at its apex. The leaves are shaped like a sheep's ears—hence the common name—and the pseudobulb somewhat resembles a clove of garlic. The inflorescence emerges from the apex of the pseudobulb and can bear 30 to 40 flowers. The genus name comes from the Greek words *stikhos*, "line," and *orkis*, "orchid," referring to the flowers appearing to be arranged in a line.

The lip surface is shiny, implying the presence of nectar, although there is none, while the color of the flowers suggests pollination by some type of fly. Pollination rate has been studied in related species, which experienced low rates of visitation, due to the lack of a pollinator reward. The same could be expected for *Stichorkis latifolia*.

The flower of the Sheep's Ear Garlic has pale orange to tan sepals and petals, the latter narrow and threadlike. The lip is darker and redder, with a glossy surface and a small, shiny bead at the base of the column, which is white to cream with a darker apex.

Actual size

SUBFAMILY	Epidendroideae
TRIBE	Neottieae
NATIVE RANGE	Tropical and subtropical Asia, from India to New Guinea and Japan
HABITAT	Open broadleaved forests, sometimes in pine forests, at 1,300–5,600 ft (400–1,700 m)
TYPE AND PLACEMENT	Terrestrial, mycoheterotrophic
CONSERVATION STATUS	Not assessed
FLOWERING TIME	June to September (midsummer to fall)

APHYLLORCHIS MONTANA
PAUPER ORCHID
REICHENBACH FILS, 1876

FLOWER SIZE
1 in (2.5 cm)

PLANT SIZE
15–28 in (38–71 cm),
with no leaves at any
point during the year

511

The leafless Pauper Orchid has a short creeping underground stem with spreading roots. It lacks chlorophyll so cannot photosynthesize and survives by parasitizing fungi from which it takes nutrients and sugars. The plant emerges from the forest floor only when flowering, and it then produces a loose raceme clothed with reflexed floral bracts subtending flowers that open widely for a short time, after which they become nodding and remain semi-closed until the production of a capsule has begun.

The genus name derives from the Greek *a*-, "without," *phyllon*, "leaf," and *orchis*, "orchid." The species name points to the mountainous habitat of this species, while its common name is a reference to the nutrient-poor soils in which it lives. Pollinators of such plants have rarely been studied—insects are limited in numbers in the deep, dark understory of tropical forests.

The flower of the Pauper Orchid has spreading, pale cream lateral sepals and petals. The upper sepal makes a hood over the arched column. The dark yellow lip is boat-shaped, and the sides envelop the base of the column.

Actual size

SUBFAMILY	Epidendroideae
TRIBE	Neottieae
NATIVE RANGE	Europe, North Africa, western Asia to Iran and Turkmenistan
HABITAT	Shady places on crumbly humus-rich clay, loess (fertile silt), or chalk soils, often in conifer or beech woods, shrubs, grassland, sometimes in stony places in sun
TYPE AND PLACEMENT	Terrestrial on alkaline soil
CONSERVATION STATUS	Not assessed, but least concern in Europe and critically endangered in England
FLOWERING TIME	June to July (early summer)

FLOWER SIZE
1 in (2.5 cm)

PLANT SIZE
6–26 × 4–11 in
(15–65 × 10–28 cm),
including inflorescence

512

CEPHALANTHERA RUBRA
RED HELLEBORINE
(LINNAEUS) RICHARD, 1817

The flower of the Red Helleborine has three spreading, pink sepals and two forward-pointing petals, which with the cup-shaped lip form a short tube leading to the pink-tipped column. A cavity is present on the base of the lip, but no nectar has been reported.

The Red Helleborine is most common in eastern Europe, where it enjoys the moderate continental climate. However, if conditions are not suitable, the plant remains below ground and lives off its fungal host, which may result in under-recording in the western and southern parts of its native range.

Climate change may be making these areas even less suitable. The species can occur in dark woodlands, where it seems unlikely to carry out sufficient photosynthesis to sustain itself. A high level of dependence on ectomychorizal fungi is thought to be taking place. Underground, the orchid does not produce a tuber or corm—just a rhizome with a mass of thick roots. It has been suggested that the flowers mimic *Campanula* and are pollinated by solitary bees of the genus *Chelosoma*.

Actual size

SUBFAMILY	Epidendroideae
TRIBE	Neottieae
NATIVE RANGE	Eurasia, east to China, south to North Africa, the Middle East, and the Himalayas; introduced and widespread across North America
HABITAT	Deciduous and coniferous woodlands, ravines, thickets, meadows, lawns, parking lots, roadside verges, stream banks, dunes, and dikes, from sea level to 8,200 ft (2,500 m)
TYPE AND PLACEMENT	Terrestrial
CONSERVATION STATUS	Not threatened
FLOWERING TIME	July (summer)

FLOWER SIZE
¾ in (2 cm)

PLANT SIZE
10–18 × 6–10 in
(25–46 × 15–25 cm),
including inflorescence

EPIPACTIS HELLEBORINE
BROAD-LEAVED HELLEBORINE
(LINNAEUS) CRANTZ, 1769

513

Actual size

The Broad-leaved Helleborine has an upright stem, slightly nodding at the top and carrying up to ten broadly ovate, pleated leaves spirally around the stem. There are as many as 50 flowers, usually all facing one side. The genus name comes from *epipaktis*, a word used by the ancient Greeks for a plant that curdled milk, possibly this species or a hellebore—the source of the species and common names. Hellebores are plants in the genus *Helleborus*, derived from the Greek *elein*, "to injure," and *bora*, "food," referring to its poisonous properties, even though it was used orally for treating sneezing.

After multiple unintended introductions to the eastern United States, possibly as a cure for gout, the species invaded the whole of North America within a century. The plant is pollinated by wasps and bees that collect nectar from its lip.

The flower of the Broad-leaved Helleborine has cupped, green to cream tinged sepals and pink petals. The cream to pink lip has a darkly colored nectar cavity with a recurved apex, and the column is broadly winged with a cream to white cap.

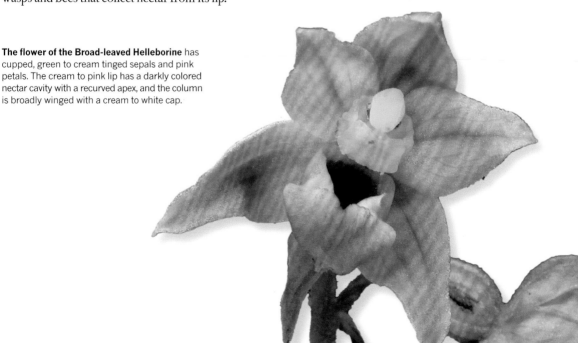

SUBFAMILY	Epidendroideae
TRIBE	Neottieae
NATIVE RANGE	Southern and central Europe, North Africa to Middle East
HABITAT	Forest margins on calcareous substrates, from sea level to 7,545 ft (2,300 m)
TYPE AND PLACEMENT	Terrestrial
CONSERVATION STATUS	Not threatened
FLOWERING TIME	April to June (spring to early summer)

FLOWER SIZE
1½ in (3.8 cm)

PLANT SIZE
No leaves, inflorescence
15–40 in (38–102 cm) tall

514

LIMODORUM ABORTIVUM

VIOLET LIMODORE

(LINNAEUS) SWARTZ, 1799

One of the larger temperate orchids, with up to 20 flowers per darkly pigmented stem, the widespread Violet Limodore is a mycoheterotroph—a leafless plant with no chlorophyll, rendering it dependent on mycorrhizal fungi for its food and minerals. The fungi to which this species is connected in turn obtain their sugar from surrounding forest trees in exchange for minerals that they can more efficiently extract from the soil. The genus name is most likely derived from *haemodorum*, a name used by the ancient Greek botanist Theophrastus for a red-flowered parasitic plant—not this species of orchid. The species name refers to the fact that in many parts of its range, the flowers never fully open and instead self-pollinate.

In some areas, the flowers do open fully and produce nectar. When this happens, nectar-seeking bees in the genera *Anthidium*, *Anthophora*, *Bombus*, and *Lasioglossum* are the pollinators.

The flower of the Violet Limodore is cream to white, overlaid with violet and usually with bright violet nectar guides on its lip, which has a nectar spur projecting at the back. Sepals and petals are lanceolate and incurving, the former wider than the latter.

Actual size

SUBFAMILY	Epidendroideae
TRIBE	Neottieae
NATIVE RANGE	Circumarctic, boreal, and temperate Eurasia and North America
HABITAT	Most frequent in moist pine forests and wet moorland, but also on beaches and dunes in northern areas; in southern regions strictly in mountains
TYPE AND PLACEMENT	Terrestrial in moss or humus
CONSERVATION STATUS	Least concern, probably under-recorded at many sites as plant is easily overlooked
FLOWERING TIME	May to August (late spring to summer)

FLOWER SIZE
¼ in (0.6 cm)

PLANT SIZE
2–6 × 1–2 in
(5–15 × 2.5–5 cm),
including inflorescence
that tops the stem

NEOTTIA CORDATA
LESSER TWAYBLADE
(LINNAEUS) RICHARD, 1817

515

Actual size

This tiny terrestrial orchid makes a short-creeping rhizome and a flowering stalk with two nearly opposite leaves, hence the "twayblade" of the common name, from the Middle-English *twain*, meaning "two," and possibly derived from Old Dutch *twee-blad*. At the base, there are a few sheaths, and the stem is topped with an upright raceme carrying between 3 and 15 flowers.

Although highly localized in its general occurrence, the plant can form large colonies in favored sites, where its bronze coloration and diminutive stature make it difficult to spot— and thus often overlooked in vegetation surveys. The species was formerly placed in the genus *Listera*, but molecular and morphological studies have shown a close relationship with the Bird's Nest Orchid (*Neottia nidus-avis*), and hence it is now placed in the same genus. The small, simple-looking flowers are pollinated by fungus gnats.

The flower of the Lesser Twayblade has an enlarged ovary. The small, spreading, greenish-brown sepals and petals are inconspicuous and often tinged purple. The lip is four-lobed with the lower two lobes elongate.

SUBFAMILY	Epidendroideae
TRIBE	Neottieae
NATIVE RANGE	Europe, Anatolia, and Caucasus to Iran, and northwestern Africa
HABITAT	Temperate shady woodland, particularly under beech, hazel, pine, oak, or rarely birch
TYPE AND PLACEMENT	Terrestrial in deep leaf litter on alkaline soil
CONSERVATION STATUS	Least concern, but locally in decline as species is an indicator of old woodland, which is rapidly declining throughout its area of occurrence
FLOWERING TIME	May to July (spring to early summer)

FLOWER SIZE
⅜ in (1 cm)

PLANT SIZE
8–15 × 3–4 in
(20–38 × 8–10 cm),
inflorescence only
(species has no leaves)

NEOTTIA NIDUS-AVIS
BIRD'S NEST ORCHID
(LINNAEUS) RICHARD, 1817

516

The thick coral-like roots of the leafless Bird's Nest Orchid form a cluster that resembles a nest—hence the common name. It lacks green leaves and has a parasitic relationship with soil fungi that obtain their carbohydrates from a mutualistic relationship with trees in the forest where the plant grows. The genus and species names refer to a "nest" and a "bird's nest," from the Greek *neottia* and the Latin *nidus-avis*.

The inflorescence, a yellowish-brown spike with up to five large scale-like clasping pale yellow bracts, appears only in years when conditions are favorable. The entire plant is easily overlooked because it has the same color as the leaf litter in the dark forests where it grows. Flies are the pollinators, but when flowers are not visited, the pollinia disintegrate and fall on the stigma, effecting self-pollination.

Actual size

The flower of the Bird's Nest Orchid has brownish-green sepals and petals that form a cup around the central column. The lip is longer, exceeds the cup, and has two outwardly curving lobes.

SUBFAMILY	Epidendroideae
TRIBE AND SUBTRIBE	Nervilieae, Nerviliinae
NATIVE RANGE	Tropical and subtropical Asia to northern Australia
HABITAT	Shaded wooded valleys and evergreen forests, at 650–5,250 ft (200–1,600 m)
TYPE AND PLACEMENT	Terrestrial in damp places
CONSERVATION STATUS	Least concern in Australia; widespread across Asia and unlikely to be threatened
FLOWERING TIME	May to June (late spring)

NERVILIA PLICATA
HAIRY TARO ORCHID
(ANDREWS) SCHLECHTER, 1911

FLOWER SIZE
1 in (2.5 cm)

PLANT SIZE
4–6 × 6–10 in
(10–15 × 15–25 cm),
excluding inflorescence
6–10 in (15–25 cm) tall

517

From a globose, hairy, whitish tuber up to ¾ in (2 cm) across, a heart-shaped, hairy, purplish or green, purple-spotted leaf emerges, typically held close to the ground. In the beginning of the wet season, after the leaf has withered away, an inflorescence just taller than the old leaf emerges with one or two flowers, which last only for a few days. The genus name comes from the Latin, *nervus*, meaning "nerve" or "vein," in reference to the prominent venation of the beautiful leaf. The folded-leaf aspect is reflected in the species name, from the Latin *plicatus*, for "folded."

Actual size

The nectarless flowers deceive solitary bees and wasps for pollination. Flowers are variable in shape and color throughout the distribution, and there may be more than one species lumped under this name. The tubers are used in parts of Asia to treat type 2 diabetes.

The flower of the Hairy Taro Orchid has spreading, lanceolate, greenish-brown sepals and petals, and a tubular, cream lip that encloses the broad, white column. In most plants there are some dark pink veins in the lip, but in others there is a yellow crest in the middle of the lip.

SUBFAMILY	Epidendroideae
TRIBE AND SUBTRIBE	Nervilieae, Epipogiinae
NATIVE RANGE	Temperate Eurasia, from Britain to Japan, south to the Caucasus and Himalayas
HABITAT	Forests, crevices, and mossy places, at 650–11,800 ft (200–3,600 m)
TYPE AND PLACEMENT	Terrestrial
CONSERVATION STATUS	Least concern, due to broad distribution, but vulnerable to critically endangered at local levels
FLOWERING TIME	May to July (late spring and summer)

FLOWER SIZE
¾ in (2 cm)

PLANT SIZE
5–10 in (13–25 cm),
only an inflorescence

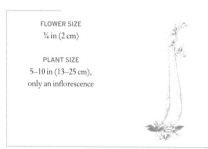

EPIPOGIUM APHYLLUM
GHOST ORCHID

SWARTZ, 1814

518

The little Ghost Orchid never produces leaves and is totally non-photosynthetic, obtaining sugars and minerals from fungi of the genus *Inocybe*, which are in turn associated with the roots of trees. Flowers grow on a white stem that has a swollen base, attached to a root mass shaped like a coral. The plants emerge above ground only to flower, especially during extremely wet summers, and populations are difficult to monitor because they may not break the surface every year.

The blooms produce a sweet vanilla or banana scent, and nectar is produced in the upward-pointing spur. Bumblebees, of the genus *Bombus*, would seem an appropriate pollinator for a flower with this scent, color, and spur, and several reports of such *Bombus* visits exist, although without pollinia removal. Bees are not frequent in such shady habitats, and production of seed capsules is rare.

Actual size

The flower of the Ghost Orchid has its lip uppermost, with narrowly lanceolate, cream sepals and petals spreading downward. The lip is white with rose-pink spots in several rows and bluntly spurred, and it forms a hood over the broad, cream column.

SUBFAMILY	Epidendroideae
TRIBE	Podochileae
NATIVE RANGE	Indonesia, from the islands of Sumatra and Borneo to Sulawesi
HABITAT	Lower montane and hill forests, usually in damp areas, at 1,970–5,250 ft (600–1,600 m)
TYPE AND PLACEMENT	Epiphytic on mossy trees, sometimes terrestrial or on rocks
CONSERVATION STATUS	Not formally assessed
FLOWERING TIME	Spring

FLOWER SIZE
¼ in (0.6 cm)

PLANT SIZE
20–45 × 5–8 in
(51–114 × 13–20 cm),
including inflorescence
produced near the
tips of the stems

APPENDICULA CRISTATA
CRESTED FERN-ORCHID
BLUME, 1825

519

Actual size

The Crested Fern-orchid can sometimes produce very long, pendent stems with flattened branches that are clad in short, oval leaves, all arranged in one plane with the flat surfaces of the leaves facing the same way. The plants can be confused with epiphytic *Huperzia* lycopods or *Grammitis* ferns, but unlike those plants these are flowering orchids. They produce terminal, often branched inflorescences up to 6 in (15 cm) long, which flower successively with small but colorful flowers.

The genus name comes from the Latin for "little appendix," which in this case is the callus on the lip. Pollination has not been observed, but many species in this genus are known to auto-pollinate if no insect visits, and this may be true of the Crested Fern-orchid as well.

The flower of the Crested Fern-orchid is yellow and pink-tinged and has a saccate spur, from which the fused sepals and petals form a funnel, the top sepal enveloping the winged column. The lip is bilobed and has an appendage.

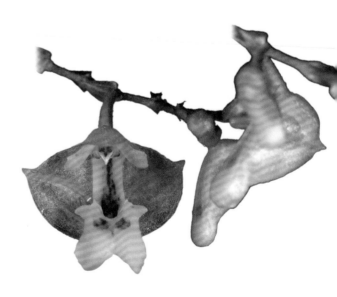

SUBFAMILY	Epidendroideae
TRIBE	Podochileae
NATIVE RANGE	Sumatra and Borneo
HABITAT	Mountain forests, at 2,300–6,600 ft (700–2,000 m)
TYPE AND PLACEMENT	Epiphytic close to the ground on tree trunks in deep shade
CONSERVATION STATUS	Not assessed
FLOWERING TIME	November to February (winter)

FLOWER SIZE
¼ in (0.8 cm)

PLANT SIZE
15–30 × 10–15 in
(38–76 × 25–38 cm),
excluding erect to arching
inflorescence, which is
shorter than the leaves,
8–12 in (20–30 cm) long

ASCIDIERIA CYMBIDIFOLIA
SNAIL ORCHID
(RIDLEY) W. SUAREZ & COOTES, 2009

520

From the Snail Orchid's small, stemlike, sheathed pseudobulbs, between two and five linear, folded leaves emerge in two ranks. Its arching to pendent densely flowered inflorescence holds many small flowers that have variable orientation—some with the lip lowermost.

The genus name is derived from the Greek word *askos*, "bladder," in reference to the inflated lip, and the genus *Eria*, a part of which is related to *Ascidieria*. The species name refers to the fact that, out of flower, it can be mistaken for a member of the orchid genus *Cymbidium*. Little is known about pollination of *A. cymbidifolia*, but the floral morphology would appear to indicate that some type of bee is the pollinator. The common name is based on the suggestion that snails might pollinate the plant, given that in many cases the inflorescence rests on the ground.

Actual size

The flower of the Snail Orchid is white to cream, with two enlarged lateral sepals that envelop the saccate lip. The upper sepal and petals are smaller, spreading, and similar in size. The lip has two appendices, and the pollinia sit on either side of the column like two eyes.

SUBFAMILY	Epidendroideae
TRIBE	Podochileae
NATIVE RANGE	Philippines
HABITAT	Lowland tropical wet forests
TYPE AND PLACEMENT	Epiphytic
CONSERVATION STATUS	Not threatened
FLOWERING TIME	Throughout the year

CERATOSTYLIS RETISQUAMA
LADY ORCHID
REICHENBACH FILS, 1857

FLOWER SIZE
¾–1¼ in (2–3 cm)

PLANT SIZE
6–10 × 3–4 in
(15–25 × 8–10 cm),
including inflorescence
1 in (2.5 cm) long

521

A large-flowered colorful species in a genus in which most species bear smaller white flowers, the pendulous Lady Orchid can eventually grow into a large plant. The terete and fleshy leaves point downward and have basal papery brown bracts—a feature to which the species name refers, from the Latin *reti*, "net," and *squama*, "scale-like." The pointed apex of the column is reflected in the genus name, which derives from the Greek words *cerato*, "horn," and *stylis*, "pillar."

Flowers are short-lived but produced repeatedly throughout the year. The plant is better known as *Ceratostylis rubra*, although the correct name predates the popular one by more than 50 years. Unlike the many red-flowered orchids that are bird-pollinated, *C. retisquama* is more likely to be bee-pollinated due to its open, spurless shape.

Actual size

The flower of the Lady Orchid is non-resupinate, with nearly equal-sized sepals and petals. They are coral red with a glistening crystalline texture. The lip is small and triangular.

SUBFAMILY	Epidendroideae
TRIBE	Podochileae
NATIVE RANGE	Eastern Himalayas, Indochina to southern China
HABITAT	Broad-leaved forests, at 5,900–7,545 ft (1,800 to 2,300 m)
TYPE AND PLACEMENT	Epiphytic on mossy branches of trees, rarely on mossy rocks
CONSERVATION STATUS	Not formally assessed
FLOWERING TIME	May to July (late spring to early summer)

FLOWER SIZE
1 in (2.5 cm)

PLANT SIZE
6–12 × 3–6 in
(16–30 × 8–15 cm),
excluding inflorescence
10–20 in (25–51 cm) long

CRYPTOCHILUS SANGUINEUS
BLOOD-RED BELL ORCHID
WALLICH, 1824

The Blood-red Bell Orchid has one globose pseudobulb per growth, enclosed in sheaths, each carrying one to three thick, upwardly projecting leaves. The inflorescence, produced from the top of the pseudobulb, is upright or arching sideways, with a stiff bract under each of the 10–20 flowers. The bell-shaped sepals cover the lip—hence the genus name (Greek, *kryptos*, "hidden," and *cheilos*, "lip")—and the flowers are held to one or two sides of the stem.

The flowers probably mimic nectar-providing members of Ericaceae (the blueberry family), and due to the shape and size of the Blood-red Bell Orchid flowers, bees that visit the species of Ericaceae are fooled and visit the orchid flowers too. Because of the red color, bird-pollination has been suggested, but observations are lacking. The blooms appear too small for Asian nectar-seeking birds to visit them. Because bees can see red colors, they seem the more likely pollinators.

The flower of the Blood-red Bell Orchid has red sepals with darker margins enclosing the flower, forming a short (empty) nectar cavity at the base. All other flower parts are deeply hidden inside.

Actual size

SUBFAMILY	Epidendroideae
TRIBE	Podochileae
NATIVE RANGE	Peninsular Malaysia
HABITAT	Forest edges, ridges, along paths, at 5,600–6,600 ft (1,700–2,000 m)
TYPE AND PLACEMENT	Epiphytic or terrestrial
CONSERVATION STATUS	Not assessed
FLOWERING TIME	March to May (spring)

FLOWER SIZE
⅜ in (1 cm)

PLANT SIZE
80–130 × 6–10 in
(203–330 × 15–25 cm),
including terminal
branching inflorescence
6–10 in (15–25 cm) long

523

DILOCHIOPSIS SCORTECHINII
FATHER BENEDETTO'S ORCHID
(HOOKER FILS) BRIEGER, 1981

With a stem and leaves more like those of a bamboo, the long-trailing Father Benedetto's Orchid only looks like an orchid when it flowers. It was named for its similarity (using the Greek ending -*opsis*, "resembling") to the distantly related genus *Dilochia*, to which the species is indeed remarkably similar. *Dilochia* itself comes from the Greek *di*, "two," and *lochos*, "file" (as in soldiers), referring to the two-rowed leaf arrangement, also a characteristic of *D. scortechinii*. The species was named in honor of Father Benedetto Scortechini (1845–86), an Italian priest who helped early settlers in Queensland, Australia, and studied the plants of that region.

Pollination has not been studied, but its color, shape, and nectar cavity would indicate some type of bee as pollinator. A nectar cavity, with or without nectar, is an attractant for bees.

Actual size

The flower of Father Benedetto's Orchid
has oval, cream sepals and petals, often with a pink suffusion. They project forward, surrounding the lip and column, the former with pink markings and trilobed. The sidelobes are upright, and the midlobe is bilobed with a central ridge of long, pinkish-purple hairs.

SUBFAMILY	Epidendroideae
TRIBE	Podochileae
NATIVE RANGE	New Guinea
HABITAT	Mid-elevation tropical wet forests in deep shade
TYPE AND PLACEMENT	Epiphytic, sometimes a creeping terrestrial in moss
CONSERVATION STATUS	Not threatened
FLOWERING TIME	Throughout the year, but mostly fall, winter, and spring

FLOWER SIZE
⅜–¾ in (1–2 cm)

PLANT SIZE
1–1½ × 1 in
(2.5–3.8 × 2.5 cm),
including short
inflorescence

524

MEDIOCALCAR DECORATUM
CANDY-CORN ORCHID
SCHUITEMAN, 1989

A miniature species with brilliant flowers similar in size and coloration to candy corn (hence its common name), the succulent creeping Candy-corn Orchid has short cylindrical pseudobulbs, each topped with three to four succulent leaves arranged like a helicopter propeller at the apex. The highly ornamental flowers appear, one or two at a time, on the newest growths and look like little inflated ornaments on a creeping plant. The plants root easily at almost every growth, and large specimens can have hundreds of growths festooned with flowers.

The genus name is derived form the Greek words *medius*, "middle," and *calcar*, "spur," referring to the shape of the middle petal, the lip, which forms a short, saclike nectar spur. The flowers of this orchid are thought to mimic those of the blueberry genus, *Vaccinium*, which are common in the same habitats as *Mediocalcar* species.

The flower of the Candy-corn Orchid is a small, inflated, reddish-orange bell. It consists of sepals and petals that form an opening and have contrastingly brilliant yellow tips. The lip is not noticeably different in size or shape from the other petals.

Actual size

SUBFAMILY	Epidendroideae
TRIBE	Podochileae
NATIVE RANGE	Indochina, Peninsular Malaysia, western Malesia east to Bali
HABITAT	Lowland and montane mossy forests, at 650–7,875 ft (200–2,400 m)
TYPE AND PLACEMENT	Epiphytic or terrestrial on mossy banks
CONSERVATION STATUS	Not assessed
FLOWERING TIME	Spring

FLOWER SIZE
¼ in (0.6 cm)

PLANT SIZE
10–15 × 4–10 in
(25–38 × 10–25 cm),
excluding erect
terminal inflorescences
6–10 in (15–25 cm) long

MYCARANTHES OBLITTERATA
FADED BAT ORCHID
BLUME, 1825

525

The elongate, arching stem of the Faded Bat Orchid carries many lanceolate, leathery leaves in two ranks, with erose leaf tips and clasping leaf bases. One to three unbranched racemes appear from the stem apex, densely set with 40–50 small flowers. The genus name is derived from Greek *mykaris*, "bat," and *anthos*, "flower," referring to the lip, which resembles a small bat. The species name refers to the pale flower color of most plants, although some have more brightly colored flowers.

Pollination has not been studied in any of the 25 species of genus *Mycaranthes*. There is no nectar or other obvious reward offered, so it has been speculated that some sort of fly pollinates the plants, although the reason why flies would be attracted to these flowers is also unclear.

Actual size

The flower of the Faded Bat Orchid has recurved, greenish-yellow lateral sepals and petals. The upper sepal makes a hood over the column. The cream lip is red-spotted and long-stalked, hinged, and three-lobed, the middle lobe thickened, woolly, and callused, with the lateral lobes outstretched.

SUBFAMILY	Epidendroideae
TRIBE	Podochileae
NATIVE RANGE	Southwestern Pacific, from New Guinea and Queensland (Australia) to Samoa and the Marianas
HABITAT	Coastal rain forests and wet forests, from sea level to 1,300 ft (400 m)
TYPE AND PLACEMENT	Epiphytic in humid airy positions on trees, often on branches overhanging streams, and less often on mossy rocks
CONSERVATION STATUS	Not assessed
FLOWERING TIME	October to February (spring to summer), sometimes at other times of the year

FLOWER SIZE
⅛ in (0.25 cm)

PLANT SIZE
8–14 × 6–10 in
(20–36 × 15–25 cm),
excluding erect-nodding
inflorescence
9–18 in (23–46 cm) long

PHREATIA MICRANTHA
PACIFIC FAN ORCHID
(A. RICHARD) LINDLEY, 1859

The short stem of the Pacific Fan Orchid is covered by up to ten linear leaves, unequally bilobed at their tip and arranged in a fan. An axillary cylindrical inflorescence holds many densely set, tiny flowers that last only two or three days. The genus name is derived from the Greek word *phrear*, a "cistern" or "well," alluding to the small cavity formed by the flower parts.

Some have speculated that raindrops pollinate *Phreatia micrantha* by dislodging the pollinia. Although this could be possible, a falling raindrop transferring pollinia from one flower into the stigmatic cavity of another seems highly implausible. A small insect is most likely involved; the nectar produced in the bilobed cavity on the lip base is also consistent with insect pollination.

Actual size

The flower of the Pacific Fan Orchid has small, translucent white sepals and petals that form a bell-shaped flower. The upper sepal is hooded, the lip is recurved, and the column is shortly winged. On the base of the lip is a nectar cavity.

SUBFAMILY	Epidendroideae
TRIBE	Podochileae
NATIVE RANGE	Southern Yunnan (China) and the Himalayas to Southeast Asia
HABITAT	Wooded slopes, rock outcrops, and valley forests, at 2,625–9,200 ft (800–2,800 m)
TYPE AND PLACEMENT	Epiphytic on trees, or epilithic
CONSERVATION STATUS	Not assessed
FLOWERING TIME	May to August (spring to summer)

FLOWER SIZE
¼ in (0.8 cm)

PLANT SIZE
6–13 × 7–12 in
(15–33 × 18–30 cm),
including arching
inflorescence
4–7 in (10–18 cm) long

527

PINALIA SPICATA
LILY OF THE VALLEY ORCHID
(D. DON) S. C .CHEN & J. J. WOOD, 2009

The Lily of the Valley Orchid has a cluster of conical pseudobulbs, basally enveloped by one or two dry sheaths and topped by up to four elliptic-oblong leaves with a short stalk and an unevenly bilobed tip. The 5–20 hairy flowers appear laxly on a slender, bract-covered inflorescence that emerges near the top of the pseudobulb. The genus name is in honor of the French botanist, Chevalier Pinal, and the species name alludes to the spikelike shape of the short-stemmed, densely flowered inflorescence.

Flowers are cupped and have a strong scent, reminiscent of Lily of the Valley (*Convallaria majalis*)—hence the common name and its former species name, *Eria convallarioides*. Pollination is likely to be by some type of bee, given the sweet scent and flower shape, although no nectar is present.

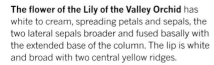

Actual size

The flower of the Lily of the Valley Orchid has white to cream, spreading petals and sepals, the two lateral sepals broader and fused basally with the extended base of the column. The lip is white and broad with two central yellow ridges.

SUBFAMILY	Epidendroideae
TRIBE AND SUBTRIBE	Podochileae
NATIVE RANGE	Southeast Asia to Borneo and Java (Indonesia)
HABITAT	Seasonally wet forests near lakes and rivers, at 985–5,250 ft (300–1,600 m)
TYPE AND PLACEMENT	Epiphytic or lithophytic
CONSERVATION STATUS	Not assessed
FLOWERING TIME	July to August (summer)

FLOWER SIZE
⅝ in (1.5 cm)

PLANT SIZE
8–15 × 5–6 in
(20–38 × 13–15 cm),
including arching-pendent
inflorescence
3–6 in (8–15 cm) long

PINALIA XANTHOCHEILA
RED-CHEEK ORCHID
(RIDLEY) W. SUAREZ & COOTES, 2009

The Red-cheek Orchid produces elongate pseudobulbs that have between two and five leaves near the tip, each oblong-lanceolate with a clasping base. Two to three inflorescences are produced from just below the leaves, under which they are often partly concealed, carrying up to 30 short-lived but colorful flowers. The genus was named for the French botanist Chevalier Pinal, who collected orchids in the nineteenth century. The species name refers to the mostly yellow lip of the plant, from the Greek *anthos*, "yellow," and *cheilos*, "lip." The common name is a reference to the red sidelobes of the lip.

Nothing is known about pollination in this species. However, the open, colorful, sweetly fragrant flowers with a shallow nectar cavity at the base of the lip suggest that a bee would be the most likely pollinator.

The flower of the Red-cheek Orchid has pale to bright yellow, spreading sepals and petals that project slightly forward, sometimes both with red stripes. The lip is three-lobed, with the sidelobe flanking the column and usually red-suffused, although pure yellow lips have also been recorded.

Actual size

SUBFAMILY	Epidendroideae
TRIBE	Podochileae
NATIVE RANGE	Peninsular Malaysia, Sumatra, Java
HABITAT	Cloud forests, at 2,950–4,920 ft (900–1,500 m)
TYPE AND PLACEMENT	Epiphytic on trees or epilithic on rocks
CONSERVATION STATUS	Not assessed
FLOWERING TIME	Throughout the year

FLOWER SIZE
¼ in (0.7 cm)

PLANT SIZE
4–8 × 1–2 in
(10–20 × 2.5–5 cm),
excluding short inflorescence,
which is terminal on the stem,
½–1½ in (1.3–4 cm) long

PODOCHILUS MURICATUS
MOSS ORCHID
(TEIJSMANN & BINNENDIJK) SCHLECHTER, 1900

529

The stems of the Moss Orchid are completely enveloped by sheathing leaves, which makes it look like a member of the genus *Peperomia* and decidedly un-orchidlike. The species is small and, when not flowering, may be mistaken for a tuft of moss or clubmoss—hence the common name. The leaves are fleshy, oval, and twist at the base, and the leaf tips are often weakly bilobed. A terminal, rarely axillary, short inflorescence with between one and three crowded flowers is subtended by spreading leaflike, concave floral bracts.

Little is known about the sexual life of this plant, but it is not self-pollinating. It has a nectar spur formed by the lip base and lateral sepals, and it seems likely that it is visited by small bees.

Actual size

The flower of the Moss Orchid is white and has sepals that form a cup covered with long hairs on the outside. The petals are narrowly pointed. The lip is broad and has a few purple blotches.

SUBFAMILY	Epidendroideae
TRIBE	Podochileae
NATIVE RANGE	Mainland tropical Asia, from Nepal to Peninsular Malaysia
HABITAT	Exposed vertical cliffs and boulders
TYPE AND PLACEMENT	Lithophytic or sometimes epiphytic
CONSERVATION STATUS	Not assessed
FLOWERING TIME	Early spring

FLOWER SIZE
½ in (1.2 cm)

PLANT SIZE
½–1 × 2–3 in
(1.2–2.5 × 5–7.5 cm),
when in leaf, excluding
terminal inflorescence
½–1 in (1.2–2.5 cm) long

530

PORPAX ELWESII
PERISCOPE ORCHID
(REICHENBACH FILS) ROLFE, 1908

Actual size

The little Periscope Orchid forms a cluster of flattened, disc-like pseudobulbs that are basally enveloped in a fibrous membrane. It flowers from the top of the disc-like naked pseudobulb with a single, down-curving bloom, and there are two elliptical, grooved, shortly petiolate leaves produced after flowering. The genus name, from the Greek *porpax*, meaning the "handle" of a shield, probably refers to the shape of the pseudobulbs, with the two leaves looking like the straps of a shield. The common name refers to the tubular flower, which looks the viewing cap of a periscope.

The dull red color of the flowers and their shape suggest that the plants are pollinated by small flies. The pollinator enters the tube, is shoved against the column by the small motile lip, and the pollinia are deposited on its back.

The flower of the Periscope Orchid has dark red sepals, fused into a saccate tube, which is surrounded basally by a large papery bract. The lip is hinged and motile, and this and the petals are red and remain hidden inside the sepal tube.

SUBFAMILY	Epidendroideae
TRIBE	Podochileae
NATIVE RANGE	Tropical Africa, from Liberia to Kenya and south to Mozambique
HABITAT	Cool forests, on mossy branches and rocks, at 5,900–7,545 ft (1,800–2,300 m)
TYPE AND PLACEMENT	Epiphytic or lithophytic
CONSERVATION STATUS	Not assessed
FLOWERING TIME	December to January (summer)

FLOWER SIZE
⅜ in (1 cm)

PLANT SIZE
1 × ½ in
(2.5 × 1.3 cm),
including erect
single-flowered
inflorescence

531

STOLZIA REPENS
AFRICAN MOSS ORCHID
(ROLFE) SUMMERHAYS, 1953

Actual size

The miniature African Moss Orchid has apically swollen, elongate pseudobulbs carrying two to three oval leaves, with each new pseudobulb produced from near the apex of the previous one. A single, relatively large flower arises from a sheathing bract that covers the base of each pseudobulb. The genus was named after a German missionary, Adolf Stolz (1871–1917), who collected in what is now Tanzania at the northern end of Lake Malawi. The common name refers to the fact that when it is not in flower, the plant looks like a moss.

Pollination involves an insect, most likely a small fly (given the unshowy colors of the flowers) that crawls onto the lip, which is attached to an extension of the column, providing it with a hinged base. After the insect passes the balance point, it is thrown into the column, removing the pollinia.

The flower of the African Moss Orchid has oblanceolate, green to reddish sepals and petals that produce a star shape. The much smaller but similarly colored lip and a short column sit in the middle, the former attached to a long extension of the latter (termed a column foot).

SUBFAMILY	Epidendroideae
TRIBE	Podochileae
NATIVE RANGE	Tropical and subtropical Asia, from the Himalayas to Taiwan, and India to the Solomon Islands
HABITAT	Old montane forests, at 650–6,600 ft (200–2,000 m)
TYPE AND PLACEMENT	Epiphytic on tree trunks and branches or lithophytic on rocks, in forests or along valleys
CONSERVATION STATUS	Not assessed
FLOWERING TIME	April to October (spring to fall)

FLOWER SIZE
¹⁄₁₆ in (0.17 cm)

PLANT SIZE
¾–1 × ½ in
(2–2.5 × 1.3 cm),
excluding erect
inflorescence
1½–2 in (3.8–5 cm) tall

532

THELASIS PYGMAEA
DWARF NIPPLE ORCHID
(GRIFFITH) LINDLEY, 1858

The tiny Dwarf Nipple Orchid has a freely branching rhizome covered with tufts of round, flattened, fleshy pseudobulbs that are partially enveloped by a small sheath. Each pseudobulb is topped by one or two narrowly lanceolate leaves that are unequally, obtusely bilobed at the tip. An inflorescence produces a slow progression of flowers, each subtended by two small sheaths spread out along the length of the conical inflorescence. There are only three or four flowers open at a time, but 30–40 eventually form on each stem.

The genus name is Greek for "with nipples," probably a reference to the shape of the leafless pseudobulb, reflected also in the common name. Pollination of the tiny flowers is highly speculative. They do not, however, self-pollinate because production of seeds is not regular.

Actual size

The flower of the Dwarf Nipple Orchid is greenish-yellow, tiny, and bell-shaped, with fleshy sepals and thinner petals and lip. The lip is a similar size and shape to the petals, and lanceolate with a sharp tip. The column is enclosed within the other flower parts.

SUBFAMILY	Epidendroideae
TRIBE	Podochileae
NATIVE RANGE	Peninsular Malaysia, Borneo, Sumatra, and Java
HABITAT	Mid-elevation wet forests
TYPE AND PLACEMENT	Epiphytic to terrestrial
CONSERVATION STATUS	Not threatened
FLOWERING TIME	Summer

FLOWER SIZE
¾ in (2 cm)

PLANT SIZE
15–30 × 6–9 in
(38–76 × 15–23 cm),
including shortly
pendent, axillary
inflorescences
3–4 in (8–10 cm) long

TRICHOTOSIA FEROX
FURRY BAMBOO ORCHID
BLUME, 1825

533

A covering of long red hair (a rare trait) and an appearance more akin to a bamboo than an orchid make the Furry Bamboo Orchid one of the most striking species. Young plants grow upright, but as they age the stems often become arching to pendent. The long, dense, rusty red hairs covering the leaves almost give the impression of spines, but in fact they are soft. Appropriately, the name for the genus is based on the Greek word for "hairy," *trichotos*. The species name means "fierce" in Latin, which is another reference to the animal-like hairiness of the species.

The flowers appear in clusters of four to eight flowers in the leaf axils. Pollination has not been studied, but the shape of the nectar cavity of the species would be compatible with pollination by some type of small bee.

The flower of the Furry Bamboo Orchid is translucent pale green to yellow with red hairs on the outer surfaces but no hairs on the inner surfaces. The trilobed lip forms a nectar cavity and is a similar color, usually with red spots.

Actual size

SUBFAMILY	Epidendroideae
TRIBE	Sobralieae
NATIVE RANGE	French Guiana, Surinam, Guyana, Venezuela, Ecuador, Brazil, and the Windward Islands
HABITAT	Seasonally wet forests, at 650–4,920 ft (200–1,500 m)
TYPE AND PLACEMENT	Epiphytic
CONSERVATION STATUS	Not threatened
FLOWERING TIME	May to September (late spring through fall)

FLOWER SIZE
⅝ in (1.5 cm)

PLANT SIZE
14–25 × 8–10 in
(36–64 × 20–25 cm),
including terminal
inflorescence

534

ELLEANTHUS CARAVATA
YELLOW CROWN ORCHID
(AUBLET) REICHENBACH FILS, 1881

The epiphytic Yellow Crown Orchid is one of the prettiest of a genus with many eye-catchingly colorful inflorescences. It is more compact than many of its sister species, with a comparatively large and showy terminal inflorescence, bearing small, tubular, yellow blooms subtended by contrasting, reddish-purple bracts. Like all *Elleanthus* species, the plant, which produces leafy stems, has no pseudobulbs. The species name is from the Greek *karabos*, a type of boat, alluding to the shape of the bracts. The genus name is from the Greek *Elle*, Helen (of Troy legend), and *anthos*, "flower."

The flowers and especially their colorful bracts make a long-lasting display that acts as a beacon for the hummingbirds that pollinate this and nearly all species of *Elleanthus*. The strikingly different pairing of colors is found in many bird-pollinated flowers.

The flower of the Yellow Crown Orchid is brilliant yellow with a purple tipped column. The sepals and petals are lanceolate and form a tube around the column together with the lip, which has a frilly margin and is positioned on the upper side of the flower away from the bract.

Actual size

SUBFAMILY	Epidendroideae
TRIBE	Sobralieae
NATIVE RANGE	Western Mexico to Costa Rica
HABITAT	Mostly shaded sites, especially steep slopes (including roadsides) with at least seasonally wet conditions
TYPE AND PLACEMENT	Terrestrial
CONSERVATION STATUS	Widespread and common, but not assessed
FLOWERING TIME	March to October

FLOWER SIZE
8–10 in (20–25 cm)

PLANT SIZE
48–60 × 10–18 in
(122–152 × 25–46 cm),
including inflorescence, which
is apical and short

SOBRALIA MACRANTHA
LARGE PURPLE DAY-ORCHID
LINDLEY, 1838

The species epithet *macrantha*, from the Greek for "large-flowered," refers to the exceptional size of its bloom. *Sobralia* species are tall, bamboo-like plants with short-lived flowers, often lasting only a few hours—an important aspect of their ecology. Tending to flower in flushes triggered by weather events, different species will often bloom en masse but on different days, thus avoiding hybridization. Pollination is by euglossine bees (males and females) and bumblebees (*Bombus*). Production of nectar has not been documented, but the high rates of visitation observed suggest that it must be present.

In recent years, the Large Purple Day-orchid, with its many color forms, has been hybridized with other *Sobralia* species. However, the large size of the plants and the short life of the flowers limit their horticultural appeal.

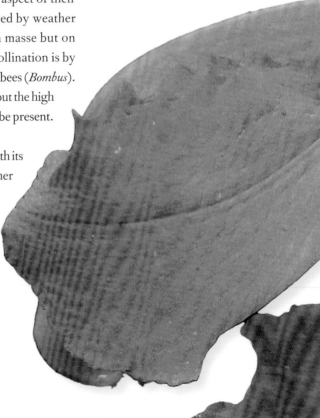

The flower of the Large Purple Day-orchid is typically a rich dark purple with a yellow and white throat. Other color forms are also common, including white with a yellow throat and white with a pink lip and a yellow throat.

Actual size

SUBFAMILY	Epidendroideae
TRIBE	Thaieae
NATIVE RANGE	Laos, northern Thailand, and China (Yunnan)
HABITAT	Crevasses between large rocks in forests, at 4,600–4,920 ft (1,400–1,500 m)
TYPE AND PLACEMENT	Terrestrial in humid shade on limestone rocks
CONSERVATION STATUS	Not assessed
FLOWERING TIME	September to October (fall)

FLOWER SIZE
⅝ in (1.5 cm)

PLANT SIZE
20–30 × 6–10 in
(51–76 × 15–25 cm),
including erect
inflorescence

THAIA SAPROPHYTICA
GREEN THAI ORCHID
SEIDENFADEN, 1975

536

Actual size

The Green Thai Orchid has no close relatives. It was originally collected flowering in a leafless condition, resulting in its species name, from the Greek *sapros*, "rotten," and *phyton*, "plant," which alludes to a plant that lives off decaying matter in the soil. In fact, no plant can do this, and non-green orchids are instead connected to soil fungi from which they obtain food and minerals. More recently, it has been discovered that this species does produce leaves that it may or may not lose. Under wetter conditions, it keeps its leaves throughout the year.

The plant, the only species in its genus—named for the country, Thailand, where it was found—forms a purple inflorescence with greenish, hooded flowers. These are distinguished by their unlobed lip and the projecting appendage at the lower margin of the stigmatic cavity.

The flower of the Green Thai Orchid is green and has triangular lateral sepals that form a small sack at the fused base. The dorsal sepal and smaller linear petals are curved forward over the arching column. The lip is tongue-shaped with two central ridges.

SUBFAMILY	Epidendroideae
TRIBE AND SUBTRIBE	Triphoreae, Triphorinae
NATIVE RANGE	Central America (Nicaragua) to northwestern South America (Ecuador)
HABITAT	Rain forests to cloud forests, at 1,300–5,600 ft (400–1,700 m)
TYPE AND PLACEMENT	Terrestrial
CONSERVATION STATUS	Not assessed
FLOWERING TIME	September to October (fall)

FLOWER SIZE
1 in (2.5 cm)

PLANT SIZE
6–10 × 4–6 in
(15–25 × 10–15 cm),
excluding apical inflorescence
3–5 in (8–13 cm) above the leaf

MONOPHYLLORCHIS MICROSTYLOIDES
STRIPED HEART
(REICHENBACH FILS) GARAY, 1962

537

The Striped Heart has a slender upright stem, carrying a single, heart-shaped, heavily veined leaf that is dark red beneath and spotted with white lines above. Underground, there is a short rhizome and many thick, hairy roots radiating out from the stem. The leaf is topped by an inflorescence with between three and eight flowers. The genus name comes from three Greek words, *mono*, "one," *phyllo*, "leaf," and *orchis*, "orchid," while the species name is Latin for "small pen," which here is a reference to the small column. *Monophyllorchis microstyloides* used to be the only species in the genus, but recently two additional species from Colombia were described.

Pollination of this relatively rarely encountered denizen of wet dark forests is unstudied, but its shape suggests pollination by a bee. No nectar is present.

The flower of the Striped Heart has tightly enclosing but free, pointed, linear, greenish sepals and two shorter, similar petals. The three-lobed white lip is recurved at its apex and has three prominent ridges at the wedge-shaped base and triangular lateral lobes. The column is slender and winged.

Actual size

SUBFAMILY	Epidendroideae
TRIBE AND SUBTRIBE	Triphoreae, Triphorinae
NATIVE RANGE	Brazil and Venezuela
HABITAT	Dwarf scrub forests, and semi-deciduous Atlantic rain forests, at 490–3,950 ft (150–1,200 m)
TYPE AND PLACEMENT	Terrestrial in leaf litter or epiphytic on lower branches of mossy trees
CONSERVATION STATUS	Not assessed
FLOWERING TIME	June to August (summer)

FLOWER SIZE
½ in (1.3 cm)

PLANT SIZE
8–15 × 3–5 in
(20–38 × 8–13),
including terminal
inflorescence

538

PSILOCHILUS MODESTUS
SHY ORCHID
BARBOSA RODRIGUEZ, 1882

Actual size

The purplish-green stems of the Shy Orchid carry obovate-lanceolate, three-veined leaves that gradually narrow below into a stalk at the base. Leaves are green above and purple below, which is a common feature of orchids growing in shady habitats. An unbranched inflorescence bears two or three successively opening flowers, subtended by overlapping floral bracts. The genus name is composed of the Greek words *psilo*, "bald," and *chilus*, "lip," referring to the hairless lip of each bloom.

Flowering time of separate plants is perfectly synchronized, with all mature flower buds open simultaneously in the morning and closing later in the day. Flowers offer nectar and pollen as rewards and are pollinated by several species of small solitary and social bees. The pollen-collecting insects perform mainly self-pollination and promote a higher fruit set than those collecting nectar. The latter are less numerous but produce an increase in cross-pollination.

The flower of the Shy Orchid has green sepals and petals, the lateral ones spreading, the upper sepals and petals curved over the column. The white to pink lip has a frilly edge, the lateral lobes shallow, clasping the column, the midlobe with three central ridges.

SUBFAMILY	Epidendroideae
TRIBE AND SUBTRIBE	Triphoreae, Triphorinae
NATIVE RANGE	North-central North America (Canada and United States) to Central America (Honduras)
HABITAT	Rich, moist forest understory with abundant humus, from sea level to 5,250 ft (1,600 m)
TYPE AND PLACEMENT	Terrestrial
CONSERVATION STATUS	Threatened throughout its range
FLOWERING TIME	July to September (summer to early fall)

TRIPHORA TRIANTHOPHOROS
THREE BIRDS ORCHID
(SWARTZ) RYDBERG, 1901

FLOWER SIZE
¾ in (1.9 cm)

PLANT SIZE
3–10 x 2–4 in
(8–25 x 5–10 cm),
including terminal
inflorescence

539

The Three Birds Orchid grows in shade and is sometimes nearly leafless, although in most cases it has two or three small leaves, which, like the stem, are usually suffused purple red. Underground, an oblong tuber produces plantlets at its tip, forming small, dense colonies. *Triphora trianthophoros* maintains a close connection to soil fungi that exchange minerals for sugars from the trees in the forests in which this orchid grows. It is thus indirectly parasitizing these trees. The genus name comes from the Greek *tri-*, "three," and *-phora*, "bearing," a reference repeated in the species name. The common name alludes to the plant looking like three small birds.

The flowers are short-lived, often no more than one to three days. Pollination is by small halictid (sweat) bees, among which the species *Augochlora pura* has been frequently cited.

Actual size

The flower of the Three Birds Orchid has lanceolate, white sepals and petals that are sometimes suffused with purple at their tips. The petals project forward along the column. The lip is white with a green central crest and three-lobed, the sidelobes partially enveloping the column.

SUBFAMILY	Epidendroideae
TRIBE	Tropidieae
NATIVE RANGE	Tropical and subtropical Asia, from India to Taiwan and New Caledonia
HABITAT	Broadleaf, evergreen, rain forests, at 650–3,300 ft (200–1,000 m)
TYPE AND PLACEMENT	Terrestrial, rocky soils in humid shady locations
CONSERVATION STATUS	Not assessed
FLOWERING TIME	June to August (summer)

FLOWER SIZE
⅜ in (1 cm)

PLANT SIZE
15–36 × 8–12 in
(38–91 × 20–30 cm),
including lateral or
terminal inflorescence

540

TROPIDIA CURCULIGOIDES
CURCULIGO ORCHID
LINDLEY, 1840

The Curculigo Orchid has a thin, woody stem covered with basally clasping, thin, tough, plicate, narrowly lanceolate, pointed leaves that are arranged in a spiral around the stem. In many cases, the stems also branch, forming a network of intertwined growths. The plant makes an axillary or terminal inflorescence, often several per stem, bearing 4–12 flowers. The genus name comes from the Greek word *tropideion*, "keel," an allusion to the boat-shaped lip. Vegetatively, the species resembles Golden Eye-grass (*Curculigo orchioides*) from the related family Hypoxidaceae, hence its species and common names.

In China, *Tropidia curculigoides* is used in traditional herbal medicine, and in Malaysia the roots are boiled in water to make a "tea" drunk to treat diarrhea or break malarial fever. No pollination studies have yet been undertaken.

Actual size

The flower of the Curculigo Orchid is greenish-white and has its lip uppermost, with broad, fleshy, recurved, lanceolate sepals and shorter, fleshy petals. The lip is short and lobed with a nectar-filled basal cavity that has a white and yellow callus in front.

SUBFAMILY	Epidendroideae
TRIBE AND SUBTRIBE	Vandeae, Adrorhizinae
NATIVE RANGE	Sri Lanka, possibly also in southern India
HABITAT	Montane forests
TYPE AND PLACEMENT	Epiphyte on mossy trunks
CONSERVATION STATUS	Endangered
FLOWERING TIME	December (end of intermonsoonal period)

ADRORHIZON PURPURASCENS
WORMROOT ORCHID
(THWAITES) J. D. HOOKER, 1898

FLOWER SIZE
¾ in (1.8 cm)

PLANT SIZE
3–4 × 2–4 in
(7.5–10 × 5–10 cm),
excluding inflorescence

541

The miniature Wormroot Orchid has narrowly cylindrical pseudobulbs, each with a single upright, rigid, linear leaf that is recurved at the margins and usually tinged purple. The roots are round, fat, and hairless. The inflorescence includes one to five attractive, relatively large flowers. *Adrorhizon purpurascens* is restricted to higher elevation wet forests, where it can be locally abundant. There is just a single species in this genus, which is closely related to *Sirhookera*, found in southern India as well as Sri Lanka. In a more general way, they are related to the vandoid orchids, especially *Polystachya*.

The name *Adrorhizon* is derived from the Greek word *hadros*, meaning "thick" or "stout," and *rhiza*, meaning "root," in reference to the vermiform (wormlike) roots of this species. Its pollinators are unknown.

Actual size

The flower of the Wormroot Orchid is white, with a short spur and three sepals, which are similar to but wider than the two petals. The lip is wider than the sepals, and the purple-tipped column has a dilated apex.

SUBFAMILY	Epidendroideae
TRIBE AND SUBTRIBE	Vandeae, Adrorhizinae
NATIVE RANGE	Indochina to the Philippines and New Guinea
HABITAT	Open scrub and montane forests, at up to 4,920 ft (1,500 m)
TYPE AND PLACEMENT	Terrestrial
CONSERVATION STATUS	Widespread and locally abundant, unlikely to be endangered
FLOWERING TIME	Throughout the year

FLOWER SIZE
3 in (8 cm)

PLANT SIZE
Single leafy stem up to
40 in (100 cm)
height with leaves up to
6 in (15 cm)
long, excluding inflorescence

542

BROMHEADIA FINLAYSONIANA
PALE REED ORCHID
(LINDLEY) MIQUEL, 1859

The Pale Reed (or Marsh) Orchid has lanceolate leaves that twist slightly to face upward, clasping at the base and shallowly bilobed at the tip. On a tall slender, leafy, upright stalk, a cluster of pale pink to white flowers is formed on a zigzag stem. A drop in temperature initiates flowering on many plants simultaneously, but only one flower on each plant is open at a time and lasts for a single day. The orchid is probably pollinated by carpenter bees (*Xylocopa*).

DNA studies suggest that the genus *Bromheadia* is related to *Sirhookera* and *Adrorhizon*—both tiny orchids from India and Sri Lanka—to which genera it bears little resemblance. Broth made from the boiled roots is consumed to treat rheumatism and joint pains, and the Indonesian Iban people used the sap to treat toothache.

The flower of the Pale Reed Orchid has three pink sepals and two slightly larger but otherwise similar petals. The lip is recurved, with a yellow, papillate callus and pinkish-purple to pink sidelobes that enclose the column.

Actual size

SUBFAMILY	Epidendroideae
TRIBE AND SUBTRIBE	Vandeae, Adrorhizinae
NATIVE RANGE	Sri Lanka and southern India
HABITAT	Semi-evergreen forests along streams and in misty, mossy montane forests, at 620–6,600 ft (190–2,000 m)
TYPE AND PLACEMENT	Epiphytic on tree trunks and large branches, forming small colonies
CONSERVATION STATUS	Not assessed
FLOWERING TIME	May to December (summer to fall)

FLOWER SIZE
⅛ in (0.4 cm)

PLANT SIZE
1–2 × 1–3 in
(2.5–5 × 2.5–8 cm)
excluding erect-arching
inflorescence
5–8 in (13–20 cm) long

SIRHOOKERA LANCEOLATA
SIR HOOKER'S ORCHID
(WIGHT) KUNTZE, 1891

543

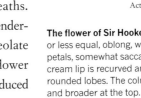

Actual size

The flower of Sir Hooker's Orchid has more or less equal, oblong, white or cream sepals and petals, somewhat saccate at base. The white to cream lip is recurved and concave, and has three rounded lobes. The column is reddish-purple and broader at the top.

The two species in the epiphytic genus *Sirhookera* have tufted stems held in place by wormlike roots. They have one or rarely two stalked leaves, surrounded by purple-veined sheaths. Young leaves are purple-spotted below. The persistent, slender-branched inflorescences extend much beyond the lanceolate leaves and produce small flowers a few at a time. They flower again the following year on many of the inflorescences produced in the previous two to four years, in addition to the new ones from that year.

The genus and common names commemorate the renowned botanist and traveler Sir Joseph Dalton Hooker (1817–1911), director of the Royal Botanic Gardens, Kew, and collector of many Indian orchids. The pollinator of *S. lanceolata* is unknown, but there is a nectar cavity on the base of the lip, so the plant is presumably visited by some type of small bee.

SUBFAMILY	Epidendroideae
TRIBE AND SUBTRIBE	Vandeae, Aeridinae
NATIVE RANGE	Tropical Asia, from India to the Philippines
HABITAT	Tropical, wet, evergreen forests on shady cliffs, karst limestone outcrops, and trees at the forest edge, at 985–2,950 ft (300–900 m)
TYPE AND PLACEMENT	Usually epilithic on rocks in forest, sometimes epiphytic on tree trunks
CONSERVATION STATUS	Not assessed
FLOWERING TIME	August to September (summer)

FLOWER SIZE
¾ in (2 cm)

PLANT SIZE
20–40 × 14–30 in
(51–102 × 36–76 cm),
including lateral inflorescence
5–8 in (13–20 cm) long

544

ACAMPE RIGIDA
STIFF TIGER ORCHID
(BUCHANAN-HAMILTON EX SMITH) P. F. HUNT, 1970

A large plant, the Stiff Tiger Orchid has a rigid, frequently branching stem with a terminal cluster of stiff, thick, linear-lanceolate, conduplicate leaves that are minutely and unequally bilobed. The old leaf bases turn brown and give the stems a ribbed appearance. The genus name derives from the Greek word *akampes*, meaning "stiff," and this is echoed in the species and common names, with the "tiger" appellation springing from the striped flowers.

Acampe rigida is listed as a medicinal plant in Hong Kong, and other species in the genus are used to treat rheumatism. Pollination has not been studied in the field, but the color and structure of the flowers, with their short spur, could suggest bee pollination.

Actual size

The flower of the Stiff Tiger Orchid is cup-shaped, with red-striped, yellow sepals and petals. The lip is similarly colored and clawed, with a short nectar spur. The flowers are not consistently oriented, and in some the column is lowermost.

SUBFAMILY	Epidendroideae
TRIBE AND SUBTRIBE	Vandeae, Aeridinae
NATIVE RANGE	Southern Indochina, Peninsular Malaysia
HABITAT	Deciduous and semi-deciduous, dry forest, at up to 2,300 ft (700 m)
TYPE AND PLACEMENT	Epiphytic on exposed branches
CONSERVATION STATUS	Not assessed
FLOWERING TIME	November to February (winter)

ADENONCOS VESICULOSA

BLISTER ORCHID

CARR, 1932

FLOWER SIZE
⅛ in (0.5 cm)

PLANT SIZE
4–8 × 1½–4 in
(10–20 × 4–10 cm),
excluding short inflorescence
1–1½ in (2.5–3.8 cm) long

545

The short, stout stem of the Blister Orchid carries swollen, linear, folded leaves and produces a short, axillary inflorescence with one or two flowers. The plant is vegetatively similar to, and often grows together with, orchids of the genus *Microsaccus*, but the leaves of the latter are not as flattened. The genus name comes from the Greek words *aden*, for "gland," and *onkos*, for "mass," referring to the callus on the base of the lip. The common name refers also to the lip callus, which resembles a watery blister.

The papillose, fleshy keel of the lip is reported as being eaten, so it could provide food for pollinating insects. Whether this is a regular part of pollination or just a coincidence is unproven. There is no nectar spur or cavity, so what would attract an insect to these small, pale flowers is unclear.

Actual size

The flower of the Blister Orchid has narrow, spreading, greenish-white sepals and narrower, spreading petals. The broad lip is greenish-white, papillose, keeled, and cupped, lacking sidelobes but with a thickened, blisterlike callus at the base.

SUBFAMILY	Epidendroideae
TRIBE AND SUBTRIBE	Vandeae, Aeridinae
NATIVE RANGE	Himalayas to Indochina (India and Nepal to Vietnam)
HABITAT	Deciduous to semi-deciduous forests, from sea level to 3,600 ft (1,100 m)
TYPE AND PLACEMENT	Epiphytic
CONSERVATION STATUS	Not threatened, but heavily collected for horticulture in some areas
FLOWERING TIME	May to June (late spring to early summer)

FLOWER SIZE
¾ in (1.9 cm)

PLANT SIZE
8–15 × 8–15 in
(20–38 × 20–38 cm),
excluding pendent
inflorescence
6–12 in (15–30 cm) long

546

AERIDES MULTIFLORA
PINK FOXTAIL ORCHID
ROXBURGH, 1820

The Pink Foxtail Orchid, which covers a broad geographical range and has many color variants, is one of the species often referred to as foxtail orchids due to their densely flowered pendent inflorescences. The genus name comes from the Latin word *aer*, "air," and the suffix *–ides*, meaning "having the nature of," a reference to the epiphytic nature of the plants. Reports of *Aerides multiflora* being of medicinal use occur in several countries, including Nepal, where it is made into a tonic for several ailments, including wounds and skin diseases.

The species has a highly sweet fragrance, and the 25 to 50 colorful flowers are waxy and long lasting. The pollinator is most likely a large nectar-seeking bee, and there is nectar present in the spur on the lip base.

The flower of the Pink Foxtail Orchid has similarly sized and shaped sepals and petals, which are white with pink suffusion and darker pink spots to solid rosy lavender. The lip is triangular, pale to dark pink to rosy lavender and with a stalked base.

Actual size

SUBFAMILY	Epidendroideae
TRIBE AND SUBTRIBE	Vandeae, Angraecinae
NATIVE RANGE	Northern Madagascar, found only on Montagne d'Ambre
HABITAT	Humid, evergreen forests, at 1,640–3,300 ft (500–1,000 m)
TYPE AND PLACEMENT	Epiphytic on *Viguieranthus alternans*
CONSERVATION STATUS	Very rare, formerly thought to be extinct, but several populations have recently been found
FLOWERING TIME	November (spring)

AMBRELLA LONGITUBA

MOUNT AMBER ORCHID

H. PERRIER, 1934

FLOWER SIZE
1¾ in (4.5 cm)

PLANT SIZE
3–5 × 4–9 in
(8–13 × 10–23 cm),
including arching-pendent
inflorescence
2–3 in (5–8 cm) long

547

The Mount Amber Orchid is the only species in a genus and has a peculiar flower. It has a short stem and five or six broad, elliptic leaves. An inflorescence arises from below the leaves with up to three flowers. The plant is known only from Amber Mountain National Park, an isolated forest surrounded by drylands, in the far north of Madagascar. This is one of the most biodiverse areas of the island, with outstandingly beautiful waterfalls and crater lakes.

The genus name is derived from the name of the mountain, as is its common name. The species name refers to the long, trumpet-shaped lip, which together with a long spur and white-green color suggest pollination by night-flying hawk moths. Due to its rarity, it has never been studied in the wild.

Actual size

The flower of the Mount Amber Orchid
has spreading, lanceolate, pale green sepals
and petals. The lip forms a long, upright,
trumpet-shaped tube with a trilobed apex,
a trilobed midlobe, and two smaller sidelobes.
There is a long, curved spur at the rear.

SUBFAMILY	Epidendroideae
TRIBE AND SUBTRIBE	Vandeae, Aeridinae
NATIVE RANGE	Mountains on islands of Luzon (Central Cordillera, Mt Mayon) and Mindoro (Mt Halcon), Philippines
HABITAT	Forested slopes, at 1,300–4,600 ft (400–1,400 m)
TYPE AND PLACEMENT	Epiphyte on mossy trees in light shade
CONSERVATION STATUS	Endangered
FLOWERING TIME	November to April (winter to spring / wet season)

FLOWER SIZE
1½ in (4.5 cm)

PLANT SIZE
1–2 × 1–2 in
(2.5–5 × 2.5–5 cm),
excluding inflorescence

548

AMESIELLA PHILIPPINENSIS
OAKES' ORCHID
(AMES) GARAY, 1972

The Oakes' Orchid is rare in the wild, where it is endangered by human destruction of its habitat. In addition, the plant is overcollected for the international orchid trade, in which it is a highly valued miniature orchid with its exceptionally large white flowers. It has fleshy roots from a short stem bearing elliptic-oblong, broad leaves. An axillary inflorescence bears up to five fragrant, long-spurred flowers. Pollination is unstudied, but the morphology and color of these flowers would seem to be adapted to moths.

The genus is named in honor of Oakes Ames (1874–1950), a renowned orchidologist at the University of Harvard who established a large herbarium and collection of orchid books and drawings. The Oakes Ames Orchid Herbarium at Harvard is still a center of orchid study today.

The flower of the Oakes' Orchid is fragrant and white. The three broadly ovate sepals are similar to the two petals. The lip is heart-shaped, yellow inside, forming a V-shaped opening to the nectar spur, which is about 1½–2⅜ in (4–6 cm) long.

Actual size

SUBFAMILY	Epidendroideae
TRIBE AND SUBTRIBE	Vandeae, Aeridinae
NATIVE RANGE	Cambodia and southern Vietnam
HABITAT	Evergreen or partly deciduous dry lowland forests, at up to 4,920 ft (1,500 m)
TYPE AND PLACEMENT	Scrambling, almost vinelike epiphytic
CONSERVATION STATUS	Not assessed
FLOWERING TIME	Spring

ARACHNIS ANNAMENSIS
SCORPION ORCHID
(ROLFE) J. J. SMITH, 1912

FLOWER SIZE
2⅛ in (5.5 cm)

PLANT SIZE
28–48 × 15–25 in
(71–122 × 38–64 cm),
including lateral, often
branched inflorescence
15–28 in (38–71 cm) long

549

The elongate, usually upright, somewhat vining stem of the Scorpion Orchid is held in place by long scrambling aerial roots. The stem has two rows of basally overlapping, leathery leaves with a shortly bilobed tip. An inflorescence emerges from one of the axils of the upper leaves with 2–12 flowers that resemble a scorpion in shape—hence the common name. The genus name, however, references another animal, the spider, from the Greek *arachne*. The species name refers to the geographical origin of this species in the Annamite Mountains of eastern Cambodia and Vietnam.

Although a musky odor is spread from scent glands on the tip of the middle sepal (the scorpion's tail), the pollinator of this unusual, spectacular flower is unknown. There is no nectar spur or an obvious reward offered but, given the shape and coloration of the blooms, a bee is the most likely candidate.

The flower of the Scorpion Orchid has narrow, red-brown-banded, yellow sepals and petals, the lateral sepals forming the "claws" of the scorpion and the upper sepal its "tail." The lip has three wings, the laterals embracing the column, and the middle lobe with a purple-striped, white base.

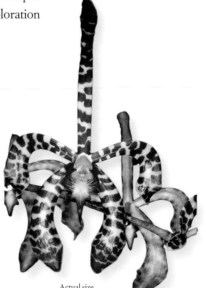

Actual size

SUBFAMILY	Epidendroideae
TRIBE AND SUBTRIBE	Vandeae, Aeridinae
NATIVE RANGE	Eastern Himalayas to Assam (India)
HABITAT	Forests, at 1,640–1,970 ft (500–600 m)
TYPE AND PLACEMENT	Epiphytic
CONSERVATION STATUS	Not assessed
FLOWERING TIME	April to July (spring to summer)

FLOWER SIZE
¼ in (0.8 cm)

PLANT SIZE
3–8 × 4–10 in
(8–20 × 10–25 cm),
excluding pendent
inflorescence
2–3 in (5–8 cm) long

550

BIERMANNIA BIMACULATA
ALMOND ORCHID
(KING & PANTLING) KING & PANTLING, 1898

Actual size

The Almond Orchid has a short stem, completely enveloped by basally sheathing leaves that are linear-oblong, somewhat curved, and unevenly bilobed at the tip. From the base of a leaf near the base of the stem, a stout inflorescence appears with two or three flowers that are strongly almond-scented. The genus is named for Adolph Biermann, a nineteenth-century German naturalist who was curator of the Calcutta Botanical Garden. He survived an attack by a tiger in the garden in 1879 but succumbed a year later to cholera. The species name refers to the two prominent yellow spots (*maculata* is Latin for "spotted") on the flower lip.

The pollinators of *Biermannia bimaculata* are unknown, but the sweet fragrance, shape, and color of the flowers suggest some sort of bee. There is a small cavity on the lip base but no nectar.

The flower of the Almond Orchid has cream to white, oblanceolate, cupped sepals with pointed, recurved tips. The petals are similar but shorter and flank the swollen column. The lip is arrowhead-shaped with upturned edges and a yellow to orange spotted, warty callus.

SUBFAMILY	Epidendroideae
TRIBE AND SUBTRIBE	Vandeae, Aeridinae
NATIVE RANGE	Close to Dalton Pass in mountains of Nueva Ecija province, Luzon Island (Philippines)
HABITAT	Lower montane forests, at 3,300–3,950 ft (1,000–1,200 m)
TYPE AND PLACEMENT	Epiphyte on tree trunks
CONSERVATION STATUS	Critically endangered
FLOWERING TIME	January to May (winter, in response to cool weather)

FLOWER SIZE
⅝–⅞ in (1.5–2.3 cm)

PLANT SIZE
2–3 × 2–3½ in
(5–8 × 5–9 cm),
excluding inflorescence

CERATOCENTRON FESSELII
HORNSPUR ORCHID
SENGHAS, 1989

551

The delightful miniature Hornspur Orchid has an upright growth with three to five leathery, dark green, elliptic leaves that clasp the stem. Inflorescences, which bear up to five bright red to red-orange flowers, are about as long as the leaves and formed in the leaf axils. It is highly threatened because it occurs only in a small area unprotected from logging or expansion of settlements and agriculture, and it is also overcollected for the illegal orchid trade.

It has been suggested that because of the orchid's bright red-orange color, bird pollination is possible, but this seems unlikely given its spur structure, which is long and tubular. No known species of bird has a beak shaped like this spur as well as one long enough to reach its base. Butterfly or bee pollination is more likely (butterflies and bees can see red pigments).

The flower of the Hornspur Orchid has similar, bright red-orange to red sepals and petals. The lip forms a red spur with a prominent pink horn to the front. The column is yellow to red-orange and hangs over the entrance to the spur.

Actual size

SUBFAMILY	Epidendroideae
TRIBE AND SUBTRIBE	Vandeae, Aeridinae
NATIVE RANGE	Peninsular Thailand, Borneo, and Java
HABITAT	Swamp forests, at 650–2,950 ft (200–900 m)
TYPE AND PLACEMENT	Epiphytic
CONSERVATION STATUS	Not assessed
FLOWERING TIME	April (spring)

FLOWER SIZE
⅛ in (0.5 cm)

PLANT SIZE
2–3 × 3–5 in
(5–8 × 8–13 cm),
including axillary
inflorescence
½–1 in (1.3–2.5 cm) long

552

CHAMAEANTHUS BRACHYSTACHYS
UNSEEN ORCHID
SCHLECHTER, 1905

The miniature Unseen Orchid, with its small, drably colored flowers, is, as the common name suggests, often overlooked, even when in full bloom. Its short stems are covered by thick, obovate leaves, from the bases of which a small stem grows, producing only one to three flowers at a time but up to 30 eventually. The genus name comes from the Greek words *chamai*, "prostrate," and *anthos*, "flower," referring to the insignificant blooms of the species. A similar allusion is repeated in the species name, which comes from the Greek words *brachy*, "short," and *stachys*, "stem." Clearly, the stature of the plant did not impress Schlechter, who first described it.

The flowers of *Chamaeanthus brachystachys* do not open widely, and so appear to form a small tube. This would require a relatively long-tongued insect as pollinator.

Actual size

The flower of the Unseen Orchid has linear, pointed, pale cream sepals and petals. The lip and column are completely hidden inside a tube created by the sepals and petals. The lip has two broad, upturned sidelobes and a short, bilobed apical lobe.

SUBFAMILY	Epidendroideae
TRIBE AND SUBTRIBE	Vandeae, Aeridinae
NATIVE RANGE	Himalayas to Indochina and on Java
HABITAT	Canopy of humid forests or on exposed rocks
TYPE AND PLACEMENT	Epiphytic or sometimes lithophytic
CONSERVATION STATUS	Unassessed, but relatively common and widespread
FLOWERING TIME	October to February (fall to winter)

FLOWER SIZE
½ in (1.3 cm)

PLANT SIZE
1–1½ × 4–8 in
(2.5–3.8 × 10–20 cm),
excluding inflorescence

CHILOSCHISTA LUNIFERA
LEAFLESS MOON ORCHID
(REICHENBACH FILS) J. J. SMITH, 1905

553

The Leafless Moon Orchid typically occurs on the trunks of deciduous trees, particularly near watercourses. This strange mass of roots attached to a short stem usually produces no leaves, as the common name suggests, although they have been occasionally reported. The crown of fleshy, greenish-white roots takes up water and nutrients and carries out photosynthesis. Dark red to plain green yellow-edged flowers that smell like vanilla grow on a 3–12 in (8–30 cm) long raceme. After flowering, the plant retains its flattened inflorescence, which is perhaps an important source of carbon.

Given the orchid's sweet fragrance and floral morphology, some form of bee is the most likely pollinator, but there are no reports of pollination in this or any other species of *Chiloschista*. No nectar has been reported.

Actual size

The flower of the Leafless Moon Orchid has similar petals and sepals, which are spreading, broad, with a rounded tip and rounded base, solid reddish-brown or spotted and edged pale yellow. The lip is cupped and has two flaring sidelobes that embrace the column.

SUBFAMILY	Epidendroideae
TRIBE AND SUBTRIBE	Vandeae, Aeridinae
NATIVE RANGE	Sumatra, Borneo, Peninsular Thailand, and Malaysia
HABITAT	Hill forests, at up to 3,300 ft (1,000 m)
TYPE AND PLACEMENT	Epiphytic
CONSERVATION STATUS	Not formally assessed but threatened in Thailand and likely to be under threat elsewhere
FLOWERING TIME	Summer

FLOWER SIZE
¼ in (0.6 cm)

PLANT SIZE
3–4 × 5–8 in
(8–10 × 13–20 cm),
including inflorescence,
which is shorter than
the leaves

554

CHRONIOCHILUS VIRESCENS
EVERLASTING ORCHID
(RIDLEY) HOLTTUM, 1960

Actual size

The Everlasting Orchid is a minute epiphyte with an erect, thick stem and spreading, fleshy leaves that are narrowly strap-shaped and unequally bilobed at the tip. The inflorescence can produce up to eight flowers that open successively (so that rarely more than one is open at any time) and smell strongly of freshly baked cupcakes. The flowers are tough and last for a long time, hence the common name. It is threatened in Thailand due to habitat conversion to agriculture. Its small size and sparse flowering makes it attractive only to connoisseur orchid growers, making collection unlikely to result in a threat.

Nothing is known about the orchid's pollination, although its floral morphology suggests some type of bee.

The flower of the Everlasting Orchid is cream to white and fleshy with three equal spreading sepals and two similar, but smaller, petals. The lip is cup-shaped.

SUBFAMILY	Epidendroideae
TRIBE AND SUBTRIBE	Vandeae, Aeridinae
NATIVE RANGE	Sabah (Borneo)
HABITAT	Mossy forests, at around 5,900 ft (1,800 m)
TYPE AND PLACEMENT	Epiphytic
CONSERVATION STATUS	Unknown, but has a restricted distribution and its forests are under threat
FLOWERING TIME	Fall

FLOWER SIZE
⅝ in (1.5 cm),
excluding spur

PLANT SIZE
10–25 × 6–10 in
(25–64 × 15–25 cm),
including lateral
arching-erect inflorescences
½–1 in (1.3–2.5 cm) long

CLEISOCENTRON GOKUSINGII
TONGUESPUR ORCHID
J. J. WOOD & A. L. LAMB, 2008

555

The elongate, often branched stems of the Tonguespur Orchid are enveloped in leafy sheaths, the almost terete blades widely spaced on the stem. A cluster of up to 20 flowers, each with a small floral bract, is produced. The genus name comes from the Greek words *kleiso*, "tongue," and *kentron*, "spur," referring to a pointed appendage inside the nectar spur, which is the also the basis for the common name of this recently described species.

Three of the six species in the genus *Cleisocentron* are blue-flowered, which is rare among orchids. Pollination has not been observed, but the highly clustered flowers and the relatively long nectar spur could indicate pollination by butterflies, which need a place to land while feeding. Long spurs are always found in species pollinated by butterflies or moths.

The flower of the Tonguespur Orchid has fleshy, forward-pointing, pale blue sepals and petals. The column has a white anther cap. The lip is short-limbed with two lateral, upturned, dark blue lobes, forming an opening to the spur.

Actual size

SUBFAMILY	Epidendroideae
TRIBE AND SUBTRIBE	Vandeae, Aeridinae
NATIVE RANGE	Tropical East Asia
HABITAT	Lowland forests, at 330–1,640 ft (100–500 m)
TYPE AND PLACEMENT	Epiphytic
CONSERVATION STATUS	Not formally assessed, but locally threatened
FLOWERING TIME	Throughout the year

FLOWER SIZE
⁵⁄₁₆ in (0.8 cm)

PLANT SIZE
15–35 × 6–10 in
(38–89 × 15–25 cm),
excluding inflorescence,
which is typically shorter
than the plant

556

CLEISOSTOMA SUBULATUM
AWL ORCHID
BLUME, 1825

Actual size

The flower of the Awl Orchid has spreading
petals and sepals, and a cup-shaped lip with
a pointed, sacklike spur. The sepals and petals
are often reddish-brown with a lighter central
stripe, and the column and lip are often white.

The Awl Orchid is an erect to arching-pendent plant that may
branch out and form a mass of stems heading off in many
directions. Its leaves are fleshy, rigid, round, and V-shaped in
cross section, with narrow tips that come to a sharp point—
hence the common name. An unbranched, stiff, pendent
inflorescence is formed, with small triangular bracts bearing
flowers that open progressively downward.

The genus name (Greek for "closed mouth") refers to the
presence of a callus (a lump of tissue) in the middle of the base
of the lip that partially blocks the entrance into the nectar spur.
The genus *Cleisostoma* is one of the largest in the subtribe, with
about 100 species, all of which occur in the Asian tropics, as far
south as northern Australia and as far east as Fiji.

SUBFAMILY	Epidendroideae
TRIBE AND SUBTRIBE	Vandeae, Aeridinae
NATIVE RANGE	India and Sri Lanka
HABITAT	Semi-evergreen to moist deciduous forests, at 985–1,970 ft (300–600 m)
TYPE AND PLACEMENT	Epiphytic
CONSERVATION STATUS	Not assessed
FLOWERING TIME	March to August (summer to early fall)

FLOWER SIZE
1 in (2.5 cm)

PLANT SIZE
20–40 × 10–20 in
(51–102 × 25–51 cm),
excluding erect-arching
inflorescence
36–60 in (91–152 cm) long

557

COTTONIA PEDUNCULARIS
INDIAN BEE ORCHID
(LINDLEY) REICHENBACH FILS, 1857

Actual size

A large epiphyte with elongate stems that carry many linear-obtuse, distantly spaced, unequally bilobed leaves, the Indian Bee Orchid produces a branched, slender inflorescence with long-lasting flowers that cluster near the apex. This is the only species in the genus *Cottonia*, named for Lieutenant Colonel Frederic Conyers Cotton (1807–1901) of the Madras Engineers, a British plant collector and orchid enthusiast.

The flowers of *Cottonia penduncularis* are thought to be pollinated through pseudocopulation. The glossy lip imitates a female insect, and a male insect, possibly a beetle, attempts to copulate with the reportedly foul-smelling flower and in doing so may transfer the pollinia. Exactly what sort of insect is involved is unknown. The flowers are remarkably similar to those of European psedocopulatory species, such as members of the genus *Ophrys*, which are, however, only distantly related.

The flower of the Indian Bee Orchid has reflexed, tan to cream, red-striped petals and sepals. The column is shortly winged, and the lip is oblong, convex, and brown with a paler central stripe, resembling the body of an insect.

SUBFAMILY	Epidendroideae
TRIBE AND SUBTRIBE	Vandeae, Aeridinae
NATIVE RANGE	Borneo, especially state of Sarawak
HABITAT	Lower montane forests, ravines, gullies, up to 5,900 ft (1,800 m)
TYPE AND PLACEMENT	Epiphytic, usually in tall trees overhanging water
CONSERVATION STATUS	Not assessed
FLOWERING TIME	During or after the monsoon (fall to early winter)

FLOWER SIZE
3 in (7.5 cm)

PLANT SIZE
25–50 × 20–36 in
(64–127 × 51–91 cm),
excluding inflorescence,
which is pendent and up to
12 ft (3.65 m) long

DIMORPHORCHIS LOWII
DIMORPHIC TIGER ORCHID
(LINDLEY) ROLFE, 1919

558

The robust, sometimes massive, Dimorphic Tiger Orchid grows an erect to slightly arching elongate stem with thick, fleshy, aerial roots near its base. Its lanceolate, folded leaves are unevenly lobed at the tip and sheathing at their bases. From a leaf axil it produces a limp, pendent, long inflorescence with many flowers.

Each inflorescence has flowers of two types: at the base of the inflorescence are two fragrant yellow flowers with small red dots, whereas the other flowers are creamy white with large red blotches and nearly scentless. Both types of flowers appear to have a complete column with pollinia and a functioning stigma, so the floral dimorphism does not appear to be associated with sexual dimorphism. Given the flower morphology, especially the lack of a nectar spur, the species should be bee-pollinated, but this has never been reported.

Actual size

The flower of the Dimorphic Tiger Orchid has spreading, wavy sepals and two forward-pointing, smaller petals. The lip is cup-shaped and forms an opening to the column.

SUBFAMILY	Epidendroideae
TRIBE AND SUBTRIBE	Vandeae, Aeridinae
NATIVE RANGE	Papua New Guinea
HABITAT	Forests on limestone ridges covered with clay, at 3,950–4,300 ft (1,200–1,300 m)
TYPE AND PLACEMENT	Epiphytic
CONSERVATION STATUS	Not assessed
FLOWERING TIME	April to May (spring)

DRYADORCHIS DASYSTELE
SPECKLED TREE NYMPH
SCHUITEMAN & DE VOGEL, 2004

FLOWER SIZE
1 in (2.5 cm)

PLANT SIZE
2–3 × 3–5 in
(5–8 × 8–13 cm),
excluding pendent
inflorescence
3–4 in (8–10 cm) long

559

The miniature Speckled Tree Nymph was described only in 2004 by the orchid specialist Ed de Vogel of Naturalis Biodiversity Center, in the Netherlands, and it has been found only once. The pendent inflorescence carries one or two short-lived flowers at a time. The genus name derives from *Dryas*, a tree nymph in Greek mythology, and the Greek word *orchis*, "orchid," while the species name is Greek for "hairy column," referring to the covering of short hairs on its column.

It has been speculated that the lip of *Dryadorchis dasystele* resembles a bee, and so pollination may occur by pseudocopulation. In this situation, a male bee is sexually attracted to and attempts to mate with the lip, in so doing removing the pollinia, which are deposited on the next flower he finds stimulating.

Actual size

The flower of the Speckled Tree Nymph has elliptic, outstretched, red-orange spotted sepals and petals. The complex lip is hairy, trilobed, white with red-orange markings, and covered with dense, short hairs. The column is yellow with red-orange spots and covered on its front with longer hairs.

SUBFAMILY	Epidendroideae
TRIBE AND SUBTRIBE	Vandeae, Aeridinae
NATIVE RANGE	New Zealand and Chatham Islands
HABITAT	Low to subalpine forests, at 330–4,920 ft (100–1,500 m)
TYPE AND PLACEMENT	Epiphytic on well-lit, lichen-covered trunks and branches of trees, occasionally on rocks or cliff faces
CONSERVATION STATUS	Not threatened
FLOWERING TIME	September to November (spring)

FLOWER SIZE
⅛ in (0.4 cm)

PLANT SIZE
2–4 × 3–6
(5–10 × 7.5–15 cm),
excluding pendent
inflorescence
2–4 in (5–10 cm) long

560

DRYMOANTHUS ADVERSUS
FLESHY TREE ORCHID
(HOOKER FILS) DOCKRILL, 1967

Actual size

The miniature Fleshy Tree Orchid has a few, glossy, lanceolate leaves in two rows, often with purple spots. There are five to eight flowers on an axillary, short inflorescence held just under the leaves, which are relatively thick and fleshy—hence the common name. The flowers often face downward, away from the plant, which gives the plant its species name, *adversus* (Latin for "contrary to"). The genus name comes from *Drymo*, one of the Nereid sea nymphs in Greek mythology, and the Greek word for flower, *anthos*.

The flowers produce nectar at the base of the lip in an open, easily accessed cavity. Ants have been reported in this cavity, but it is unlikely that they function as proper pollinators because they do not to range far enough to cross-pollinate plants. The flowers are slightly fragrant, which may suggest bee pollination.

The flower of the Fleshy Tree Orchid has thick, bell-shaped flowers with pale green, red-spotted sepals and petals of similar size and shape. The lip is white to cream and cupped, and the column is short with a white cap.

SUBFAMILY	Epidendroideae
TRIBE AND SUBTRIBE	Vandeae, Aeridinae
NATIVE RANGE	Borneo
HABITAT	Riverine, seasonal swamp forests, at 1,640–2,625 ft (500–800 m)
TYPE AND PLACEMENT	Epiphytic
CONSERVATION STATUS	Not assessed
FLOWERING TIME	May to July (spring to summer)

DYAKIA HENDERSONIANA

PINK SNAIL ORCHID

(REICHENBACH FILS) CHRISTENSON, 1986

FLOWER SIZE
¾ in (2 cm)

PLANT SIZE
4–8 × 5–8 in
(10–20 × 13–20 cm),
excluding erect
inflorescence
5–8 in (13–20 cm) tall

561

The Pink Snail Orchid, a small epiphytic species with a short stem carrying lanceolate, unequally bilobed leaves, has a distinctive spur that hangs down like a long tongue from an otherwise nonexistent lip. The spur and column resemble a white slug that is crawling on the flower. The name *Dyakia* is also a genus of land snails, and both they and the orchid species were named for the Dyaks or Dayaks, a native people of Borneo. The Henderson of the species name was an English nurseryman in the mid-nineteenth century.

Dyakia hendersoniana produces a densely flowered spike with up to 40 showy, fragrant flowers. Pollination has not been studied, but the color and fragrance of the flowers, shape of the narrow long nectar spur, and position of the column close to its top would be consistent with butterfly pollination.

The flower of the Pink Snail Orchid has broad, spreading, bright pink sepals and petals, with the lateral sepals the largest. The column is short and sits directly above the opening to the long, sacklike, pale pink to white nectar cavity.

Actual size

SUBFAMILY	Epidendroideae
TRIBE AND SUBTRIBE	Vandeae, Aeridinae
NATIVE RANGE	Japan, Korea, Taiwan, Hong Kong
HABITAT	Mountain forests, at 650–6,600 ft (200–2,000 m)
TYPE AND PLACEMENT	Epiphytic on branches of pines and broad-leaved trees
CONSERVATION STATUS	Not assessed
FLOWERING TIME	September to November (fall)

FLOWER SIZE
½–¾ in (1.3–1.8 cm)

PLANT SIZE
4–8 × 6–10 in
(10–20 × 15–25 cm)
including inflorescence

562

GASTROCHILUS JAPONICUS
YELLOW PINE ORCHID
(MAKINO) SCHLECHTER, 1913

Actual size

In cool to even cold habitats the miniature Yellow Pine Orchid produces a short stem that is densely set with fleshy, linear-falcate leaves that have sheathing bases and unequally bilobed tips. Pendent inflorescences, which grow from the base of one or two leaves, are shorter than the leaves and hold up to seven flowers. In some cases, several short inflorescences are produced, making this a showy species, prized by horticulturalists interested in micro-orchids.

The floral fragrance has a high percentage of eugenol, a spicy and clove-like chemical compound. The lip has a deep cavity that is devoid of nectar. The general shape of the lip is similar to those of some species of *Epipactis*, which are pollinated by wasps, but the sweet fragrance of this species suggests a bee pollinator (eugenol not being generally attractive to wasps).

The flower of the Yellow Pine Orchid has curved, yellowish-green sepals and petals. The white lip with red spots and other markings is deeply cupped. The cup is yellow with red spots and has a broad terminal lobe with a yellow, red-spotted center.

SUBFAMILY	Epidendroideae
TRIBE AND SUBTRIBE	Vandeae, Aeridinae
NATIVE RANGE	Taiwan
HABITAT	Forests, at 3,300–7,200 ft (1,000–2,200 m)
TYPE AND PLACEMENT	Epiphytic
CONSERVATION STATUS	Not assessed
FLOWERING TIME	August to March (summer to spring)

FLOWER SIZE
¾ in (1.9 cm)

PLANT SIZE
2–3 × 3–5 in
(5–8 × 8–13 cm),
excluding pendent
inflorescence
3–5 in (8–13 cm) long

GASTROCHILUS RETROCALLUS
TAIWANESE BEE ORCHID
(HAYATA) HAYATA, 1917

563

Actual size

The miniature Taiwanese Bee Orchid has a short, often somewhat pendent stem clothed in 4–10 curved-lanceolate leaves with clasping bases. From near the base of the stem, there can be between one and four pendent inflorescences, each with three to five flowers that are large, given the small size of the plant. The origin of the species name, from the Latin *retro* and *callus*, meaning "backward callus," is unclear. The common name refers to the darkly pigmented areas of the lip, which appear to be mimicking the body of an insect, like a bee.

Pollination of *Gastrochilus retrocallus* has yet to be studied, but this may be a case of pseudocopulation, where a male insect mistakes the lip for a female of its species. In its attempts to mate with the fake female, pollination takes place.

The flower of the Taiwanese Bee Orchid has similar, yellow green sepals and petals that spread widely. The pale green lip has three lobes, the middle one larger and fringed. Covering all three lobes are reddish-purple blotches that vaguely resemble the body of an insect.

SUBFAMILY	Epidendroideae
TRIBE	Collabieae
NATIVE RANGE	Madagascar
HABITAT	Seasonally wet, semi-deciduous mossy forests, at 3,950–6,600 ft (1,200–2,000 m)
TYPE AND PLACEMENT	Terrestrial
CONSERVATION STATUS	Not assessed
FLOWERING TIME	May to October (winter to fall)

FLOWER SIZE
2 in (5 cm)

PLANT SIZE
20–30 × 18–25 in
(51–76 × 46–64 cm),
excluding erect inflorescence
28–40 in (71–102 cm),
which is longer than the leaves

GASTRORCHIS HUMBLOTII
BEAUTY OF THE FOREST
(REICHENBACH FILS) SCHLECHTER, 1924

The Beauty of the Forest has underground, corm-like round pseudobulbs and upright stems enveloped in large, basally sheathing, stalked, pleated, three-veined, lanceolate-ovate leaves. An upright pyramidal inflorescence grows from the base of a pseudobulb, carrying 6–14 beautiful, long-lasting flowers clustered near the apex. Their color is variable, and several of the forms have been given taxonomic recognition. The genus name comes from the Greek words *gaster*, "stomach," and *orchis*, "orchid," which allude to the broad circular base of the lip.

Pollination of the species has not been studied. Its flowers, however, are similar in shape to those of the closely related genus *Phaius*, a member of which, *P. delaveyi*, has been recorded as pollinated by bumblebees (*Bombus*). A similar system is probably operating in *Gastrorchis humblotii*, given its shape and color. No nectar is present.

The flower of the Beauty of the Forest varies considerably, with white to pink spreading sepals and petals of similar size and broad shape. The lip varies from rusty red purple to bright pink, sometimes with darker spotting, ruffled edges, and lateral wings that form a cup with a two-ridged, yellow, central callus.

Actual size

SUBFAMILY	Epidendroideae
TRIBE AND SUBTRIBE	Vandeae, Aeridinae
NATIVE RANGE	Tropical East Asia, from Hainan island (China) and Myanmar to Sulawesi and Java (Indonesia)
HABITAT	Evergreen broad-leaved forests, at up to 3,950 ft (1,200 m)
TYPE AND PLACEMENT	Epiphytic on tree trunks
CONSERVATION STATUS	Not assessed
FLOWERING TIME	August (summer)

GROSOURDYA APPENDICULATA

TORCH ORCHID

(BLUME) REICHENBACH FILS, 1868

FLOWER SIZE
¼ in (0.8 cm)

PLANT SIZE
3–5 × 4–6 in
(8–13 × 10–15 cm),
including arching inflorescences
2–4 in (5–10 cm) long,
which are shorter than
the leaves

565

The miniature Torch Orchid has a short stem with fleshy aerial roots and three to six, lanceolate, fleshy, curved leaves in two rows. The leaf tips are unequally bilobed. The threadlike inflorescence stalk carries black hairs and appears from the base of the stem. It carries up to five, short-lived and sequentially produced flowers subtended by a minute bract. The genus is named, ironically, for René de Grosourdy (1836–64), who was a medicinal plant hunter in the tropical Americas. The species name *appendiculata*, meaning "with appendages" in Latin, refers to the fat nectar cavity on the base of the lip, whereas the common name springs from the bright orange red-spotted flowers.

Pollination has not been studied in nature. However, the open shape of the colorful flowers and the short nectar spur with a wide opening suggest some type of bee as pollinator.

Actual size

The flower of the Torch Orchid has broad-spreading, yellow-orange sepals and petals. All parts except the lip and column are red-spotted. The white lip is elaborately three-lobed, the lobe tips orange or yellow, with a short spur under the protruding white column with a yellow apex.

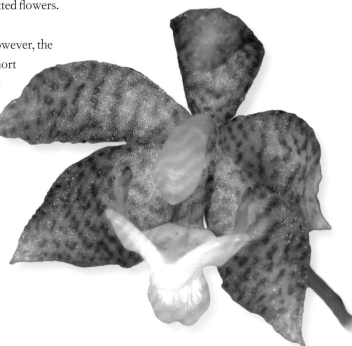

SUBFAMILY	Epidendroideae
TRIBE AND SUBTRIBE	Vandeae, Aeridinae
NATIVE RANGE	Myanmar, Thailand, Laos, and Yunnan province (China)
HABITAT	Open areas of forests on rocks or trees, at 3,300–4,920 ft (1,000–1,500 m)
TYPE AND PLACEMENT	Epiphytic and lithophytic
CONSERVATION STATUS	Not threatened
FLOWERING TIME	Spring

FLOWER SIZE
1⅜–2 in (3.5–5 cm)

PLANT SIZE
10–30 × 8–16 in
(25–76 × 20–41 cm),
excluding erect to arching
inflorescence
12–24 in (30–61 cm) long,
which is longer than the leaves

566

HOLCOGLOSSUM KIMBALLIANUM
ROYAL BUTTERFLY ORCHID
(REICHENBACH FILS) GARAY, 1972

The plants of the Royal Butterfly Orchid are either erect or pendent with terete leaves (round in cross section). The inflorescences, holding up to 20 flowers, project away from the plants, making them a spectacular sight, reminiscent of a flight of butterflies—hence the common name (the "royal" due to the rich purple coloration of the lip). The genus name is based on the Greek words *holkos*, "strap," and *glossa*, "tongue," referring to the prominent nectar spur, which projects behind the lip.

Pollination of related species with a similar floral morphology, such as *Holcoglossum nujiangense*, has been demonstrated in China to be by wild bee species related to the honeybee. The nectar spur, which in *H. kimballianum* has a wide opening, does not seem to contain nectar, so these beautiful flowers may be deceiving their pollinators.

Actual size

The flower of the Royal Butterfly Orchid is white to pale pink with a small dorsal sepal and large, winglike laterals. The petals twist and recurve markedly, and the three-lobed lip, notched in the center, is a bright rose-purple or red to white with red-purple veins.

SUBFAMILY	Epidendroideae
TRIBE AND SUBTRIBE	Vandeae, Aeridinae
NATIVE RANGE	West Java and Pahang (Peninsular Malaysia)
HABITAT	Moist rain forests, at 2,950–3,300 ft (900–1,000 m)
TYPE AND PLACEMENT	Epiphytic on tall exposed trees
CONSERVATION STATUS	Not assessed
FLOWERING TIME	February to August (spring and summer)

FLOWER SIZE
½ in (1.3 cm)

PLANT SIZE
2–3 × 4–6 in
(5–8 × 10–15 cm),
excluding short pendent
inflorescence

HYMENORCHIS JAVANICA
MEMBRANE ORCHID
(TEIJSMANN & BINNENDIJK) SCHLECHTER, 1913

567

The lovely little Membrane Orchid has a short stem densely packed with four to ten dark green, leathery leaves that have finely toothed margins. It produces a short, pendulous inflorescence with up to 12 small, cup-shaped, crystal-white, translucent flowers. The common name is a direct translation of the genus name, from the Greek, *hymen*, "membrane," and *orchis*, "orchid," referring to the thin texture of the flowers, which are opaque but permit light to pass through.

The genus is related to the large genus *Pteroceras*, from which it may not be distinct. This species is an epiphyte on thin, lichen-covered branches and twigs in humid forests. No pollinator has been reported, but given the pale color of the flowers and the presence of a nectar spur a moth would be expected.

Actual size

The flower of the Membrane Orchid has thin, forward-pointing, translucent white petals and sepals with finely toothed margins. The lip forms a short, flaring tube around the column with a green center. There is a short, blunt spur at the back of the flower.

SUBFAMILY	Epidendroideae
TRIBE AND SUBTRIBE	Vandeae, Aeridinae
NATIVE RANGE	Eastern Himalayas, and Assam to Bangladesh
HABITAT	Monsoon forests
TYPE AND PLACEMENT	Epiphytic
CONSERVATION STATUS	Not assessed
FLOWERING TIME	March to June (spring to early summer)

FLOWER SIZE
¾ in (2 cm)

PLANT SIZE
8–15 × 4–8 in
(20–38 × 10–20 cm),
excluding inflorescence,
which is arching-pendent,
6–10 in (15–25 cm) long

568

MICROPERA ROSTRATA
BEAKED
WALLET ORCHID
(ROXBURGH) N. P. BALAKRISHNAN, 1970

The Beaked Wallet Orchid has an arching, sparsely branching stem attached to larger branches and trunks of large trees by long, thick, wiry roots. The folded leaves clasp the stem on both sides. In the axils of a leaf a pendent to arching raceme bears 10–25 fragrant, pink flowers that are held upside down, with the pouched lip uppermost.

The genus name refers to the small open nectar pouch on the front of the lip (from the Greek for "small," *mikros*, and "pouch," *pera*), as does the "wallet" element of its common name. "Beaked" and *rostrata* (from the Latin *rostrum*, "beak") describe the hooked, forward-projecting column. The color, fragrance, and open nectar cavity might suggest pollination by a bee, but no pollinator has been reported.

The flower of the Beaked Wallet Orchid is bright pink with spreading sepals and petals all the same shape and size. The pink lip is pouched and flattened, with a short nectar spur at the base. The column is long and pointed.

Actual size

SUBFAMILY	Epidendroideae
TRIBE AND SUBTRIBE	Vandeae, Aeridinae
NATIVE RANGE	Vietnam, Thailand, Peninsular Malaysia, and Indonesia
HABITAT	Open swamps and wet grasslands, from sea level to 1,640 ft (500 m)
TYPE AND PLACEMENT	Terrestrial, but also epiphytic
CONSERVATION STATUS	Not threatened
FLOWERING TIME	March to July (spring to summer)

PAPILIONANTHE HOOKERIANA

HOOKER'S BUTTERFLY ORCHID

(REICHENBACH FILS) SCHLECHTER, 1915

FLOWER SIZE
2½ in (6.5 cm)

PLANT SIZE
25–50 × 10–18 in
(64–127 × 25–46 cm),
including lateral
erect-arching inflorescence
10–15 in (25–38 cm) long

569

The sun-loving Hooker's Butterfly Orchid bears 2–12 showy, long-lasting flowers, and its pencil-shaped leaves alternate on slender climbing stems with plentiful roots. The plants often grow next to stiff grasses or shrubs, which provide support for their scrambling stems. The genus name is from the Latin *papilio*, "butterfly," and the Greek *anthos*, "flower," allusions also made in the common name. The species name, and part of the common name, both refer to the renowned botanist, geographer, and explorer Joseph D. Hooker (1817–1911).

This species has been used extensively in horticultural hybridization and is a parent of the hybrid *Papilionanthe* Miss Joaquim, the national flower of Singapore. Pollination is by carpenter bees (genus *Xylocopa*) seeking nectar produced in the short lip spur. The leaves are also used in the treatment of rheumatism and other bone or joint pains.

The flower of the Hooker's Butterfly Orchid has pale lavender, broadly lanceolate sepals and petals with a large, trilobed lip. The purple sidelobes frame the purple column, and the large, flat, broadly triangular, white midlobe, with dark purple spots and stripes, has a short, cone-shaped nectar spur at the back.

Actual size

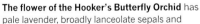

SUBFAMILY	Epidendroideae
TRIBE AND SUBTRIBE	Vandeae, Aeridinae
NATIVE RANGE	Indonesia (West Kalimantan province)
HABITAT	Seasonally wet swamp forests in partial shade, up to 3,300 ft (1,000 m)
TYPE AND PLACEMENT	Epiphytic or lithophytic on mossy branches or rocks
CONSERVATION STATUS	Not assessed, but has a narrow distribution
FLOWERING TIME	Spring

FLOWER SIZE
1 in (2.5 cm)

PLANT SIZE
12–18 × 15–30 in
(30–46 × 38–76 cm),
excluding inflorescence,
which is mostly upright and
shorter than the leaves

PARAPHALAENOPSIS SERPENTILINGUA
SERPENT ORCHID
(J. J. SMITH) A. D. HAWKES, 1963

The short, sometimes pendent stem of the Serpent Orchid carries between four and ten cylindrical leaves and an upcurved inflorescence crowded with up to seven, long-lasting, fragrant flowers. The genus name refers to the similarity of these flowers to those of the moth orchid, *Phalaenopsis*, in which genus they were previously included, although *Paraphalaenopsis* species are more closely related to *Vanda*. The epithet *serpentilingua* refers to the split midlobe of the lip, which resembles the tongue of a snake. Sometimes, the plant is called the Rat's Tail Orchid or Rat-tail Phalaenopsis because of its terete leaves.

Pollination has not been studied. However, the similarity of the flowers to those of the larger-flowered species of *Phalaenopsis*, which are pollinated by carpenter bees (*Xylocopa*), would suggest something similar for *Paraphalaenopsis* species.

The flower of the Serpent Orchid has clawed, broad, white petals and sepals that are slightly recurved and undulating. The lip has a white stalk and bright red to pink lobes—two lateral spreading and the middle one strap-shaped with reddish barring and a forked tip.

Actual size

SUBFAMILY	Epidendroideae
TRIBE AND SUBTRIBE	Vandeae, Aeridinae
NATIVE RANGE	Southern Thailand to Peninsular Malaysia
HABITAT	Riverine forests, at 1,640 ft (500 m)
TYPE AND PLACEMENT	Epiphytic
CONSERVATION STATUS	Not assessed
FLOWERING TIME	June to August (summer)

PENNILABIUM STRUTHIO
OSTRICH ORCHID
CARR, 1930

FLOWER SIZE
1 in (2.5 cm)

PLANT SIZE
3–4 × 3–6 in
(8–10 × 8–15 cm),
excluding pendent
inflorescence
1–2 in (2.5–5 cm) long

571

The tiny Ostrich Orchid has a short stem held in place on the side of a tree trunk by fleshy roots. It has a few closely set elliptical leaves, carried in two rows, below which an unbranched, flattened inflorescence, with one or two flowers, emerges. *Pennilabium* means "feathered lip" in Latin (*penna* and *labium*), and the common and species names allude to an ostrich—*struthio* in Latin. The long spur hanging below the relatively wide flower suggests an ostrich standing on one leg.

The flowers are lightly fragrant, and the erect keels at the spur entrance are transparent and could serve as guides for the tongue of visiting moths to reach the nectar below. The sticky disc that attaches the pollinia to the pollinator projects over the opening to the nectar spur, where it would easily come into contact with the moth.

Actual size

The flower of the Ostrich Orchid has cream to yellow, cupped, broad sepals and petals of similar size and shape. The white lip is elaborate and two-lobed, the lobes broad with a lacy margin. Between the lobes is a long, yellow spur with a thickened tip.

SUBFAMILY	Epidendroideae
TRIBE AND SUBTRIBE	Vandeae, Aeridinae
NATIVE RANGE	Borneo and the Philippines to Queensland (Australia)
HABITAT	Wet evergreen forests, often overhanging streams and swamps, from sea level to 1,970 ft (600 m)
TYPE AND PLACEMENT	Epiphytic
CONSERVATION STATUS	Not assessed, but threatened in some parts of its range by collection for horticulture
FLOWERING TIME	April to July (spring to summer, winter in Australia)

FLOWER SIZE
3 in (8 cm)

PLANT SIZE
3–5 × 8–15 in
(8–13 × 20–38 cm),
excluding erect to
arching inflorescence
20–35 in (51–89 cm) long

572

PHALAENOPSIS AMABILIS
WHITE MOTH ORCHID
(LINNAEUS) BLUME, 1825

The White Moth Orchid has between four and six thick, ovate-elliptic leaves arranged in two ranks on a short stem. The inflorescence emerges from near the base of the stem and can carry up to 40 large, sweetly fragrant flowers. The species was one of the few tropical Asian orchids known to Linnaeus, and he placed it with all other epiphytic orchids in the genus *Epidendrum*, on the simple basis that they grew on trees. The genus name is Greek for "like a moth" (*phalaina*, "moth"), and its member species, along with many horticultural hybrids, are collectively known as moth orchids.

In spite of its mothlike shape, the species is pollinated by carpenter bees (genus *Xylocopa*). There is no reward offered, but the combination of the large showy flowers with nectar guides (spots and stripes) and sweet fragrance is enough to attract these large insects.

Actual size

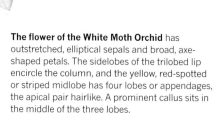

The flower of the White Moth Orchid has outstretched, elliptical sepals and broad, axe-shaped petals. The sidelobes of the trilobed lip encircle the column, and the yellow, red-spotted or striped midlobe has four lobes or appendages, the apical pair hairlike. A prominent callus sits in the middle of the three lobes.

SUBFAMILY	Epidendroideae
TRIBE AND SUBTRIBE	Vandeae, Aeridinae
NATIVE RANGE	Bangladesh, Myanmar, Thailand, Laos, Vietnam, Nicobar Islands, Malaysia, Java, Borneo, Sumatra, and the Philippines
HABITAT	Dense riparian forests, often high in the canopy, at 650–2,625 ft (200–800 m)
TYPE AND PLACEMENT	Epiphytic
CONSERVATION STATUS	Not threatened
FLOWERING TIME	Throughout the year, most profusely in summer

FLOWER SIZE
2 in (5 cm)

PLANT SIZE
4–8 × 8–14 in
(10–20 × 20–36 cm),
excluding arching-pendent
inflorescence
5–16 in (13–41 cm) long

573

PHALAENOPSIS CORNU-CERVI
DEER-ANTLERED MOTH ORCHID
(BREDA) BLUME & REICHENBACH FILS, 1860

The Deer-antlered Moth Orchid blooms repeatedly, often for years on the same persistent, flattened, sometimes branching, stem, which resembles a deer's antlers. The species name is Latin for "deer antler." The flowers are variable in color and shape but usually glossy and fleshy; red, yellow, brown, and bicolor forms are common. The genus name derives from the Greek *phalaina*, "moth," and *-opsis*, "looking like," although this species is less mothlike than many others in the genus. *Phalaenopsis* is the most important commercially of all orchid genera (as hybrids), but *P. cornu-cervi* has seen limited use in hybridization.

Growing at a lower elevation than most species in the genus, the plant makes copious aerial roots in its highly humid treetop habitat and grows steadily throughout the year. Pollination is by bees, but there is no reward.

The flower of the Deer-antlered Moth Orchid is most often yellow or chartreuse, overlaid with reddish spots or bars. In some plants, these spots coalesce to appear a solid color. The lip projects forward out of the plane of the rest of the flower and is white with purple spots and three lobes.

Actual size

SUBFAMILY	Epidendroideae
TRIBE AND SUBTRIBE	Vandeae, Aeridinae
NATIVE RANGE	Peninsular Malaysia, Borneo, and the Philippines
HABITAT	Dipterocarp forests, from sea level to 3,300 ft (1,000 m)
TYPE AND PLACEMENT	Epiphytic
CONSERVATION STATUS	Not assessed, but under pressure from logging and agriculture
FLOWERING TIME	May to August (spring to summer)

FLOWER SIZE
1½ in (3.8 cm)

PLANT SIZE
2–4 × 10–15 in
(5–10 × 25–38 cm),
excluding lateral
erect-arching inflorescence
12–20 in (30–51 cm) long

PHALAENOPSIS FUSCATA
DARK-SPOTTED MOTH ORCHID
REICHENBACH FILS, 1874

The Dark-spotted Moth Orchid prefers shady sites near streams, where it never gets too dry. Although its leaves are thick and can hold water, it does not have pseudobulbs and so needs more humidity than species that do have them. The genus name comes from the Greek word *phalaina*, "moth," and the ending *–opsis*, "likeness," alluding to the mothlike appearance of its members, which also gives them their common name, moth orchids. The species and common names refer to the large brown spots on the sepals and petals, from the Latin *fuscatus*, meaning "darkened."

The flower of the Dark-spotted Moth Orchid has oblong, brown-blotched, yellow, outstretched sepals and petals. The lip is yellow with brown stripes and trilobed, with a broad, cupped midlobe and short, raised sidelobes, with a complex, lobed callus between them.

In spite of their "moth orchid" name, *Phalaenopsis* species are not pollinated by moths, and are visited instead by various species of bee, although *P. fuscata* has not been specifically studied. It is fragrant during the daytime but produces no nectar.

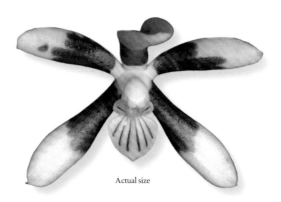

Actual size

SUBFAMILY	Epidendroideae
TRIBE AND SUBTRIBE	Vandeae, Aeridinae
NATIVE RANGE	Southern China, southern Korea, and southern Japan (including Okinawa Island)
HABITAT	Tree trunks in open forests or along valleys, at 1,640–4,600 ft (500–1,400 m)
TYPE AND PLACEMENT	Epiphytic
CONSERVATION STATUS	Not assessed, but potentially endangered due to collection for cultivation throughout its range
FLOWERING TIME	May to August (spring to summer)

FLOWER SIZE
1 in (2.5 cm)

PLANT SIZE
3–4 × 5–6 in
(8–10 × 13–15 cm),
including arching-pendent
inflorescence
5–8 in (13–20 cm) long

PHALAENOPSIS JAPONICA
NAGO ORCHID
(REICHENBACH FILS) KOCYAN & SCHUITEMAN, 2014

575

The Nago Orchid is a small species that produces a short stem clothed with closely spaced elliptic leaves that basally wrap around the stem. In many cases, two or more inflorescences are produced at the same time, and when each of them holds between six and ten flowers the mass of blooms can almost completely obscure the leaves. The common name comes from the city on Okinawa island, where this species was originally collected. It was formerly considered a member of genus *Sedirea* until DNA studies showed it to be a member of *Phalaenopsis*, with species of which hybrids have been produced.

Pollination is by small bees that are attracted by the highly fragrant, orange flower-scented blooms. The colorful flowers appear to the bees to contain nectar, but no reward is offered—another example of deceit pollination.

Actual size

The flower of the Nago Orchid has white sepals and petals that project slightly forward. The lateral sepals often have deep purple-red bars. The lip is white with lavender-purple marking and a large, flaring, apical lobe with two short lateral lobes that flank the opening into the spur.

SUBFAMILY	Epidendroideae
TRIBE AND SUBTRIBE	Vandeae, Aeridinae
NATIVE RANGE	Eastern and southeastern Australia
HABITAT	Rain forests along gullies and swampy areas, from sea level to 3,300 ft (1,000 m)
TYPE AND PLACEMENT	Epiphytic
CONSERVATION STATUS	Least concern in Queensland and New South Wales, where abundant and widespread, but near threatened in Victoria
FLOWERING TIME	September to January (spring to summer)

FLOWER SIZE
¼ in (0.6 cm)

PLANT SIZE
6–15 × 6–9 in
(15–38 × 15–23 cm),
excluding pendent-arching
inflorescence
6–10 in (15–25 cm) long

576

PLECTORRHIZA TRIDENTATA
TANGLE ORCHID
(LINDLEY) DOCKRILL, 1967

Actual size

The flower of the Tangle Orchid has yellow
to green, usually densely mottled sepals and
petals. The lip is white with three sharply pointed
lobes and yellow spots. The column is green with
brown spots and a yellow to white cap, and it
forms a nectar cavity with the lip.

The Tangle Orchid produces an elongate, erect to pendent stem
with two rows of lanceolate leaves that are 1 in (2.5 cm) apart.
It typically grows on twigs and other small branches, in which
its long roots form a tangled mass—hence the common name.
This feature is also the reference for the genus name, which
comes from the Greek *plektos*, "twisted," and *rhiza* "root." The
species name refers to the three lobes of the lip callus, from the
Latin for "three-toothed."

The shape, color, and lemony fragrance of *Plectorrhiza tridentata*
are consistent with bee pollination, and native bees have
been reported as pollinators, although the species of bee
observed was not recorded. Natural hybrids of this
species with some of the other Australian native
species, such as members of genus *Sarcochilus*,
have also been reported.

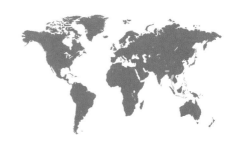

SUBFAMILY	Epidendroideae
TRIBE AND SUBTRIBE	Vandeae, Aeridinae
NATIVE RANGE	Patchily distributed across southern tropical Asia, Sri Lanka, the Nicobar and Andaman islands, eastern Himalayas, Laos, Thailand, Peninsular Malaysia, Java, and the Philippines
HABITAT	Tropical lowland wet forests, up to 1,970 ft (600 m), usually in shady places and about 3–20 ft (1–6 m) above the ground
TYPE AND PLACEMENT	Epiphytic on tree trunks, branches of small trees, or shrubs
CONSERVATION STATUS	Least concern due to its broad distribution, but may be under threat locally due to habitat destruction and collecting for the orchid trade
FLOWERING TIME	March to May and October to November (spring and fall)

FLOWER SIZE
¼ in (0.8 cm)

PLANT SIZE
22–48 × 10–18 in
(56–122 × 25–46 cm),
excluding inflorescences,
which are erect and
12–20 in (30–51 cm) long

577

POMATOCALPA MACULOSUM
SPOTTED PITCHER ORCHID
(LINDLEY) J. J. SMITH, 1912

Actual size

The Spotted Pitcher Orchid has a rambling habit, with elongate stems and fleshy roots, clambering over other vegetation and forming tangled masses. The leathery, dark-green leaves are spread along the nodes, and up to three, stiffly erect, branched inflorescences appear together once or twice a year. The flowers cluster at the tips of the inflorescence branches and are pollinated by the small stingless sugarbag or carbonaria bee (*Tetragonula*), which picks up pollinaria on its head.

The flowers vary over the orchid's wide distribution, and therefore the nomenclature for this species has been confusing. A thorough morphological revision has reunited all segregates into one species and accepted the minor changes as two geographically isolated subspecies. The genus name is derived from the Greek words *pomatos*, meaning "lid," and *kalpis*, "jug," referring to the lip shape.

The flower of the Spotted Pitcher Orchid is variable in position, with the saccate, yellow lip uppermost in many flowers and the opening at the base, which is nearly closed by the short column. The yellow to cream, red-spotted petals and sepals are strap-shaped and spreading.

SUBFAMILY	Epidendroideae
TRIBE AND SUBTRIBE	Vandeae, Angraecinae
NATIVE RANGE	Tropical Africa
HABITAT	*Brachystegia* woodland, often on rocks, at 1,970–7,200 ft (600–2,200 m)
TYPE AND PLACEMENT	Epiphytic or occasional lithophytic
CONSERVATION STATUS	Not formally assessed, but locally common
FLOWERING TIME	January to March (fall to winter)

FLOWER SIZE
½ in (1.3 cm), with a spur
up to 3 in (8 cm) long

PLANT SIZE
6–10 × 4–8 in
(15–25 × 10–20 cm),
excluding lateral inflorescence,
which is longer than the leaves

RANGAERIS MUSCICOLA
BLUSHING COMET ORCHID
(REICHENBACH FILS) SUMMERHAYES, 1936

578

The stout stem of the Blushing Comet Orchid is topped with a fan of up to ten stiff, curved leaves and held upright by its fleshy roots. It can produce up to four lateral racemes with as many as 15 or even 20 fragrant flowers that open white but become apricot orange tinged as they age—they "blush" pale orange, as in the common name. White-flowered orchids with a long nectar spur are often referred to as "comet orchids."

The species name is Greek for "moss-loving," referring to the mosses often present at the sites where this species grows. Natural pollination has not been observed, but the white flowers with long spurs suggest that hawk moths are likely to be involved. Pollination by two hawk moths, *Agrius convolvuli* and *Coelonia fulvinotata*, was observed in a related species, *Rangaeris amaniensis*.

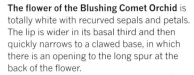

The flower of the Blushing Comet Orchid is totally white with recurved sepals and petals. The lip is wider in its basal third and then quickly narrows to a clawed base, in which there is an opening to the long spur at the back of the flower.

Actual size

SUBFAMILY	Epidendroideae
TRIBE AND SUBTRIBE	Vandeae, Aeridinae
NATIVE RANGE	Assam, Myanmar, southern China, Laos, and Vietnam
HABITAT	Mid-elevation tropical wet forests, often near or in gorges
TYPE AND PLACEMENT	Epiphytic
CONSERVATION STATUS	Threatened due to collection for horticulture
FLOWERING TIME	April to May

FLOWER SIZE
2⅜–2¾ in (6–7 cm)

PLANT SIZE
25–40 × 8–14 in
(64–102 × 20–36 cm),
including lateral,
erect-arching inflorescence
15–30 in (38–76 cm) long

RENANTHERA IMSCHOOTIANA
RED VANDA
ROLFE, 1891

579

In a genus of giant species, the Red Vanda is a moderately sized species that climbs trees with a vinelike stem. The long-lasting flowers are produced in great profusion on side inflorescences near the apex of the new growth. Remarkably beautiful when in flower and bearing 20–30 large flowers, plants have been stripped from the wild in the past for garden use in the tropics and the horticultural trade.

The genus name comes from the Latin words *ren*, "kidney," and *anthera*, "anther," referring to the shape of the pollen masses. Although commonly called Red Vanda, in reference to the similarity of its flowers to those of the genus *Vanda*, this species is not closely related to that genus genetically. The color, nectar cavity on the lip, and fruity fragrance indicate pollination by bees.

The flower of the Red Vanda has relatively small dorsal sepals and petals that are yellow with orange red spots. The small lip is dark red with a white margin around the entrance to the nectary. The large, ovoid lateral sepals are brilliant scarlet to red orange.

Actual size

SUBFAMILY	Epidendroideae
TRIBE AND SUBTRIBE	Vandeae, Aeridinae
NATIVE RANGE	Southern Indochina
HABITAT	Semi-deciduous, dry woodland and savannas, at up to 2,300 ft (700 m)
TYPE AND PLACEMENT	Epiphytic
CONSERVATION STATUS	Not assessed
FLOWERING TIME	July to November (summer to fall)

FLOWER SIZE
¾ in (2 cm)

PLANT SIZE
8–12 × 10–18 in
(20–30 × 25–46 cm),
excluding erect
lateral inflorescence
10–18 in (25–46 cm) long

580

RHYNCHOSTYLIS COELESTIS
BLUE FOXTAIL ORCHID
(REICHENBACH FILS) A. H. KENT, 1891

The Blue Foxtail Orchid continues to grow from its apex over many years, and its upright stem bears several strap-like, fleshy leaves. From the bases of these leaves, it produces many waxy, fragrant flowers on one to several inflorescences borne simultaneously, with possible color variants including pink and white forms. The genus name is derived from Greek and refers to the beaked (*rhynchos*) column (*stylis*).

The generally blue color of the flowers makes them popular in horticulture; hybrids with *Vanda*, to which the species is closely related, are popular. Species of *Rhynchostylis*, such as *R. retusa*, are used medicinally in Nepal, India, and Sri Lanka to treat wounds and rheumatism. The dried flowers are also reportedly useful as an insect repellant.

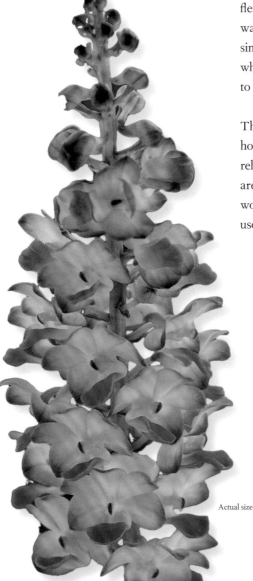

The flower of the Blue Foxtail Orchid
has spreading, white to pale blue petals
and sepals, and a lip with a darker apex
that forms a prominent spur at the back.
In some forms, the tips of the sepals
and petals have a dark splash.

Actual size

SUBFAMILY	Epidendroideae
TRIBE AND SUBTRIBE	Vandeae, Aeridinae
NATIVE RANGE	Philippines
HABITAT	Tropical rain forests, at 1,150 ft (350 m)
TYPE AND PLACEMENT	Epiphytic
CONSERVATION STATUS	Not threatened
FLOWERING TIME	Fall and winter

FLOWER SIZE
¼ in (0.8 cm)

PLANT SIZE
10–30 × 5–10 in
(25–76 × 13–25 cm),
excluding pendent
inflorescences
4–6 in (10–15 cm) long

ROBIQUETIA CERINA
WAXY CONE ORCHID
(REICHENBACH FILS) GARAY, 1972

581

With its tightly clustered, pendent inflorescences, the Waxy Cone Orchid does not at first even appear to be an orchid. The flowers do not open fully, and at a glance the cone-like inflorescences might be mistaken for a cluster of berries rather than flowers. A climbing plant that grows on tree trunks, *Robiquetia cerina*, like many other eventually tall orchids, can become pendent as it matures. The genus name honors Pierre-Jean Robiquet (1780–1840), the French chemist who discovered codeine and the first amino acid, asparagine. The scientific and common names both refer to the appearance of the flowers, from the Latin *cerinus*, meaning "wax colored."

Color variants range from purplish red to orange yellow. This and the shape of the flowers, particularly the short nectar spur, suggest that pollination by butterflies or bees might be expected.

The flower of the Waxy Cone Orchid has cupped sepals and petals that open only slightly and a trilobed lip with a short fleshy spur, often with nectar visible in its mouth. Typical colors are maroon red through pink and coral to yellow.

Actual size

SUBFAMILY	Epidendroideae
TRIBE AND SUBTRIBE	Vandeae, Aeridinae
NATIVE RANGE	Papua New Guinea, northeastern Australia (Queensland)
HABITAT	Coastal scrub and seasonally dry rain forests, from sea level to 650 ft (200 m)
TYPE AND PLACEMENT	Epiphytic
CONSERVATION STATUS	Least concern in Australia, not assessed for Papua New Guinea
FLOWERING TIME	September to January (spring)

FLOWER SIZE
⅛ in (0.5 cm)

PLANT SIZE
3–4 × 4–6 in
(7.5–10 × 10–15 cm),
excluding pendent
inflorescence
1–4 in (2.5–10 cm) long

582

SACCOLABIOPSIS ARMITII
SPOTTED PITCHER ORCHID
(FERDINAND VON MUELLER) DOCKRILL, 1967

The small Spotted Pitcher Orchid forms small clumps of short hanging stems, each with three to six lanceolate leaves and a pendant raceme with up to 50 tiny flowers. The common name refers to the red anther cap and the white apex of the lip, which create a spotted appearance. The lip forms a nectar cavity that is cupped or pitcher-shaped. The genus name refers to the shape of the lip, from the Latin *saccus*, "bag," and *labium*, "lip." The ending—*opsis*—means "looking like."

Saccolabiopsis armitii is a truly tropical species that occurs at low elevations. Its flowers last for many days and are pollinated by native bees in Queensland, Australia. The cavity on the lip does not appear to produce nectar, which indicates a method of pollination by deceit.

The flower of the Spotted Pitcher Orchid
has yellow to yellow-green, forward-pointing, concave sepals and clawed petals. The lip is pitcher-shaped with a white, apical lobe, and the column has a small red cap.

Actual size

SUBFAMILY	Epidendroideae
TRIBE AND SUBTRIBE	Vandeae, Aeridinae
NATIVE RANGE	Northeastern Australia, from southeastern Queensland to northeastern New South Wales
HABITAT	Rain forests, sea level to 1,300 ft (400 m)
TYPE AND PLACEMENT	Epiphytic
CONSERVATION STATUS	Least concern in Queensland, but endangered in New South Wales
FLOWERING TIME	January to May (summer to fall)

SARCOCHILUS DILATATUS
BROWN BUTTERFLY ORCHID
F. VON MUELLER, 1859

FLOWER SIZE
¾ in (1.9 cm)

PLANT SIZE
3–6 × 4–8 in
(8–15 × 10–20 cm),
excluding pendent
inflorescence
1½–3 in (3.8–8 cm) long

583

The small Brown Butterfly Orchid is often found clinging to the sides of large Hoop Pines (*Araucaria cunninghamii*), where it forms large colonies in shaded sites. The plants produce a short stem covered by lanceolate, sometimes darkly pigmented leaves, from the bases of which a pendent inflorescence with 2–12 flowers grows. The genus name, from the Greek *sarkos*, "flesh," and *cheilos*, "lip," alludes to the fleshy lip of its member species. The species name refers to the way the lip becomes broader, or dilates, near its top, while the common name is simply a fanciful reference to the flowers.

Pollination is presumed to be by small bees, such as those of the stingless genus *Trigona*, which have been observed to pollinate species with similarly shaped flowers in eastern Australia. No reward awaits the insects.

Actual size

The flower of the Brown Butterfly Orchid has the petals and dorsal sepal pointing upward and the other sepals pointing downward, all colored yellow with brown spots and tips. The white lip has some brown stripes and three lobes, the middle pointing downward and the sidelobes surrounding the column.

SUBFAMILY	Epidendroideae
TRIBE AND SUBTRIBE	Vandeae, Aeridinae
NATIVE RANGE	Eastern Australia
HABITAT	Temperate and subtropical rain forests, from sea level to 4,600 ft (1,400 m)
TYPE AND PLACEMENT	Epiphytic, rarely on rocks
CONSERVATION STATUS	Vulnerable, but in the past considered endangered due to the extensive extraction for horticulture, which has declined in recent years
FLOWERING TIME	June to October (winter to spring)

FLOWER SIZE
1 in (2.5 cm)

PLANT SIZE
3–8 × 6–10 in
(8–20 × 15–25 cm),
excluding arching-pendent
inflorescence
5–9 in (13–23 cm) long

SARCOCHILUS FALCATUS
ORANGE BLOSSOM ORCHID
R. BROWN, 1810

584

Actual size

The flower of the Orange Blossom Orchid has pure white, broadly oval, stalked sepals and petals. The lip is trilobed, the sidelobes curling over the column and the midlobe forming a saclike cavity. The lip varies from white with just yellow to white with red and purple stripes.

The extremely common Orange Blossom Orchid produces a short stem clothed in curving, broadly lanceolate leaves with a finely toothed margin. The plants start out erect but with age become lax to pendent. The curving leaves inspire the species name, which is Latin for "scythe-shaped," while the genus name comes from the Greek words *sarkos*, "flesh," and *cheilos*, "lip," referring to the thickness of the lip. The common name reflects both the appearance and fragrance of the flowers, which are like those of an orange blossom.

Pollination is by native bees, although the species involved is not known. Natural hybrids between *Sarcochilus* and other Australian native genera are reported. Propagation from seed has relieved the pressure previously experienced by many populations of the species, and their numbers have largely recovered.

SUBFAMILY	Epidendroideae
TRIBE AND SUBTRIBE	Vandeae, Aeridinae
NATIVE RANGE	Indochina, Thailand to Vietnam
HABITAT	Evergreen rain forests, at 650–985 ft (200–300 m)
TYPE AND PLACEMENT	Epiphytic
CONSERVATION STATUS	Not assessed
FLOWERING TIME	April to August (spring to summer)

FLOWER SIZE
⅜ in (1 cm)

PLANT SIZE
3–5 × 3–5 in
(8–13 × 8–13 cm),
excluding arching-pendent
inflorescence
1–3 in (2.5–8 cm) long

SARCOGLYPHIS MIRABILIS
PERCHING BIRD
(REICHENBACH FILS) GARAY, 1972

585

The miniature Perching Bird has a short stem with ligulate, obtusely bilobed leaves arranged in two rows. It produces a side inflorescence with between five and ten fleshy flowers. The genus name comes from the Greek words *sarx*, "flesh," and *glyphe*, "carving," alluding to the sculptured nature of the flower column, which is actually no fleshier than it is in most species of this group of tropical Asian orchids. The species name is Latin for "wonderful," and the common name a fanciful reference to the column looking like the head of a bird as it sits on a branch.

Pollination of *Sarcoglyphis mirabilis* has not been studied, but its color, fragrance, and nectar spur are consistent with the involvement of some type of bee. Nectar is produced by surface glands along the walls of the nectar cavity.

Actual size

The flower of the Perching Bird has cream to white, spreading sepals and petals, often with some purplish-brown blotches on their base. The pink to lavender, trilobed lip bears a spur. The column is upright with a bird-head-like anther cap.

SUBFAMILY	Epidendroideae
TRIBE AND SUBTRIBE	Vandeae, Aeridinae
NATIVE RANGE	Assam (India), Myanmar, Thailand, Vietnam, and Yunnan province (China)
HABITAT	Semi-deciduous forests, at 1,640–3,950 ft (500–1,200 m)
TYPE AND PLACEMENT	Epiphytic
CONSERVATION STATUS	Not threatened
FLOWERING TIME	Summer

FLOWER SIZE
⅜ in (1 cm)

PLANT SIZE
¼–1½ × 1 in
(2–4 × 2.5 cm),
including short, pendent
axillary inflorescences
½ in (1.3 cm) long

586

SCHOENORCHIS FRAGRANS
FRAGRANT PENNY ORCHID
(PARISH & REICHENBACH FILS) SEIDENFADEN & SMITINAND, 1963

Actual size

The beautifully proportioned miniature Fragrant Penny Orchid, one of the most endearing of all orchids, resembles a tiny succulent moth orchid (genus *Phalaenopsis*). Growing in fairly high light situations in semi-deciduous forests, its tough, fleshy leaves create a contrasting backdrop for the pretty spurred flowers. Older plants will often branch basally and form clumps of growths over time. The genus name comes from the Greek *schoenus*, "reed," and *orchis*, "orchid," referring to the reedlike leaf characteristics of other species in the genus, the opposite of those exhibited by *Schoenorchis fragrans*. The common name is a reflection of its highly fragrant flowers and small size (about the diameter of a small coin).

Pollination has not been documented. However, the color, fragrance, and shape of the flowers (particularly the nectar spur) would be compatible with some type of bee as pollinator.

The flower of the Fragrant Penny Orchid
appears in tight clusters that open progressively. Sepals and petals are white to pale pink, tipped with darker pink to purple. The mostly pink to purple lip is long, narrow, and subtly trilobed, with a broad-mouthed nectar spur behind.

SUBFAMILY	Epidendroideae
TRIBE AND SUBTRIBE	Vandeae, Aeridinae
NATIVE RANGE	Myanmar and Thailand
HABITAT	Forests, at 330–2,625 ft (100–800 m)
TYPE AND PLACEMENT	Epiphytic
CONSERVATION STATUS	Not assessed, but more common in the past
FLOWERING TIME	April to May (spring)

FLOWER SIZE
½–¾ in (1.25–1.9 cm)

PLANT SIZE
8–15 × 6–8 in
(20–38 × 15–20 cm),
including inflorescence,
6–10 in (15–25 cm) long,
which is shorter than the leaves

587

SEIDENFADENIA MITRATA
BISHOP'S MITRE ORCHID
(REICHENBACH FILS) GARAY, 1972

The Bishop's Mitre Orchid is a small epiphyte that has short stems topped with cylindrical, centrally grooved, pendent leaves and an axillary spike with more than ten, sweetly scented flowers subtended by small triangular bracts. The species produces a spur at the back of the flower, a structure that is usually associated with pollination by a bee, although in this case no observations of pollination have been published.

Seidenfadenia mitrata is the sole species in its genus but is closely related to the better-known genus *Rhynchostylis*. The species name refers to the anther cap, which is shaped somewhat like a bishop's mitre. The genus name honors diplomat and orchidologist Gunnar Seidenfaden (1908–2001), who was Danish ambassador to Thailand in the 1950s and arranged several orchid expeditions, during which more than 120 new species from Southeast Asia were discovered and described.

The flower of the Bishop's Mitre Orchid has reflexed-spreading, white petals and sepals. The purple-pink lip is tongue-shaped and opens to a saccate, forward-pointing spur. The column is flattened from the sides with two sidelobes.

Actual size

SUBFAMILY	Epidendroideae
TRIBE AND SUBTRIBE	Vandeae, Aeridinae
NATIVE RANGE	Himalayan valleys, Bengal, Indochina, Peninsular Malaysia, Borneo
HABITAT	Lower montane, riverine forests on limestone or ultramafic soils, usually at elevations below 985 ft (300 m) but sometimes up to 4,300 ft (1,300 m)
TYPE AND PLACEMENT	Pendently epiphytic
CONSERVATION STATUS	Not assessed but widespread
FLOWERING TIME	June to December (summer to winter)

FLOWER SIZE
⅜ in (1.1 cm)

PLANT SIZE
8–12 × 4–8 in
(20–30 × 10–20 cm),
including short side
inflorescences
2–7 in (5–18 cm) long

588

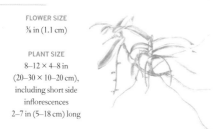

SMITINANDIA MICRANTHA
LITTLE ROSE ORCHID
(LINDLEY) HOLTTUM, 1969

Actual size

The Little Rose Orchid has an almost vinelike pendent or arching, branching stem, topped with many distantly spaced, linear, apically bilobed leaves and a lateral, densely flowered, bract-bearing, arching-pendent inflorescence. The flowers are arranged spirally and open from the base, with several open at once. They have a broad, short cavity, but no nectar has been reported. Pollination is most likely to be by bees, for which this type of morphology seems to be adapted.

The species name, *micrantha*, refers to the small size of the flowers, while the genus is named after Thai forest botanist Tem Smitinand (1920–95), who worked on the orchid flora of Thailand with the Danish orchidologist Gunnar Seidenfaden. The stems are reported to have antiseptic properties, and in Nepal the root is powdered to prepare a tonic.

The flower of the Little Rose Orchid has spreading, pale pink petals and sepals with a saccate spur at the back. The lip has three darker pink lobes and a callus that partially blocks the opening to the spur.

SUBFAMILY	Epidendroideae
TRIBE AND SUBTRIBE	Vandeae, Aeridinae
NATIVE RANGE	China (Yunnan province), Thailand, and Vietnam
HABITAT	Semi-deciduous to evergreen forests, at 985–5,900 ft (300–1,800 m)
TYPE AND PLACEMENT	Epiphytic
CONSERVATION STATUS	Not assessed, but likely to be threatened due to habitat conversion
FLOWERING TIME	April to May (spring)

FLOWER SIZE
¼ in (0.8 cm)

PLANT SIZE
3–5 × 2–3 in
(8–13 × 5–8 cm),
excluding erect to
arching inflorescence
3–6 in (8–15 cm) long

STEREOCHILUS DALATENSIS
SOLID-LIP ORCHID
(GUILLAUMIN) GARAY, 1972

589

The Solid-lip Orchid is a miniature species with a short stem covered by closely spaced thick leaves in two ranks. The inflorescence is produced from the base of one of the lower leaves and bears up to ten sweetly scented colorful flowers. The common name refers to the lip, which is sturdy and appears thick, although it has a concealed nectar cavity at its base. The genus name is based on the Greek words *stereo*, "solid," and *cheilos*, "lip," referring again to the same feature. The species name refers to the town in Vietnam, Da Lat, where this orchid was first found.

Pollination is likely to be carried out by a small bee, attracted by the bright colors and sweet fragrance. It is not known if the cavity on the lip in fact contains nectar.

Actual size

The flower of the Solid-lip Orchid has similar, reflexing, pale pink to white, lanceolate sepals and petals. The bright pink to lavender lip has sidelobes that surround the entrance to the spur and a much larger concave midlobe. The column is shaped like a long-beaked bird's head.

SUBFAMILY	Epidendroideae
TRIBE AND SUBTRIBE	Vandeae, Aeridinae
NATIVE RANGE	Java to southwestern China (Yunnan)
HABITAT	Swampy, primary forests, on trees and in coffee plantations, at 490–5,250 ft (150–1,600 m)
TYPE AND PLACEMENT	Epiphytic
CONSERVATION STATUS	Not assessed
FLOWERING TIME	April to October (spring to fall)

FLOWER SIZE
¼ in (0.8 cm)

PLANT SIZE
1 × 6–10 in
(2.5 × 15–25 cm),
excluding inflorescence
1–1½ in (2.5–3.8 cm) long

TAENIOPHYLLUM PUSILLUM
SMALL
TAPEWORM ORCHID
(WILLDENOW) SEIDENFADEN & ORMEROD, 1995

Actual size

The unattractive genus name of this leafless epiphyte comes from the Greek words *taenia*, "tapeworm," and *phyllos*, "leaf," despite the lack of leaves. Its versatile roots have developed into flat, long, tapeworm-like shapes, twisting and turning over the branch on which it sits. The roots are green and have taken on the function of a leaf in photosynthesis to produce food for the orchid. The redundant leaves have been reduced to tiny brown scales on the short stem that carries the flowers.

The zigzag inflorescence is covered with tubercles, and carries one or two flowers successively, subtended by warty floral bracts. Pollination is most likely by small bees. The lip is almost entirely composed of a nectar spur, with the column buried inside the tube leading to the spur.

The flower of the Small Tapeworm Orchid is
fleshy with spreading, linear, greenish-brown to
tan sepals and petals of similar shape and size.
The lip is white and pouch-like. The column is
short, and the anther cap has two pinkish-purple
spots, resembling eyes.

SUBFAMILY	Epidendroideae
TRIBE AND SUBTRIBE	Vandeae, Aeridinae
NATIVE RANGE	Malay Peninsula, Sumatra, Borneo, Thailand, and Vietnam
HABITAT	Open areas and bluffs in and adjacent to evergreen and semi-deciduous forests, at 165–3,950 ft (50–1,200 m)
TYPE AND PLACEMENT	Epiphytic or terrestrial
CONSERVATION STATUS	Not threatened
FLOWERING TIME	Summer

THRIXSPERMUM CALCEOLUS

CLIMBING SHOE ORCHID

(LINDLEY) REICHENBACH FILS, 1868

FLOWER SIZE
1½ in (3.8 cm)

PLANT SIZE
8–30 × 6–8 in
(20–76 × 15–20 cm),
including short inflorescences
1–2 in (2.5–5 cm)
long that emerge
opposite the leaves

591

The vinelike Climbing Shoe Orchid is equally at home in trees and on rocky outcrops. The pretty, crystalline-textured flowers, although short-lived (usually only for a few hours on one day), are produced repeatedly, one or two at a time, from the same inflorescences on most plants in the population in coordinated flushes. This mass flowering is usually initiated by a sharp temperature drop associated with a rainstorm. The genus name, from the Greek words *thrix*, "hair," and *sperma*, "seeded," refers to the thin, elongate seeds, although these are found in most orchids.

The common and species names both reflect the pouch-like or shoe-like lip—*calceolus* means "like a small shoe." Bees are the pollinators, attracted by a sweet floral fragrance and nectar. They become trapped by the lip and remove the pollinia as they struggle to leave.

The flower of the Climbing Shoe Orchid has crystalline white, broadly lanceolate sepals and petals. The lip, which can point in any direction, bears a pouch often with yellow patches near its opening and underneath.

Actual size

SUBFAMILY	Epidendroideae
TRIBE AND SUBTRIBE	Vandeae, Aeridinae
NATIVE RANGE	Philippines
HABITAT	Rain forests, from sea level to 985 ft (300 m)
TYPE AND PLACEMENT	Epiphytic
CONSERVATION STATUS	Not assessed
FLOWERING TIME	May to July (spring to summer)

FLOWER SIZE
2½ in (6.5 cm)

PLANT SIZE
40–160 × 15–25 in
(102–406 × 38–64 cm),
including lateral
inflorescences
2–4 in (5–10 cm) long

TRICHOGLOTTIS ATROPURPUREA
PURPLE BEARD ORCHID
REICHENBACH FILS, 1876

592

The cane-like Purple Beard Orchid can reach great length, its stem clothed in oval leaves in two ranks, each wrapping its base around the stem and spaced about 1 in (2.5 cm) apart. The plant is held in position by thick roots produced from all along the stem, and it bears short, one or two flowered inflorescences that can occur along a large portion of its length simultaneously, making it incredibly showy when in flower. The genus name derives from the Greek words *trichos*, "hair," and *glotta*, "tongue," the latter referring to the lip. The midlobe of the lip is decidedly hairy, which is responsible for the common name.

Pollination of *Trichoglottis atropurpurea* is likely to involve bees, given the color, shape, and fragrance of its flowers. However, there have so far been no observations of pollination in the wild.

Actual size

The flower of the Purple Beard Orchid has dark reddish-purple, broadly lanceolate, outstretched sepals and petals. The lip has two basal lobes that together with the column form a yellow-edged nectar cavity, and the midlobe is itself trilobed with a central, bearded callus.

SUBFAMILY	Epidendroideae
TRIBE AND SUBTRIBE	Vandeae, Aeridinae
NATIVE RANGE	Borneo and Sumatra
HABITAT	*Gymnostoma* (she-oak) forests on sandstone, at 650–4,300 ft (200–1,300 m)
TYPE AND PLACEMENT	Epiphytic on trunks in light shade
CONSERVATION STATUS	Not assessed
FLOWERING TIME	April to June, October to November (spring and fall)

FLOWER SIZE
1 in (2.5 cm)

PLANT SIZE
40–80 × 3–5 in
(102–203 × 8–13 cm),
including short
lateral inflorescences
1–2 in long (2.5–5 cm) long

TRICHOGLOTTIS SMITHII

STRIPED VINE ORCHID

CARR, 1935

593

A vinelike climber that forms clusters of twisted stems with fat white roots, the Striped Vine Orchid has short inflorescences, producing one or two flowers at a time, blooming simultaneously along the stem. The leaves are oblong with a notched tip and spaced ½ in (1.3 cm) apart. The plant was named in honor of Dutch botanist Johannes Jacobus Smith (1867–1947), Director of the Buitenzorg (now Bogor, Indonesia) Botanical Gardens, 1913–24. Smith traveled extensively and discovered and described hundreds of orchid species native to the Dutch East Indies. The genus name is derived from the Greek *trichos*, "hair," and *glotta*, "tongue," the latter referring to the lip.

Some type of bee is the expected pollinator of *Trichoglottis smithii*, given its color, fragrance, and nectar cavity on the lip. There does not appear to be any nectar produced.

Actual size

The flower of the Striped Vine Orchid
has white to cream, reddish-brown-barred, oblanceolate sepals and petals, the lateral sepals the largest. The trilobed lip, with two rounded, upright sidelobes and a pointed middle lobe, is also white, but lavender-purple-spotted and covered in long hairs.

SUBFAMILY	Epidendroideae
TRIBE AND SUBTRIBE	Vandeae, Aeridinae
NATIVE RANGE	Sulawesi (Indonesia)
HABITAT	Forests and plantations, at 3,950–8,200 ft (1,200–2,500 m)
TYPE AND PLACEMENT	Epiphytic
CONSERVATION STATUS	Not assessed
FLOWERING TIME	April to May (spring)

FLOWER SIZE
¼ in (0.8 cm)

PLANT SIZE
5–10 × 3–5 in
(13–25 × 7.6–13),
including lateral
inflorescence
⅜ in (1 cm) long

594

TRACHOMA CELEBICUM
SULAWESI
RED-SPOTTED ORCHID
(SCHLECHTER) GARAY, 1972

Actual size

The small Sulawesi Red-spotted Orchid has an erect to arching, slightly flattened stem with narrowly elliptic, thick leaves that are sheathing at the base and spaced about ⅜ in (1 cm) apart. It produces one or two inflorescences opposite a leaf, each carrying one or two strongly scented flowers at a time, but eventually producing 10–15 blooms on each. The genus name means "toughness" in Greek, a reference to the dried bracts left on an inflorescence after older flowers have fallen off. Celebes, referenced in the species name, is the former name of Sulawesi.

A halictid bee (sweat bee) has been recorded as the pollinator of a closely related species, which would be logical for this species also. The distinction between the genera *Trachoma* and *Tuberolabium* has long been problematic, and assignment of this species to the former is tentative.

The flower of the Sulawesi Red-spotted Orchid
has fleshy, forward-pointing, yellow sepals and
petals, each with a reddish basal blotch. The
white lip forms a sacklike spur and has two
hornlike appendages and two yellow raised
blotches. Its sidelobes enfold the column.

SUBFAMILY	Epidendroideae
TRIBE AND SUBTRIBE	Vandeae, Aeridinae
NATIVE RANGE	From Assam and the Khasi Hills (India) to Yunnan province (China), Myanmar, and northern Thailand
HABITAT	Dry-deciduous forests, at 2,625–5,600 ft (800–1,700 m)
TYPE AND PLACEMENT	Epiphytic on exposed deciduous trees, primarily dwarf oak
CONSERVATION STATUS	Originally listed as highly endangered, and it remains so in Assam, where it was first found, but is widespread and locally common in eastern Himalayas
FLOWERING TIME	September to November (fall)

VANDA COERULEA
BLUE VANDA
GRIFFITH EX LINDLEY, 1847

FLOWER SIZE
4–4¾ in (10–12 cm)

PLANT SIZE
20–75 × 18–30 in
(51–191 × 46–76 cm),
excluding inflorescence,
which is erect to arching,
10–30 in (25–76 cm) long

595

When the Blue Vanda, with its stout stems, fans of folded leathery leaves, and beautifully checkered blue flowers, was first discovered by William Griffith in 1847, it caused a big stir because such a large blue orchid, offering spikes of flat long-lasting flowers, was an orchid grower's dream. The discovery resulted in several expeditions to Assam to collect this and other remarkable plants for the hothouses of Europe.

Vanda coerulea is responsible for the vibrant blues and purples of many cultivated *Vanda* hybrids. The scientific name is derived from *vandaar*, the vernacular Sanskrit name for epiphyte. Juice from the flower has been used to create eyedrops for treating glaucoma and cataracts. Laboratory research has also indicated that extracts of this blue orchid may have potential for use in antiaging skin treatments.

The flower of the Blue Vanda has clawed, broad, spreading sepals and petals, the latter often with a twisted stalk. The lip is short and three-lobed, and the column has a white cap.

Actual size

SUBFAMILY	Epidendroideae
TRIBE AND SUBTRIBE	Vandeae, Aeridinae
NATIVE RANGE	Himalayas to southwestern China and Indochina
HABITAT	Moss-covered branches in seasonally wet forests, at 1,970–7,545 ft (600–2,300 m)
TYPE AND PLACEMENT	Epiphytic
CONSERVATION STATUS	Not assessed
FLOWERING TIME	April to July (spring to summer)

FLOWER SIZE
2 in (5 cm)

PLANT SIZE
20–45 × 12–20 in
(51–114 × 30–51 cm),
including axillary
inflorescence
5–8 in (13–20 cm) long

VANDA CRISTATA
FORKED BEETLE ORCHID
WALLICH EX LINDLEY, 1833

596

The strangely forked and darkly colored lip of the Forked Beetle Orchid has always made it stand out from the other species of the genus *Vanda*. The plant has a long stem clothed on opposite sides in linear leaves, from which the inflorescences emerge, bearing between two and six fragrant flowers. The genus name comes from the Sanskrit word for a plant that grows on a tree.

The peculiarly colored lip of *V. cristata* made pollination a subject of speculation until, in 1983, it was reported that an unidentified beetle was the pollinator. Beetles were subsequently also reported as pollinators of a sister species, *V. griffithii*, in Bhutan. Why a beetle would visit these apparently rewardless flowers is an unsolved mystery.

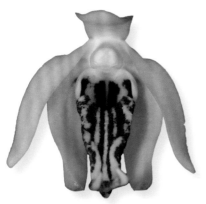

Actual size

The flower of the Forked Beetle Orchid has lanceolate, green to yellow green sepals and petals. The lip is white to cream with nearly black to maroon strips and is trilobed—the middle with a forked apex and the laterals forming a circular basal cavity under the column.

SUBFAMILY	Epidendroideae
TRIBE AND SUBTRIBE	Vandeae, Aeridinae
NATIVE RANGE	Eastern Himalayas to Thailand
HABITAT	Semi-deciduous and deciduous dry forests, from sea level to 2,300 ft (700 m)
TYPE AND PLACEMENT	Epiphytic on deciduous trees
CONSERVATION STATUS	Not assessed
FLOWERING TIME	March to May (late spring to summer)

VANDA CURVIFOLIA
VERMILION BOTTLEBRUSH VANDA
(LINDLEY) L. M. GARDINER, 2012

FLOWER SIZE
1 in (2.5 cm)

PLANT SIZE
6–20 × 8–15 in
(15–51 × 20–38 cm),
excluding lateral, erect
inflorescence
5–12 in (13–30 cm) tall

597

The short often basally branching, stout stems of the Vermilion Bottlebrush Vanda are covered by two rows of narrowly linear, strongly curving leaves that are two-toothed at the tip and have sheathing bases. The species produces densely flowered, upright inflorescences of 20–60 bright orange-red flowers. This species was well known in the previously recognized genus *Ascocentrum*, but is now included in a broader concept of the genus *Vanda*, which gets its name from the Sanskrit word for an epiphyte.

The flowers are scentless, have dark pollinia caps and produce nectar in a relatively large spur lined by secretory hairs in its middle section. These traits, plus the color, are typical for bird pollination, although in this case it has not been confirmed by studies in the wild.

The flower of the Vermilion Bottlebrush Vanda is red orange and has spreading, broad, oblanceolate sepals and petals of similar size and shape. The lip is simple, lanceolate, and curved backward. At its base, in the middle of a pair of yellow knobs, is an opening to the nectary.

Actual size

SUBFAMILY	Epidendroideae
TRIBE AND SUBTRIBE	Vandeae, Aeridinae
NATIVE RANGE	China, Japan, and Korea
HABITAT	Deciduous forests, at 1,640–3,300 ft (500–1,000 m)
TYPE AND PLACEMENT	Epiphytic and lithophytic
CONSERVATION STATUS	Not threatened
FLOWERING TIME	Late spring and summer

FLOWER SIZE
1½ in (3.8 cm)

PLANT SIZE
5–8 × 6–10 in
(13–20 × 15–25 cm),
excluding axillary,
mostly erect inflorescence
2–5 in (5–13 cm) long

598

VANDA FALCATA
WIND ORCHID
(THUNBERG) BEER, 1854

The Wind Orchid has been treasured in East Asia for centuries as a cultivated plant. It has a revered cultural significance in its native range, despite being cultivated all over the world in recent decades. As recently as 2014, the species was treated as a member of the genus *Neofinetia*, but this small epiphyte was then transferred back to the much larger genus *Vanda*, in which it had been included in 1854. *Vanda* is the Sanskrit word for "epiphyte," and its common name is a translation of its name in Japanese, *fuuran*, reflecting the graceful form of the flower. Many hybrids between this species and other genera in subtribe Aeridinae have been produced.

In its native habitats, *V. falcata* is pollinated by hawk moths. They are attracted by the orchid's sweet scent, only produced at night, and the nectar in its long spur.

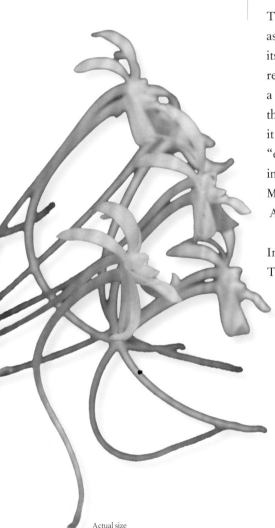

The flower of the Wind Orchid is generally white, although pale purple, pink, yellow, and green-tinted forms exist, most with an entirely white lip. Sepals and petals are narrowly lanceolate and reflexed, and the trilobed lip leads to a long, curved nectar spur.

Actual size

SUBFAMILY	Epidendroideae
TRIBE AND SUBTRIBE	Vandeae, Aeridinae
NATIVE RANGE	Java and the Lesser Sunda Islands (Indonesia), naturalized in the Philippines and Hawaii
HABITAT	Seasonally wet forests, from sea level to 1,640 ft (500 m)
TYPE AND PLACEMENT	Epiphytic or occasionally terrestrial
CONSERVATION STATUS	Not assessed
FLOWERING TIME	December to January (winter)

FLOWER SIZE
2½ in (6.5 cm)

PLANT SIZE
20–70 × 25–35 in
(51–178 × 64–89 cm),
including axillary
inflorescence
10–20 in (25–51 cm) long

VANDA TRICOLOR
SWEET PURPLE TONGUE
LINDLEY, 1847

599

The Sweet Purple Tongue is highly fragrant and has long been cultivated for its showy flowers in the lowland tropics around the world, from where it has started to invade local forests. Plants can grow to a large size, forming an erect main stem clothed by linear succulent leaves. From the bases of the leaves, lateral inflorescences carry 6–15 flowers. There are a number of color forms, but all of them have a lavender purple lip, which gives this species its common name.

Large bees of many types are attracted to the colorful and fragrant flowers wherever they are grown throughout the tropics. In Panama, where the plant is cultivated outside in gardens, euglossine bees have been observed as visitors, and in the wild in Java, a carpenter bee, *Xylocopa latipes*, has been recorded as a pollinator.

The flower of the Sweet Purple Tongue has pale yellow to white, spoon-shaped sepals and petals with brown to purple spots, the petals twisting almost 180 degrees. The lip has three pale to dark purple lobes. The column is white with an expanded base that surrounds the lip base.

Actual size

SUBFAMILY	Epidendroideae
TRIBE AND SUBTRIBE	Vandeae, Aeridinae
NATIVE RANGE	China (Yunnan and Guangxi provinces) through Indochina to Peninsular Malaysia
HABITAT	Evergreen to semi-deciduous forests, usually on rocky outcrops, at 985–3,600 ft (300–1,100 m)
TYPE AND PLACEMENT	Terrestrial, lithophytic
CONSERVATION STATUS	Not assessed, but threatened in some parts of its range by collection for horticulture
FLOWERING TIME	April to July (spring to summer)

FLOWER SIZE
3 in (8 cm)

PLANT SIZE
12–40 × 12–25 in
(30–102 × 30–64 cm),
including arching
inflorescence
15–25 in (38–64 cm) long

600

VANDOPSIS GIGANTEA
PORPOISE-HEAD ORCHID
(LINDLEY) PFITZER, 1889

The flower of the Porpoise-head Orchid has outstretched, broadly ovate sepals and petals. The lip is trilobed, with the sidelobes encircling the base of the column. A callus sits in the middle of the three lobes. Flower color is generally yellow to cream with brown-red to purple spotting on all parts.

The Porpoise-head Orchid can form massive plants, with a long stem clothed by two ranks of tough, strap-shaped leaves. The inflorescence is produced from the base of a leaf and can have up to 20 flowers. The size of the plants bestows the species name, while the genus name refers to the similarity of the member species to the genus *Vanda*. The common name refers to the shape of the column, which resembles the head of a porpoise with two purple spots for its eyes.

As in other species of this tribe, pollination is likely to be by large bees, most likely carpenter bees (genus *Xylocopa*). The flowers have a faintly sweet fragrance during the daytime, and bees could mistake the empty chamber on the base of the lip as a nectar cavity.

Actual size

SUBFAMILY	Epidendroideae
TRIBE AND SUBTRIBE	Vandeae, Aeridinae
NATIVE RANGE	Himalayas to northern Myanmar and China (Yunnan province)
HABITAT	Trunks of pine and oak trees or rocks, at 4,920–7,545 ft (1,500–2,300 m)
TYPE AND PLACEMENT	Epiphytic or lithophytic
CONSERVATION STATUS	Not assessed
FLOWERING TIME	May to June (late spring to early summer)

VANDOPSIS UNDULATA

WHITE QUASI VANDA

(LINDLEY) J. J. SMITH, 1912

FLOWER SIZE
1¼ in (3 cm)

PLANT SIZE
25–40 × 7–12 in
(64–102 × 18–30 cm),
excluding lateral,
arching inflorescence
6–10 in (15–25 cm) long

601

The common name for this orchid is a translation of its Chinese name. The plant is almost vinelike, with a long stem—erect or nearly pendent, depending on where it is growing—and oblong leaves spaced about 2–3 in (5–8 cm) apart. On slopes, it can form tangled clumps of stems that clamber over shrubby vegetation. The inflorescence bears between three and eight sweetly fragrant flowers. The genus name parallels the Chinese name, the Greek ending *-opsis*, meaning "likeness," alluding to its similarity to *Vanda*, with which it has no close genetic relationship. The species name, Latin for "wavy," reflects the undulating margins of the petals and sepals.

The large sweet-smelling flowers have a colorful lip and nectar guides, but no nectar spur. They are probably pollinated by bees, attracted by their general similarity to reward-offering flowers.

Actual size

The flower of the White Quasi Vanda
has spoon-shaped, white sepals and petals, the lateral sepals the largest. The trilobed lip has a yellow, pointed median lobe with purple stripes, a swollen tip, and small teeth on its sides, with the yellow sidelobes encircling the column base.

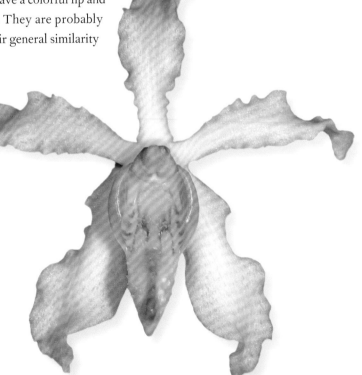

SUBFAMILY	Epidendroideae
TRIBE AND SUBTRIBE	Vandeae, Angraecinae
NATIVE RANGE	Madagascar
HABITAT	Rain forests, at up to 4,920 ft (1,500 m)
TYPE AND PLACEMENT	Epiphytic and lithophytic
CONSERVATION STATUS	Not threatened
FLOWERING TIME	Spring to fall (May to September)

FLOWER SIZE
¾–1¼ in (2–3 cm)

PLANT SIZE
4–5 × 3–4 in
(10–13 × 8–10 cm),
excluding erect to
arching inflorescences
10–18 in (25–46 cm) long

602

AERANGIS CITRATA
YELLOW-SPOTTED ORCHID
(THOUARS) SCHLECHTER, 1914

The miniature Yellow-spotted Orchid has glossy leaves and long pendent to arching spikes of up to 20 elegant, gleaming white flowers, named for the yellow-colored spots on its anther cap. These are the two pollen masses showing through the thin-walled anther cap. Some have reported a faint lemony fragrance, but most people detect no scent. Other species of the genus that have been studied produce nectar in their spur and attract night-flying hawk moths as pollinators.

The genus name is derived from the Greek words *aer*, "air," and *angos*, "vessel," a reference to the mostly epiphytic nature of this species, found throughout Madagascar, wherever there are ponds or streams to keep the atmosphere humid. Like the rest of its subtribe, the plant does not produce pseudobulbs, making it more dependent on environmental water.

The flower of the Yellow-spotted Orchid is small and white to creamy white with narrow sepals and much broader petals. The anther cap bears two yellow bumps on its surface, and the lip often has a pale violet flush and bears a curving nectar spur at the back.

Actual size

SUBFAMILY	Epidendroideae
TRIBE AND SUBTRIBE	Vandeae, Angraecinae
NATIVE RANGE	Comoro Islands
HABITAT	Humid evergreen and semi-deciduous forests, from sea level to 2,460 ft (750 m)
TYPE AND PLACEMENT	Epiphytic
CONSERVATION STATUS	Not assessed
FLOWERING TIME	September to October (spring)

AERANGIS HARIOTIANA
RUSTY STARS
(KRAENZLIN) P. J. CRIBB & CARLSWARD, 2012

FLOWER SIZE
⅛ in (0.3 cm)

PLANT SIZE
2–3 × 2–4 in
(5–8 × 5–10 cm),
excluding pendent
axillary inflorescence
5–8 in (13–20 cm) long

603

The miniature Rusty Stars has a short stem clothed in oblong, unequally bilobed leaves and often produces between three and five inflorescences at a time. Unusually, the flowers open at the end of the stem first, which is the opposite of most plants. The genus name is from the Greek for *aer*, "air," and *angos*, "vessel," an allusion to the epiphytic nature of the plants. Some authors have placed this and other similarly small-flowered members of the genus in *Microterangis*, but the size of the flowers alone does not warrant this special treatment.

In spite of their small size, the flowers have a deep nectar spur, so apparently there is an insect, probably a moth, on the Comoro Islands that can reach into the spur. The major nocturnal pollinators are hawk moths, which are rare on small islands such as the Comoros, but common in nearby Madagascar.

Actual size

The flower of Rusty Stars has pale to rusty orange, triangular, outstretched sepals, petals, and lip, the latter with a central groove that leads to the entrance to the nectar spur, which projects directly backward. The column is perched immediately over this entrance.

SUBFAMILY	Epidendroideae
TRIBE AND SUBTRIBE	Vandeae, Angraecinae
NATIVE RANGE	Madagascar
HABITAT	Wet forests, at 2,950–4,600 ft (900–1,400 m)
TYPE AND PLACEMENT	Epiphytic
CONSERVATION STATUS	Threatened by deforestation
FLOWERING TIME	Mostly summer to fall

FLOWER SIZE
2 in (5 cm)

PLANT SIZE
10–16 × 8–12 in
(25–41 × 20–30 cm),
excluding wiry,
pendent inflorescences
8–20 in (20–51 cm) long

AERANTHES RAMOSA
GREEN SPIDER ORCHID
ROLFE, 1901

604

The Green Spider Orchid has between five and twelve elongate
lanceolate leaves on opposing sides of the stem, covering it with
their bases. From the base of the stem, the plant produces dark,
wiry inflorescences carrying one to three flowers that appear
in succession and seem to be suspended in midair. The genus
name is derived from the Greek *aer*, "air," and *anthos*, "flower,"
referring to the apparent lack of connection between the plant
and its flower. The species name refers to the branching nature
of the inflorescences (from the Latin *ramosus*), and the common
name to the spidery appearance of the flowers.

The faintly sweet nocturnal fragrance and long nectar spur
suggest that it has a moth pollination syndrome, although this
has not been studied in the wild. Green or white starry flowers
are usually associated with such pollination.

Actual size

The flower of the Green Spider Orchid is a
translucent green with spidery petals and sepals.
The lip is sometimes whitish, with a green,
club-shaped nectary hanging behind the flower.
The anther cap is usually white to whitish green.

SUBFAMILY	Epidendroideae
TRIBE AND SUBTRIBE	Vandeae, Angraecinae
NATIVE RANGE	Ghana, Guinea, Ivory Coast, Nigeria, Liberia, Sierra Leone, and Togo
HABITAT	Wet evergreen forests, at 1,300–3,300 ft (400–1,000 m)
TYPE AND PLACEMENT	Epiphytic
CONSERVATION STATUS	Not threatened
FLOWERING TIME	October to November (fall)

FLOWER SIZE
⅝ in (1.5 cm)

PLANT SIZE
6–10 × 6–8 in
(15–25 × 15–20 cm),
including short side
inflorescence
1–2 in (2.5–5 cm) long

ANCISTRORHYNCHUS CEPHALOTES
SNOWBALL ORCHID
(REICHENBACH FILS) SUMMERHAYES, 1944

605

Dense clusters of sweetly scented blooms arising at the base of the plant are the hallmark of the genus *Ancistrorhynchus*. The Snowball Orchid produces 15–20 flowers in a cone-like inflorescence. The genus name comes from the Greek words *ankistron*, "fishhook," and *rhynchos*, "snout," referring to the hooked apex of the column. The common name is from the Greek *kephalos*, "head," alluding to the ball-shaped head of flowers.

The flowers have short, curving nectar spurs, and these, plus their white color, are usually associated with hawk moth pollination. However, hawk moth pollinated plants, unlike *A. cephalotes*, produce flowers on long stems that project away from stems and foliage so that the hovering moths can easily reach the flowers. Perhaps this genus has switched to pollination by butterflies, which need a place to land while they feed.

The flower of the Snowball Orchid is white with similar, broadly lanceolate petals and sepals. The lip is also white and much broader, surrounding the column, often with a green blotch or stripe leading to the short curving spur.

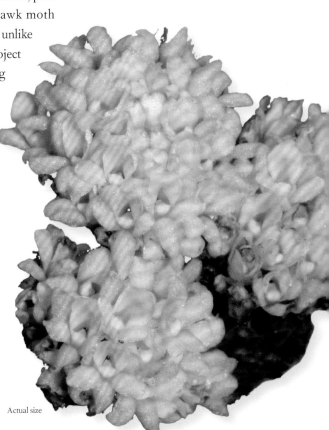

Actual size

SUBFAMILY	Epidendroideae
TRIBE AND SUBTRIBE	Vandeae, Angraecinae
NATIVE RANGE	Tropical Africa, Madagascar, and Mascarene Islands
HABITAT	Lower montane forests at 1,900–5,600 ft (580–1,700 m)
TYPE AND PLACEMENT	Epiphytic
CONSERVATION STATUS	Not assessed, but widespread
FLOWERING TIME	January to March (summer to fall), but can flower sporadically in other months

FLOWER SIZE
¼ in (0.5 cm)

PLANT SIZE
3–6 × 3–5 in
(8–15 × 8–13 cm),
excluding inflorescence

ANGRAECOPSIS PARVIFLORA
DWARF GREEN COMET ORCHID
(THOUARS) SCHLECHTER, 1914

The Dwarf Green Comet Orchid, which often grows pendently, has a short stem attached to the substrate by wiry roots and a set of closely spaced, linear leaves with unequally bilobed tips on opposite sides of the stem. The orchid produces numerous, many-flowered stems with green flowers that open—usually simultaneously—clustered to the tips of the stems. The wiry inflorescences are longer than the leaves and, like the whole plant, may be pendent.

The green flowers with their long spur suggest pollination by moths, but this has never been studied or observed. The genus name refers to an overall similarity (-*opsis*, appearance) to the genus *Angraecum*, which is based on the Indonesian word for orchid—*anggrek*. However, phylogenetic studies indicate that *Angraecopsis* is not closely related to *Angraecum*, but rather to the genera *Mystacidium* and *Sphyrarhynchus*.

Actual size

The flower of the Dwarf Green Comet Orchid has similar, spreading sepals and petals. The lip has three long lobes and is grooved at the base. The spur can be up to ⅝ in (1.5 cm) long and opens at the base of the lip.

SUBFAMILY	Epidendroideae
TRIBE AND SUBTRIBE	Vandeae, Angraecinae
NATIVE RANGE	Réunion and Mauritius
HABITAT	Wet inland forests, at 1,300–3,950 ft (400–1,200 m)
TYPE AND PLACEMENT	Epiphytic on mossy trunks and branches
CONSERVATION STATUS	Least concern, but of local concern
FLOWERING TIME	December to April (winter to spring)

FLOWER SIZE
½ in (1.25 cm)

PLANT SIZE
6–10 × 8–14 in
(15–25 × 20–36 cm).
excluding arching-pendent
inflorescences
3–8 in (8–20 cm) long

ANGRAECUM CADETII
CRICKET ORCHID
BOSSER, 1988

607

The lovely little Cricket Orchid has a short stem completely clothed in overlapping, clasping leaves, forming a fan. It attaches to mossy branches by wiry roots. The name of the genus comes from the Indonesian word for orchid, *anggrek*, but the species of *Angraecum* are found only in Africa, Madagascar, and the Mascarene Islands.

This is the only species known to be pollinated by a cricket, filmed in the act at night on the island of Réunion with infrared cameras. Normally crickets eat flowers (and other plant parts), but here juvenile crickets drink nectar produced in the sacklike spur and the pollinia become attached to their heads. The cricket involved also turned out to be a species unknown to science and was named *Glomeremus orchidophilus* (which means orchid-loving cricket).

The flower of the Cricket Orchid is pale green when it first opens and cup-shaped, with the pointed sepals and petals projecting forward and the lip forming a sacklike, nectar-filled spur. As the flowers age, they become whiter.

Actual size

SUBFAMILY	Epidendroideae
TRIBE AND SUBTRIBE	Vandeae, Angraecinae
NATIVE RANGE	Madagascar
HABITAT	Rain forests, at up to 330 ft (100 m)
TYPE AND PLACEMENT	Epiphytic and lithophytic
CONSERVATION STATUS	Threatened by poaching
FLOWERING TIME	Fall to spring

FLOWER SIZE
6–8 in (15–20 cm)

PLANT SIZE
18–36 × 24–40 in
(46–91 × 61–102 cm),
including arching
inflorescences
15–18 in (38–46 cm) tall

608

ANGRAECUM SESQUIPEDALE
COMET ORCHID
THOUARS, 1822

The remarkable Comet Orchid bears one of the most extraordinary flowers of any plant. The species epithet, from the Latin words *sesqui*, meaning "one and a half," and *pedalis*, for "foot," refers to the prodigious length of the nectary at the back of the flower. The large fan-shaped plants, with thick, coarse aerial roots, grow epiphytically on larger tree branches near sea level. The genus name comes from the Indonesian word for orchid, *anggrek*.

When Charles Darwin observed the structure of these flowers in 1862, he hypothesized the existence of a then undescribed moth species with a proboscis long enough to reach the full length of the nectar spur, up to 12 in (30 cm). Darwin was vindicated decades later (1903) when just such a moth was discovered in Madagascar, *Xanthopan morgana*. Actual pollination has not, however, been observed.

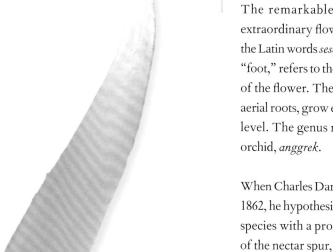

Actual size

The flower of the Comet Orchid is star-shaped and pure glistening white, with a triangular lip. When they first open, the flowers have a greenish tinge. The most noticeable feature is the exceptionally long nectar spur, in which nectar can often be observed in the bottom portion.

SUBFAMILY	Epidendroideae
TRIBE AND SUBTRIBE	Vandeae, Angraecinae
NATIVE RANGE	Madagascar, Comoros, Réunion, and Mauritius
HABITAT	Coastal and humid, evergreen forests, at 1,300–6,200 ft (400–1,900 m)
TYPE AND PLACEMENT	Epiphytic
CONSERVATION STATUS	Not formerly assessed and not threatened in Réunion, but critically endangered and declining in Mauritius
FLOWERING TIME	December (summer)

FLOWER SIZE
1½ in (3.8 cm)

PLANT SIZE
6–15 × 7–12 in
(15–38 × 18–30 cm),
excluding erect-arching
inflorescence
8–15 in (20–38 cm) long

609

BECLARDIA MACROSTACHYA
WHITE-EYE ORCHID
(THOUARS) A. RICHARD, 1828

The White-eye Orchid has a short stem carrying up to a dozen linear-oblong leaves and produces a raceme partially enveloped by bracts with up to 12 strongly scented flowers. The genus name is in honor of Pierre Auguste Béclard (1785–1825), a French anatomist and surgeon who was also interested in orchids and other plants. The species name derives from the Greek words *makros*, "large," and *stachys*, "spike," referring to the long, pointed leaves.

The white, spurred flowers are typically pollinated by night-flying hawk moths. In Réunion, however, it has been reported that two birds, the Olive White-eye (*Zosterops olivaceus*) and Gray White-eye (*Z. borbonicus*), visit the blooms, although they do not appear to be well suited to the shape of the flowers, especially as their short bills cannot easily reach into the long nectar spur.

The flower of the White-eye Orchid has three similar, white, recurved sepals and two broader petals. The white lip wraps around the column and is hairy inside with a green center, wavy edges, and broad, often bilobed apex. It also has a green-tipped spur pointing backward.

Actual size

SUBFAMILY	Epidendroideae
TRIBE AND SUBTRIBE	Vandeae, Angraecinae
NATIVE RANGE	Tropical America, from southern Mexico and the Antilles to Bolivia
HABITAT	Swamps in hot tropical forests, from sea level to 4,600 ft (1,400 m)
TYPE AND PLACEMENT	Loosely attached epiphyte on twigs in shade with high humidity
CONSERVATION STATUS	Not assessed but with a broad distribution and elevational range, so not likely to be of conservation concern
FLOWERING TIME	Spring and fall

FLOWER SIZE
½ in (1.3 cm)

PLANT SIZE
6–15 × 3–5 in
(15–38 × 8–13 cm),
including arching
lateral inflorescences
2–4 in (5–10 cm) long

610

CAMPYLOCENTRUM MICRANTHUM
FAIRY BENT-SPUR ORCHID
(LINDLEY) ROLFE, 1901

Actual size

The small, almost vinelike Fairy Bent-spur Orchid has an elongate stem loosely held among the smaller branches by aerial roots. It bears elliptical leaves in two ranks that clasp the stem and has a fleshy inflorescence covered in small floral bracts with minute flowers on two sides of the thickened stem. The genus name comes the Greek words *kampylox*, "crooked," and *kentron* "spur," which is reflected in the common name ("bent-spur"). This genus is one of two American genera of the subtribe Angraecinae, the members of which are otherwise found in Africa, Madagascar, and the Mascarene Islands.

Pollination by small stingless halictid bees has been reported, but in some areas automatic self-pollination takes place. There is nectar present in the relatively long nectar spur, even in those plants in which self-pollination occurs. A sweet scent has also been reported.

The flower of the Fairy Bent-spur Orchid is white and starlike, with fused sepals that have a spur at the back. Petals are similar and also spreading. The lip encloses the column and is similar to the sepals and petals.

SUBFAMILY	Epidendroideae
TRIBE AND SUBTRIBE	Vandeae, Angraecinae
NATIVE RANGE	Southern Florida and Mesoamerica, the Caribbean, and northern South America
HABITAT	Moist montane forests, cypress swamps, and wet hammocks, up to 4,600 ft (1,400 m)
TYPE AND PLACEMENT	Epiphytic or, rarely, terrestrial
CONSERVATION STATUS	Rare in Florida, but not assessed elsewhere; easily overlooked because of small flowers and leaflessness, so may be more common than assumed
FLOWERING TIME	November to February (fall to early spring)

CAMPYLOCENTRUM PACHYRRHIZUM

LEAFLESS BENT-SPUR ORCHID

(REICHENBACH FILS) ROLFE, 1903

FLOWER SIZE
¼ in (0.6 cm)

PLANT SIZE
2–3 × 2–3 in
(5–8 × 5–8 cm),
including arching-pendent
inflorescence

611

The thick, fleshy, green roots of the Leafless Bent-spur Orchid spread widely on tree trunks from a short rhizome out of which a densely flowered spike appears, with small, nearly stalk-less flowers emerging in two rows on either side of the stem. The plant lacks leaves throughout its life; its strongly flattened green roots perform photosynthesis. The species name refers to these thick roots (Greek *pachy*, "thick," and *riza*, "root").

The shape of the flower lip and short spur suggest bee pollination, but no observation data are available. The short curving nectar spur projects behind the lip, and it is the curving spur for which the genus is named (Greek for "curved" or "crooked," *kampylox*, and *centron*, "spur"). Further study is needed on all aspects of the life of this enigmatic orchid.

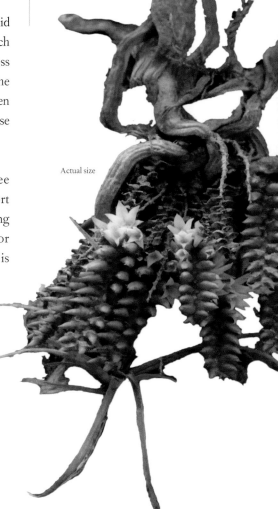

Actual size

The flower of the Leafless Bent-spur Orchid is small with similarly shaped white petals and sepals. The lip forms a small cup around the column and a short, curved spur at the back, the opening of the flower almost filled by the yellow anther cap.

SUBFAMILY	Epidendroideae
TRIBE AND SUBTRIBE	Vandeae, Angraecinae
NATIVE RANGE	Tropical West Africa, from Liberia to Cameroon
HABITAT	Humid forests, at 1,970–6,900 ft (600–2,100 m)
TYPE AND PLACEMENT	Epiphytic
CONSERVATION STATUS	Not formally assessed
FLOWERING TIME	March to May (spring)

FLOWER SIZE
¾ in (1.8 cm)

PLANT SIZE
5–9 × 6–12 in
(13–23 × 15–30 cm),
excluding inflorescence
4–7 in (10–18 cm) long

CRIBBIA CONFUSA
MUDDLED ORCHID
P. J. CRIBB, 1996

612

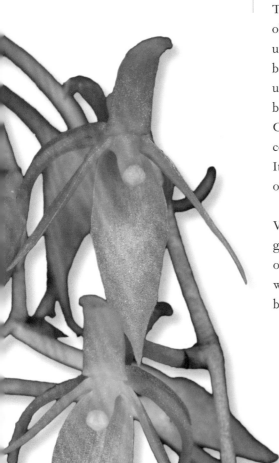

The short, vertical stems of the Muddled Orchid carry linear-oblanceolate, folded leaves in two ranks. The leaf tips are unequally bilobed, and the leaves are relatively thick, with leaf bases that clothe the stems. A basal or axillary inflorescence has up to a dozen flowers that are supported by oval, acute-tipped bracts. The genus, named in honor of Kew orchidologist Philip Cribb, a specialist in African and Asian orchids, was originally confused with *Cribbia brachyceras*, hence the species epithet. It differs from *C. brachyceras* in having larger, pale greenish-orange flowers.

Very little is known about the pollination of species in this genus, but the similarity in shape and color to other members of subtribe Angraecinae, such as *Angraecum*—in particular the well-developed nectar spur—would seem to indicate pollination by moths.

Actual size

The flower of the Muddled Orchid has greenish-orange, narrow, pendent lateral sepals, and a spreading upper sepal and petals. The green lip is cupped and heart-shaped at the base, surrounding the opening to a nectar spur.

SUBFAMILY	Epidendroideae
TRIBE AND SUBTRIBE	Vandeae, Angraecinae
NATIVE RANGE	Indian Ocean islands of Mauritius and Réunion
HABITAT	Wet, open forests and cloud forests, from sea level up to 2,625 ft (800 m)
TYPE AND PLACEMENT	Epiphytic on tree trunks, lower branches of bushes, and sometimes on mossy rocks
CONSERVATION STATUS	Near threatened
FLOWERING TIME	December to March (spring)

FLOWER SIZE
2½ in (6.4 cm)

PLANT SIZE
12–30 × 8–14 in
(30–76 × 20–36 cm),
excluding inflorescence
12–24 in (30–61 cm) long

CRYPTOPUS ELATUS
STATELY MOTH-ORCHID
(PETIT-THOUARS) LINDLEY, 1825

613

The long stem of the Stately Moth-orchid is attached upright by fleshy roots to the tree trunk on which it grows. The roots appear at random from among folded, distantly spaced leaves that clasp the stem and are irregularly bilobed at their tip. From the base of a leaf, a simple or sparsely branched inflorescence grows, carrying up to 15 white flowers that turn apricot-orange to reddish in their center with age. The orchids are often not apparent until they bloom, when their showy white flowers on wiry stems attract attention.

The spur, which is 10–12 in (25–30 cm) long, does not contain nectar, and the flower lacks a scent. Pollination has not been documented, but floral morphology indicates pollination by long-tongued moths—a case of deceit as the insect gets no reward for its visit.

Actual size

The flower of the Stately Moth-orchid has narrow, recurved sepals. The white petals are fan-shaped and four-lobed at the top with a clawed base. The elaborate white lip is five-lobed, often with a yellow to red blotch in the center, and a long, curved spur.

SUBFAMILY	Epidendroideae
TRIBE AND SUBTRIBE	Vandeae, Angraecinae
NATIVE RANGE	Sub-Saharan Africa
HABITAT	Woodland and riverine forests, at up to 8,200 ft (2,500 m)
TYPE AND PLACEMENT	Epiphytic or lithophytic
CONSERVATION STATUS	Not formally assessed but widespread and likely to be of little conservation concern
FLOWERING TIME	March to May (fall)

FLOWER SIZE
1½ in (3.75 cm)

PLANT SIZE
8–15 × 10–16 in
(20–38 × 25–41 cm),
excluding pendent
inflorescence
8–12 in (20–30 cm) long

CYRTORCHIS ARCUATA
CURVED ORCHID
(LINDLEY) SCHLECHTER, 1914

614

Actual size

The flower of the Curved Orchid has linear,
recurved, white to greenish-white sepals, petals,
and lip, all of the same shape, although the
petals and lip are somewhat smaller. A long
nectar spur is curved forward and down.

The Curved Orchid produces an elongate stout stem clothed with persistent, overlapping leaf bases. Among these, aerial roots form, each about ¼ in (0.7 cm) thick, which attach the plant to the branch for support. The apically bilobed leaves are fleshy and form on opposite sides of the stem. In a leaf axil, a raceme emerges and bears 10–20 waxy flowers that are sweetly fragrant at night and turn orange once the pollinia are removed.

Pollination by moths is the default in flowers of this shape, color, and nocturnal scent. Chrysomelid beetles have been observed as frequent visitors but could not be the effective pollinators given the long length of the nectar spur. An extract of one unspecified species of *Cyrtorchis* has been used in the treatment of malaria.

SUBFAMILY	Epidendroideae
TRIBE AND SUBTRIBE	Vandeae, Angraecinae
NATIVE RANGE	Southern Florida, Cuba, and the Bahamas
HABITAT	Lower elevation rain forests in swamps
TYPE AND PLACEMENT	Epiphytic
CONSERVATION STATUS	Endangered by poaching, but now specifically protected by law in Florida
FLOWERING TIME	July to September (summer to fall)

DENDROPHYLAX LINDENII
GHOST ORCHID
(LINDLEY) BENTHAM EX ROLFE, 1888

FLOWER SIZE
3½–6 in (9–15 cm)

PLANT SIZE
Leafless, the mass of roots
often 3–5 × 8–15 in
(8–13 × 20–38 cm),
excluding arching
inflorescence
3–12 in (8–30 cm) long

615

The rare Ghost Orchid is completely leafless, with photosynthetic roots, and therefore difficult to observe when not in bloom. The ghostly flowers appear to be dangling in space, unconnected physically to any support. The genus name comes from the Greek words *dendro*, "tree," and *phylax*, "guard," referring to the mass of roots "guarding" the host tree. The species name is in honor of the Belgian botanist, Jean Jules Linden (1817–98), who revolutionized the cultivation of orchids in Europe. Before his time, orchids were grown in conditions too hot for them, killing nearly all imported plants.

The flowers are similar in shape to their African and Madagascan cousins, *Angraecum* and related Angraecinae genera. It has been suggested that the Ghost Orchid is pollinated by the Giant Sphinx Moth (*Cocytius antaeus*), based on the length of its proboscis (tongue).

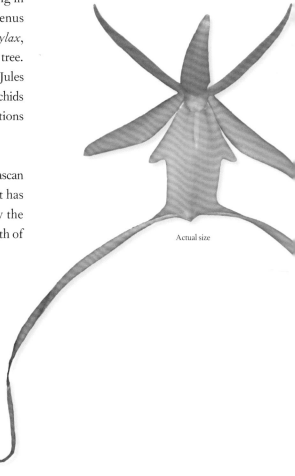

Actual size

The flower of the Ghost Orchid has pale, greenish-white, lanceolate sepals and petals and a remarkable four-lobed white lip. The two terminal lobes are long, leg-like, and bent, giving the flower the appearance of a frog.

SUBFAMILY	Epidendroideae
TRIBE AND SUBTRIBE	Vandeae, Angraecinae
NATIVE RANGE	Tropical Africa from Sierra Leone to Uganda
HABITAT	Dense shade in evergreen forests, at 1,970–4,920 ft (600–1,500 m)
TYPE AND PLACEMENT	Epiphytic in moist pockets on tree trunks and lower branches
CONSERVATION STATUS	Not assessed
FLOWERING TIME	September to December

FLOWER SIZE
½ in (1.3 cm)

PLANT SIZE
15–30 × 12–25 in
(38–76 × 30–64 cm),
excluding pendent
inflorescence
8–20 in (20–51 cm) long

616

DIAPHANANTHE PELLUCIDA
TRANSLUCENT ORCHID
(LINDLEY) SCHLECHTER, 1914

Actual size

The flower of the Translucent Orchid has similar,
pale cream, spreading, lanceolate sepals and
petals. The paler, deflexed lip has fringed
margins with a tooth in the mouth of the
spur, which is about ⅜ in (1 cm) long.

The Translucent Orchid has short, often arching to slightly
pendent stems that bear many, pendent, oblanceolate, fleshy
leaves with bases that cover the stem. The leaf apex is unequally
bilobed, and the blade has obvious netlike venation. From the
stem, a pendent raceme arises, densely covered with translucent
flowers subtended by small, papery bracts.

The genus name refers to the translucent texture of the flowers,
from the Greek *diaphanes*, meaning "translucent" and *anthos*,
"flower," while *pellucida* refers to the small, more opaque dots on
the blooms. The exact pollinator of this species is uncertain, but
Euchromia moths are thought to be frequent visitors, fitting
well with the floral morphology, color, and long
nectar spur. The roots of other *Diaphananthe*
species are reportedly used for
basket making.

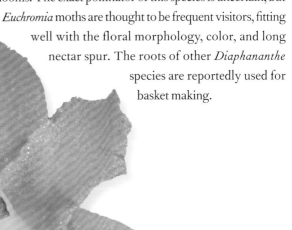

SUBFAMILY	Epidendroideae
TRIBE AND SUBTRIBE	Vandeae, Angraecinae
NATIVE RANGE	Madagascar
HABITAT	Evergreen forests on sandstone, at 820–3,300 ft (250–1,000 m)
TYPE AND PLACEMENT	Epiphytic
CONSERVATION STATUS	Not assessed
FLOWERING TIME	October to December (summer)

ERASANTHE HENRICI
NIGHT HAWK
(SCHLECHTER) CRIBB, HERMANS & ROBERTS, 2007

FLOWER SIZE
4 in (10 cm)

PLANT SIZE
6–10 × 7–12 in
(15–25 × 18–30 cm),
excluding pendent
inflorescence
5–16 in (13–41 cm) long

617

The Night Hawk is better known in the genus *Aeranthes*, in which it was included until 2007. *Erasanthe* is an anagram of *Aeranthes*, although the two genera are not closely related. *Erasanthe henrici* produces an erect stem clothed with four to six wavy-margined leaves, and, from the base of a leaf, a pendent stem carries 4–12 flowers with a long spur at the back. The plants are denizens of the darker portions of humid forests, growing on sandstone massifs. The manner in which the inflorescence holds the large flowers makes them appear to float in the air like a bird of prey, hence the common name.

Pollination has not been studied. However, the coloration, long nectar spur, and suspension away from the plant are consistent with pollination by a night-flying hawk moth.

The flower of the Night Hawk has white, outstretched, lanceolate, pointed sepals and petals. The lip is white and fringed and has a deep, central, green cavity leading to the nectar spur at the back of the flower. The pale green column hangs over this cavity.

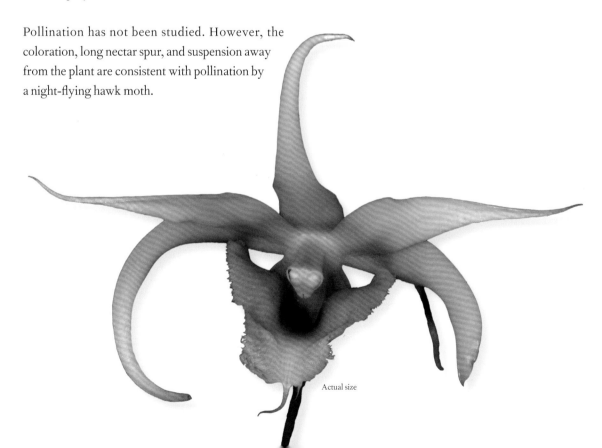

Actual size

SUBFAMILY	Epidendroideae
TRIBE AND SUBTRIBE	Vandeae, Angraecinae
NATIVE RANGE	Tropical Africa, from Guinea to Uganda
HABITAT	Moist evergreen forests near streams, at 3,600–3,950 ft (1,100–1,200 m)
TYPE AND PLACEMENT	Epiphytic on mossy branches, occasionally on rocks
CONSERVATION STATUS	Not assessed
FLOWERING TIME	Throughout the year, but more flowers during warmer months

FLOWER SIZE
2½ in (6 cm)

PLANT SIZE
4–8 × 6–10 in
(10–20 × 15–25 cm),
including inflorescence,
which is shorter than
the leaves

EURYCHONE ROTHSCHILDIANA
GREEN THROATED ORCHID
(O'BRIEN) SCHLECHTER, 1918

The small Green Throated Orchid has a short stem, fleshy roots, and leaves—in a plane (distichous) and bilobed apically—with an undulating margin. It is the larger of the two species in this genus, closely related to the genus *Aerangis*. From the base of a leaf, the plant produces a pendent inflorescence bearing one to seven flowers, with triangular bracts protecting the flower buds. The flowers are large, showy, and fragrant, with the scent of cinnamon.

The orchid produces its scent after dark; that plus its color and floral morphology with a spur suggests that the pollinator is a moth, probably a hawk moth. It is unusual, however, for a hawk moth-pollinated orchid to have such a wide entrance to the spur, so it may have an entirely different pollinator.

The flower of the Green Throated Orchid is white with slightly twisted sepals and petals that are both green-centered. Its large "horn of plenty"-shaped white lip has a spur and a dark green center.

Actual size

SUBFAMILY	Epidendroideae
TRIBE AND SUBTRIBE	Vandeae, Angraecinae
NATIVE RANGE	Central Madagascar
HABITAT	Humid forests, at 1,640–3,950 ft (500–1,200 m)
TYPE AND PLACEMENT	Lithophytic, occasionally epiphytic
CONSERVATION STATUS	Not threatened
FLOWERING TIME	Spring

JUMELLEA DENSIFOLIATA
DOG-EARED ORCHID
SENGHAS, 1964

FLOWER SIZE
1⅜ in (3.5 cm)

PLANT SIZE
8–12 × 8–14 in
(20–30 × 20–36 cm),
including single-flowered
lateral inflorescences
3–5 in (8–13 cm) long

619

The Dog-eared Orchid has a dense cluster of leathery, dark green leaves that form a starkly contrasting background for the elegant, long-spurred flowers. Nestling mostly among mossy rocks, the plants are protected from the sometimes harsh conditions and send fleshy roots deep into the fissures and cracks among the rocks. The genus was named in honor of Henri Jumelle (1866–1935), Professor of Botany at the University of Marseille and Director of the Botanical Garden in the same French city. The species name refers to the relatively densely arranged leaves of this species, and the common name to its unusual drooping petals.

Blessed with a most intoxicating sweet nocturnal perfume, the species attracts hawk moths, which pollinate the flowers at dusk. The moths probe the long nectar-containing spur, and while doing this come into contact with the apex of the column and pick up the pollinia.

Actual size

The flower of the Dog-eared Orchid has brilliant white, starry, narrowly lanceolate sepals, with the dorsal sepal erect, while the laterals flare sideways. Petals are white, hang down, and often curve behind the lip, which is arrowhead-shaped and bears a long and extremely thin nectar spur at its rear.

SUBFAMILY	Epidendroideae
TRIBE AND SUBTRIBE	Vandeae, Angraecinae
NATIVE RANGE	Central and eastern Madagascar
HABITAT	Humid evergreen forests, at 6,600–7,200 ft (2,000–2,200 m)
TYPE AND PLACEMENT	Epiphytic
CONSERVATION STATUS	Not assessed, but its habitat is under great threat from fire and human activities
FLOWERING TIME	December to January (late summer/wet season)

FLOWER SIZE
⅛ in (0.5 cm)

PLANT SIZE
25–35 × 12–20 in
(64–89 × 31–51 cm),
excluding inflorescence,
which is apically pendent and
shorter than the leaves

620

LEMURORCHIS MADAGASCARIENSIS
GHOST LEMUR ORCHID
KRAENZLIN, 1893

The relatively large and rare Ghost Lemur Orchid forms a fan of up to 15 upright, strap-shaped leaves—each one up to 20 in (50 cm) long with a bilobed tip. No pseudobulb is present, and growth is continuous at the apex of the stem. From the bases of leaves, the plant produces one to three inflorescences that are initially upright, but the apical half is pendent, reminiscent of a lemur's tail, and covered with large, sheathing, brown bracts and numerous small, long-spurred flowers.

It is the only species in the genus, the name of which refers to the endemic Madagascan primate, the lemur (a word that in turn comes from the Latin *lemures* meaning "malevolent ghosts"), and *orchis*, Greek for "orchid." Pollination has not been observed, but the pale color and presence of a long spur suggest that a small moth may be the pollinator.

Actual size

The flower of the Ghost Lemur Orchid has three spreading, blunt-tipped, yellow sepals and two pointed petals. The yellow lip has a white center and is short, enclosing the column and forming a long spur at the back of the flower.

SUBFAMILY	Epidendroideae
TRIBE AND SUBTRIBE	Vandeae, Angraecinae
NATIVE RANGE	Tropical West and Central Africa, from Liberia to Gabon
HABITAT	Evergreen tropical forests, at 1,640–1,970 ft (500–600 m)
TYPE AND PLACEMENT	Epiphytic
CONSERVATION STATUS	Not assessed
FLOWERING TIME	June

FLOWER SIZE
¼ in (0.6 cm)

PLANT SIZE
8–15 × 8–20 in
(20–38 × 20–51 cm),
excluding lateral,
arching inflorescence
10–15 in (25–38 cm) long

LISTROSTACHYS PERTUSA
POLONAISE ORCHID
(LINDLEY) REICHENBACH FILS, 1852

621

The Polonaise Orchid has a short stem and 10–15 closely set,
linear, stiff leaves in two rows. From just above a leaf, it produces
a stem with many, densely set flowers in two opposing ranks.
The long spike of little delicate flowers resembles noble ladies
dancing the polonaise as the petals touch each other as if holding
hands. The Latin species name *pertusa*, meaning "perforated,"
refers to the deep depression in the lip. The genus name comes
from the Latin *lustra*, "to illuminate," and the Greek *stachys*, "ear
of corn," referring to the way the inflorescence appears to light
up with bright flowers.

Pollination of this small-flowered species is a mystery. The
color, shape, and presence of a relatively long nectar spur
suggest hawk moth pollination, but it would have to be an
especially small species.

Actual size

The flower of the Polonaise Orchid has short
sepals and larger, transparent, spreading petals,
both white. The column is enlarged at the top,
and the lip has a large, green-edged depression
in the middle. There is a green nectar spur below
the flower.

SUBFAMILY	Epidendroideae
TRIBE AND SUBTRIBE	Vandeae, Angraecinae
NATIVE RANGE	Kenya to Zimbabwe (tropical East Africa)
HABITAT	Wet evergreen forests, at 2,625–8,200 ft (800–2,500 m)
TYPE AND PLACEMENT	Epiphytic
CONSERVATION STATUS	Not assessed, but probably not threatened
FLOWERING TIME	April to May (spring)

FLOWER SIZE
⅛ in (0.5 cm)

PLANT SIZE
2–3 × 6–10 in
(5–8 × 15–25 cm),
excluding arching to
pendent inflorescence
4–6 in (10–15 cm) long

MICROCOELIA STOLZII
YELLOW-CAPPED ORCHID
(SCHLECHTER) SUMMERHAYES, 1943

622

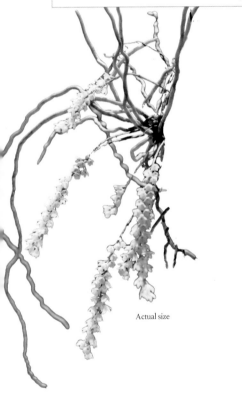

Actual size

The Yellow-capped Orchid is generally leafless and carries out photosynthesis in its masses of long, flattened gray green roots. When actively growing, there may be a small, short-lived bract produced on the erect central stem, but this soon falls off. A single plant can have 50 to 80 roots and up to 15 inflorescences present at a single time. The genus name is from the Greek words *mikros*, "small," and *koilos*, "hollow," referring to the small spur found in the member plants. The species name honors Adolph Stolz (1871–1917), a German missionary and plant collector in the Tanzanian Highlands.

Microcoelia stolzii produces masses of small but sweetly fragrant flowers with a relatively long nectar spur, and it is likely to be pollinated by butterflies. Other species in the genus with similar morphology produce fragrance at night, consistent with moth pollination.

The flower of the Yellow-capped Orchid has similar, bright white sepals and petals that form a cup around the lip and column. The lip is white and trilobed, the two lateral lobes surrounding the entrance to the spur, with the midlobe larger. The column is yellow and the spur brown-tipped.

SUBFAMILY	Epidendroideae
TRIBE AND SUBTRIBE	Vandeae, Angraecinae
NATIVE RANGE	Southern Africa
HABITAT	Savannas and evergreen forests in shade, at up to 2,626 ft (800 m)
TYPE AND PLACEMENT	Epiphytic
CONSERVATION STATUS	Not threatened
FLOWERING TIME	Late spring and summer

MYSTACIDIUM CAPENSE
SOUTHERN STAR
(LINNAEUS FILS) SCHLECHTER, 1914

FLOWER SIZE
1¼ in (3 cm)

PLANT SIZE
3–5 × 4–5 in
(8–13 × 10–13 cm),
excluding arching-pendent
inflorescence
8–15 in (20–38 cm) long,
which is longer than the leaves

623

The miniature Southern Star can branch as it matures, each branch producing multiple flower stems, completely obscuring the leaves with dainty and starry long-spurred blooms. The genus name comes from the Greek word *mystax*, "mustache," a reference to the fringed edge of the column apex. The species name refers to the Cape of Good Hope, although the species itself is not found near the Cape but generally in southern Africa.

The perfume fills the evening air with a scent reminiscent of jasmine, attracting the hawk moth *Hippotion celerio*, which has a tongue long enough to reach into the deep nectar spur. *Mystacidium capense* was one of the first epiphytes described from South Africa, as far back as 1781, when it was placed in the genus *Epidendrum*, in which all epiphytic orchids were originally included.

The flower of the Southern Star is crystalline white with a starlike form and graceful nectar spur that extends about three times the length of the bloom behind the lip. When opening, the flowers can be greenish-white, but this soon fades to pure white.

Actual size

SUBFAMILY	Epidendroideae
TRIBE AND SUBTRIBE	Vandeae, Angraecinae
NATIVE RANGE	Northeastern-central Madagascar
HABITAT	Shady forests, at 330–660 ft (100–200 m)
TYPE AND PLACEMENT	Epiphytic
CONSERVATION STATUS	Not assessed
FLOWERING TIME	Late fall

FLOWER SIZE
1 in (2.5 cm)

PLANT SIZE
3–6 × 4–8 in
(8–15 × 10–20 cm),
excluding inflorescence,
which is arching-pendent,
2–3 in (5–8 cm) long

624

NEOBATHIEA KERAUDRENAE
LITTLE COMET ORCHID
TOILLIEZ-GENOUD & BOSSER, 1964

The Little Comet Orchid has a short stem that carries four or five fleshy, spathulate leaves, usually shortly bilobed at the tip. On a short inflorescence, one or two long-spurred flowers are formed. The common name references Darwin's Comet Orchid, *Angraecum sesquipedale*, also found on Madagascar. The genus was named in honor of the eminent French botanist, Henri Perrier de la Bâthie (1873–1958), who studied the orchids of Madagascar.

The flower type suggests pollination by a hawk moth with a tongue long enough to obtain nectar from the tip of the long spur. Pollinaria of related species were found on *Panogea lingens* hawk moths, with the pollinia attached dorsally on the proboscis of the moth. This hawk moth visits various species as a number of pollinia from different angraecoid orchids were found on the proboscis of a single insect.

Actual size

The flower of the Little Comet Orchid has spreading, spathulate, cream to white sepals and petals with greenish bases, and a larger, strap-like, white lip, which has a long, descending nectar spur.

SUBFAMILY	Epidendroideae
TRIBE AND SUBTRIBE	Vandeae, Angraecinae
NATIVE RANGE	Réunion, Mauritius, and eastern coastal forests of Madagascar
HABITAT	Humid, mossy evergreen forests, from sea level to 6,600 ft (2,000 m)
TYPE AND PLACEMENT	Epiphytic
CONSERVATION STATUS	Threatened by deforestation
FLOWERING TIME	September to November (southern spring)

FLOWER SIZE
1 in (2.5 cm)

PLANT SIZE
10–20 × 4–8 in
(25–51 × 10–20 cm),
excluding arching
lateral inflorescence
10–16 in (25–41 cm) long

625

OEONIA VOLUCRIS
SNOWY VINE ORCHID
(THOUARS) SPRENGEL, 1826

The Snowy Vine Orchid produces an elongate stem with roots along its length, which scrambles through the canopy of forest trees. It produces an inflorescence with three to eight attractive flowers. The genus name is from the Greek *oionos*, for a bird of prey, while the species name, *volucris*, is another Greek word for a type of bird. Both refer to the beak-like apex of the column. The common name alludes to the flower color and the almost vinelike nature of the plant.

The long inflorescence, with flowers clustered at its apex, the white color, sweet fragrance, and upwardly curving nectar spur are all typical features associated with hawk moth pollination. The majority of genera and species in subtribe Angraecinae exhibit these same features, and although *Oeonia volucris* has never been studied it is likely that it, too, is pollinated by hawk moths.

The flower of the Snowy Vine Orchid has pure white, oblanceolate, spreading sepals and petals. The lip is trilobed, with the sidelobes wrapping around the column with a nectar spur at the back, its entrance marked by a pale olive throat, and a midlobe with a deep central notch.

Actual size

SUBFAMILY	Epidendroideae
TRIBE AND SUBTRIBE	Vandeae, Angraecinae
NATIVE RANGE	Madagascar, Comoros, and Mascarene Islands
HABITAT	Evergreen forests in deep shade, at sea level to 330 ft (100 m)
TYPE AND PLACEMENT	Epiphytic
CONSERVATION STATUS	Not threatened
FLOWERING TIME	Spring and fall, often twice a year

FLOWER SIZE
Up to 1¼–2⅜ in (3–6 cm)

PLANT SIZE
8–20 × 6–12 in
(20–51 × 15–30),
excluding arching
inflorescence
8–12 in (20–30 cm) long,
which is longer than the leaves

OEONIELLA POLYSTACHYS
AWL-LIPPED ORCHID
(THOUARS) SCHLECHTER, 1918

626

A common species that occurs in Madagascar and most of the nearby island chains, the erect, somewhat climbing Awl-lipped Orchid bears copious aerial roots along its stem and grows in bright, highly humid situations—in some cases in mangrove swamps at sea level. The genus was named in reference to its similarity (*-ella*, a "lesser" version) to the related genus *Oeonia*, which was named from the Greek word *oionos*, "bird of prey." The exact reason for this reference is not clear, but probably the author thought that some part of the flower looked like such a bird. The common name derives from the awl-like shape of the apex of the lip.

The flowers are arranged in two ranks. Pollination has not been studied, but the short spur and tubular shape of the lip would be compatible with some type of bee as the pollinator.

Actual size

The flower of the Awl-lipped Orchid is pale green to crystalline white with narrowly lanceolate sepals and petals. The white lip has a green base and a distinctive, sharply tipped, awl-like midlobe. The sidelobes enfold the column.

SUBFAMILY	Epidendroideae
TRIBE AND SUBTRIBE	Vandeae, Angraecinae
NATIVE RANGE	Widespread in tropical and subtropical West and Central Africa
HABITAT	Low on large trees in forests, at sea level to 3,300 ft (1,000 m)
TYPE AND PLACEMENT	Epiphytic
CONSERVATION STATUS	Not threatened
FLOWERING TIME	November to January, but can bloom anytime

PLECTRELMINTHUS CAUDATUS

WORM ORCHID

(LINDLEY) SUMMERHAYES, 1949

FLOWER SIZE
3 in (8 cm)

PLANT SIZE
10–30 × 10–30 in
(25–76 × 25–76 cm),
excluding arching-pendent
inflorescence
15–25 in (38–64 cm) long

627

The leathery-leaved Worm Orchid has stunning, non-resupinate flowers with astonishing twisting, wormlike nectar spurs, which give rise to both its common and genus names (Greek for "spur," *plektron*, and "worm," *minthion*). The species name comes from the Latin word for "tailed," also a reference to the narrow spur.

The flowers appear alternately in two ranks, with the blooms oddly facing slightly inward rather than outward. The prodigious curling spurs and nocturnal fragrance indicate a large, long-tongued hawk moth as the likely pollinator, but it is difficult to imagine how the tongue of the moth deals with the curves in the spur. The apex of the column projects into the mouth of the nectar spur, contacting the underside of the body of the moth as it approaches to insert its tongue.

The flower of the Worm Orchid has narrow, olive green to almost orange sepals and petals with pointed tips. The flowers alternate on opposite sides and have a pure white, spade-shaped lip with a long, narrow, lance-shaped midlobe pointing straight up.

Actual size

SUBFAMILY	Epidendroideae
TRIBE AND SUBTRIBE	Vandeae, Angraecinae
NATIVE RANGE	Widespread in tropical and subtropical West Africa to Angola and Tanzania
HABITAT	Tropical wet forests, at sea level to 3,950 ft (1,200 m)
TYPE AND PLACEMENT	Epiphytic
CONSERVATION STATUS	Not threatened
FLOWERING TIME	Can bloom anytime

FLOWER SIZE
½ in (1.3 cm)

PLANT SIZE
3–5 × 3–5 in
(8–13 × 8–13 cm),
including axillary
inflorescence
1–2 in long (2.5–5 cm)

628

PODANGIS DACTYLOCERAS
EMERALD FOOT ORCHID
(REICHENBACH FILS) SCHLECHTER, 1918

Actual size

The little fan-shaped plants of the Emerald Foot Orchid have slightly curving, thick, flattened leaves. The inflorescences emerge from the tightly overlapping leaf bases and are compressed, creating a tight cluster of blooms. The genus name is derived from the Greek *pous*, "foot," and *angos*, "vessel," referring to the foot-shaped nectar spur. The common name also uses this idea, to which is added a reference to the bright greenish-yellow, emerald-like apex of the column. The species name, from the Greek words *dactylos*, "finger," and *keras*, "horn," refers as well to the unusual shape of the spur.

Pollination has not been studied, but the clustered nature of the inflorescence plus the color and shape (a long spur with a narrow mouth) could be an indication that butterflies pollinate the plant. Unlike hawk moths, butterflies require a landing site in order to feed.

The flower of the Emerald Foot Orchid is round and translucent white with a prominent, greenish-yellow anther cap. The flowers are densely clustered near the apex of the inflorescence and are held high by their long individual stem.

SUBFAMILY	Epidendroideae
TRIBE AND SUBTRIBE	Vandeae, Aeridinae
NATIVE RANGE	Borneo
HABITAT	Wet forests, on smaller branches, at 1,300–5,250 ft (400–1,600 m)
TYPE AND PLACEMENT	Epiphytic
CONSERVATION STATUS	Not assessed
FLOWERING TIME	March to May

FLOWER SIZE
1 in (2.5 cm)

PLANT SIZE
2–5 × 3–7 in
(5–13 × 8–18 cm),
including inflorescence
3–6 in (8–15 cm) long

629

PTEROCERAS FRAGRANS
SWOOPING BIRD ORCHID
(RIDLEY) GARAY, 1972

The Swooping Bird Orchid is a miniature species with a short stem completely covered in lanceolate leaves on two sides. From the base of a leaf, a slowly elongating inflorescence holds two to three flowers at a time, usually facing downward. The genus name, from the Greek, *pteron*, "wing," and *keras*, "horn," refers to the two upright lobes of the lip. The common name is a fanciful allusion to the column, which looks like a bird—the apex its head, the petals its wings—swooping down on unsuspecting prey.

Pollination is thought to be by some sort of bee, which uses the lip as a landing platform to gain access to the nectar spur formed by the lip base. The sweet scent and bright color of the flowers also fit this mode of pollination.

The flower of the Swooping Bird Orchid has broadly lanceolate, cream sepals and smaller, spathulate, cream petals, both often with a broad concentric brown stripe. The complex white lip has a nectar spur with a swollen tip and upright sidelobes, over which the white column hangs.

Actual size

SUBFAMILY	Epidendroideae
TRIBE AND SUBTRIBE	Vandeae, Aeridinae
NATIVE RANGE	Tropical Asia, from the eastern Himalayas to Indochina, western Malesia, and the Philippines
HABITAT	Lowland and mid-elevation forests, at up to 4,920 ft (1,500 m)
TYPE AND PLACEMENT	Pendant epiphytic
CONSERVATION STATUS	Not formally assessed
FLOWERING TIME	March to May (spring)

FLOWER SIZE
⅜ in (1 cm)

PLANT SIZE
10–25 × 5–10 in
(25–64 × 13–25 cm),
excluding pendent
inflorescence
6–14 in (15–36 cm) long

PTEROCERAS TERES
ASIAN TIGER ORCHID
(BLUME) HOLTTUM, 1960

The unbranched stem of the Asian Tiger Orchid is clothed with many linear-lanceolate leaves that are obliquely bilobed at the tip. Between one and four, unbranched, pendent inflorescences grow from the bases of the leaves and have many flowers, which open several at a time. The inflorescence stalk is thick and winged at the nodes, and the short-lived, fragrant flowers are subtended by scale-like bracts. The genus name comes from the Greek words *pteron*, for "wing," and *keras*, "horn," in reference to the hornlike structures on the base of the lip.

Pollination of the species is not studied, but it appears that carpenter bees (*Xylocopa*) may play a role. The spur of the lip, flower color, and sweet scent produced during the daytime are all classic features of bee-pollinated flowers.

Actual size

The flower of the Asian Tiger Orchid has spreading, spotted, leathery, cream-yellow petals and sepals with reddish spots. The white lip is stalked, with two small, yellow sidelobes with red marking, and a midlobe with a nectar cavity and a dark red tip.

SUBFAMILY	Epidendroideae
TRIBE AND SUBTRIBE	Vandeae, Angraecinae
NATIVE RANGE	Tropical Africa, from Senegal to Sudan and south to Mozambique and Angola
HABITAT	Seasonally wet evergreen forests, at 1,800–6,600 ft (550–2,000 m)
TYPE AND PLACEMENT	Epiphytic
CONSERVATION STATUS	Not assessed
FLOWERING TIME	September to October

RHIPIDOGLOSSUM RUTILUM
RED FOXTAIL ORCHID
(REICHENBACH FILS) SCHLECHTER, 1918

FLOWER SIZE
⅛ in (0.5 cm)

PLANT SIZE
10–25 × 8–15 in
(25–64 × 20–38 cm),
excluding pendent
inflorescence
3–10 in (8–25 cm) long

631

The Red Foxtail Orchid can produce a long stem with linear-lanceolate leaves down two sides with little spacing between them. Its large white roots form a suspending network surrounding the plants, often branching to form large clumps that appear not to be attached to a single branch. The genus name, from the Greek words *rhipis*, "fan," and *glossa*, "tongue," alludes to the broad, fan-shaped lip. The species name is Latin for "red," although the flowers are more brownish purple.

At night, the tiny blooms produce a sweet gardenia-like scent, which, together with their color and long nectar spur, make pollination by night-flying hawk moths likely. There is nectar produced in the spurs, and the inflorescences stand clear of the leaves so that the pollinators can easily reach them while in flight.

Actual size

The flower of the Red Foxtail Orchid has reddish to greenish, oval sepals and petals. The lip has two small sidelobes that project upward around the column and a midlobe that is wider at the apex. The column hangs over the opening to the long, forward-curving nectar spur.

SUBFAMILY	Epidendroideae
TRIBE AND SUBTRIBE	Vandeae, Angraecinae
NATIVE RANGE	Western Madagascar
HABITAT	Seasonally dry forests and scrub, at 4,920–6,600 ft (1,500–2,000 m)
TYPE AND PLACEMENT	Epiphytic or lithophytic
CONSERVATION STATUS	Not assessed
FLOWERING TIME	November to January (spring to summer)

FLOWER SIZE
2¼ in (5.6 cm)

PLANT SIZE
10–30 × 10–22 in
(25–76 × 25–56 cm),
excluding lateral,
arching-erect inflorescence
12–24 in (30–61 cm) long

632

SOBENNIKOFFIA ROBUSTA
DOVE IN FLIGHT
(SCHLECHTER) SCHLECHTER, 1925

The flower of the Dove in Flight is pure white
with outstretched, lanceolate sepals and petals,
the laterals sickle-shaped. The lip has a pale
green center and a long spur and is trilobed,
the sidelobes rounded at their apex and the
midlobe narrow and sharply pointed.

The Dove in Flight produces an erect stem, clothed in linear,
fleshy, closely spaced leaves with an unequally bilobed tip.
The plants grow on the bases of trees and on rocks, often
reaching a large size and producing up to 20 highly fragrant
flowers on an inflorescence. The genus name honors Alexandra
Sobennikoff, the wife of the German taxonomist, Rudolf
Schlechter (1872–1925), who described the genus. The species
name is Latin for "strong growing," and the common name
refers to the fanciful appearance of the large white flowers.

Pollination has not been studied but, as in most of the other
genera in the subtribe Angraecinae, *Sobennikoffia robusta* is likely
to be pollinated by hawk moths. The plant has a long, upwardly
curving nectar spur, so the moth that pollinates it must have an
equally long tongue.

Actual size

SUBFAMILY	Epidendroideae
TRIBE AND SUBTRIBE	Vandeae, Angraecinae
NATIVE RANGE	Sub-Saharan Africa
HABITAT	A wide variety, from dry forests and woodlands to wet forests, from sea level to 8,200 ft (2,500 m)
TYPE AND PLACEMENT	Epiphytic and sometimes on rocks
CONSERVATION STATUS	Not threatened, but removal of forest habitat for agriculture, housing, and other uses could result in the plant becoming infrequent
FLOWERING TIME	November to March (summer)

TRIDACTYLE BICAUDATA

TWO TAILS—THREE FINGERS

(LINDLEY) SCHLECHTER, 1914

FLOWER SIZE
½ in (1.2 cm)

PLANT SIZE
8–30 × 9–15 in
(20–76 × 23–38 cm),
excluding pendent
inflorescence
4–8 in (10–20 cm) long

633

The Two Tails–Three Fingers has a tight cluster of slender, erect to hanging stems, each carrying up to two rows of 14 spreading linear leaves with unequally bilobed tips. A dense inflorescence bears 8–16 flowers arranged in two rows, all facing the same way and producing a strong fragrance. The genus name comes from the Greek *dactylos*, "finger," and *tri*, "three," referring to the three-parted lip, which in this species has two parts that are larger, hence the common name and the species name, from the Latin *bi*, "two," and *cauda*, "tail." The plant is called *iphamba* by the Zulus in South Africa, who use it as a charm.

The flowers release a strong scent of vanilla at night and have a long nectar spur. The likely pollinators are, therefore, night-flying hawk moths.

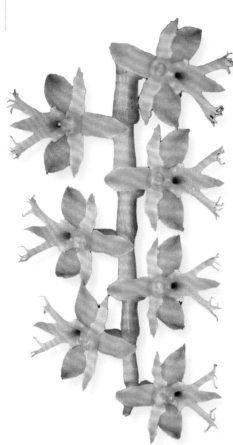

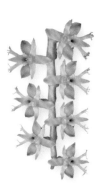

Actual size

The flower of the Two Tails–Three Fingers is green to tan and has spreading sepals and petals, the sepals broadly triangular and the petals lanceolate. The lip is elaborately five-lobed, with two short lateral lobes, and the limb has two fimbriate lobes and a central simple one.

SUBFAMILY	Epidendroideae
TRIBE AND SUBTRIBE	Vandeae, Angraecinae
NATIVE RANGE	Southeast Africa to the Cape, north to Tanzania
HABITAT	Miombo (*Brachystegia*) woodland and coniferous (*Juniperus, Widdringtonia*) forests on rocky slopes, at 3,300–6,900 ft (1,000–2,100 m)
TYPE AND PLACEMENT	Epiphytic or occasionally lithophytic
CONSERVATION STATUS	Not globally assessed, but least concern in South Africa
FLOWERING TIME	March to April (fall)

FLOWER SIZE
⅝ in (1.6 cm),
excluding spur

PLANT SIZE
10–18 × 8–15 in
(25–46 × 20–38 cm),
excluding arching-
pendent inflorescence
10–22 in (25–56 cm) long

634

YPSILOPUS ERECTUS
MIOMBO COMET ORCHID
(P. J. CRIBB) P. J. CRIBB & J. STEWART, 1985

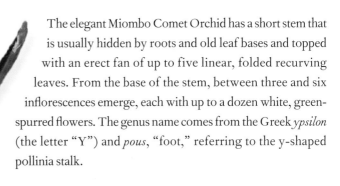

The elegant Miombo Comet Orchid has a short stem that is usually hidden by roots and old leaf bases and topped with an erect fan of up to five linear, folded recurving leaves. From the base of the stem, between three and six inflorescences emerge, each with up to a dozen white, green-spurred flowers. The genus name comes from the Greek *ypsilon* (the letter "Y") and *pous*, "foot," referring to the y-shaped pollinia stalk.

Pollination of the species has not been studied in the field, but on the basis of the shape (especially the long nectar spur) and color of the flower, a hawk moth is the likely pollinator. The common name refers to the type of woodland (miombo) in which the orchid occurs and its similarity to the Comet Orchid, *Angraecum sesquipedale*, from Madagascar.

The flower of the Miombo Comet Orchid is white and has similar, fleshy, recurved sepals and petals. The lip is diamond-shaped with a long point and a green nectar spur about three times longer than the size of the flower.

Actual size

SUBFAMILY	Epidendroideae
TRIBE AND SUBTRIBE	Vandeae, Polystachyinae
NATIVE RANGE	Northern and central Madagascar
HABITAT	Evergreen forests, at 2,300–3,120 ft (700–950 m)
TYPE AND PLACEMENT	Epiphytic or terrestrial
CONSERVATION STATUS	Not assessed
FLOWERING TIME	December (summer)

POLYSTACHYA CLAREAE
CLARE'S CORN ORCHID
HERMANS, 2003

FLOWER SIZE
¼ in (0.8 cm)

PLANT SIZE
6–10 × 3–5 in
(15–25 × 8–13 cm),
excluding terminal
inflorescence
6–8 in (15–20 cm) tall

635

Clare's Corn Orchid occurs both in trees and on the ground and has cylindrical pseudobulbs bearing three to six leaves with bases that wrap around the pseudobulbs. There are two to four narrow bracts that protect the developing terminal inflorescence, which has one to three short side branches. The genus name derives from the Greek *polys*, "many," and *stachys*, "ear of grain," a reference to the stems looking like heads of wheat (or corn, as it is in England, which gives the common name). The species name is for Clare Hermans, an English orchid grower and expert.

Most species of *Polystachya* are pollinated by bees, which are attracted by non-rewarding hairs on the lip that resemble pollen. Such hairs are present in *P. clareae*, so it is assumed that the same system is operating here.

Actual size

The flower of Clare's Corn Orchid has the lip uppermost and bright red-orange sepals and petals that frame the bonnet-shaped yellow orange lip and column. The lip bears yellow hairs, thought to mimic pollen, and sits directly above the column.

SUBFAMILY	Epidendroideae
TRIBE AND SUBTRIBE	Vandeae, Polystachyinae
NATIVE RANGE	Central to southern Africa, Madagascar, Mauritius, Réunion, and the Seychelles
HABITAT	Trees in a variety of different forest types and on rocks in leaf litter, from sea level to 9,850 ft (3,000 m)
TYPE AND PLACEMENT	Epiphytic, rarely lithophytic
CONSERVATION STATUS	Not assessed, but frequent over a large range
FLOWERING TIME	April to August (spring to summer)

FLOWER SIZE
¾ in (1.9 cm)

PLANT SIZE
6–10 × 1–2½ in
(15–25 × 2.5–6.5 cm),
excluding arching
inflorescence
4–15 in (10–38 cm) long

636

POLYSTACHYA CULTRIFORMIS
POWDER ORCHID
(THOUARS) LINDLEY EX SPRENGEL, 1826

Actual size

The Powder Orchid has thin, spindle-like pseudobulbs, each capped by a single leaf that can be elliptic to almost broadly oblong. The inflorescence is produced from the point where the leaf and stem meet, and it can be highly branched and bear up to 20 flowers. The genus name comes from the Greek *polys*, "many," and *stachys*, "an ear of grain," referring to the appearance of a cluster of its swollen stems.

Pollination is by small bees that are attracted by the highly fragrant and colorful flowers. The species produces a powdery material on its lip—hence the common name—and bees collect this, thinking that it is pollen. They are victims of deceit, as the powder has no food value and so is worthless to the insects. The pseudobulbs are used in West Africa to make an aphrodisiac.

The flower of the Powder Orchid is highly variable in color, ranging from cream to deep rose pink, and the usually upside-down flowers have a hooded lip on which the tip is darker and bears a powdery substance. The lip base and column form an empty cavity.

SUBFAMILY	Epidendroideae
TRIBE AND SUBTRIBE	Vandeae, Polystachyinae
NATIVE RANGE	South Africa to eastern Zimbabwe
HABITAT	Forests, scrub, or sandstone outcrops, at up to 4,920 ft (1,500 m)
TYPE AND PLACEMENT	Epiphytic on tree trunks or branches, or sometimes epilithic on sandstone
CONSERVATION STATUS	Locally common
FLOWERING TIME	October to December (spring)

FLOWER SIZE
¾ in (2 cm)

PLANT SIZE
5–10 × 2–3 in
(13–25 × 5–8 cm),
excluding inflorescence,
which is erect and
7–12 in (18–30 cm) tall

637

POLYSTACHYA PUBESCENS
LUCKY CHARM ORCHID
(LINDLEY) REICHENBACH FILS, 1863

The narrowly conical pseudobulbs of the Lucky Charm Orchid are clustered on a clump of fleshy roots. Two or three lanceolate, unequally lobed leaves top a pseudobulb that also carries an upright, pubescent raceme with up to a dozen fragrant flowers. The blooms lack nectar but produce "food hairs" on the lip that are collected by small bees as pseudopollen. Little evidence exists that these hairs are nutritious, so this is another case of deceit.

There are records of the plant's use as a protective charm by the amaZulu people, and the pseudobulbs of many species of *Polystachya*, including *P. pubescens*, have been used in the production of aphrodisiacs. The genus name comes from the Greek words *polys*, meaning "many," and *stachys*, "an ear of grain," in reference to the clustered stems of the first species ascribed to this genus.

Actual size

The flower of the Lucky Charm Orchid has the lip uppermost. The yellow perianth is spreading, with the lateral sepals boat-shaped and red-striped at the base and the yellow lip forming a small, pointed cap over the column.

APPENDICES

GLOSSARY

Acuminate With a pointed tip.

Adnate Fused.

Anther Pollen-bearing structure of a flower.

Anthesis Period of flowering.

Apex (adj. apical) Point at the upper end of a structure.

Apomictic Producing seeds without pollination.

Auricle Earlike appendage.

Axil (adj. axillary) Upper side of the point where a leaf attaches to a stem.

Basal Portion of a structure nearest to the point of attachment.

Boreal Subarctic regions.

Bract Small leaflike structure.

Bracteole Small bract.

Buzz pollination Form of pollination in which the anthers have apical pores and are gathered together such that the tips all touch; the pollinating bee grabs these anthers and buzzes its flight muscles, making the pollen erupt from the anthers, at which point the bee gathers the pollen.

Calcareous Chalky.

Callus (pl. calli) Swollen structure on a flower, usually helping to direct the movement of the pollinator or the pollinator's tongue.

Canaliculated With a canal.

Cilia (adj. ciliate) Large hairs.

Circumboreal Occurring throughout the boreal (subarctic) zone.

Cleistogamous Not opening, in reference to a flower that self-pollinates without opening.

Clonal Asexual, reproducing by runners or stolons (see stolon).

Commensal relationship Living together with benefit to one of the partners.

Conduplicate Folded into two parts.

Cordate Heart-shaped.

Corm Swollen stem that is dormant during the winter or dry season.

Cristate With a crest.

Deflexed Downwardly flexed (bent).

Dimorphic With two forms.

Distal End farthest away from the point of attachment.

Distichous Double-ranked, in reference to leaves.

Ectomycorrhizal Connected by the roots to a fungus.

Elaiophore Gland that produces a compound (scent or oil, for example).

Elfin woodland Short montane woodland created by windy, wet conditions.

Emarginated With a blunt (unpointed) apex.

Epilithic Growing on rocks.

Epiphytic Growing on a tree, but not a parasite.

Erose With a fringed margin.

Exsert[ed] Sticking out.

Extirpated Made extinct, at least locally.

Falcate Curved like a sickle.

Foliar Relating to a leaf.

Fractiflex Zigzag.

Frons Upper portion of the head of an insect.

Fusiform Tapering at both ends, spindle-shaped.

Glaucous With a powdery appearance.

Globose Globe-shaped.

Hammock Raised area in a swamp covered with hardwoods, specifically in the southern United States.

Hastate With a narrow base.

Inflorescence Stem that bears the flowers.

Keel Elongate raised area.

Labellum Lip, the modified petal of an orchid.

Lamella (pl. lamellae, adj. lamellate) Elongate raised area.

Lamina Flat part of a leaf (not the stem).

Lanceolate Lance-shaped.

Lateritic Soil type rich in iron and aluminum.

Ligule (adj. ligulate) Bearing a tonguelike piece of tissue.

Lithophytic Growing on rocks.

Lycopod Fernlike plant that bears spores and looks like a small pine tree.

Mesic Not wet, not dry, with a balanced amount of moisture.

Monopodial Growing from the same point over many years.

Monotypic Genus with a single species.

Montane Growing on mountains.

Morphology Shape and structure of a plant.

Mycoheterotrophic Orchid that derives its nutrition from an association with a fungus.

Mucilage Slime.

Mycorrhizal Referring to a beneficial relationship with fungus.

Non-resupinate With the lip uppermost, achieved by a twisting of the stem holding the flower or the flower simply flipping over on its axis.

Oblanceolate (also linear-oblanceolate) Leaf or flower part with the basal end narrower than the apical end.

Obligate Epiphyte growing only on trees.

Obovate (also obovoid) Leaf or floral part that is wider at the apical end; a synonym of oblanceolate.

Osmophore Gland that releases fragrance compounds.

Outbreeding Crossing with an unrelated member of the same species, as opposed to inbreeding, which is crossing with a closely related individual.

Ovate Leaf or flower part with the broader part at the base.

Pandurate (also panduriform) Violin-shaped (thinner in the middle).

Panicle (adj. paniculate) Branched inflorescence in which the oldest flower is at the base.

Papillae (adj. papillate) Nipple-shaped protuberances.

Pedicel Stalk that supports a single flower.

Peloric Where an orchid's lip is the same size as the petals, producing an abnormal symmetry.

Perianth Sepals and petals collectively.

Petiole (adj. petiolate) Stalk that supports a leaf.

Picotee With the border colored differently from the rest.

Pilose Hairy.

Plicate (of leaves) Folded, like a lady's fan.

Pollinarium Structure, consisting of the pollinia, stipe, and viscidium (see definitions below), associated with the pollen in an orchid.

Pollinium (pl. pollinia) Mass of non-powdery pollen.

Pseudobulb Swollen part of an orchid stem.

Pseudocopulation Copulation in which one of the partners is a flower, not a member of the opposite sex.

Pseudopollen Structures produced by a flower to resemble pollen.

Raceme Unbranched flower stem in which the oldest flowers are at the base.

Racemose Resembling a raceme (see above).

Rachis Stem bearing flower stalks.

Recurved Curved back toward the base.

Resupinate With the lip lowermost, achieved by a twisting of the stalk.

Reticulation Pattern resembling a net.

Rhizome Underground stem that grows horizontally.

Rostellum Structure holding the pollen in an orchid and separating it from the female receptive surface (the stigma).

Rugose Wrinkled.

Saccate Sack-shaped.

Scape (adj. scapose) Leafless stem.

Sclerophyll (of forests) Hard-leaved.

Sepal Outermost part of a flower.

Sidelobe Lobe other than the apical lobe of a flower part.

Spatulate (or spathulate) Having a structure with a broad, rounded apex.

Stamen Pollen-bearing structure.

Staminodium Sterile stamen.

Stelidia Fingerlike structure on the side of the column.

Stipe Stalk, part of the pollinarium attaching the pollinia to viscidium.

Stolon Stemlike structure that produces a plantlet at its tip.

Style Stem that bears the stigma (the structure in a flower that receives the pollen).

Suborbicular Almost round in shape.

Substrate Material in which a plant grows.

Subtend To sit below.

Subulate Awl-shaped, slender, and tapering to a point.

Sulcate With a groove (sulcus).

Sympatric Growing together.

Synsepal Fused structure composed of two sepals.

Tepal Petals and sepals.

Terete Round in cross section.

Terrestrial Growing in soil.

Thrips Small insects (order Thysanoptera) that feed on plants, especially flowers.

Tridentate With three teeth.

Tuber Swollen underground stem (like a potato).

Tubercle (adj. tuberculate) Swollen spot.

Ultrabasic Soil derived from igneous (volcanic) rocks with a low silica content.

Ultramafic See ultrabasic.

Umbel Flower stem in which the flowers are all attached to a central point.

Understory Ground section of a forest.

Unifacial With one side facing out, used for leaves that have been folded so that the inner surfaces are fused together.

Vandoid Genus or species of orchid that looks like the orchid genus *Vanda*.

Venation Collective system of veins in a leaf or floral part.

Ventral Lower or on the underside.

Vermiform Worm-shaped.

Villous Shaggy.

Viscidium (also viscid disc) Sticky structure in an orchid flower that adheres to the pollinator and is attached to the pollen, such that the pollinator's action removes the pollen.

641

CLASSIFICATION
of the ORCHIDACEAE

Below is a classification of the orchids, listing genera followed by their authors, from Chase et al. "An updated classification of Orchidaceae" (*Botanical Journal of the Linnean Society*, 2015).

Number of species are shown in brackets

SUBFAMILY APOSTASIOIDEAE
Apostasia Blume [6]
Neuwiedia Blume [8]

SUBFAMILY VANILLOIDEAE

Tribe Vanilleae [14 genera]
SUBTRIBE POGONIINAE:
Cleistes Rich. ex Lindl. [64]
Cleistesiopsis Pansarin & F. Barros [2]
Duckeella Porto & Brade [3]
Isotria Raf. [2]
Pogonia Juss. [5]

SUBTRIBE VANILLINAE:
Clematepistephium N. Hallé [1]
Cyrtosia Blume [5]
Epistephium Kunth [21]
Eriaxis Rchb. f. [1]
Erythrorchis Blume [2]
Galeola Lour. [6]
Lecanorchis Blume [20]
Pseudovanilla Garay [8]
Vanilla Plum. ex Mill. [105]

SUBFAMILY CYPRIPEDIOIDEAE
[five genera]

Cypripedium L. [51]
Mexipedium V.A. Albert & M.W. Chase [1]
Paphiopedilum Pfitzer [86]
Phragmipedium Rolfe [26]
Selenipedium Rchb. f. [5]

SUBFAMILY ORCHIDOIDEAE

Tribe Codonorchideae
Codonorchis Lindl. [1]

Tribe Cranichideae [100 genera]
SUBTRIBE CHLORAEINAE:
Bipinnula [11]
Chloraea Comm. ex Juss. [52]
Gavilea Poepp. [15]

SUBTRIBE CRANICHIDINAE:
Aa Rchb. f. [25]
Altensteinia Kunth [7]
Baskervilla Lindl. [10]
Cranichis Sw. [53]
Fuertesiella Schltr. [1]
Galeoglossum A. Rich & Galeotti [3]
Gomphichis Lindl. [24]
Myrosmodes Rchb. f. [12]
Ponthieva R. Br. in W.T. Aiton [66]
Porphyrostachys Rchb. f. [2]
Prescottia Lindl. [26]
Pseudocentrum Lindl. [7]

Pterichis Lindl. [20]
Solenocentrum Schltr. [4]
Stenoptera C. Presl [7]

SUBTRIBE GALEOTTIELLINAE:
Galeottiella Schltr. [6]

SUBTRIBE GOODYERINAE:
Aenhenrya Gopalan [1]
Anoectochilus Blume [43]
Aspidogyne Garay [45]
Chamaegastrodia Makino & F. Maek. [3]
Cheirostylis Blume [53]
Cystorchis Blume [21]
Danhatchia Garay & Christenson [1]
Dossinia C. Morren [1]
Erythrodes Blume [26]
Eurycentrum Schltr. [7]
Gonatostylis Schltr. [2]
Goodyera R. Br. in W.T. Aiton [98]
Halleorchis Szlach. & Olszewski [1]
Herpysma Lindl. [1]
Hetaeria Blume [29]
Hylophila Lindl. [7]
Kreodanthus Garay [14]
Kuhlhasseltia J.J. Sm. [9]
Lepidogyne Blume [1]
Ligeophila Garay [12]
Ludisia A. Rich. [1]
Macodes Lindl. [11]
Microchilus C. Presl [137]
Myrmechis Blume [17]
Odontochilus Blume [25]
Orchipedum Breda [3]
Pachyplectron Schltr. [3]
Papuaea Schltr. [1]
Platylepis A. Rich. [17]
Platythelys Garay [13]
Rhamphorhynchus Garay [1]
Rhomboda Lindl. [22]
Schuitemania Ormerod [1]
Stephanothelys Garay [5]
Vrydagzynea Blume [43]
Zeuxine Lindl. [74]

SUBTRIBE MANNIELLINAE:
Manniella Rchb. f. [2]

SUBTRIBE PTEROSTYLIDINAE:
Achlydosa M.A. Clem. & D.L. Jones [1]
Pterostylis R. Br. [211]

SUBTRIBE DISCYPHUSINAE:
Discyphus Schltr. [1]

SUBTRIBE SPIRANTHINAE:
Aracamunia Carnevali & I. Ramírez [1]
Aulosepalum Garay [7]

Beloglottis Schltr. [7]
Brachystele Schltr. [21]
Buchtienia Schltr. [3]
Coccineorchis Schltr. [7]
Cotylolabium Garay [1]
Cybebus Garay [1]
Cyclopogon C. Presl [83]
Degranvillea Determann [1]
Deiregyne Schltr. [18]
Dichromanthus Garay [4]
Eltroplectris Raf. [13]
Eurystyles Wawra [20]
Funkiella Schltr. [27]
Hapalorchis Schltr. [10]
Helonoma Garay [4]
Kionophyton Garay [4]
Lankesterella Ames [11]
Lyroglossa Schltr. [2]
Mesadenella Pabst & Garay [7]
Mesadenus Schltr. [7]
Nothostele Garay [2]
Odontorrhynchus M.N. Correa [6]
Pelexia Poit. ex Rich. [77]
Physogyne Garay [3]
Pseudogoodyera Schltr. [1]
Pteroglossa Schltr. [11]
Quechua Salazar & L. Jost [1]
Sacoila Raf. [7]
Sarcoglottis C. Presl [48]
Sauroglossum Lindl. [11]
Schiedeella Schltr. [24]
Skeptrostachys Garay [13]
Sotoa Salazar [1]
Spiranthes Rich. [34]
Stalkya Garay [1]
Stenorrhynchos Rich. ex Spreng. [5]
Svenkoeltzia Burns-Bal. [3]
Thelyschista Garay [1]
Veyretia Szlach. [11]

Tribe Diurideae [39]
SUBTRIBE ACIANTHINAE:
Acianthus R. Br. [20]
Corybas Salisb. [132]
Cyrtostylis R. Br. [5]
Stigmatodactylus Maxim. ex Makino [10]
Townsonia Cheeseman [2]

SUBTRIBE CALADENIINAE:
Adenochilus Hook. f. [2]
Aporostylis Rupp & Hatch [1]
Caladenia R. Br. [267]
Cyanicula Hopper & A.P. Brown [10]
Elythranthera [Endl.] A.S. George [2]
Ericksonella Hopper & A.P. Br. [1]
Eriochilus R. Br. [9]
Glossodia R. Br. [2]
Leptoceras [R. Br.] Lindl. [1]

Pheladenia D.L. Jones & M.A. Clem. [1]
Praecoxanthus Hopper & A.P. Brown [1]

SUBTRIBE CRYPTOSTYLIDINAE:
Coilochilus Schltr. [1]
Cryptostylis R. Br. [23]

SUBTRIBE DIURIDINAE:
Diuris Sm. [71]
Orthoceras R. Br. [2]

SUBTRIBE DRAKAEINAE:
Arthrochilus F. Muell. [15]
Caleana R. Br. [1]
Chiloglottis R. Br. [23]
Drakaea Lindl. [10]
Paracaleana Blaxell [13]
Spiculaea Lindl. [1]

SUBTRIBE MEGASTYLIDINAE:
Burnettia Lindl. [1]
Leporella A.S. George [1]
Lyperanthus R. Br. [2]
Megastylis [Schltr.] Schltr. [7]
Pyrorchis D.L. Jones &
  M.A. Clements [2]
Rimacola Rupp [1]
Waireia D.L. Jones, Molloy & M.A.
  Clements [1]

SUBTRIBE PRASOPHYLLINAE:
Genoplesium R. Br. [47]
Microtis R. Br. [19]
Prasophyllum R. Br. [131]

SUBTRIBE RHIZANTHELLINAE:
Rhizanthella R.S. Rogers [3]

SUBTRIBE THELYMITRINAE:
Calochilus R. Br. [27]
Epiblema R. Br. [1]
Thelymitra J.R. Forest. & G. Forest. [110]

Tribe Orchideae [59 genera]
SUBTRIBE BROWNLEEINAE:
Brownleea Harv. ex Lindl. [8]
Disperis Sw. [78]

SUBTRIBE CORYCIINAE:
Ceratandra Lindl. [6]
Corycium Sw. [15]
Evotella Kurzweil & H.P. Linder [1]
Pterygodium Sw. [19]

SUBTRIBE DISINAE:
Disa P.J. Bergius [182]
Huttonaea Harv. [5]
Pachites Lindl. [2]

643

SUBTRIBE ORCHIDINAE:
Aceratorchis Schltr. [1]
Anacamptis Rich. [11]
Androcorys Schltr. [10]
Bartholina R. Br. [2]
Benthamia A. Rich. [29]
Bhutanthera J. Renz [5]
Bonatea Willd. [13]
Brachycorythis Lindl. [36]
Centrostigma Schltr. [3]
Chamorchis Rich. [1]
Cynorkis Thouars [156]
Dactylorhiza Neck. ex Nevski [40]
Diplomeris D. Don [3]
Dracomonticola H.P. Linder
    & Kurweil [1]
Galearis Raf. [10]
Gennaria Parl. [1]
Gymnadenia R. Br. [23]
Habenaria Willd. [835]
Hemipilia Lindl. [13]
Hsenhsua X.H. Jin, Schuit.
    & W.T. Jin [1]
Herminium L. [19]
Himantoglossum Spreng. [11]
Holothrix Rich. ex Lindl. [45]
Megalorchis H. Perrier [1]
Neobolusia Schltr. [3]
Neotinea Rchb. f. [4]
Oligophyton H.P. Linder [1]
Ophrys L. [34]
Orchis Tourn. ex L. [21]
Pecteilis Raf. [8]
Peristylus Blume [103]
Physoceras Schltr. [12]
Platanthera Rich. [136]
Platycoryne Rchb. f. [19]
Ponerorchis Rchb. f. [55]
Porolabium Tang & F.T. Wang [1]
Pseudorchis Ség. [1]
Roeperocharis Rchb. f. [5]
Satyrium L. [86]
Schizochilus Sond. [11]
Serapias L. [13]
Silvorchis J.J. Sm. [3]
Sirindhornia H.A. Pedersen
    & Suksathan [3]
Stenoglottis Lindl. [7]
Steveniella Schltr. [1]
Thulinia P.J. Cribb [1]
Traunsteinera Rchb. [2]
Tsaiorchis Tang & F.T. Wang [1]
Tylostigma Schltr. [8]
Veyretella Szlach.
    & Olszewski [2]

SUBFAMILY EPIDENDROIDEAE
Tribe Neottieae [six genera]
Aphyllorchis Blume [22]
Cephalanthera Rich. [19]
Epipactis Zinn [49]
Limodorum Boehm. [3]
Neottia Guett. [64]
Palmorchis Barb. Rodr. [21]

Tribe Sobralieae [four genera]
Elleanthus C. Presl [111]
Epilyna Schltr. [2]

Sertifera Lindl. [7]
Sobralia Ruiz & Pav. [149]

Tribe Triphoreae [five genera]
SUBTRIBE DICERATOSTELINAE:
Diceratostele Summerh. [1]

SUBTRIBE TRIPHORINAE:
Monophyllorchis Schltr. [1]
Pogoniopsis Rchb. f. [2]
Psilochilus Barb. Rodr. [7]
Triphora Nutt. [18]

Tribe Tropidieae [two genera]
Corymborkis Thouars [6]
Tropidia Lindl. [31]

Tribe Xerorchideae
Xerorchis Schltr. [2]

Tribe Wullschlaegelieae
Wullschlaegelia Rchb. f. [2]

Tribe Gastrodieae [six genera]
Auxopus Schltr. [4]
Didymoplexiella Garay [8]
Didymoplexis Griff. [17]
Gastrodia R. Br. [60]
Neoclemensia Carr [1]
Uleiorchis Hoehne [2]

Tribe Nervilieae [three genera]
SUBTRIBE NERVILIINAE:
Nervilia Comm. ex Gaudich. [67]

SUBTRIBE EPIPOGIINAE:
Epipogium Borkh. [3]
Stereosandra Blume [1]

Tribe Thaieae
Thaia Seidenf. [1]

Tribe Arethuseae [26 genera]
SUBTRIBE ARETHUSINAE:
Anthogonium Wall. ex Lindl. [9]
Arethusa L. [1]
Arundina Blume [2]
Calopogon R. Br. [5]
Eleorchis Maek. [1]

SUBTRIBE COELOGYNINAE:
Aglossorrhyncha Schltr. [13]
Bletilla Rchb. f. [5]
Bracisepalum J.J. Sm. [2]
Bulleyia Schltr. [1]
Chelonistele Pfitzer [13]
Coelogyne Lindl. [200]
Dendrochilum Blume [278]
Dickasonia L.O. Williams [1]
Dilochia Lindl. [8]
Entomophobia de Vogel [1]
Geesinkorchis de Vogel [4]
Glomera Blume [131]
Gynoglottis J.J. Sm. [1]
Ischnogyne Schltr. [1]
Nabaluia Ames [3]
Neogyna Rchb. f. [1]
Otochilus Lindl. [5]

Panisea Lindl. [11]
Pholidota Lindl. [39]
Pleione D. Don [21]
Thunia Rchb. f. [5]

Tribe Malaxideae [16 genera]
SUBTRIBE DENDROBIINAE:
Bulbophyllum Thouars [1867]
Dendrobium Sw. [1509]

SUBTRIBE MALAXIDINAE:
Alatiliparis Marg. & Szlach. [5]
Crepidium Blume [260]
Crossoglossa Dressler
    & Dodson [26]
Crossoliparis Marg. [1]
Dienia Lindl. [6]
Hammarbya Kuntze [1]
Hippeophyllum Schltr. [10]
Liparis Rich. [426]
Malaxis Sol. ex Sw. [182]
Oberonia Lindl. [323]
Oberonioides Szlach. [2]
Orestias Ridl. [4]
Stichorkis Thouars [8]
Tamayorkis Szlach. [1]

Tribe Cymbidieae [174 genera]
SUBTRIBE CYMBIDIINAE:
Acriopsis Reinw. ex Blume[9]
Cymbidium Sw. [71]
Grammatophyllum Blume [12]
Porphyroglottis Ridl. [1]
Thecopus Seidenf. [2]
Thecostele Rchb. f. [1]

SUBTRIBE EULOPHIINAE:
Acrolophia Pfitzer [7]
Ansellia Lindl. [1]
Claderia Hook. f. [2]
Cymbidiella Rolfe [3]
Dipodium R. Br. [25]
Eulophia R. Br. [200]
Eulophiella Rolfe [5]
Geodorum Jacks. [12]
Grammangis Rchb. f. [2]
Graphorkis Thouars [4]
Imerinaea Schltr. [1]
Oeceoclades Lindl. [38]
Paralophia P.J. Cribb & Hermans [2]

SUBTRIBE CATASETINAE:
Catasetum Rich. ex Kunth [176]
Clowesia Lindl. [7]
Cyanaeorchis Barb. Rodr. [3]
Cycnoches Lindl. [34]
Dressleria Dodson [11]
Galeandra Lindl. [38]
Grobya Lindl. [5]
Mormodes Lindl. [80]

SUBTRIBE CYRTOPODIINAE:
Cyrtopodium R. Br. [47]

SUBTRIBE COELIOPSIDINAE:
Coeliopsis Rchb. f. [1]
Lycomormium Rchb. f. [5]
Peristeria Hook. [13]

SUBTRIBE ERIOPSIDINAE:
Eriopsis Lindl. [5]

SUBTRIBE MAXILLARIINAE:
Anguloa Ruiz & Pav. [9]
Bifrenaria Lindl. [21]
Guanchezia G.A. Romero & Carnevali [1]
Horvatia Garay [1]
Lycaste Lindl. [32]
Maxillaria Ruiz & Pav. [658]
Neomoorea Rolfe [1]
Rudolfiella Hoehne [6]
Scuticaria Lindl. [11]
Sudamerlycaste Archila [42]
Teuscheria Garay [7]
Xylobium Lindl. [30]

SUBTRIBE ONCIDIINAE:
Aspasia Salisb. [7]
Brassia R. Br. [64]
Caluera Dodson & Determann [3]
Capanemia Barb. Rodr. [9]
Caucaea Schltr. [9]
Centroglossa Barb. Rodr. [5]
Chytroglossa Rchb. f. [3]
Cischweinfia Dressler
    & N.H. Williams [11]
Comparettia Poepp. & Endl. [78]
Cuitlauzina La Lllave & Lex. [7]
Cypholoron Dodson & Dressler [2]
Cyrtochiloides N.H. Williams
    & M.W. Chase [3]
Cyrtochilum Kunth [137]
Dunstervillea Garay [1]
Eloyella P. Ortiz [10]
Erycina Lindl. [7]
Fernandezia Ruiz & Pav. [51]
Gomesa R. Br. [119]
Grandiphyllum Docha Neto [7]
Hintonella Ames [1]
Hofmeisterella Rchb. f. [2]
Ionopsis Kunth [6]
Leochilus Knowles & Westc. [12]
Lockhartia Hook. [28]
Macradenia R. Br. [11]
Macroclinium Barb. Rodr. [42]
Miltonia Lindl. [12]
Miltoniopsis God.-Leb. [5]
Notylia Lindl. [56]
Notyliopsis P. Ortiz [2]
Oliveriana Rchb. f. [6]
Oncidium Sw. [311]
Ornithocephalus Hook. [55]
Otoglossum [Schltr.] Garay & Dunst. [13]
Phymatidium Lindl. [10]
Platyrhiza Barb. Rodr. [1]
Plectrophora H. Focke [10]
Polyotidium Garay [1]
Psychopsiella Lückel & Braem [1]
Psychopsis Raf. [4]
Pterostemma Kraenzl. [3]
Quekettia Lindl. [4]
Rauhiella Pabst & Braga [3]
Rhynchostele Rchb. f. [17]
Rodriguezia Ruiz & Pav. [48]
Rossioglossum [Schltr.] Garay
    & G.C. Kenn. [9]
Sanderella Kuntze [2]

Saundersia Rchb. f. [2]
Schunkea Senghas [1]
Seegeriella Senghas [2]
Solenidium Lindl. [3]
Suarezia Dodson [1]
Sutrina Lindl. [2]
Systeloglossum Schltr. [5]
Telipogon Kunth [205]
Thysanoglossa Porto & Brade [3]
Tolumnia Raf. [27]
Trichocentrum Poepp. & Endl. [70]
Trichoceros Kunth [10]
Trichopilia Lindl. [44]
Trizeuxis Lindl. [1]
Vitekorchis Romowicz & Szlach. [4]
Warmingia Rchb. f. [4]
Zelenkoa M.W. Chase
    & N.H. Williams [1]
Zygostates Lindl. [22]

SUBTRIBE STANHOPEINAE:
Acineta Lindl. [17]
Braemia Jenny [1]
Cirrhaea Lindl. [7]
Coryanthes Hook. [59]
Embreea Dodson [2]
Gongora Ruiz & Pav. [74]
Horichia Jenny [1]
Houlletia Brongn. [9]
Kegeliella Mansf. [4]
Lacaena Lindl. [2]
Lueckelia Jenny [1]
Lueddemannia Linden & Rchb. f. [3]
Paphinia Lindl. [16]
Polycycnis Rchb. f. [17]
Schlimia Planch. & Linden [7]
Sievekingia Rchb. f. [16]
Soterosanthus F. Lehm. ex Jenny [1]
Stanhopea J. Frost ex Hook. [61]
Trevoria F. Lehm. [5]
Vasqueziella Dodson [1]

SUBTRIBE ZYGOPETALINAE:
Aetheorhyncha Dressler [1]
Aganisia Lindl. [4]
Batemannia Lindl. [5]
Benzingia Dodson [9]
Chaubardia Rchb. f. [3]
Chaubardiella Garay [8]
Cheiradenia Lindl. [1]
Chondrorhyncha Lindl. [7]
Chondroscaphe [Dressler]
    Senghas & G. Gerlach [14]
Cochleanthes Raf. [4]
Cryptarrhena R. Br. [3]
Daiotyla Dressler [4]
Dichaea Lindl. [118]
Echinorhyncha Dressler [5]
Euryblema Dressler [2]
Galeottia A. Rich. [12]
Hoehneella Ruschi [2]
Huntleya Bateman ex Lindl. [14]
Ixyophora Dressler [5]
Kefersteinia Rchb. f. [70]
Koellensteinia Rchb. f. [17]
Neogardneria Schltr. ex Garay [1]
Otostylis Schltr. [4]
Pabstia Garay [5]

Paradisanthus Rchb. f. [4]
Pescatoria Rchb. f. [23]
Promenaea Lindl. [18]
Stenia Lindl. [22]
Stenotyla Dressler [9]
Vargasiella C. Schweinf. [1]
Warczewiczella Rchb. f. [11]
Warrea Lindl. [3]
Warreella Schltr. [2]
Warreopsis Garay [4]
Zygopetalum Hook. [14]
Zygosepalum [Rchb. f.] Rchb. f. [8]

Tribe Epidendreae [99 genera]
SUBTRIBE BLETIINAE:
Basiphyllaea Schltr. [7]
Bletia Ruiz & Pav. [33]
*Chysis*Lindl. [10]
Hexalectris Raf. [10]

SUBTRIBE LAELIINAE:
Acrorchis Dressler [1]
Adamantinia van den Berg
    & C.N. Conç [1]
Alamania Llave & Lex. [1]
Arpophyllum Llave & Lex. [3]
Artorima Dressler & G.E. Pollard [1]
Barkeria Knowl. & Westc. [17]
Brassavola R. Br. [22]
Broughtonia R. Br. [6]
Cattleya Lindl. [112]
Cattleyella van den Berg
    & M.W. Chase [1]
Caularthron Raf. [4]
Constantia Barb. Rodr. [6]
Dimerandra Schltr. [8]
Dinema Lindl. [1]
Domingoa Schltr. [4]
Encyclia Hook. [165]
Epidendrum L. [1413]
Guarianthe Dressler & W.E. Higgins [4]
Hagsatera R. González [2]
Homalopetalum Rolfe [8]
Isabelia Barb. Rodr. [3]
Jacquiniella Schltr. [12]
Laelia Lindl. [23]
Leptotes Lindl. [9]
Loefgrenianthus Hoehne [1]
Meiracyllium Rchb. f. [2]
Microepidendrum Brieger
    ex W.E. Higgins [1]
Myrmecophila Rolfe [10]
Nidema Britton & Millsp. [2]
Oestlundia W.E. Higgins [4]
Orleanesia Barb. Rodr. [9]
Prosthechea Knowles & Westc. [117]
Pseudolaelia Porto & Brade [18]
Psychilis Raf. [14]
Pygmaeorchis Brade [2]
Quisqueya Dod [4]
Rhyncholaelia Schltr. [2]
Scaphyglottis Poepp. & Endl. [69]
Tetramicra Lindl. [14]

SUBTRIBE PLEUROTHALLIDINAE:
Acianthera Scheidw. [118]
Anathallis Barb. Rodr. [152]
Andinia [Luer] Luer [13]

645

Barbosella Schltr. [19]
Brachionidium Lindl. [75]
Chamelophyton Garay [1]
Dilomilis Raf. [5]
Diodonopsis Pridgeon & M.W. Chase [5]
Draconanthes [Luer] Luer [2]
Dracula Luer [127]
Dresslerella Luer [13]
Dryadella Luer [54]
Echinosepala Pridgeon
    & M.W. Chase [11]
Frondaria Luer [1]
Kraenzlinella Kuntze [9]
Lepanthes Sw. [1085]
Lepanthopsis [Cogn.] Ames [43]
Masdevallia Ruiz & Pav. [589]
Myoxanthus Poepp. & Endl. [48]
Neocogniauxia Schltr. [2]
Octomeria D. Don [159]
Pabstiella Brieger & Senghas [29]
Phloeophila Hoehne & Schltr. [11]
Platystele Schltr. [101]
Pleurothallis R. Br. [552]
Pleurothallopsis Porto & Brade [18]
Porroglossum Schltr. [43]
Restrepia Kunth [53]
Restrepiella Garay & Dunst. [2]
Scaphosepalum Pfitzer [46]
Specklinia Lindl. [135]
Stelis Sw. [879]
Teagueia [Luer] Luer [13]
Tomzanonia Nir [1]
Trichosalpinx Luer [111]
Trisetella Luer [23]
Zootrophion Luer [22]

SUBTRIBE PONERINAE:
Helleriella A.D. Hawkes [2]
Isochilus R. Br. [13]
Nemaconia Knowles & Westc. [6]
Ponera Lindl. [2]

SUBTRIBE CALYPSOINAE:
Aplectrum Nutt. [1]
Calypso Salisb. [1]
Changnienia S.S. Chien [1]
Coelia Lindl. [5]
Corallorhiza Gagnebin [11]
Cremastra Lindl. [4]
Dactylostalix Rchb. f. [1]
Danxiaorchis J.W. Zhai,
    F.W. Xing & Z.J. Liu [1]
Ephippianthus Rchb. f. [2]
Govenia Lindl. [24]
Oreorchis Lindl. [16]
Tipularia Nutt. [7]
Yoania Maxim. [4]

SUBTRIBE AGROSTOPHYLLINAE:
Agrostophyllum Blume [100]
Earina Lindl. [7]

Tribe Collabieae [20 genera]
Acanthephippium Blume [13]
Ancistrochilus Rolfe [2]
Ania Lindl. [11]
Calanthe R. Br. [216]
Cephalantheropsis Guillaumin [4]

Chrysoglossum Blume [4]
Collabium Blume [14]
Diglyphosa Blume [3]
Eriodes Rolfe [1]
Gastrorchis Thouars [8]
Hancockia Rolfe [1]
Ipsea Lindl. [3]
Nephelaphyllum Blume [11]
Pachystoma Blume [3]
Phaius Lour. [45]
Pilophyllum Schltr. [1]
Plocoglottis Blume [41]
Risleya King & Pantl. [1]
Spathoglottis Blume [48]
Tainia Blume [23]

Tribe Podochileae [27 genera]
Appendicula Blume [146]
Ascidieria Seidenf. [8]
Bryobium Lindl. [8]
Callostylis Blume [5]
Campanulorchis Brieger in F.R.R.
    Schlechter [5]
Ceratostylis Blume [147]
Conchidium Griff. [10]
Cryptochilus Wall. [5]
Dilochiopsis [Hook.] Brieger in F.R.R.
    Schlechter [1]
Epiblastus Schltr. [22]
Eria Lindl. [237]
Mediocalcar J.J. Sm. [17]
Mycaranthes Blume [36]
Notheria P. O'Byrne and J.J. Verm. [15]
Octarrhena Thwaites [52]
Oxystophyllum Blume [36]
Phreatia Lindl. [211]
Pinalia Lindl. [105]
Poaephyllum Ridl. [6]
Podochilus Blume [62]
Porpax Lindl. [13]
Pseuderia Schltr. [20]
Ridleyella Schltr. [1]
Sarcostoma Blume [5]
Stolzia Schltr. [15]
Thelasis Blume [26]
Trichotosia Blume [78]

Tribe Vandeae [137 genera]:
SUBTRIBE ADRORHIZINAE:
Adrorhizon Hook. f. [1]
Bromheadia Lindl. [30]
Sirhookera Kuntze [2]

SUBTRIBE POLYSTACHYINAE:
Hederorkis Thouars [2]
Polystachya Hook. [234]

SUBTRIBE AERIDINAE:
Acampe Lindl. [8]
Adenoncos Blume [17]
Aerides Lour. [25]
Amesiella Schltr. ex Garay [3]
Arachnis Blume [14]
Biermannia King & Pantl. [11]
Bogoria J.J. Sm. [4]
Brachypeza Garay [10]
Calymmanthera Schltr. [5]
Ceratocentron Senghas [1]

Chamaeanthus Schltr. [3]
Chiloschista Lindl. [20]
Chroniochilus J.J. Sm. [4]
Cleisocentron Brühl [6]
Cleisomeria Lindl. ex D. Don in Loud. [2]
Cleisostoma Blume [88]
Cleisostomopsis Seidenf. [2]
Cottonia Wight [1]
Cryptopylos Garay [1]
Deceptor Seidenf. [1]
Dimorphorchis Rolfe [5]
Diplocentrum Lindl. [2]
Diploprora Hook. f. [2]
Dryadorchis Schltr. [5]
Drymoanthus Nicholls [4]
Dyakia Christenson [1]
Eclecticus P. O'Byrne [1]
Gastrochilus D. Don [56]
Grosourdya Rchb. f. [11]
Gunnarella Senghas [9]
Holcoglossum Schltr. [14]
Hymenorchis Schltr. [12]
Jejewoodia Szlach. [6]
Luisia Gaudich. [39]
Macropodanthus L.O. Williams [8]
Micropera Lindl. [21]
Microsaccus Blume [12]
Mobilabium Rupp [1]
Omoea Blume [2]
Ophioglossella Schuit. & Ormerod [1]
Papilionanthe Schltr. [11]
Papillilabium Dockrill [1]
Paraphalaenopsis A.D. Hawkes [4]
Pelatantheria Ridl. [8]
Pennilabium J.J. Sm. [15]
Peristeranthus T.E. Hunt [1]
Phalaenopsis Blume [70]
Phragmorchis L.O. Williams [1]
Plectorrhiza Dockrill [3]
Pomatocalpa Breda [25]
Porrorhachis Garay [2]
Pteroceras Hassk. [27]
Renanthera Lour. [20]
Rhinerrhiza Rupp [1]
Rhinerrhizopsis Ormerod [3]
Rhynchogyna Seidenf. & Garay [3]
Rhynchostylis Blume [3]
Robiquetia Gaudich. [45]
Saccolabiopsis J.J. Sm. [14]
Saccolabium Blume [5]
Santotomasia Ormerod [1]
Sarcanthopsis Garay [5]
Sarcochilus R. Br. [25]
Sarcoglyphis Garay [12]
Sarcophyton Garay [3]
Schistotylus Dockrill [1]
Schoenorchis Reinw. ex Blume [25]
Seidenfadenia Garay [1]
Seidenfadeniella C.S. Kumar [2]
Singchia Z.J. Liu & L.J. Chen [1]
Smithsonia C.J. Saldanha [3]
Smitinandia Holttum [3]
Spongiola J.J. Wood & A.L. Lamb [1]
Stereochilus Lindl. [7]
Taeniophyllum Blume [185]
Taprobanea Christenson [1]
Thrixspermum Lour. [161]
Trachoma Garay [14]

646

Trichoglottis Blume [69]
Tuberolabium Yaman. [11]
Uncifera Lindl. [6]
Vanda R. Br. [73]
Vandopsis Pfitzer in Engler & Prantl [4]

SUBTRIBE ANGRAECINAE:
Aerangis Rchb. f. [58]
Aeranthes Lindl. [43]
Ambrella H. Perrier [1]
Ancistrorhynchus Finet [17]
Angraecopsis Kraenzl. [22]
Angraecum Bory [221]
Beclardia A. Rich. [2]
Bolusiella Schltr. [6]
Calyptrochilum Kraenzl. [2]
Campylocentrum Benth. [65]
Cardiochilos P.J. Cribb [1]
Chauliodon Summerh. [1]
Cribbia Senghas [4]
Cryptopus Lindl. [4]
Cyrtorchis Schltr. [18]
Dendrophylax Rchb. f. [14]
Diaphananthe Schltr. [33]
Dinklageella Mansf. [4]
Distylodon Summerh. [1]
Eggelingia Summerh. [3]
Erasanthe P.J. Cribb, Hermans
     & D.L. Roberts [1]

Eurychone Schltr. [2]
Jumellea Schltr. [59]
Lemurella Schltr. [4]
Lemurorchis Kraenzl. [1]
Listrostachys Rchb. f. [1]
Margelliantha P.J. Cribb [6]
Microcoelia Lindl. [30]
Mystacidium Lindl. [10]
Neobathiea Schltr. [5]
Nephrangis Summerh. [2]
Oeonia Lindl. [5]
Oeoniella Schltr. [2]
Ossiculum P.J. Cribb & Laan [1]
Plectrelminthus Raf. [1]
Podangis Schltr. [1]
Rangaeris [Schltr.] Summerh. [6]
Rhaesteria Summerh. [1]
Rhipidoglossum Schltr. [35]
Sobennikoffia Schltr. [4]
Solenangis Schltr. [8]
Sphyrarhynchus Mansf. [1]
Summerhayesia P.J. Cribb [2]
Taeniorrhiza Summerh. [1]
Triceratorhynchus Summerh. [1]
Tridactyle Schltr. [47]
Ypsilopus Summerh. [5]

INCERTAE SEDIS [IN EPIDENDROIDEAE]:
Devogelia Schuit. [1]

RESOURCES

BOOKS AND JOURNALS

Chase, M. W., K. M. Cameron, J. V. Freudenstein, A. M. Pridgeon, G. Salazar, C. van den Berg, and A. Schuiteman. An updated classification of Orchidaceae. *Botanical Journal of the Linnean Society* 177: 151–174 (2015).

A list of orchid genera and the number of species in each with a review of recently published phylogenetic papers on orchids.

Van der Cingel, N. H. *An Atlas of Orchid Pollination: European Orchids.* CRC PRESS, 2001.

Van der Cingel, N. H. *An Atlas of Orchid Pollination: America, Africa, Asia and Australasia.* BALKEMA, 2001.

Davy, A. and D. Gibson. Virtual issue: Charismatic Orchids. *Journal of Ecology*, 2015: www.journalofecology.org/view/0/orchidVI.html.

This is a compilation of orchid articles published in recent years in the *Journal of Ecology*; the topics covered include biological flora of the British Isles, demographic studies, mycorrhizal associations, and reproductive ecology.

Fay, M. F. and M. W. Chase. Orchid biology: from Linnaeus via Darwin to the 21st century. *Annals of Botany* 104: 359–364 (2009).

A review article from a volume of this journal dedicated to orchids, covering many areas of orchid biology.

Kull, T., J. Arditti, and S. M. Wong. *Orchid biology: reviews and perspectives* X (2009). SPRINGER.

This is the tenth volume in a series that began in 1977, most volumes edited by Arditti; it is comprised of solicited chapters on various aspects of orchid biology, history, cultivation, and even orchids in space.

Pridgeon, A. M., P. J. Cribb, M. W. Chase, and F. N. Rasmussen. *Genera Orchidacearum, Vol. 1: General Introduction, Apostasioideae and Cypripedioideae.* OXFORD UNIVERSITY PRESS, 1999.

Pridgeon, A. M., P. J. Cribb, M. W. Chase, and F. N. Rasmussen. *Genera Orchidacearum, Vol. 2: Orchidoideae (Part one).* OXFORD UNIVERSITY PRESS, 2001.

Pridgeon, A. M., P. J. Cribb, M. W. Chase, and F. N. Rasmussen. *Genera Orchidacearum, Vol. 3: Orchidoideae (Part two) and Vanilloideae.* OXFORD UNIVERSITY PRESS, 2003.

Pridgeon, A. M., P. J. Cribb, M. W. Chase, and F. N. Rasmussen. *Genera Orchidacearum, Vol. 4: Epidendroideae (Part one).* OXFORD UNIVERSITY PRESS, 2005.

Pridgeon, A. M., P. J. Cribb, M. W. Chase, and F. N. Rasmussen. *Genera Orchidacearum, Vol. 5: Epidendroideae (Part two).* OXFORD UNIVERSITY PRESS, 2009.

Pridgeon, A. M., P. J. Cribb, M. W. Chase, and F. N. Rasmussen. *Genera Orchidacearum, Vol. 6: Epidendroideae (Part three).* OXFORD UNIVERSITY PRESS, 2014.

648

USEFUL WEBSITES

World checklist of selected plant families (Orchidaceae)

apps.kew.org/wcsp/home.do

This has information about each published orchid species name and synonym and the geographical distribution for each accepted; there are also links to other online resources, such as Google images, etc.

An online resource for monocot plants: e-Monocot

e-monocot.org

This has all the information from the *Genera Orchidacearum* series and can be queried by genus.

Internet orchid species photo encyclopaedia

www.orchidspecies.com

This has images of about half of all orchid species with habitat and cultural information.

World orchid iconography/Bibliorchidea

orchid.unibas.ch

This site contains the archives of the Swiss Orchid Foundation, including herbarium specimens, drawings, and images for over 11,000 orchids.

Epidendra: the global orchid taxonomic network

www.epidendra.org

This site has a variety of types of information and resources about orchids, including images, national park information, floras, and history.

First nature: nature and biology of orchids

www.first-nature.com/flowers/~nature-orchids.php

A website with a set of commonly asked questions about orchids with answers written by a non-biologist.

ABBREVIATIONS
in AUTHOR NAMES

It is common botanical practice to follow the scientific name of a plant with the name of its author and the latter is frequently abbreviated. For example, many species were first described by Carl Linnaeus, the father of scientific nomenclature, whose name is usually shortened to L. In this book, we have given the surname of the author rather than the abbreviation, for ease of reference. However, the following abbreviations have been used:

ex The Latin for "from." Smith ex Jones, for example, indicates that Jones was the first to publish a name validly while recognizing that the name was first given but not published by an earlier author, Smith.

fils An abbreviation of the Latin filius, meaning "son." Useful in cases where a father and son are both authors, as for example, Linnaeus and Linnaeus fils, who were both called Carl Linnaeus.

649

INDEX *of* COMMON NAMES

While botanical names must be unique, this is not the case for common names, and some duplication is inevitable. Where two species share a common name, the botanical name for each is given in brackets.

652

INDEX *of* SCIENTIFIC NAMES

653

654

ACKNOWLEDGMENTS

PICTURE CREDITS

The publisher would like to thank the following individuals and organizations for their kind permission to reproduce the images in this book. All reasonable efforts have been made to acknowledge the images, however we apologize if there are any unintentional omissions and would be grateful if notified of any corrections that should be incorporated in future reprints or editions of this book.

656

Alamy/©Krystyna Szulecka Photography: 60. Alpsdake/CC BY-SA 4.0: 407. Manolo Arias: 370. Prof. Leonid V. Averyanov/www.binran. ru: 43, 595. Prashant Awale: 181. Tom Ballinger: 208. Dalton Holland Baptista/CC BY-SA 3.0: 278. Dot Potter Barnett: 50, 80. Ella Baron: 307. Guillermo Barreto, Terrestrial Orchid Collector: 104. Alejandro Bayer Tamayo/CC BY-SA 2.0: 223. Cássio van den Berg/CC BY-SA 2.5: 410. Gavin Campbell: 431. André Cardoso: 331. Prof. Sahut Chanta-naorrapint: 44. Jason Marcus Chin, Nature Guide (TG11678), Cameron Highlands, Malaysia: 523. © Maarten Christenhusz: 143, 179, 298, 608. © Mark A. Clements: 88. Jim Cootes: 85. Alan Cressler: 163, 237, 402, 513, 539. Norbert Dank/www.flickr.de/nurelias: 15 (top), 291, 306, 328, 345, 386. Wiel Driessen: 330, 346, 452. Maja Dumat/CC BY 2.0: 546. Ecuagenera: 366, 454. Felix/CC BY-SA 2.0: 211. Suranjan Fernando: 541, 543. Branka Forscek: 260, 585. Brett Francis/CC BY-SA 2.0: 460. Elena Andrews Gaillard: 103, 166. Juan Galarza: 459. Stephan Gale: 78. Mark A. Garland/U. S. Department of Agriculture: 87. Brian Gratwicke/CC BY 2.0: 149. Lourens Grobler: 57, 64, 156, 267, 310, 349, 357, 360, 362, 494, 535, 633. Martin Guenther: 550. Claudine et Pierre Guezennec: 75. Roger L. Hammer: 398. Jörg Hempel/CC BY-SA 2.0: 172. Benoît Henry: 41. Frédéric Henze: 84, 247, 607. John Henry Hills: 124. Johan Hermans: 145, 151, 547, 609, 627, 632. VanLap Hoàng/CC BY 2.0: 588. Tim Hodges, Melbourne Australia: 570. Jason Hollinger/CC BY 2.0: 29, 62, 405. Jean and Fred Hort: 119, 121. Richard C. Hoyer: 91. Eric Hunt: 1, 3, 6-11, 12 (top, bottom center, bottom right), 14 (top), 15 (bottom), 16, 17 (top), 18, 21 (top, bottom), 27 (top), 32, 42, 46, 51-55, 65, 72, 74, 79, 158-159, 167, 169, 192-193, 196, 200, 202, 206, 209-210, 212-214, 222, 226, 229-234, 242, 248-252, 254, 258, 263, 265 266, 268, 271-273, 276, 279-281, 284-285, 289, 292-293, 297, 300-303, 309, 311-317, 321-322, 324, 326, 332, 337-338, 341, 348, 353-356, 358, 361, 364-365, 372, 376, 378-382, 384, 387-390, 394, 396-397, 403-404, 408, 411-413, 417-418, 420, 423-429, 433, 435-442, 444-445, 447, 450-451, 455-456, 462-466, 469-472, 474, 476, 478, 485-488, 491-493, 495, 497-499, 501, 503, 505, 509, 519, 521, 524-525, 527-529, 534, 544, 548-549, 551, 555-556, 558, 562-567, 573-575, 579-580, 583, 589, 591-593, 596-598, 601-604, 610, 612-620, 625-626, 628-629, 631, 635. Rudolf Jenny: 359, 581, 605. Carlos Jerez: 282. Daniel Jiménez: 93, 255, 304, 319, 475, 507. Bart Jones, Memphis, TN, USA: 66. Kevin B. Jones, Charleston Southern University Biology Department: 335. Marie Gyslene Kamdem Meikeu: 531. Dominique Karadjoff: 624 Mikael Karlbom: 457. Andreas Kay, Ecuador Megadiverso: www.flickr.com/andreaskay/albums: 70. Ron Kinsey: 90, 138. Ryan Kitko/CC BY 2.0: 236. Jacques Klein: 102, 538. Pablo Leautaud/CC BY 3.0: 100. Rich Leighton: 611. Michael Lo/www.junglemikey.blogspot.co.uk: 12 (bottom left), 35, 241, 569. David Lochli/CC BY 2.0: 243. Carlos Velazco Macias: 96. Malcolm M. Manners: 105, 600. Roberto Martins: 69, 203, 262. Peter Matthews: 139. Buddhika Mawella: 219, 557. Joshua Mayer/CC BY-SA 2.0: 197. David R. McAdoo: 36, 37, 99, 195, 401. Warren McCleland: 23 (right), 155, 184, 244. Cameron McMaster: 147. João Medeiros/CC BY 2.0: 168. Mauricio Mercadante: 40. Guilmin Micheline: 467. Juan Sebastián Moreno: 253. Margaret Morgan: 137. Fabien Naneix: 374. Philip Norton: 28 (bottom), 116, 339. Patricio Novoa. National Botanic Gardens, Chile/CC BY 2.0: 61, 92. Dr Henry Oakeley: 239, 245, 256, 257, 264, 290, 296, 320, 323, 344, 363, 406, 443, 568. P. O'Byrne: 86, 594. Jin-Yao Ong: 511. P. T. Ong: 218. Orchi/CC BY-SA 3.0: 216. Stefano Pagnoni: 73, 190, 416 Guillaume Paumier/CC BY-SA 2.0: 553. Marcelo Pedron: 108. P.B. Pelser & J.F. Barcelona: 34. Luis Pérez/CC BY-2.0: 342. Andreas Philipp, Colombia/andreas.philipp@quimbaya.me: 13 (right), 391. Udai C. Pradhan: 162. Michael Pratt: 560. Qwert1234/CC BY-SA 3.0: 39, 409. rduta/CC BY 2.0: 586. Rebecca E. Repasky/AABP Atrium. Atrium Biodiversity Information System for the Andes to Amazon Biodiversity Program at the Botanical Research Institute of Texas/atrium. andesamazon.org: 71 Rexness/CC BY-SA 2.0: 89 Richard/CC BY-SA 2.0: 13 (left), 58, 59, 183. Mauro Rosim - Brazil: 351, 448. Colin & Mischa Rowan/www.RetiredAussies.com: 112, 117, 122, 123, 125, 127, 129,

133, 490, 576, 584. Björn S…/CC BY 2.0: 165, 175, 178, 187, 191. Rich Sajdak: 81. Gerardo A. Salazar/Instituto de Biologia, Universidad Nacional Autonoma de Mexico: 95, 299. David Scherberich: 537. André Schuiteman: 207, 526, 530, 545. Antje Schultner/CC BY-ND 2.0: 2, 47. Eerika Schulz: 368, 480, 561, 587. Michael Schwerdtfeger/University of Goettingen: 295. Shikoku Garden Inc.: 198. Shutterstock: 14 (center left), 20, 22 (top, bottom), 23 (left), 24 (left, right), 25 (top, bottom), 28 (top), 160-161, 164. Sociedad Colombiana de Orquideologia: 98, 228, 308, 333, 393. Species Orchids/www.flickr.com/photos/54925614@ N08: 148, 329. Herbert Stärker: 144, 146, 153, 578, 623, 634. Hans Stieglitz/CC BY-SA 3.0: 45. StingrayPhil/CC BY 2.0: 109. Jeremy Storey: 19 (top), 130-131, 136, 140. Reproduced with kind permission from the Swiss Orchid Foundation at the Herbarium Jany Renz, University of Basel/ W. Bachmann: 327, 622; /H. Baumann: 515; /P. Bernet: 606; /P. Bertaux: 269; /J. Blättler: 48; /G. Chiron: 106, 395; /J. F. Christians: 504, 508; /J. B Comber: 552; /P. J. Cribb: 49, 141; /E. la Croix: 173; /M. Erijri: 419; /D. Gerhard: 400; /R. Jenny: 270, 325, 334, 352, 367, 468, 554; /K. Keller: 76, 194, 536; /H. Kretzschmar: 17 (bottom), 152; /C. A. J. Kreutz: 174; /R. Kuehn: 171, 176, 177; /J. Levy: 5, 225, 496; /W. Löderbusch: 26, 399; /C. Luer: 94, 97; /S. Manning: 204, 461, 479, 482, 484; /G. Meyer: 369; /M.N./BBG: 189; /Th. Nordhausen: 120, 134; /R. Parsons: 215, 318, 336, 371; /J. Renz: 188; /D. Rückbrodt: 182; /A. Schuiteman: 502; /K. Senghas: 83, 118, 132, 224, 240, 275, 287, 305, 347, 350, 373, 375, 377, 385, 414-415, 477, 522, 630, 637; /S. Sprunger: 514; /R. van Vugt: 19 (center), 38, 68, 110-111, 113-114, 128, 180, 185, 221, 238, 277, 421-422, 500, 512, 516-517, 533, 572. Hisanori Takeuchi: 115. Edgardo Varela Torres, M.D. from Puerto Rico: 235. Luiz Filipe Klein Varella/www. orquideasgauchas.net: 101, 107, 227, 259, 294, 392, 446, 473. John Varigos: 246, 274, 283, 286, 340, 343, 383, 430, 434, 453, 458, 481, 483, 510, 532, 542, 577, 582, 599, 621, 636. Miguel Vieira/CC BY 2.0: 63. Sebastián Vieira: 288. Ed de Vogel: 520. Rogier van Vugt: 67, 77, 142, 150, 154, 157, 170, 186, 199, 201, 205, 217, 220, 261, 540, 559, 571, 590. Ming-I Weng, Taiwan: 82, 489. Chinthaka Wijesinghe: 506. Scott Wilson/CC BY-ND 2.0: 449. Len Worthington: 518. Gary Yong Gee: 126. Scott Zona: 56, 135.

Thanks also to the following for the botanical illustrations: David Anstey, Bibliothèque de l'Université de Strasbourg, Naturalis Biodiversity Centre, New York Botanical Garden, Pennsylvania Horticultural Society, Peter H Raven Library/Missouri Botanical Garden, the Swiss Orchid Foundation at the Herbarium Jany Renz, University of Basel.